Lecture Notes in Computer Science 1376
Edited by G. Goos, J. Hartmanis and J. van Leeuwen

Springer
Berlin
Heidelberg
New York
Barcelona
Budapest
Hong Kong
London
Milan
Paris
Santa Clara
Singapore
Tokyo

Francesco Parisi Presicce (Ed.)

Recent Trends in Algebraic Development Techniques

12th International Workshop, WADT'97
Tarquinia, Italy, June 3-7, 1997
Selected Papers

Springer

Series Editors

Gerhard Goos, Karlsruhe University, Germany
Juris Hartmanis, Cornell University, NY, USA
Jan van Leeuwen, Utrecht University, The Netherlands

Volume Editor

Francesco Parisi Presicce
DSI-Dipartimento di Scienze dell'Informazione
Universitá di Roma "La Sapienza"
Via Salaria 113, I-00198 Roma, Italy
E-mail: parisi@dsi.uniroma1.it

Cataloging-in-Publication data applied for

Die Deutsche Bibliothek - CIP-Einheitsaufnahme

Recent trends in algebraic development techniques : 12th international workshop ; selected papers / WADT '97, Tarquinia, Italy, June 3 - 7, 1997. Francesco Parisi Presicce (ed.). - Berlin ; Heidelberg ; New York ; Barcelona ; Budapest ; Hong Kong ; London ; Milan ; Paris ; Santa Clara ; Singapore ; Tokyo : Springer, 1998
(Lecture notes in computer science ; Vol. 1376)
ISBN 3-540-64299-4

CR Subject Classification (1991): D.2.1-2, D.2.4, D.2-m, D.3.1,F.3.1-2

ISSN 0302-9743
ISBN 3-540-64299-4 Springer-Verlag Berlin Heidelberg New York

Printed in Germany

Typesetting: Camera-ready by author
SPIN 10631942 06/3142 – 5 4 3 2 1 0 Printed on acid-free paper

Preface

Since the early 1970s, the algebraic specification of data types has been an important area of research which has provided foundations, methodologies, and tools for the formal development of software. The algebraic approach to the specification and development of systems, born as a formal method for abstract data types, encompasses today the formal design of integrated hardware and software systems, new specification frameworks, and a wide range of applications.

The 12th WADT Workshop on Algebraic Development Techniques was held in Tarquinia (approximately 100 km north of Rome) from June 3 to June 7 1997, and was organized by Francesco Parisi Presicce.

The main topics addressed during the workshop were:

- algebraic and other approaches to formal specifications
- algebraic structures and their logics
- specification languages and their associated methods and tools
- algebraic specification of concurrent systems
- term rewriting and theorem proving
- applications to reverse engineering, object systems, and compiler optimization.

The program consisted of 40 presentations describing ongoing research, three invited lectures by Hartmut Ehrig, José Meseguer, and Ugo Montanari surveying different topics and presenting recent results and direction for future work, and a tutorial by Peter D. Mosses on CoFI (Common Framework Initiative), part of a collaborative effort to design a Common Framework for Algebraic Specification and Development of Software (http://www.brics.dk/Projects/CoFI).

A Selection Committee consisting of

Egidio Astesiano, Michel Bidoit, Hartmut Ehrig, Hans-Jörg Kreowski, José Meseguer, Ugo Montanari, Fernando Orejas, Peter D. Mosses, Francesco Parisi Presicce, Don T. Sannella, Andrzej Tarlecki, and Martin Wirsing

selected a number of papers based on the abstracts and the presentations at the workshop and invited their authors to submit a written version of their talks for possible publication in the proceedings. All the submissions underwent a careful refereeing process and were discussed (by e-mail) by the Selection Committee during a final acceptance/rejection round. This volume contains the final versions of the 21 accepted papers and the written versions of the three invited lectures.

We are very grateful to all the workshop participants, to the members of the Selection Committee, and to the following (additional) referees:

M. Cerioli, A. Corradini, M. Gogolla, B. Graves, M. Große-Rhode, J. E. Hannay, R. Heckel, B. Jacobs, C. Kirchner, A. Labella, N. Marti-Oliet, E. Moggi, T. Mossakowski, O. Owe, P. Padawitz, J. Padberg, W. Pawłowski, A. Pierantonio, A. Piperno, G. Reggio, J. J .M .M. Rutten, M. Simeoni, U. Wolter for their contribution to the scientific quality of the workshop and of these proceedings.

We also wish to thank M. Boreale, S. De Simoni, M. Große-Rhode, A. Pierantonio, A. Piperno and M. Simeoni for their invaluable help in the organization before, during and after the workshop, and Springer-Verlag for agreeing to publish this volume.

The Workshop was sponsored by the Dipartimento di Scienze dell'Informazione of the Universitá di Roma La Sapienza and received financial support from the Consiglio Nazionale delle Ricerche through GNIM (Gruppo Nazionale per l'Informatica Matematica) and the Comitato Nazionale Scienza e Tecnologia dell'Informazione and the Comitato Nazionale per le Scienze di Ingegneria ed Architettura.

January 1998 Francesco Parisi Presicce

Table of Contents

Invited Papers

Contributed Papers

From Abstract Data Types to Algebraic Development Techniques: A Shift of Paradigms

H. Ehrig, M. Gajewsky, U. Wolter
e-mail: {ehrig, gajewsky, wolter}@cs.tu-berlin.de

Technical University of Berlin

Abstract. The concept of abstract data types and the corresponding series of ADT-workshops have been most fruitful for the development of algebraic specification techniques within the last two decades. Since abstract data types by now are well-established in all areas of Computer Science and algebraic specification techniques go far beyond their classical roots of universal algebra the name of the ADT-workshop has been changed to "Workshop on Algebraic Development Techniques", where the acronym ADT has now a new interpretation. In this paper we discuss this shift of paradigms from "Abstract Data Types" to "Algebraic Development Techniques". We show how the scope of algebraic development techniques has been extended from algebraic specification to several other formal specification approaches already and we make a proposal how the scope of this area can be seen today and in the future.

Contents

1 Introduction

The concept of abstract data types was developed in the 70'ies in order to have an abstract and implementation-independent definition of data types. The original idea of using algebraic specification as a formal specification technique for abstract data types was quite successful leading to a series of workshops on abstract data types starting in the early 80'ies.

The main topics of these ADT-workshops (for a survey see [GC95]) have been the following

- models and semantics, including initial, terminal, loose and behavioural semantics
- different specification formalisms, including equational, first and higher order axioms in combination with total, partial, order-sorted, and continuous models
- abstract theory of specification formalisms, including categorical frameworks like institutions, specification frames, parchments, and general logics
- structuring and refinement, including horizontal structuring, vertical refinement and compatibility of parameterization with implementation
- correctness by verification and transformation, including completeness and consistency results based on the theory of term rewriting and correctness preserving transformations
- specification executability, language and tool support, including the development of several algebraic specification languages with tool support for software development

More recently also the following topics have been considered in more detail

- development methodology, including the KORSO Methodology and the Common Framework Initiative (CoFI)
- algebraic specification of concurrent, object-oriented and distributed systems, including process algebraic and class-oriented approaches and integration of algebraic specification with other specification techniques
- dynamic abstract data types, aiming at a general integration framework including data type, process and time-oriented techniques
- algebraic development techniques, aiming at an extension of structuring and development techniques from algebraic specification and category theory to other formal specification techniques.

In this paper we want to discuss the corresponding shift of paradigms during the last two decades. This shift is summarized in four slogans presented already in our contribution [EM97] for TAPSOFT'97, which will be reviewed and discussed in section 2. In section 3 we give an overview of algebraic development techniques in different projects of the Berlin group going beyond algebraic specification to algebraic graph transformation, Petri nets, and statecharts. As an example of algebraic development techniques based on category theory we review in section 4 how union and rule based modification concepts and results can be formulated in the categorical framework of high-level replacement systems and applied to all of these different specification approaches. In the conclusion we try to summarize the new scope and future trends of ADT in the sense of algebraic development techniques from our point of view.

2 Shift of Paradigms

The shift of topics presented at the ADT-workshops - as discussed in the introduction - has been caused by a shift of paradigms which will be discussed in this section.

2.1 From Diversity of Mathematical Concepts to Unification of Computational Models and Semantic Theories

The main aim of Theoretical Computer Science in the 70'ies and 80'ies was the presentation of suitable mathematical models and solutions for different concepts and problems in Practical Computer Science. The classical topics "Automata Theory" and "Formal Languages" at the beginning of the 70'ies were extended by a diversity of mathematical concepts and computational models. In the area of specification and semantics different styles of semantic theories for specification and programming have been developed, like operational, algebraic, denotational and axiomatic semantics. In the area of algebraic specification different kinds of logics have been considered, like equational, universal Horn and first order predicate logic, leading to a variety of different theories. Starting already in the 80'ies the notions of "institution" and "specification frame" have been developed as a unification for different logics and semantics. Today most approaches concerning horizontal structuring, vertical refinement and implementation of algebraic specification are presented in an institution-independent way. In a similar way rewrite logic in the sense of Meseguer [Mes92] and the tile model in the sense of Montanari [GM97, GM96] are unified computational models for different kinds of calculi for the mathematical modelling of concurrent, distributed and reactive systems. This means that the current and future trend is the unification of computational models and semantic theories using abstract general models, especially from category theory.

2.2 From Algebraic Specification to Integration of Formal Techniques

The original idea of algebraic specification in the 70'ies was to provide a specification technique for basic abstract data types, like integers, stacks and queues.

It turned out that algebraic specifications and corresponding specification morphisms are most adequate to define various structuring and refinement techniques leading to the specification of parameterized data types and the development of modular software systems in the 80'ies. Since an algebraic specification mainly supports the functional and data type view of software systems these techniques were extended to specify concurrent, distributed and reactive systems. On the other hand several other formal specification techniques, like Petri nets, process algebraic techniques, graph transformation, statecharts, Z, and temporal logic, have been developed and applied to specify all kinds of communication based systems. It turned out, that it is not always useful to apply only one formal specification technique to model all aspects of a complex software system. In fact, there are different views of a complex system, like functional, process, and temporal view, which are important to be considered separately for different purposes. While algebraic specification techniques are especially suitable for the functional or data type view, other techniques are more suitable for the process and temporal view.

The use of different views to specify a complex system raises, of course, the problem of consistency and integration of views, which is one important research topic in software engineering and related areas. Meanwhile several process specification techniques, like CCS, Petri nets, graph transformation and statecharts, have been extended by data type specification techniques. In this way several integrated formalisms have been developed, like LOTOS [LOT87], algebraic high-level nets [PER95], attributed graph transformation [LKW93], where algebraic specification techniques are used for the data type aspects. A different example is μSZ [BDG$^+$96], an integration of statecharts and Z, where Z is used for functional aspects. The current and future trend is the integration of different specification techniques, which are suitable for the functional, process and temporal view of communication based systems, where algebraic specifications are especially useful for the functional or date type view.

2.3 From Trees to Graphs, Graph Transformation and Visual Languages

In the 70'ies and 80'ies the concept of trees has been advocated to model various structures in Computer Science, especially in automata theory and programming within the conference series CAAP. When the last CAAP was celebrated as part of TAPSOFT'97 in Lille it was not because trees are no longer useful. On the contrary — similar to abstract data types — trees are well-established in all areas of Computer Science, especially in connection with terms and term rewriting. On the other hand graphs and graphical representation of structures have been even more used in all areas of Computer Science, but mainly on an informal level. Although graph grammars were developed as a generalization of Chomsky grammars on strings on one hand and of tree and term rewriting systems on the other hand already in the 70'ies, graph grammars and transformations have been developed as a specification technique only in the late 80'ies and 90'ies.

The algebraic theory of graph transformation in contrast to other graph grammar approaches [Roz97] is based on the categorical concept of pushouts. This allows to apply the corresponding pushout techniques not only to graphs, but also to hypergraphs and other kinds of high-level structures used in other specification techniques, like algebraic specifications, Petri nets and statecharts. The corresponding theory of high-level replacement systems can be formulated in a purely categorical way [EHKP91]. The application of this theory to the category of graphs leads to the well-known theory of graph transformation, while the application to the categories of algebraic specifications, Petri nets and statecharts has opened up a new dimension for the corresponding specification techniques. Important issues are the rule-based modification, short transformation, of the corresponding specification document, and compatibility of horizontal structuring, like union, with transformation of documents. The corresponding technical concept and results will be briefly discussed in section 4. On the other hand graphs and graph transformations have turned out to be most useful as a formal basis to model visual languages and visual design of distributed systems (see section 3.2). Since visualization and animation are most important for the intuitive understanding of all kinds of specifications, it is certainly a current and future trend to support visualization and visual languages by formal techniques, especially by algebraic graph transformation.

2.4 From Abstract Data Types to Object-Oriented Techniques and Continuous Software Engineering

In the 70'ies the main objective was the development of suitable mathematical models for abstract data types. In the 80'ies the main focus was on parameterized abstract data types and modules with import and export interfaces. In the late 80'ies and 90'ies, when object-oriented techniques became more and more popular, the concepts of ADT's and algebraic specification were extended to model at least certain aspects of object-orientation. At the same time it became apparent that for state-oriented systems ADT's should be extended by a dynamic component to take care of state changes in an adequate way. This was reflected by the concept of DOID's by Astesiano [AZ95] and an informal concept of "Dynamic Abstract Data Types" [EO94] with several instantiations based on algebraic specification and algebraic graph transformation techniques.

By now, there are several formal concepts, like classical and dynamic abstract data types, process algebras, co-algebras, actor systems and attributed graph transformation systems, which have the capability of modelling certain aspects of object-oriented techniques. But it is still open and hence an important future topic to develop a widely accepted formal model for the object-oriented design and programming paradigm. Another important aspect in the area of software engineering, database and information systems as well as communication technology and computer networks is the problem of continuous change of requirements for already existing software systems in all areas of administration, commercial services and industry. This means that today maintenance of software includes re-engineering and hence continuous software engineering. Although this problem is known and faced in practice since the very beginning it has become a matter of research only recently. Unfortunately, formal methods for software development have been almost neglected in the area of re-engineering up to now. An important problem is that the adaption of the software means to change the design patterns online, because shut down of the system is highly undesirable for commercial reasons. Research concerning continuous software engineering and — even more general — for evolutionary systems in different areas of science is certainly an important future trend. Most recently it has been shown [Löw97] that among all existing semi-formal and formal methods the algebraic theory of graph transformation has the greatest potential to solve the problems of re-engineering and continuous software engineering mentioned above. Since algebraic graph transformation is within the scope of "Algebraic Development Techniques" (see section 5) the development of formal models for continuous software engineering seems to be a promising future topic for the new ADT-workshops.

3 Algebraic Development Techniques in different Projects of the Berlin Group

In this section we show how algebraic development techniques in the projects of the Berlin group have evolved from classical algebraic specification in the 70'ies and 80'ies to categorical techniques in other formal specification approaches in the 90'ies.

3.1 ACT - COMPASS - KORSO - CoFI

Motivated by the ADJ-group in the mid 70'ies [GTW78] the Berlin group has at the end of the 70'ies started to study abstract data types using algebraic specification with initial semantics. In joint work with the ADJ-group we studied parameterized algebraic specifications and developed the semantics for the algebraic specification language LOOK. Within the DFG-project ACT (1982-1987) [CEW93] we developed the algebraic specification language ACT-ONE in the early 80'ies which was later integrated into the language LOTOS. In this period we studied algebraic specifications with equational axioms in the sense of classical universal algebra, while other groups allowed already first order logical axioms. During the first ADT-workshops there were hot debates about the question, whether predicate logic should be allowed in algebraic specifications and what kind of semantics to take: Initial, final, loose or behavioural semantics? In our languages ACT-ONE and ACT-TWO we decided to take initial semantics for the constructive parts of the specifications and loose semantics with constraints for the interfaces. Moreover, we shifted from total to partial algebras. Within the COMPASS-project (1989-1996) [BKL$^+$91] we started in cooperation with the Barcelona group to extend our classical theory of algebraic specification — as documented in [EM85] and [EM90] — to the case of behavioural semantics and to develop an institution-independent version of this theory. More precisely, we considered specification frames (indexed categories) instead of institutions, where the behavioural case could be considered as one specific instantiation. In the KORSO-project (1992-1994), a German project concerning the development of correct software [BJ95], we supported the development of the language SPECTRUM, an algebraic specification language with three-valued first order logic, polymorphism, type classes, higher-order functions and infinite objects. However, in spite of the great potentials of this language, other specification formalisms, like Petri nets, were used in the main KORSO-case study HDMS in addition to SPECTRUM in order to improve understandability. The Common Framework Initiative (CoFI) [Mos97], started in 1996, is a promising approach to integrate the best of all algebraic specification techniques and languages developed so far. The first result is the new language CASL [cas], which was tentatively approved by the IFIP WG 1.3 in June 1997. The idea is, that other well-known algebraic specification languages, developed by specific groups, like ACT-ONE in Berlin, become sublanguages of CASL, and several extensions of CASL should be defined in view of specific application areas. The Berlin group is especially interested to integrate suitable sublanguages of CASL with other specification formalisms, like graph transformation, Petri nets and statecharts.

3.2 COMPUGRAPH - GETGRATS - APPLIGRAPH

The algebraic approach of graph transformation developed within the COMPUGRAPH-project (1989-1996) [EL93] has its roots in the algebraic theory of graph grammars developed already in the 70'ies and 80'ies. This algebraic theory [Ehr79] is based on the construction of pushouts in the category of graphs. The keyword "algebraic" was chosen, because algebraic and categorical techniques are used in contrast to other "set theoretical" or "logic" approaches to graph grammars [Roz97]. In the early 90'ies the algebraic approach of graph transformation was reformulated in a pure categorical framework leading to the concept of high-level-replacement systems [EHKP91] (see section 3.3). In the late phase of COMPUGRAPH and in the GETGRATS-project (1996-1999) [CK97] an algebraic semantics for graph transformation based on free and a coalgebraic semantics based on cofree constructions have been developed. Another example of algebraic development techniques in this area is the integration of different views of systems, which has been studied based on pushout resp. pullback constructions in suitable categories of graph transformation systems. A good example of transfer of concepts from the area of algebraic specification to that of algebraic graph transformation is the development of a module concept for graph transformation systems with import and export interfaces [EE96] motivated by the algebraic module concept in [EM90]. For the interfaces of modules it was necessary to develop a concept of loose semantics for graph transformations in analogy to loose semantics of algebraic specifications. The new concept of double-pullback derivations for graph transformations is the basis of our loose semantics. Another variant of the module concept is based on distributed graph transformations. The concept of distributed graph transformations is closely related to amalgamation of graph transformation rules and derivations, which was motivated by the amalgamation concept in the area of algebraic specification. In our APPLIGRAPH-project (1997-2000) [CK97] one main task is to develop a specification language for graph transformation systems where the experience with ACT-ONE and ACT-TWO in the ACT-project (see 3.1) will be most valuable.

3.3 PETRI-NET-TECHNOLOGY

An algebraic version of Petri nets was provided by Meseguer and Montanari in the famous article "Petri Nets are Monoids" [MM90]. In fact, this algebraic version is an excellent basis to introduce morphisms and to study Petri nets from a categorical point of view. In two projects concerning Petri nets in Berlin a similar algebraic version was also provided for algebraic high-level Petri nets, where place-transition nets are combined with algebraic specifications. For a suitable choice of morphisms the category of Petri nets in the low-level case (i.e. without data types) as well as in the high-level case (i.e. with data types) has pushouts, coproducts and coequalizers and hence general colimits. This allows to transfer concepts of horizontal structuring and rule-based modification (see 2.3) from the areas of algebraic specification and graph transformation to Petri nets. Another important example of algebraic development techniques in the area of Petri nets is the concept of "Abstract Petri Nets" [Pad96], which provides a unified framework for different kinds of low-level and high-level Petri nets. This concept is based on adjoint functors and institutions allowing to express different kinds of net structures and data type specification formalisms. In our PETRI-NET-TECHNOLOGY-project (1996-1999) [WRE95] the concept of abstract Petri nets is the semantical basis for a Petri net construction kit. The general idea is that this construction kit allows to tailor a suitable Petri net technique for specific application domains considered in this project. Petri net technology — in contrast to most of the classical theory of Petri nets — is considered to be a system development technology in the sense of software engineering, where structuring and refinement as well as integration with other specification techniques play a fundamental role. In particular, these aspects can be supported by algebraic development techniques as mentioned above. This shows that algebraic development techniques in the sense of algebraic and categorical constructions and transfer of concepts from other formal specification areas are very important in the area of Petri net technology.

3.4 ESPRESS-INTEGRATION PROJECTS

The ESPRESS-project (1995-1998) [ESP] aims at an engineering driven development of safety critical embedded systems. In addition to TU Berlin and two R&D-institutions it includes the industrial partners Daimler Benz and Bosch in Germany. The industrial case studies considered in this project are an intelligent car cruise control and a traffic light control system. From the very beginning it was a requirement of the industrial partners to use statecharts and Z as formal specification techniques in this project. The reason to use these techniques instead of other ones, like algebraic specification and graph transformation, was the fact that commercial tools for statecharts and Z are available and already used in some other projects of Daimler Benz. Hence one main task of TU-Berlin in this project was to design a suitable integration of Z and statecharts, called μSZ [BDG+96], and an integration with temporal logic is under development. Although statecharts are very convenient for the specification of embedded reactive systems in general suitable techniques for structuring and refinement are missing. Motivated by similar techniques for algebraic specification, graph transformation and Petri nets in previous projects (see 3.1 – 3.3) we were able to transfer concepts for horizontal structuring and rule-based modification and to develop new refinement techniques for statecharts which are compatible with their behaviour [EGKP97]. In spring of 1997 the German Research Council (DFG) has approved a frame program (DFG-Schwerpunktprogramm) [spp] with the title "Integration of Techniques for Software Specification Concerning Applications in Engineering". Within this frame program TU Berlin is going to apply for specific projects concerning applications to production automation and traffic guidance systems together with leading experts in engineering. In one of these projects statecharts will be integrated with object-oriented techniques and in another one Petri nets with E/R-techniques. At least in the second project, which is closely related to the important EU-project "European Train Control Systems", algebraic development techniques will be applied to Petri nets. In analogy to abstract data types we are going to develop new concepts for Petri nets, called open processes and abstract process types. These new techniques are suitable to model and integrate different scenarios in the area of traffic guidance systems, especially for European train control systems.

Finally let us point out that the concept of dynamic abstract data types, discussed in 2.4, is a good starting point for a general scheme to integrate different kinds of specification formalisms, especially for the data type and process view of systems. This suggests that algebraic development techniques will also play an important role for the integration of formal specification techniques.

4 Union and Rule-Based Modification in Different Specification Approaches

The concept of union has been studied for algebraic specifications and that of rule-based modification for algebraic graph grammars already in the 70'ies and 80'ies. The key idea in both cases is to use the categorical concept of pushouts. Meanwhile it has turned out that both concepts are important also in several other specification approaches. The main concepts and results can be formulated already on a categorical level, called high-level structures and replacement systems, and instantiated to the different approaches [EHKP91]. In our view this is an interesting example of algebraic development techniques supporting the transfer of concepts and results between different specification approaches.

In this section we review the idea of high-level structures and replacement systems including concepts and results for union and rule-based modification, where the latter is called transformation in this context. We discuss the instantiations of this categorical framework to graphs, algebraic specifications, Petri nets, action nets and statecharts. Corresponding results in all these different specification approaches have been presented separately in literature.

4.1 High-Level Structures and Replacement Systems

The algebraic approach to graph grammars (introduced in [Ehr79]) has been generalized to high-level replacement systems [EHKP91]. High-level replacement systems are formulated for an arbitrary category with a distinguished class $\mathcal{M}$ of morphisms which is used to classify different types of productions or rules. Like other kinds of formal grammars, high-level replacement systems consist of a start item and a set of productions or rules. The notion of rules and transformations in high-level replacement systems is based on morphisms and pushouts in a category **CAT**, where the objects can be regarded as high-level structures, for example graphs, and the morphisms as structure preserving functions between these objects, for example graph morphisms. Transformation of a high-level structure is achieved by the application of a rule leading to another high-level structure. The rule is split into a deleting part L, an adding part R and an interface K which is preserved, such that the rule[1] p is given by $p = (L \overset{l}{\leftarrow} K \overset{r}{\rightarrow} R)$ where l and r are morphisms of the corresponding category. Deleted are those parts of the object L that are not in the image of the morphism $l : K \rightarrow L$. Analogously, the part that is not in the image of $r : K \rightarrow R$ is added to the interface. More precisely we have:

Definition 4.1.1 (Rules and Transformations)

1. A *rule* $p = (L \overset{l}{\leftarrow} K \overset{r}{\rightarrow} R)$ in **CAT** consists of the objects L, K and R, called left hand side, interface (or gluing object) and right hand side respectively, and two morphisms $K \overset{l}{\rightarrow} L$ and $K \overset{r}{\rightarrow} R$ both in $\mathcal{M}$.
2. Given a rule $p = (L \overset{l}{\leftarrow} K \overset{r}{\rightarrow} R)$ a *direct transformation* $G \overset{p}{\Longrightarrow} H$, from an object G to an object H is given by the following two pushout diagrams **(1)** and **(2)** in the category **CAT** as shown below:

$$\begin{array}{ccccc} L & \overset{l}{\longleftarrow} & K & \overset{r}{\longrightarrow} & R \\ {\scriptstyle g_1}\downarrow & (1) & \downarrow{\scriptstyle g_2} & (2) & \downarrow{\scriptstyle g_3} \\ G & \underset{c_1}{\longleftarrow} & C & \underset{c_2}{\longrightarrow} & H \end{array}$$

The morphisms $L \overset{g1}{\rightarrow} G$ and $R \overset{g3}{\rightarrow} H$ are called occurrences of L in G and R in H, respectively. By an occurrence of rule $p = (L \overset{l}{\leftarrow} K \overset{r}{\rightarrow} R)$ in a structure G we mean an occurrence of the left hand side L in G.

3. A *transformation sequence* $G \overset{*}{\Longrightarrow} H$, short transformation, between objects G and H means G is isomorphic to H or there is a sequence of $n \geq 1$ direct transformations

$$G = G0 \overset{p1}{\Longrightarrow} G1 \overset{p2}{\Longrightarrow} \ldots \overset{pn}{\Longrightarrow} Gn = H$$

[1] The rule is called p traditionally, due to the original notion of production.

For this sequence we may also write $G \Longrightarrow H$ via $(p1, \ldots, pn)$. $\square$

The idea of a direct transformation is that the left hand side L mapped to G is replaced by the right hand side R mapped to H, where the interface K together with its mappings to L, R and to the context C designates corresponding interfaces in L, R and C.

Definition 4.1.2 (Union)

The union A of two high-level structures A_1 and A_2 via some interface I, denoted by $(A_1, A_2) \overset{I}{\Longrightarrow} A$, is given by the following pushout in the category **CAT**:

$$
\begin{array}{ccc}
I & \xrightarrow{in_1} & A_1 \\
{\scriptstyle in_2}\downarrow & (PO) & \downarrow \\
A_2 & \longrightarrow & A
\end{array}
$$

$\square$

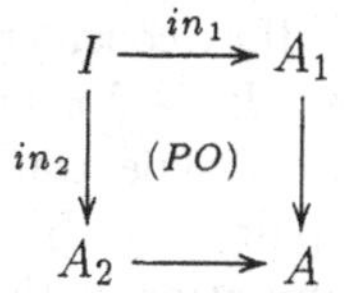

To achieve compatibility between horizontal (union) and vertical (transformation) structuring, we require the following independence condition which intuitively means that the interface I for the union is not changed by the transformations.

Definition 4.1.3 (Independence of Union and Transformation)

A union $(A_1, A_2) \overset{I}{\Longrightarrow} B$ in **CAT** is called independent from transformations $A_i \overset{p_i}{\Longrightarrow} B_i$ for $i \in \{1, 2\}$ given by the pushouts **(1)** and **(2)** if there are morphisms $I \to C_i$ so that the triangle **(3)** commutes for $i \in \{1, 2\}$.

$$
\begin{array}{ccccc}
L_i & \longleftarrow & K_i & \longrightarrow & R_i \\
\downarrow & & \downarrow & & \downarrow \\
& (1) & & (2) & \\
A_i & \longleftarrow & C_i & \longrightarrow & B_i
\end{array}
$$

$\square$

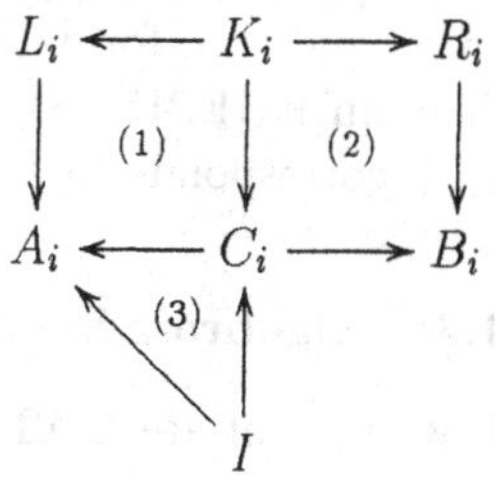

Theorem 4.1.4 (Compatibility of Union and Transformation)

If the category **CAT** has pushouts and finite coproducts we have:

Given a union $(A_1, A_2) \overset{I}{\Longrightarrow} A$ independent from the transformations $A_i \overset{p_i}{\Longrightarrow} B_i$ for $i \in \{1, 2\}$, then we have compatibility of union and transformation in the following sense:

There is an object B obtained by the union $(B_1, B_2) \overset{I}{\Longrightarrow} B$ and by the transformation $A \overset{p_1 + p_2}{\Longrightarrow} B$ via the parallel rule[2] $p_1 + p_2$, illustrated by the following picture:

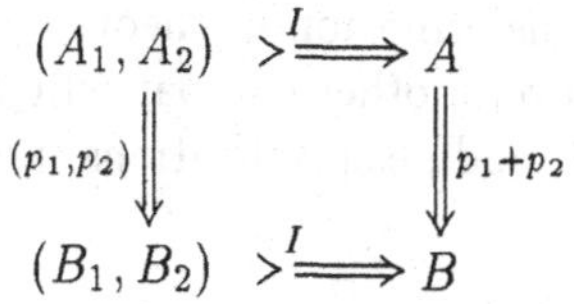

$$
\begin{array}{ccc}
(A_1, A_2) & \overset{I}{\Longrightarrow} & A \\
{\scriptstyle (p_1, p_2)}\Big\Vert\downarrow & & \Big\Vert\downarrow{\scriptstyle p_1 + p_2} \\
(B_1, B_2) & \overset{I}{\Longrightarrow} & B
\end{array}
$$

Proof: See [PER95] $\square$

---

[2] The parallel rule $p_1 + p_2 = (L_1 + L_2 \leftarrow K_1 + K_2 \to R_1 + R_2)$ of rules p_1 and p_2 is defined by the coproducts in each component, where the class $\mathcal{M}$ is supposed to be closed under coproducts.

4.2 Graphs and Graph Transformations

A graph in the algebraic theory of graph grammars and transformation systems is given by $G = (N, E, source, target)$, where N is a set of nodes, E a set of edges and $source, target : E \to N$ are functions assigning to each edge its source and target node respectively:

$$E \underset{target}{\overset{source}{\rightrightarrows}} N$$

If we consider labelled or typed graphs we have in addition two label functions $l_E : E \to \Sigma_E$ and $l_N : N \to \Sigma_N$ or a graph morphism $f : G \to TG$ from G into a type graph TG. In general a graph morphism $f : G_1 \to G_2$ consists of a pair $f = (f_E, f_N)$ of functions $f_E : E_1 \to E_2$ and $f_N : N_1 \to N_2$ which are compatible with *source* and *target* functions of G_1 and G_2. The corresponding category **GRAPH** of graphs and graph morphisms has pushouts and coproducts which are constructed componentwise in the category **SET** of sets. This allows to instantiate the category **CAT** in 4.1 by **GRAPH** such that high-level structures become graphs and high-level-replacement systems are graph grammars in the sense of the algebraic approach [Ehr79], also called double-pushout-approach in [Roz97]. In particular, rules and transformations considered in 4.1.1 are equal to graph rules and graph transformations in this case. The compatibility of union and transformation in this case corresponds to the distributed parallelism theorem in [EBHL88], where the independence of union and transformation in 4.1.3 corresponds to I-strictness of local direct derivations in [EBHL88].

4.3 Algebraic Specifications

If we instantiate **CAT** in section 4.1 by the category **ALGSPEC** of algebraic specifications (see e.g. [EM85]) we have pushouts and coproducts and obtain algebraic specification grammars as considered in [EP91]. In this case the interface part $IMP \leftarrow PAR \to EXP$ of an algebraic module specification

$$\begin{array}{ccc} PAR & \longrightarrow & EXP \\ \downarrow & & \downarrow \\ IMP & \longrightarrow & BOD \end{array}$$

in the sense of [EM90] can be considered as a rule in a corresponding algebraic specification grammar. This allows to apply rule-based modification techniques to the development of modular software systems [PP90]. In this case the compatibility of union and transformation in theorem 4.1.4 has been considered in [PP91], (thm. 3.14), and several other compatibility results, like distributivity of composition over union of module specifications, are presented in [EM90].

4.4 Petri Nets

The general theory can be applied to low-level and high-level Petri nets (see [Pad96]): For simplicity we have a closer look only at the low-level case of place/transition nets.

A place/transition net, short PT-net, is given by $N = (P, T, pre, post)$, where P is a set of places, T a set of transitions and $pre, post : T \to P^{\oplus}$ are functions assigning to each transition an element of the free commutative monoid $P^{\oplus}$ over P, its *pre*- and *post*-domain. N can be represented by the following diagram:

$$T \underset{post}{\overset{pre}{\rightrightarrows}} P^{\oplus}$$

This algebraic notation of place/transition nets has been introduced by Meseguer and Montanari in [MM90]. It is equivalent to the traditional one, where arcs are given as elements of a flowrelation and equipped with a weight function.

A morphism $f : N_1 \to N_2$ between two PT-nets $N_i = (P_i, T_i, pre_i, post_i)$ for $i \in \{1, 2\}$ is a pair of functions $f_P : P_1 \to P_2$ and $f_T : T_1 \to T_2$ such that the following diagram commutes componentwise for *pre*- and *post*-function:

$$\begin{array}{ccc} T_1 & \underset{post_1}{\overset{pre_1}{\rightrightarrows}} & P_1^{\oplus} \\ f_T \downarrow & = & \downarrow f_P^{\oplus} \\ T_2 & \underset{post_2}{\overset{pre_2}{\rightrightarrows}} & P_2^{\oplus} \end{array}$$

The corresponding category of place/transition nets is denoted by **PT**.

The category **PT** has pushouts and coproducts which can be constructed componentwise in **SET**. (Note, that this would fail if we would allow arbitrary monoid morphisms $f_P; P_1^{\oplus} \to P_2^{\oplus}$ instead of freely generated ones.) The compatibility of union and transformation has been shown in [PER95] for the case of algebraic high-level nets and applied in a case study of medical information systems [EPE96].

4.5 Action Nets

Another formalism for modelling concurrent systems are action nets, introduced in [EGP97]. They build a link between Petri nets, statecharts and graphs as they combine elements of all of them.

An action net $AN = (GS, T, E, source, dest, cond, action)$ consists of a graphical state space[3] $GS = (S, sub : S \to \mathbb{F}S)$, a graph without multiple edges and finite outdegree of all nodes S. Furthermore, we have a set of transitions T, a set of events E, two functions $source, dest : T \to \mathbb{F}_1 S$, denoting the source resp. destination states for every transition, a function[4] $cond : T \to \mathcal{T}_{\mathbb{B}}E$ assigning to each transition a boolean expression called trigger condition, and a function $action : T \to \mathbb{F}E$ assigning to each transition a finite set of events generated when the transition is taken.

An AN-morphism $f : AN_1 \to AN_2$ is a triple $f = (f_S, f_T, f_E)$, where $f_S : GS_1 \to GS_2$ is a GS-morphism, i.e. satisfying the formula $\mathbb{F}f(sub_1(s_1)) \subseteq sub_2(f(s_1))$ for $s_1 \in GS_1$, and $f_T : T_1 \to T_2$, $f_E : E_1 \to E_2$ are functions such that the following diagram commutes componentwise

---

[3] $\mathbb{F}$ stands for the finite powerset, $\mathbb{F}_1$ for the nonempty finite powerset.

[4] $\mathcal{T}_{\mathbb{B}}E$ denotes boolean expressions over E

$$
\begin{array}{ccccccc}
\mathbb{F}_1 S_1 & \overset{source_1}{\underset{dest_1}{\Longleftarrow}} & T_1 & \xrightarrow{(cond_1, action_1)} & T_{\mathbb{B}} E_1 \times \mathbb{F}E_1 & & GS_1 \\
\Big\downarrow {\scriptstyle \mathbb{F}_1 f_S} & = & \Big\downarrow {\scriptstyle f_T} & = & \Big\downarrow {\scriptstyle T_{\mathbb{B}} f_E \times \mathbb{F} f_E} & & \Big\downarrow {\scriptstyle f_S} \\
\mathbb{F}_1 S_2 & \overset{source_2}{\underset{dest_2}{\Longleftarrow}} & T_2 & \xrightarrow[(cond_2, action_2)]{} & T_{\mathbb{B}} E_2 \times \mathbb{F}E_2 & & GS_2
\end{array}
$$

The corresponding category of action nets is denoted by **AN**.

According to [EGP97] the category **AN** has pushouts of injective morphisms and finite coproducts leading to compatibility of union and transformation as formulated in theorem 4.1.4

4.6 Statecharts

Motivated by corresponding results for Petri nets horizontal and vertical structuring techniques for statecharts have been studied in [EGKP97]. Considering statecharts in the sense of [Har87] without initial state and without a completeness requirement for transitions we obtain the notion of abstract statecharts considered in [EGKP97]. Since we have already introduced action nets (see 4.5) we can define abstract statecharts as a special case of action nets. In fact, an abstract statechart $ASC = (HS, T, E, source, dest, cond, action)$ is an action net $AN = (GS, T, E, source, dest, cond, action)$ where the graphical state space $GS = (S, sub : S \to \mathbb{F}S)$ is specialized to be a hierarchical state space $HS = (S, root, sub : S \to \mathbb{F}S, dec : S \to \{and, or, basic\})$, i.e. a tree where all nodes are labelled by the function *dec*. This means that the nodes are states, which are either *basic*, *or*- or *and*- states. Moreover, the *source*- resp. *destination* states of each transition in abstract statecharts are required to be parallel.

Unfortunately, the full subcategory **ASC** (category of abstract statecharts) of **AN** (category of action nets) is not closed under pushouts in general. But if we restrict the morphisms to be injective, one of them *root*-preserving and the other one leaf-preserving, then the compatibility theorem 4.1.4 holds also in the case of abstract statecharts (see [EGP97, EGKP97] for more detail).

5 Conclusion

In this paper we have discussed the shift of paradigms leading from abstract data types as considered in the 70'ies and 80'ies to algebraic development techniques in the 90'ies. In the following we try to summarize the scope of algebraic development techniques as discussed in this paper.

1. Algebraic development techniques include first of all the study of abstract data types and algebraic specification techniques in the sense of the COMPASS-project [BKL+91]. In particular the Common Framework Initiative CoFI, the development of the new algebraic specification language CASL and the corresponding software development techniques are important cases of algebraic development techniques.

2. The transfer of categorical based concepts for structuring, modularity, refinement, and compositionality from the area of algebraic specification to other formal specification approaches is a second important branch of algebraic development techniques.
3. A third branch, as outlined in section 4, consists in generalizing concepts and results from the area of algebraic graph transformation to the categorical theory of high-level replacement systems and to apply them to other areas of formal specification techniques.
4. Finally, we claim that based on an abstract theory of specification formalisms a general scheme for the integration of different specification formalisms and formal techniques for re-engineering and continuous software engineering can be developed using algebraic and categorical techniques. This will lead to an important new branch of algebraic development techniques.

References

[AZ95] E. Astesiano and E. Zucca. D-oids: A model for dynamic data types. *Math. Struct. in Comp. Sci.*, 5(2):257–282, 1995.

[BDG+96] Robert Büssow, Heiko Dörr, Robert Geisler, Wolfgang Grieskamp, and Marcus Klar. μSZ – Ein Ansatz zur systematischen Verbindung von Z und Statecharts. Technical Report 96-32, Technische Universität Berlin, February 1996.

[BJ95] M. Broy and S. (Eds.) Jähnichen. KORSO*: Methods, Languages, and Tools for the Construction of Correct Software, Lect. Notes in Comp. Sci. no.*1009 . Springer, 1995.

[BKL+91] M. Bidoit, H. J. Kreowski, P. Lescanne, F. Orejas, and D. Sannella. *Algebraic System Specification and Development, Lect. Notes in Comp. Sci. no.* 501. Springer, 1991.

[cas] Available via URL: "http://www.brics.dk/Projects/CoFI/".

[CEW93] I. Claßen, H. Ehrig, and D. Wolz. *Algebraic Specification Techniques and Tools for Software Development - The Act Approach.* AMAST Series in Computing Vol. 1. World Scientific Publishing Co., 1993.

[CK97] A. Corradini and H.-J. Kreowski. GETGRATS and APPLIGRAPH: Theory and Application of Graph Transformation Systems. In *Bull. EATCS*, November 1997. to appear.

[EBHL88] H. Ehrig, P. Böhm, U. Hummert, and M. Löwe. Distributed parallelism of graph transformation. In *13th Int. Workshop on Graph Theoretic Concepts in Computer Science, Lect. Notes in Comp. Sci. no. 314*, pages 1–19, Berlin, 1988. Springer Verlag.

[EE96] H. Ehrig and G. Engels. Pragmatic and semantic aspects of a module concept for graph transformation systems. In *Lect. Notes in Comp. Sci. no. 1073 , Proc. Williamsburg, U.S.A.*, pages 137–154. Springer Verlag, 1996.

[EGKP97] H. Ehrig, R. Geisler, M. Klar, and J. Padberg. Horizontal and Vertical structuring Techniques for Statecharts. In A. Mazurkiewicz and J. Winkowski, editors, In *Lect. Notes in Comp. Sci. no. 1243, CONCUR'97: Concurrency Theory, 8th International Conference, Warsaw, Poland*, pages 181 – 195. Springer Verlag, July 1997.

[EGP97] H. Ehrig, M. Gajewsky, and J. Padberg. Action Nets, Hierarchical State Spaces and Abstract Statecharts as High-Level Structures. Technical Report TR 97 - 14, Technical University Berlin, 1997.

[EHKP91] H. Ehrig, A. Habel, H.-J. Kreowski, and F. Parisi-Presicce. From graph grammars to High Level Replacement Systems. In H. Ehrig, H.-J. Kreowski, and G. Rozenberg, editors, *4th Int. Workshop on Graph Grammars and their Application to Computer Science, Lect. Notes in Comp. Sci. no. 532*, pages 269–291. Springer Verlag, 1991.

[Ehr79] H. Ehrig. Introduction to the algebraic theory of graph grammars. In V. Claus, H. Ehrig, and G. Rozenberg, editors, *1st Graph Grammar Workshop, Lect. Notes in Comp. Sci. no. 73*, pages 1–69. Springer Verlag, 1979.

[EL93] H. Ehrig and M. Löwe. The ESPRIT BRWG COMPUGRAPH Computing by Graph Transformations : A survey. In *TCS 109*, pages 3 – 6. North-Holland, 1993.

[EM85] H. Ehrig and B. Mahr. *Fundamentals of Algebraic Specification 1: Equations and Initial Semantics*, volume 6 of *EATCS Monographs on Theoretical Computer Science*. Springer, Berlin, 1985.

[EM90] H. Ehrig and B. Mahr. *Fundamentals of Algebraic Specification 2: Module Specifications and Constraints*, volume 21 of *EATCS Monographs on Theoretical Computer Science*. Springer, Berlin, 1990.

[EM97] H. Ehrig and B. Mahr. Future Trends of TAPSOFT. In *TAPSOFT'97: Theory and Practice of Software Development, Lect. Notes in Comp. Sci. no. 1214*, pages 6–10. Springer Verlag, 1997.

[EO94] H. Ehrig and F. Orejas. Dynamic abstract data types: An informal proposal. *Bull. EATCS 53*, pages 162–169, 1994.

[EP91] H. Ehrig and F. Parisi-Presicce. Algebraic specification grammars: a junction between module specifications and graph grammars. *Lect. Notes in Comp. Sci. no.* 532, pages 292–310, Springer Verlag, 1991.

[EPE96] C. Ermel, J. Padberg, and H. Ehrig. Requirements Engineering of a Medical Information System Using Rule-Based Refinement of Petri Nets. In D. Cooke, B.J. Krämer, P. C-Y. Sheu, J.P. Tsai, and R. Mittermeir, editors, *Proc. Integrated Design and Process Technology*, pages 186 – 193. Society for Design and Process Science, 1996. Vol.1.

[ESP] ESPRESS: Ingenieurmäßige Entwicklung sicherheitsrelevanter eingebetteter Systeme, BMBF-Projektantrag, Berlin 1994.

[GC95] M. Gogolla and M. Cerioli. What is an Abstract Data Type after all? In *ADT-Workshop'94, Lect. Notes in Comp. Sci. no. 906*, pages 499–523. Springer Verlag, 1995.

[GM96] F. Gadducci and U. Montanari. The Tile Model. Technical Report TR-96-27, Università de Pisa, Dipartimento di Informatica, June 1996.

[GM97] F. Gadducci and U. Montanari. The Tile Model. In Colin Stirling Gordon Plotkin and Mads Tofte, editors, *Proof, Language and Interaction: Essays in Honour of Robin Milner*. MIT Press, 1997. to appear.

[GTW78] J. A. Goguen, J. W. Thatcher, and E. G. Wagner. An initial algebra approach to the specification, correctness and implementation of abstract data types. In R. Yeh, editor, *Current Trends in Programming Methodology IV: Data Structuring*, pages 80–144. Prentice Hall, 1978.

[Har87] D. Harel. Statecharts: a visual formalism for complex systems. *Science of Computer Programming*, 8:231–274, 1987.

[LKW93] M. Löwe, M. Korff, and A. Wagner. An algebraic framework for the transformation of attributed graphs. In M.R. Sleep, M.J. Plasmeijer, and M.C. van Eekelen, editors, *Term Graph Rewriting: Theory and Practice*, chapter 14, pages 185–199. John Wiley & Sons Ltd, 1993.

[LOT87] LOTOS - A formal description technique based on temporal ordering of observational behaviour. Information Processing Systems - Open Systems Interconnection **ISO DIS 8807**, jul. 1987. (ISO/TC 97/SC 21 N).

[Löw97] M. Löwe. Evolution Patterns, 1997. Habilitation Thesis, Technical University Berlin, to appear.

[Mes92] J. Meseguer. Conditional rewriting logic as a unified model of concurrency. *TCS*, 96:73–155, 1992.

[MM90] J. Meseguer and U. Montanari. Petri nets are monoids. *Information and Computation*, 88(2):105–155, 1990.

[Mos97] P. D. Mosses. CoFI: The Common Framework Initiative for Algebraic Specification and Development. In *TAPSOFT'97: Theory and Practice of Software Development, Lect. Notes in Comp. Sci. no. 1214*, pages 115–137. Springer Verlag, 1997.

[Pad96] J. Padberg. *Abstract Petri Nets: A Uniform Approach and Rule-Based Refinement.* PhD thesis, Technical University Berlin, 1996. Shaker Verlag.

[PER95] J. Padberg, H. Ehrig, and L. Ribeiro. Algebraic high-level net transformation systems. *Math. Struct. in Comp. Sci.*, 5:217–256, 1995.

[PP90] F. Parisi-Presicce. A Rule-Based Approach to Modular System Design. In *Proc. of the 12th Int. Conf. on Software Engeneering, Nice*, pages 202–211. IEEE Computer Science Press, 1990.

[PP91] F. Parisi-Presicce. Foundations for Rule-based Design of Modular Systems. In *TCS, no. 83*, pages 131 - 155, 1991.

[Roz97] G. Rozenberg, editor. *Handbook of Graph Grammars and Computing by Graph Transformations, Volume 1: Foundations.* World Scientific, 1997.

[spp] Available via URL: "http://tfs.cs.tu-berlin.de/SPP/".

[WRE95] H. Weber, W. Reisig, and H. Ehrig. Concept, Theoretical Foundation, and Validation of an Application Oriented Petri Net Technology. Proposal for a "Forschergruppe" to the German Research Council (DFG), 1995. (Accepted as a DFG-project from April 1996 to March 1999).

Membership Algebra as a Logical Framework for Equational Specification*

José Meseguer

SRI International, Menlo Park, CA 94025, USA
meseguer@csl.sri.com

A mi madre, Fuensanta Guaita de Meseguer,
con todo cariño en su 85 cumpleaños

Abstract. This paper proposes *membership equational logic*—a Horn logic in which the basic predicates are equations $t = t'$ and membership assertions $t : s$ stating that a term t belongs to a sort s—as a logical framework in which a very wide range of total and partial equational specification formalisms can be naturally represented. Key features of this logic include: simplicity, liberality and equational character; generality and expressiveness in supporting subsorts, overloading, errors and partiality; and efficient implementability in systems such as Maude. The paper presents the basic properties of the logic and its models, and discusses in detail how many total and partial equational specification formalisms, including order-sorted algebra and *partial* membership equational logic, can be represented in it, as well as the practical benefits in terms of tool reusability that this opens up for other languages, including CASL.

Table of Contents

* Supported by Office of Naval Research Contracts N00014-95-C-0225 and N00014-96-C-0114, National Science Foundation Grant CCR-9633363, and by the Advanced Software Enrichment Project of the Information-Technology Promotion Agency, Japan (IPA).

1 Introduction

This work is an intermediate stage in a long-term quest for ever better equational specification formalisms. The fact that many of us have been at it for quite some time and that this has already produced a rich collection of formalisms should not be an excuse to lose nerve and give up the quest. We should press on.

In fact, the time may be ripe for a task of reflection and unification to try to pull together and understand in a more holistic way the best ideas that have been developed in the algebraic specification research program. This program, both in its core ideas and in its ever wider applications to other areas such as object-oriented programming, concurrency, software engineering, theorem proving, and logical frameworks, has yielded very powerful techniques and has shown sufficient maturity and promise as to make Joseph Goguen state that we may be entering "a golden era of algebraic specification." One may indeed wonder whether the next millennium might not in fact usher in such a marvelous event.

Whatever our degree of optimism may be about such predictions, it seems undeniable that strong winds, of a syncretic and ecumenical nature, are blowing around us and are stimulating much activity and excitement in projects such as the Common Algebraic Specification Language (CASL) [9]. There are good pragmatic reasons for this. It takes long years of hard work to develop well-finished and well-documented *systems* and *tools* that are essential to transfer our techniques and ideas to the world of engineering practice. An atomization of our energies into countless formalisms and experimental systems without a way to integrate them and to have good tool support for them would be a sad waste of efforts, and perhaps a lack of scientific responsibility.

All this makes the idea of a *universal language* or *logical framework* for equational specification[2] quite attractive. This paper is in fact a proposal for a specific logical framework, *membership equational logic*, in which a very wide range of total and partial equational specification formalisms can be naturally represented. Whoever makes a proposal of this kind should give good reasons for it. What are the good characteristics of the framework? In which sense is it better than other options? Is it in competition with other proposals?

Of course, statements of optimality can be provisional at best. What is really important is to make clear the intended *goals*, so that the adequacy and success of the proposed formalism to meet such goals can then be judged. In this work the quest is for an equational specification formalism meeting as best as possible the following goals:

1. It should be as *simple* as possible.
2. It should have *all the good properties* of *equational* logics, such as soundness and completeness, initial, free and relatively free algebras, good modularity properties for theory composition and parameterization, and so on.
3. It should be as *general* as possible, so as to encompass the widest possible range of total and partial equational specification formalisms.
4. It should be as *expressive* and as *convenient to use* as possible. In particular, it should directly support features such as
   - sorts and subsorts,
   - operator overloading,
   - errors and error recovery,
   - partiality,

   that are very useful in practical specifications. Therefore, a formalism that would "compile away" such features while still keeping a comparable expressive power at the mathematical level would utterly fail to meet this goal.
5. It should be *efficiently implementable* by rewriting, so that executable specification and declarative programming using this logic become practical and attractive possibilities once the appropriate interpreters and compilers are built. This means that a natural path from specifications to efficient programs exists within the logic. It also means that many other formalisms, once represented in it, can be executed, with enormous savings in tool-building effort.

It should be obvious that the above goals exercise a dynamic and creative *tension* among each other, similar to that of two groups of people pulling a rope in opposite directions, with, say, simplicity, efficient implementability and equational character pulling on one side, and with generality and expressiveness pulling on the other. This makes the task more interesting, because the best

[2] Since the expression "algebraic specification" is now often used in a quite latitudinarian way, so as to mean almost anything that uses an algebraic technique, such as pushouts, in a possibly nonalgebraic setting, I have chosen "equational specification" to mean the hardcore stuff of the equational logics and algebras that we all know and love.

solutions should keep the rope taut, not allowing any particular goal to win at the expense of the others.

Judging whether membership equational logic does indeed meet the above goals is left as an exercise, perhaps the most important exercise, for the reader. It should in any case be pointed out that membership equational logic is in fact very efficiently implementable under usual Church-Rosser assumptions about the specifications. The Maude interpreter [7] can reach up to 590,000 rewrites per second on such specifications running on a Pentium PC, and a compiler would easily achieve several million rewrites per second.

While sharing with CASL its ecumenical leanings, it should be obvious that the goals and criteria involved in each project are quite different, and that they are in fact complementary. Thus, CASL is algebraic in the latitudinarian sense of the word, but, being based on full first-order logic, does not enjoy the usual initiality, freeness and relative freeness properties of equational logics; and it is not reasonable to expect or demand any efficient executability of arbitrary CASL specifications. This is because the goals of generality and expressiveness of CASL's specification formalism completely dominate over those of efficient implementability and equational character, a perfectly legitimate choice for the purposes intended for CASL, but a very different one from those of membership equational logic and Maude. It is however reasonable to expect that a sizable number of CASL specifications will in fact fall within its equational subset, and that one would then desire to benefit from the executability and the powerful automated deduction techniques available in the equational world through appropriate tools. Since the equational logic subset of CASL can be mapped in a conservative way into membership equational logic, this opens up the possibility of using Maude as a tool for CASL, both for execution and—as several of us at SRI are currently doing, by developing a Church-Rosser checker tool and an inductive theorem prover for the Cafe language—as a metatool for building a formal environment of proving tools written in Maude.

Membership equational logic extends in a very natural way both order-sorted algebra with sort constraints [14, 25], and Wadge's classified algebra [34, 33] which can be regarded as the one-kinded[3] subcase of membership equational logic. In addition, membership equational logic is naturally a special case of, and as we shall see in Section 3 actually equivalent to, many-sorted Horn logic with equality, a logic that Joseph Goguen and I studied in detail as the natural setting for unifying equational and relational programming [15]. Therefore, there is nothing particularly surprising about it, except perhaps for how naturally it can be used as an equational logical framework, how well it satisfies the goals listed above, and how long it has taken some of us to recognize it for what it is worth.

The adequacy of membership algebra as a logical framework for equational

[3] In membership equational logic data is first classified in *kinds*, with each kind then having a set of different *sorts*. Kind checking is decidable, whereas sort checking is in general only semidecidable; however, it becomes decidable under adequate Church-Rosser conditions [4, 3].

specification is supported by the fact, discussed in detail in the paper, that—besides the formalisms already mentioned—other well-known total equational approaches to types such as many-sorted equational logic, equational type logic [21], unified algebras [28], and heterogeneous unified algebras [30] can be faithfully embedded in it. Similarly, a wide range of partial equational specification formalisms, including essentially algebraic theories [11, 1], hierarchical equational partial logic [32], left-exact sketches [2], limit theories [10], (conditional) many-sorted partial equational logic with existence equations [32], and partial equational logic with strong equations [5] can all be faithfully embedded in *partial membership equational logic*, a partial analogue of membership equational logic also studied in this paper. Partial membership equational logic can be mapped conservatively—and preserving key model-theoretic constructions such as initial, free, and relatively free algebras—into (total) membership equational logic. In this way, it is shown that membership equational logic has good properties as a logical framework for representing both total and partial equational specification formalisms. In addition, as shown in joint work with Adel Bouhoula and Jean-Pierre Jouannaud [4, 3] membership equational logic enjoys quite attractive automated deduction methods—such as completion and inductive theorem proving—and operational semantics using rewriting techniques. Claus Hintermeier and Claude and Hélène Kirchner have developed similar rewriting and completion techniques in a partial order-sorted setting [17, 18]. However, it seems fair to say that the total approach of membership algebra leads to simpler solutions and methods.

This work should be placed in context by explaining that it is part of a larger collective effort. Membership equational logic was first implemented in the Maude interpreter in joint work with Manuel Clavel, Steven Eker and Patrick Lincoln [7]. Shortly afterwards, Adel Bouhoula, Jean-Pierre Jouannaud and I started a joint effort to investigate in a systematic way its rewriting, proof-theoretic, and automated deduction aspects [4, 3]. At the same time I undertook a systematic study of both its model-theoretic aspects and its properties as an equational logical framework. The fruits of this last study were first reported at Dagstuhl in July of 1996 [23], distributed in long manuscript form during the Asilomar rewriting logic workshop in September of 1996, and, with more advances, presented at WADT'97 in Tarquinia and further spelled out in this version.

The structure of the paper is as follows. After giving the basic definitions in Section 2, we prove in Section 3 that membership equational logic and many-sorted Horn logic with equality define the same classes of models. Then, sound and complete rules of deduction, as well as constructions for initial and free algebras, and for free extensions along theory maps are given in Sections 4–7. A first discussion of membership equational logic as a logical framework for *total* equational specification formalisms is then presented in Section 8. But this raises the question of how, and how well, can partial equational logics be represented. Answering this question properly requires introducing some new technical concepts from the theory of institutions, namely a generalized notion of map of institutions and the more specific notion of an extension map of institutions in Section 10. The rest of the paper studies in detail two key extension maps of this kind representing order-sorted algebra (Sections 11 and 12) and *partial* membership equational logic (Sections 13 and 14) in membership equational logic. The overall properties of membership equational logic as an equational logical framework in which one can represent a wide range of logics are then discussed in Section 15. The paper finishes with some concluding remarks in Section 16.

2 Basic Definitions

Definition 1. A *signature* Ω in membership equational logic is a triple (K, Σ, π) where K is a set whose elements are called *kinds*, $\Sigma = \{\Sigma_{w,k}\}_{(w,k)\in K^*\times K}$ is a $K^* \times K$-indexed family of function symbols[4], and π is a function $\pi : S \to K$ that assigns to each element of a set S of *sorts* its corresponding kind. We denote by S_k the set $\pi^{-1}(k)$ for $k \in K$.

Note that in the above definition Σ is a K-sorted signature in the usual many-sorted sense. However, since in our terminology the elements of K are called kinds, and they are different from the sorts in S, we will throughout refer to Σ as a *K-kinded* signature, and to its usual many-sorted Σ-algebras as *many-kinded Σ-algebras.*

Definition 2. Given a signature Ω, an *Ω-algebra* consists of

1. a many-kinded Σ-algebra A, together with
2. an assignment to each sort $s \in S$ of a subset $A_s \subseteq A_{\pi(s)}$.

Given two Ω-algebras A and B, an *Ω-homomorphism* $h : A \to B$ is a Σ-homomorphism $h : A \to B$ such that for each sort $s \in S$ we have $h_{\pi(s)}(A_s) \subseteq B_s$.
With the usual composition of homomorphisms this defines a category $\mathbf{Alg}_\Omega$.

In membership equational logic there are two types of *atomic formulas*, namely:

1. *equations* of the form $t = t'$ where $t, t' \in T_\Sigma(X)_k$ for some $k \in K$, for X a K-kinded set of variables and $T_\Sigma(X)$ the free Σ-kinded algebra on X, and
2. *membership assertions* of the form $t : s$, where $s \in S$ and $t \in T_\Sigma(X)_{\pi(s)}$.

[4] As usual, K^* denotes the free monoid on the set K with unit λ. Each $f \in \Sigma_{w,k}$ is denoted $f : w \to k$. Moreover, to avoid ambiguous terms—so that each term has a unique kind—we will assume that Σ satisfies the property that whenever $f \in \Sigma_{w,k} \cap \Sigma_{w',k'}$ with w and w' of same length, then $w = w' \;\Leftrightarrow\; k = k'$. Any K-kinded signature has an isomorphic signature with this property. Note that this requirement does not rule out subsort overloading of function symbols. The point is that the same operator at the level of kinds may have several different overloadings at the level of sorts, which, as we shall see, can be expressed as axioms in the logic.

The *sentences* that we shall consider are universally quantified Horn clauses on these atomic formulas, that is, sentences of the form

$$(i)\ (\forall X)\, t = t' \Leftarrow u_1 = v_1 \wedge \ldots \wedge u_n = v_n \wedge w_1 : s_1 \wedge \ldots \wedge w_m : s_m$$
$$(ii)\ (\forall X)\, t : s \Leftarrow u_1 = v_1 \wedge \ldots \wedge u_n = v_n \wedge w_1 : s_1 \wedge \ldots \wedge w_m : s_m$$

where X is a K-kinded set containing all the variables appearing in t, t' (resp. t) and the u_i, v_i, and w_j.

Given an Ω-algebra A and a K-kinded set of variables X such that $t, t' \in T_\Sigma(X)$ (resp. $t \in T_\Sigma(X)$), then *satisfaction* of an atomic formula $t = t'$ (resp. $t : s$) relative to a K-kinded function $a : X \to A$, called an *assignment* of values in A to the variables X, is defined in the obvious way, that is,

$$A, a \models_\Omega t = t' \text{ iff } \overline{a}(t) = \overline{a}(t')$$
$$A, a \models_\Omega t : s \text{ iff } \overline{a}(t) \in A_s$$

where $\overline{a} : T_\Sigma(X) \to A$ is the unique K-kinded Σ-homomorphism from $T_\Sigma(X)$ to A as a Σ-algebra extending the assignment a.

Similarly, for φ a Horn sentence of type (i) (resp. (ii)), we say that A *satisfies* φ, written $A \models_\Omega \varphi$, iff for all K-kinded assignments $a : X \to A$ such that $A, a \models_\Omega u_i = v_i$, $1 \leq i \leq n$, and $A, a \models_\Omega w_j : s_j$, $1 \leq j \leq m$, we have $A, a \models_\Omega t = t'$ (resp. $A, a \models_\Omega t : s$).

For Γ a set of Horn sentences we write $A \models_\Omega \Gamma$ iff for all $\varphi \in \Gamma, A \models_\Omega \varphi$.

A *theory* is a pair (Ω, Γ), with Γ a set of Horn Ω-sentences. Such a theory defines a subclass of Ω-algebras, and therefore a full subcategory $\mathbf{Alg}_{\Omega,\Gamma} \subseteq \mathbf{Alg}_\Omega$, namely, exactly those Ω-algebras A such that $A \models_\Omega \Gamma$.

Example 1. The basic features of membership algebra can be illustrated by the following parameterized specification in Maude-like notation[5] for paths in a graph.

```
fth GRAPH is
  sorts Node Edge .
  ops s t : Edge -> Node .
efth

fmod PATH[G :: GRAPH] is
  sorts Path .
  subsorts Node Edge < Path .
  ops s t : Path -> Node .
  op _;_ : [Path] [Path] -> [Path] [assoc] .
  var E : Edge .
  var N : Node .
```

[5] The parameter theory GRAPH, introduced by the keyword fth, has a *loose* semantics; whereas the parameterized module PATH, introduced by the keyword fmod, has a *free algebra semantics*, sending each graph to the free PATH-algebra that it generates. Such free algebras exist by Theorem 10.

```
  vars P Q : Path .
  cmb E ; P : Path if t(E) == s(P) .
  ceq N ; P = P if s(P) == N .
  ceq P ; N = P if t(P) == N .
  eq s(N) = N .
  eq t(N) = N .
  ceq s(E ; P) = s(E) if t(E) == s(P) .
  ceq t(P ; E) = t(E) if t(P) == s(E) .
efm
```

The above specification is Church-Rosser and terminates with canonical terms that have the least possible sort among all their equivalent terms; therefore, it can be executed in Maude by term rewriting. Note that some syntactic sugar is used in this notation. There is only one kind in the module `PATH`, which is identified with the single connected component {`Node`, `Edge`, `Path`} generated by the subsort ordering, and is denoted by the equivalence class notation `[Path]`. Therefore, strictly speaking all variables are of kind `[Path]`; the sorts declared for the variables are in fact conditions left implicit in the axioms in which the variables appear. For example, the equation `s(N) = N` is syntactic sugar for

```
eq s(N) = N if N : Node .
```

Both subsort declarations and operator declarations are syntactic sugar for the corresponding conditional membership axioms. For example, `Edge < Path` and the operator declaration for the source function `s` mean, respectively,

```
cmb X : Path if X : Edge .
cmb s(X) : Node if X : Path .
```

for `X` a variable of kind `[Path]`. The attribute `[assoc]` of the path composition operator declares that the axiom of associativity holds for it and permits matching and rewriting modulo such an axiom. Path composition is in fact the only operator declared explicitly at the kind level; that the sort and target functions are unary functions of sort `[Path]` can be trivially inferred from their (sorted) operator declarations.

Note that, because of the conditional membership axiom for path composition, in the free `PATH`-algebra generated by a given graph a composition of edges will have sort `Path` if and only if the respective target and source of each pair of contiguous edges coincide. If this condition is not satisfied, such an expression will have kind `[Path]`, but will have no sort assignable to it. That is, it will be an *error expression*, corresponding to the fact that composition of paths is a partial function that is defined exactly when the respective sources and targets coincide.

At parse time we cannot in general determine whether an expression has a sort or not, since this requires performing logical inferences. Therefore, it is possible to give dubious expressions the benefit of the doubt, so that if in the end their canonical form is sortable, we know that the expression is not an error and denotes a well-defined value.

3 The Equivalence with Many-Sorted Horn Logic with Equality

With the above definition of sentences and satisfaction, membership equational logic is indeed an institution[6] [13], that we shall denote *MEqtl*. Furthermore, it is also quite clear that membership equational logic is a special type of Horn formalism and should therefore be viewed as a sublogic [24] of many-sorted Horn logic with equality, denoted $MSHorn^{=}$. That is, there is a sublogic inclusion $I : MEqtl \to MSHorn^{=}$.

Indeed, recall that signatures in $MSHorn^{=}$ are triples (L, Σ, Π) with L a set of sorts, $\Sigma = \{\Sigma_{w,l}\}_{(w,l)\in L^* \times L}$ a family of function symbols, and $\Pi = \{\Pi_w\}_{w \in L^*}$ a family of predicate symbols, and that models are L-sorted Σ-algebras A together with the assignment of a subset $P_A \subseteq A^w$ to each $P \in \Pi_w, w \in L^*$. Then, the inclusion I sends a signature $\Omega = (K, \Sigma, \pi : S \to K)$ to the signature $I(\Omega) = (K, \Sigma, \overline{S})$, where $\overline{S}_k = S_k$ for $k \in K$ and $\overline{S}_w = \emptyset$ for any $w \in K^* - K$, and where for each predicate symbol $s \in S_k$ we adopt the postfix notation $\_ : s$ to match our previous notation.

With this postfix convention the translation of sentences from *MEqtl* to $MSHorn^{=}$ is indeed the identity function, and the corresponding model categories are not only isomorphic, but in fact identical, that is, $\mathbf{Alg}_{\Omega} = \mathbf{Mod}_{I(\Omega)}$ and similarly $\mathbf{Alg}_{\Omega,\Gamma} = \mathbf{Mod}_{I(\Omega),\Gamma}$.

Conversely, we can regard many-sorted Horn logic with equality as a special case of membership equational logic by defining a map of institutions[7] [24] $J : MSHorn^{=} \to MEqtl$. The map J assigns to each signature (L, Σ, Π) the theory whose signature

1. has kind set $K = L \uplus \{p(w) \mid w \in L^* - L \text{ and } \Pi_w \neq \emptyset\}$;
2. for each kind $k \in K$ the set of sorts S_k is
   - if $k \in L$ then $S_k = \Pi_k$,
   - if $k = p(w)$ then $S_{p(w)} = \Pi_w$;
3. the signature of function symbols is

$$\begin{aligned}\Delta = \Sigma &\cup \{\langle \_, \ldots, \_ \rangle : l_1 \ldots l_n \to p(l_1 \ldots l_n) \mid p(l_1 \ldots l_n) \in K - L\} \\ &\cup \{\pi_i : p(l_1 \ldots l_n) \to l_i, 1 \leq i \leq n \mid p(l_1 \ldots l_n) \in K - L\} \ .\end{aligned}$$

   In particular, in the case $\Pi_\lambda \neq \emptyset$, we just have a constant $\langle\rangle : \lambda \to p(\lambda)$ and no projections.

The axioms of the theory $J(L, \Sigma, \Pi)$ are then the equations

$$\begin{aligned}&(\forall x_1 : l_1 \ldots x_n : l_n)\ \pi_i(\langle x_1 \ldots x_n \rangle) = x_i, \ \ 1 \leq i \leq n \\ &(\forall y : p(l_1 \ldots l_n))\ y = \langle \pi_1(y), \ldots, \pi_n(y) \rangle\end{aligned}$$

[6] However, the notion of maps of signatures and the corresponding preservation of satisfaction will be dealt with in Section 7.

[7] This map of institutions is furthermore an *embedding* in the sense explained in Section 8.

for each $p(l_1 \ldots l_n) \in K - L$, stating that the kind $p(l_1 \ldots l_n)$ is, up to isomorphism, the cartesian product of the kinds $l_1, \ldots, l_n$ with projection functions $\pi_1, \ldots, \pi_n$. In particular, in the case $p(\lambda) \in K - L$ these equations become the equation $(\forall y : p(\lambda))\ y = \langle\rangle$, stating that the kind $p(\lambda)$ is a one-point set.

The translation α of sentences sends each equation $t = t'$ to itself, each predicate atomic formula $P(t_1, \ldots, t_n)$ to the membership assertion $\langle t_1, \ldots, t_n \rangle : P$, and each Horn clause

$$(\forall X)\ at \Leftarrow u_1 = v_1 \wedge \ldots \wedge u_n = v_n \wedge P_1(\overline{w_1}) \wedge \ldots \wedge P_m(\overline{w_m})$$

to the Horn sentence

$$(\forall X)\ \alpha(at) \Leftarrow u_1 = v_1 \wedge \ldots \wedge u_n = v_n \wedge \langle \overline{w_1} \rangle : P_1 \wedge \ldots \wedge \langle \overline{w_m} \rangle : P_m \ .$$

A general Horn theory (L, Σ, Π, Γ) is then mapped by J to the theory obtained by adding to $J(L, \Sigma, \Pi)$ all the axioms $\alpha(\Gamma)$. It is then easy to show that there is an equivalence of categories

$$\mathbf{Mod}_{(L,\Sigma,\Pi)} \approx \mathbf{Alg}_{J(L,\Sigma,\Pi)} \ ,$$

and more generally,

$$\mathbf{Mod}_{(L,\Sigma,\Pi,\Gamma)} \approx \mathbf{Alg}_{J(L,\Sigma,\Pi,\Gamma)} \ .$$

This shows that, as institutions, *MSHorn*$^=$ and *MEqtl* have essentially the same expressive power, since theories can be translated in both directions by maps of institutions, and the corresponding categories of models are equivalent.

4 Rules of Deduction, Soundness and Completeness

The above equivalence with Horn logic also shows that, since we can regard *MEqtl* as a sublogic of many-sorted Horn logic with equality, all the well-known results for many-sorted Horn logic with equality hold true when restricted to the sublogic *MEqtl*. In particular, the proof of soundness and completeness for the rules of deduction of order-sorted Horn logic with equality in [15], of which many-sorted Horn logic with equality is a special case, yields the soundness and completeness of the following rules of deduction for a theory (Ω, Γ) in membership equational logic as an immediate corollary (where all the quantifications are assumed to involve sets of K-kinded variables containing those occurring in all terms of the given sentence).

1. **Reflexivity.**

$$\frac{}{\Gamma \vdash_\Omega (\forall X)\ t = t}$$

2. **Symmetry.**

$$\frac{\Gamma \vdash_\Omega (\forall X)\ t = t'}{\Gamma \vdash_\Omega (\forall X)\ t' = t}$$

3. **Transitivity.**

$$\frac{\Gamma \vdash_{\Omega} (\forall X)\, t = t' \qquad \Gamma \vdash_{\Omega} (\forall X)\, t' = t''}{\Gamma \vdash_{\Omega} (\forall X)\, t = t''}$$

4. **Congruence.**

$$\frac{\Gamma \vdash_{\Omega} (\forall X)\, t_1 = t_1' \quad \ldots \quad \Gamma \vdash_{\Omega} (\forall X)\, t_n = t_n'}{\Gamma \vdash_{\Omega} (\forall X)\, f(t_1, \ldots, t_n) = f(t_1', \ldots, t_n')}$$

where we assume that $f : k_1 \ldots k_n \to k$ is in Σ, and the terms $t_i, t_i' \in T_{\Sigma}(X)_{k_i}$, $1 \leq i \leq n$.

5. **Membership.**

$$\frac{\Gamma \vdash_{\Omega} (\forall X)\, t = t' \qquad \Gamma \vdash_{\Omega} (\forall X)\, t : s}{\Gamma \vdash_{\Omega} (\forall X)\, t' : s}$$

6. **Modus ponens.** Given a sentence

$$(\forall X)\, t = t' \Leftarrow u_1 = v_1 \wedge \ldots \wedge u_n = v_n \wedge w_1 : s_1 \wedge \ldots \wedge w_m : s_m$$
$$(\text{resp. } (\forall X)\, t : s \Leftarrow u_1 = v_1 \wedge \ldots \wedge u_n = v_n \wedge w_1 : s_1 \wedge \ldots \wedge w_m : s_m)$$

in the set Γ of axioms, and given a K-kinded assignment $\theta : X \to T_{\Sigma}(Y)$, then

$$\frac{\Gamma \vdash_{\Omega} (\forall Y)\, \overline{\theta}(u_i) = \overline{\theta}(v_i) \;\; 1 \leq i \leq n \qquad \Gamma \vdash_{\Omega} (\forall Y)\, \overline{\theta}(w_j) : s_j \;\; 1 \leq j \leq m}{\Gamma \vdash_{\Omega} (\forall Y)\, \overline{\theta}(t) = \overline{\theta}(t') \quad (\text{resp. } (\forall Y)\, \overline{\theta}(t) : s)}$$

5 Initial and Free Models

Also as an immediate corollary of the existence of initial and free models for order-sorted Horn logic with equality [15], we get both a concrete construction for initial and free algebras as well as their corresponding universal properties.

Theorem 3. *Let X be a K-kinded set, $\Omega = (K, \Sigma, \pi)$ a signature, and Γ a set of Ω-sentences. Then, the quotient $\mathcal{T}_{\Omega,\Gamma}(X)$ of $T_{\Sigma}(X)$ obtained by imposing the Σ-congruence*[8]

$$t \equiv_{\Gamma(X)} t' \iff \Gamma \vdash_{\Omega} (\forall X)\, t = t' \;,$$

together with the assignment of a subset[9]

$$\mathcal{T}_{\Omega,\Gamma}(X)_s = \{[t] \in \mathcal{T}_{\Omega,\Gamma}(X)_k \mid \Gamma \vdash_{\Omega} (\forall X)\, t : s\} \;,$$

is a free (Ω, Γ)-algebra on X, that is, $\mathcal{T}_{\Omega,\Gamma}(X) \in \mathbf{Alg}_{\Omega,\Gamma}$ and given $A \in \mathbf{Alg}_{\Omega,\Gamma}$, for each assignment $a : X \to A$ there exists a unique Ω-homomorphism $\overline{a} : \mathcal{T}_{\Omega,\Gamma}(X) \to A$ such that the diagram

[8] This is trivially a Σ-congruence by rules 1–4.
[9] Which does not depend on the choice of a representative by rule 5.

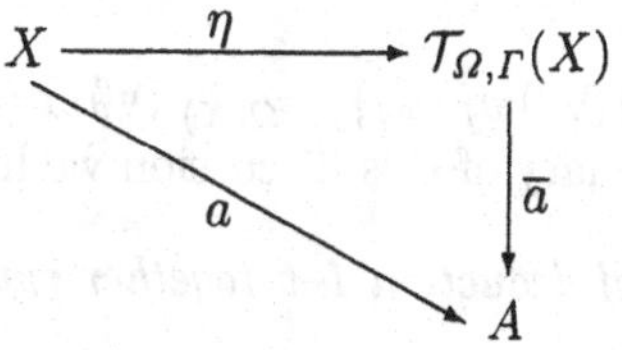

commutes, where η is the assignment mapping each variable x to the equivalence class $[x]$.

Corollary 4. *The algebra $\mathcal{T}_{\Omega,\Gamma} = \mathcal{T}_{\Omega,\Gamma}(\emptyset)$, for $\emptyset$ the empty K-kinded set, is initial in $\mathbf{Alg}_{\Omega,\Gamma}$.*

6 The Theorem of Constants

The rules of deduction 1–6 can only deduce atomic sentences. What about deducing other implications from those in the set of axioms Γ? By viewing variables as extra constants, we can justify the soundness and completeness of adding an implication introduction rule for this purpose to our original rules 1–6.

Given a signature $\Omega = (K, \Sigma, \pi)$ and a K-kinded set X which, without loss of generality, we can assume disjoint from Σ and with $X_k \cap X_{k'} = \emptyset$ if $k \neq k'$, we define $\Omega(X) = (K, \Sigma(X), \pi)$, where $\Sigma(X)$ is the signature obtained by adding the elements of X as constants of the respective kind.

Notice that an Ω-algebra together with an assignment $a : X \to A$, with X a K-kinded set, determines an $\Omega(X)$-algebra structure that can be denoted by the pair (A, a), namely, the extra constants in X are interpreted according to the assignment a. Furthermore, any $\Omega(X)$-algebra can be so described by such a pair (A, a), with A its underlying Ω-algebra structure. Notice also that, by the definition of satisfaction, $A, a \models_\Omega t = t'$ (resp. $t : s$) iff $(A, a) \models_{\Omega(X)} (\forall\emptyset)\, t = t'$ (resp. $(\forall\emptyset)\, t : s$).

Let now

$$\varphi = (\forall X)\, t = t' \Leftarrow u_1 = v_1 \wedge \ldots \wedge u_n = v_n \wedge w_1 : s_1 \wedge \ldots \wedge w_m : s_m$$
$$(\text{resp. } \varphi = (\forall X)\, t : s \Leftarrow u_1 = v_1 \wedge \ldots \wedge u_n = v_n \wedge w_1 : s_1 \wedge \ldots \wedge w_m : s_m)$$

be a Horn sentence. Can we characterize when all algebras for a theory (Ω, Γ) satisfy such a sentence?

By reinterpreting the definition of satisfaction as satisfaction for the corresponding $\Omega(X)$-algebras, we can do this as follows:

For all $A \in \mathbf{Alg}_{\Omega,\Gamma}$, $A \models_\Omega \varphi$

iff

For all $(A, a) \in \mathbf{Alg}_{\Omega(X),\Gamma}$, if $(A, a) \models_{\Omega(X)} (\forall\emptyset)\, u_i = v_i$, $1 \leq i \leq n$ and $(A, a) \models_{\Omega(X)} (\forall\emptyset)\, w_j : s_j$, $1 \leq j \leq m$, then $(A, a) \models_{\Omega(X)} (\forall\emptyset)\, t = t'$ (resp. $(A, a) \models_{\Omega(X)} (\forall\emptyset)\, t : s$)

iff

For all $(A, a) \in \mathbf{Alg}_{\Omega(X),\Gamma\cup\{(\forall\emptyset)\, u_i = v_i\}_i \cup\{(\forall\emptyset)\, w_j : s_j\}_j}$ we have $(A, a) \models_{\Omega(X)} (\forall\emptyset)\, t = t'$ (resp. $(A, a) \models_{\Omega(X)} (\forall\emptyset)\, t : s$)

iff (by completeness)
$\Gamma \cup \{(\forall\emptyset)u_i = v_i\}_i \cup \{(\forall\emptyset)w_j : s_j\}_j \vdash_{\Omega(X)} (\forall\emptyset)\, t = t'$ (resp. $(\forall\emptyset)\, t : s$) .
As an immediate corollary of this discussion we have

Theorem 5. *The rules of deduction 1–6 together with the rule*

7. **Implication introduction.**

$$\frac{\Gamma \cup \{(\forall\emptyset)\; u_i = v_i\}_i \cup \{(\forall\emptyset\; w_j : s_j\}_j \vdash_{\Omega(X)} (\forall\emptyset)\; t = t'}{\Gamma \vdash_\Omega (\forall X)\; t = t' \Leftarrow \bigwedge_i u_i = v_i \wedge \bigwedge_j w_j : s_j}$$

$$\frac{\Gamma \cup \{(\forall\emptyset)\; u_i = v_i\}_i \cup \{(\forall\emptyset)\; w_j : s_j\}_j \vdash_{\Omega(X)} (\forall\emptyset)\; t : s}{\Gamma \vdash_\Omega (\forall X)\; t : s \Leftarrow \bigwedge_i u_i = v_i \wedge \bigwedge_j w_j : s_j}$$

are sound and complete for deducing all Horn sentences that are logical consequences of some given set of Horn sentences Γ, where now we allow some of the deductions to take place in theories extending (Ω, Γ) by additional constants and additional hypotheses.

7 Free Extensions along Theory Maps

We now define signature morphisms and theory morphisms, and show that membership equational logic is a liberal institution.

Definition 6. Given two signatures $\Omega = (K, \Sigma, \pi)$ and $\Omega' = (K', \Sigma', \pi')$, a *signature morphism* $H : \Omega \to \Omega'$ is a triple $H = (H_0, H_1, H_2)$, where $H_0 : K \to K'$ and $H_1 : S \to S'$ are functions such that the diagram

$$\begin{array}{ccc} S & \xrightarrow{H_1} & S' \\ \downarrow{\scriptstyle \pi} & & \downarrow{\scriptstyle \pi'} \\ K & \xrightarrow[H_0]{} & K' \end{array}$$

commutes, and where H_2 is a $K^* \times K$-indexed function

$$H_2 = \{H_{2,w,k} : \Sigma_{w,k} \to \Sigma'_{H_0(w),H_0(k)}\}_{(w,k)\in K^*\times K}$$

with $H_0(w)$ the homomorphic extension of H_0 to strings in K^*.

We can therefore describe signature morphisms $(H_0, H_1, H_2) : (K, \Sigma, \pi) \longrightarrow (K', \Sigma', \pi')$ as ordinary many-kinded signature morphisms $(H_0, H_2) : (K, \Sigma) \longrightarrow (K', \Sigma')$ together with a map $H_1 : S \longrightarrow S'$ commuting the square above.

Note that a theory map $H : \Omega \to \Omega'$ induces a forgetful functor

$$\mathcal{U}_H : \mathbf{Alg}_{\Omega'} \to \mathbf{Alg}_\Omega$$

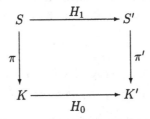

sending each algebra $A' \in \mathbf{Alg}_{\Omega'}$ to the Ω-algebra $\mathcal{U}_H(A')$ with $\mathcal{U}_H(A')_k = A'_{H_0(k)}$, $\mathcal{U}_H(A')_s = A'_{H_1(s)}$, and for each $f : w \to k$ in Σ, $\mathcal{U}_H(A')_f = A'_{H_2(f)}$.

In particular, for X a K-kinded set with $X_{k_1} \cap X_{k_2} = \emptyset$ if $k_1 \neq k_2$, denoting by $H_0(X)$ the K'-kinded set with $H_0(X)_{k'} = \bigcup_{H_0(k)=k'} X_k$, we can consider the algebra $\mathcal{U}_H(\mathcal{T}_{\Omega'}(H_0(X)))$ and define the K-kinded assignment

$$j : X \to \mathcal{U}_H(\mathcal{T}_{\Omega'}(H_0(X)))$$

sending each variable $x \in X_k$ to itself in $H_0(X)_{H_0(k)} \subseteq \mathcal{T}_{\Omega'}(H_0(X))_{H_0(k)} = \mathcal{U}_H(\mathcal{T}_{\Omega'}(H_0(X)))_k$. By the freeness of $\mathcal{T}_\Omega(X)$, j then extends uniquely to an Ω-homomorphism

$$H(X) : \mathcal{T}_\Omega(X) \to \mathcal{U}_H(\mathcal{T}_{\Omega'}(H_0(X)))$$

which provides the obvious translation from Ω-terms to Ω'-terms induced by H. We can extend this translation to Horn sentences by defining

$$\begin{aligned} &H((\forall X)\, t = t' \ \ (\text{resp. } t : s) \Leftarrow \textstyle\bigwedge_i u_i = v_i \wedge \bigwedge_j w_j : s_j) = \\ &(\forall H_0(X))\ H(X)(t) = H(X)(t') \ \ (\text{resp. } H(X)(t) : H_1(s)) \Leftarrow \\ &\qquad \textstyle\bigwedge_i H(X)(u_i) = H(X)(v_i) \wedge \bigwedge_j H(X)(w_j) : H_1(s_j)\ . \end{aligned}$$

We leave for the reader to check the following

Lemma 7 (Satisfaction Lemma). *Let $H : \Omega \to \Omega'$ be a signature morphism and let φ be a Horn Ω-sentence. Then, for each algebra $A' \in \mathbf{Alg}_{\Omega'}$ we have*

$$\mathcal{U}_H(A') \models_\Omega \varphi \iff A' \models_\Omega H(\varphi)\ .$$

Theory morphisms are then defined in the usual way.

Definition 8. Given two theories (Ω, Γ) and (Ω', Γ'), a *theory morphism* $H : (\Omega, \Gamma) \to (\Omega', \Gamma')$ is a signature morphism $H : \Omega \to \Omega'$ such that for each $\varphi \in \Gamma$ we have $\Gamma' \vdash_{\Omega'} H(\varphi)$.

Signature morphisms compose in the obvious way and form a category **Sign**. Similarly, theory morphisms compose as well, forming a category **Th**. Furthermore, by a general result about institutions, the category **Th** is finitely cocomplete iff **Sign** is so. Indeed, we have

Lemma 9. **Sign** *and therefore* **Th** *are finitely cocomplete.*

Proof sketch. The initial object in **Sign** is the signature $\emptyset = (\emptyset, \emptyset, 1_\emptyset)$. To see that **Sign** has pushouts, notice that many-kinded signatures have pushouts, and obtain the function π for the pushout as the unique function from the pushout of the sorts to the pushout of the kinds. □

Thanks to the satisfaction lemma, a theory morphism $H : (\Omega, \Gamma) \to (\Omega', \Gamma')$ induces a forgetful functor

$$\mathcal{U}_H : \mathbf{Alg}_{\Omega', \Gamma'} \to \mathbf{Alg}_{\Omega, \Gamma}\ .$$

We now show that membership equational logic is a *liberal* institution in the sense that any such $\mathcal{U}_H$ always has a left adjoint.

Theorem 10. *For any theory morphism $H : (\Omega, \Gamma) \to (\Omega', \Gamma')$ the functor $\mathcal{U}_H$ has a left adjoint $\mathcal{F}_H$.*

Proof sketch. Given $A \in \mathbf{Alg}_{\Omega,\Gamma}$, let $Diag_H(A)$ be the following set of ground $\Omega'(H_0(A))$-sentences:

1. $(\forall\emptyset)\ H(f)(a_1, \dots, a_n) = A_f(a_1, \dots, a_n)$
   for each $f : k_1 \dots k_n \to k$ in Σ, and $a_i \in A_{k_i}$, $1 \le i \le n$, where $A_f(a_1, \dots, a_n)$ denotes the result of the operation f applied to $a_1, \dots, a_n$ in the algebra A.
2. $(\forall\emptyset)\ a : H_1(s)$ for each $s \in S$ and $a \in A_s$ in A.

Then define $\mathcal{F}_H(A)$ as the underlying Ω'-algebra of the initial algebra

$$\mathcal{T}_{\Omega'(H_0(A)), \Gamma' \cup Diag_H(A)} .$$

□

8 Comparison with Some Total Equational Logics

The claim that membership equational logic has good properties as a logical framework for equational specification has to be substantiated in practice by showing how other equational specification formalisms can be represented inside membership equational logic. To begin with, we should consider *total* equational specification formalisms; partial specification formalisms will be discussed later, in Section 15.

The total equational specification formalisms that we shall consider constitute what might be called a representative collection of such formalisms, in the sense that, although not being an exhaustive list of the formalisms that have been proposed, seems to cover the most commonly used ones[10]. In all cases, the representation maps translating each formalism into membership equational logic are *embeddings* of institutions, that is, a particular kind of maps of institutions generalizing sublogic inclusions [24] and having especially nice properties. We assume familiarity with the notion of maps of institutions in [24] and give here a definition of embedding maps that generalizes one due to Mossakowski [27].

Definition 11. A map of institutions $(\Phi, \alpha, \beta) : \mathcal{I} \to \mathcal{I}'$ is called an *embedding* iff for each $T \in \mathbf{Th}_{\mathcal{I}}$ the functor $\beta_T : Mod'(\Phi(T)) \to Mod(T)$ is an equivalence of categories. We then use the notation $(\Phi, \alpha, \beta) : \mathcal{I} \hookrightarrow \mathcal{I}'$ to denote an embedding map. It is easy to show that embedding maps are closed under composition.

Remark that Φ maps theories in the first institution into theories in the second. As it is well-known for maps of institutions, this allows more flexible mappings than those mapping signatures to signatures.

[10] For a good survey of equational specification formalisms see [29].

Mossakowski [27] has generalized his notion of embedding to that of *weak embedding*, denoted $\mathcal{I} \leftarrow\!-\!\!\rightarrow \mathcal{I}'$. We shall encounter such weak embeddings later, in Section 15.

Embedding maps $\mathcal{I} \hookrightarrow \mathcal{I}'$ have very good properties. If $\mathcal{I}$ and $\mathcal{I}'$ are liberal institutions, they preserve relatively free model constructions, and if isomorphic models in $\mathcal{I}$ satisfy the same sentences then, assuming that $\mathcal{I}$ and $\mathcal{I}'$ have entailment systems that are complete relative to their satisfaction relations, the embedding is a *conservative* map of logics. All this follows from more general results in Section 10. In summary, an embedding $\mathcal{I} \hookrightarrow \mathcal{I}'$ indicates that $\mathcal{I}$ can be regarded as essentially a sublogic of $\mathcal{I}'$, but gives more syntactic flexibility than a sublogic inclusion by allowing maps Φ from theories to theories, and by not requiring the translation of sentences to be an inclusion, or even injective.

Note that we have already encountered in Section 3 embeddings $\mathit{MSHorn}^= \hookrightarrow \mathit{MEqtl}$ and $\mathit{MEqtl} \hookrightarrow \mathit{MSHorn}^=$, demonstrating that both logics have essentially the same expressive power.

Since *unsorted* Horn logic with equality $\mathit{USHorn}^=$ is a special case sublogic of many-sorted Horn logic with equality, we obtain by composition an embedding

$$\mathit{USHorn}^= \hookrightarrow \mathit{MSHorn}^= \hookrightarrow \mathit{MEqtl} \ .$$

A number of important total equational specification formalisms are naturally embedded in $\mathit{USHorn}^=$, often as sublogics, including the following:

- Unsorted (conditional) equational logic *USEqtl*.
- The *equational type logic* (*EqtlTypL*) of Manca, Sallibra and Scollo [21], which is the special case of unsorted Horn logic with equality obtained by allowing only a binary "typing" predicate $\_ : \_$ besides equality.
- As already pointed out in the introduction, the *classified algebra* (*ClassEqtl*) of Wadge [34, 33] is the sublogic of membership equational logic obtained by allowing only a single kind. In spite of its very good properties, such a restriction to a single kind makes classified algebra strictly more limited in expressive power—it cannot for example represent the usual many-sorted formalisms by means of embeddings. Furthermore, the useful type checking allowed by many-kinded specifications, by which we can throw away nonsensical expressions such as `true/7` that do not have a parse at the kind level, is not available in the one-kinded case.
- The *unified algebra* formalism (*UnifEqtl*) of Mosses [28] can also be regarded as essentially a sublogic of unsorted Horn logic with equality. Indeed, Mosses [28] defines in detail an embedding $\mathit{UnifEqtl} \hookrightarrow \mathit{USHorn}^=$ showing that the models of a unified algebra specification are precisely the models of a corresponding theory in unsorted Horn logic with equality having a typing predicate $\_ : \_$ and a partial order predicate $\_ \leq \_$ in addition to the equality predicate, and satisfying suitable axioms that characterize the lattice-theoretic and typing properties of unified algebras. Similarly, the *heterogeneous* version of unified algebras (*HUnifEqtl*) proposed by Parisi-Presicce and Veglioni [30] has an embedding into into many-sorted Horn logic with

equality $HUnifEqtl \hookrightarrow MSHorn^=$, provided that the signature morphisms are restricted to be injective (see [30, Props. 18–21]).

Of course, the well-known case of (conditional) many-sorted equational logic ($MSEqtl$) is also a sublogic of membership equational logic, namely the sublogic obtained by making the set S of sorts empty. Finally, it is also well-known that (conditional) many-sorted equational logic is a sublogic of (conditional) order-sorted equational logic ($OSEqtl$) as formulated in [16], namely the case when the poset of sorts is discrete, and that, conversely, there is also an embedding of logics $OSEqtl \hookrightarrow MSEqtl$ [16], which composed with the embedding $MSEqtl \hookrightarrow MEqtl$ gives us an embedding $OSEqtl \hookrightarrow MEqtl$. However, as we shall see later (in Sections 11 and 12), this is not the only possible—nor the most natural—way of representing order-sorted equational logic in membership equational logic.

We can summarize our present discussion by means of the following diagram of embeddings:

$$
\begin{array}{ccccccc}
EqtlTypL & \hookrightarrow & USHorn^= & \hookrightarrow & MSHorn^= & \hookleftarrow & \boxed{MEqtl} \\
& \nearrow & \uparrow & & \uparrow & & \uparrow \\
UnifEqtl & & ClassEqtl & & HUnifEqtl & & \\
& & \uparrow & & & & \\
& & USEqtl & \longrightarrow & & MSEqtl & \rightleftarrows OSEqtl
\end{array}
$$

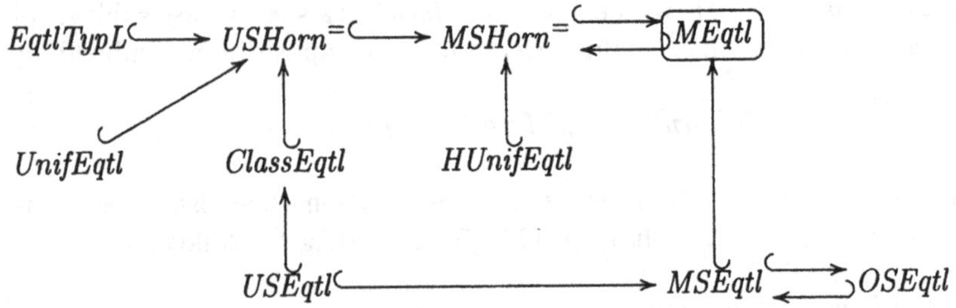

9   A Partiality Puzzle

The very fact that membership equational logic has the same expressive power as many-sorted Horn logic with equality seems to point to a fundamental limitation, and to cast serious doubts about its adequacy as a logical framework for equational specification. After all, from a model-theoretic and categorical characterization of categories of models defined by increasingly more expressive logics we know that:

1. Each *variety*, that is, each category of many-sorted algebras definable by a set of unconditional equations, is a special case of a *semivariety*, that is, a category of many-sorted algebras definable by conditional equations, but that there are semivarieties that cannot be specified by unconditional equations alone in the very strong sense that they can never be equivalent as categories to a variety, regardless of how such a variety is specified. A simple such counterexample is the category of graphs (as sets endowed with a binary relation) and graph homomorphisms [1, 3.21.2].

2. Similarly, each semivariety is a category of many-sorted models definable in Horn logic with equality but the converse is not true. A well-known counterexample is the category of partially ordered sets [1, 3.21.1].

3. The real trouble comes with ***partial semivarieties***, that is, categories of partial algebras definable by conditional equations. Indeed, such categories are exactly the finitely locally presentable categories (see [1, Theorem 3.36] and Section 13) but there are finitely locally presentable categories that are not Horn definable [1, Example 5.13].

In summary, we have

Varieties $\subsetneq$ *Semivarieties* $\subsetneq$ *Horn*$^{=}$*-Classes* $\subsetneq$ *Partial Semivarieties*

and it is clear that membership equational logic can specify, up to equivalence of categories, exactly the categories of many-sorted models definable by Horn clauses with equality. Therefore, doesn't it then follow that membership equational logic, while clearly adequate as a logical framework for total algebras, is also clearly inadequate as a logical framework for partial ones?

The answer to this puzzle is a rotund *no*. It only shows that (conditional) partial equational logics cannot be mapped into *MEqtl* by means of *embeddings*. But what seems at first sight a limitation is actually the strongest and most attractive feature of membership equational logic as a logical framework, namely:

1. Its logic and model theory are quite simple, in fact strictly simpler than the more complex logic and model categories of (conditional) partial algebras, yet
2. This more complex world of partiality can be faithfully embedded in the simpler world of membership equational logic by means of *extension maps of institutions* such that:
   (a) they leave intact the categories of partial algebras, embedding them in categories of membership algebras, from which they can be recovered back by an adjoint construction,
   (b) they preserve all free constructions in both directions, so that for all practical purposes we can reason in the simpler world of membership algebras,
   (c) the simpler logic of membership algebra can be *borrowed* to endow the corresponding categories of partial algebras with *sound* and *complete* rules of logical deduction in the framework logic.

10 Extension Maps of Institutions

Therefore, before discussing further the logical framework properties of *MEqtl* we introduce some new general notions in the theory of institutions, namely, generalized maps of institutions and extension maps; and we show that extension maps enjoy very good properties.

Definition 12. Given two institutions $\mathcal{I}$ and $\mathcal{I}'$, a functor $\Phi : \mathbf{Th} \to \mathbf{Th}'$ is called *signature-preserving* iff there is a functor $\Phi^\diamond : \mathbf{Sign} \to \mathbf{Sign}'$ such that the diagram

$$\begin{array}{ccc} \mathbf{Th} & \xrightarrow{\Phi} & \mathbf{Th}' \\ {\scriptstyle sign}\downarrow & & \downarrow{\scriptstyle sign'} \\ \mathbf{Sign} & \xrightarrow[\Phi^\diamond]{} & \mathbf{Sign}' \end{array}$$

commutes. It is then easy to see that $\Phi^\diamond$ is necessarily unique.

The following definition of generalized maps of institutions provides a more general and flexible way of mapping one institution to another than the usual maps of institutions in [24]. This extra generality will in fact be needed in Section 14 to map partial algebras into membership algebras. The extra generality consists in allowing one sentence in the first institution to be translated into a *set* of sentences in the second, and in relaxing the natural transformation requirement for the translation of sentences α to a weaker condition on the corresponding categories of models.

Definition 13. Given two institutions $\mathcal{I}$ and $\mathcal{I}'$, a *generalized map of institutions* $(\Phi, \alpha, \beta) : \mathcal{I} \to \mathcal{I}'$ is given by

- a signature-preserving functor $\Phi : \mathbf{Th} \to \mathbf{Th}'$,
- a family of functions $\alpha = \{\alpha_\Sigma : sen(\Sigma) \to \mathcal{P}(sen'(\Phi^\diamond(\Sigma)))\}_{\Sigma \in \mathbf{Sign}}$, and
- a natural transformation $\beta : Mod \circ \Phi^{op} \Rightarrow Mod$,

such that:

(i) Denoting $\Phi(\Sigma, \Gamma) = (\Sigma', \Gamma')$ and $\Phi(\Sigma, \emptyset) = (\Sigma', \emptyset'_\Sigma)$, we have

$$Mod'(\Sigma', \Gamma') = Mod'(\Sigma', \emptyset'_\Sigma \cup \bigcup \alpha_\Sigma(\Gamma)) \ .$$

(ii) For each $H : \Sigma_1 \to \Sigma_2$ in **Sign**, and each $\varphi \in sen(\Sigma_1)$, we have

$$Mod'(\Sigma'_2, \emptyset'_{\Sigma_2} \cup \alpha_{\Sigma_2}(H(\varphi))) = Mod'(\Sigma'_2, \emptyset'_{\Sigma_2} \cup \Phi^\diamond(H)(\alpha_{\Sigma_1}(\varphi))) \ .$$

(iii) For each $\Sigma \in \mathbf{Sign}, \varphi \in sen(\Sigma)$, and $M' \in Mod'(\Sigma', \emptyset'_\Sigma)$, we have

$$M' \models'_{\Sigma'} \alpha_\Sigma(\varphi) \iff \beta_{(\Sigma,\emptyset)}(M') \models_\Sigma \varphi \ .$$

Where we have shortened $sen(H)$ to H, and $sen'(\Phi^\diamond(H))$ to $\Phi^\diamond(H)$.

It is not hard to show that generalized maps of institutions form a category that contains the usual maps of institutions as a subcategory.

The following proposition gives a particularly compact way of defining a generalized map of institutions.

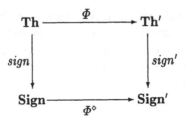

Proposition 14. *Given two institutions $\mathcal{I}$ and $\mathcal{I}'$, a functor $\Phi_0 : \mathbf{Sign} \to \mathbf{Th}'$ (where we then denote by $\Phi^\diamond : \mathbf{Sign} \to \mathbf{Sign}'$ the functor $\Phi^\diamond = sign' \circ \Phi_0$), a family of functions $\alpha = \{\alpha_\Sigma : sen(\Sigma) \to \mathcal{P}(sen'(\Phi^\diamond(\Sigma)))\}_{\Sigma \in \mathbf{Sign}}$, and a natural transformation $\beta : Mod \circ \Phi_0^{op} \Rightarrow Mod$, satisfying conditions (ii) (with the slight change of $\beta_{(\Sigma,\emptyset)}$ by β_Σ) and (iii), can be extended*[11] *in a unique way to a generalized map of institutions $(\Phi, \alpha, \beta) : \mathcal{I} \to \mathcal{I}'$ whose restriction to signatures is $\Phi^\diamond$ and such that if $\Phi_0(\Sigma) = (\Sigma', \emptyset'_\Sigma)$, then $\Phi(\Sigma, \Gamma) = (\Sigma', \emptyset'_\Sigma \cup \bigcup \alpha_\Sigma(\Gamma))$.*

We are now ready to define extension maps.

Definition 15. A generalized map of institutions $(\Phi, \alpha, \beta) : \mathcal{I} \to \mathcal{I}'$ is called an *extension* iff for each $T \in \mathbf{Th}_\mathcal{I}$ the functor $\beta_T : Mod'(\Phi(T)) \to Mod(T)$ has a left adjoint ρ_T such that the unit of the adjunction is a natural isomorphism

$$1_{\mathrm{Mod}} \simeq \beta_T \circ \rho_T \ .$$

By general results of category theory [20], the functor ρ is then full and faithful. The functor ρ_T is called the *extension* functor, whereas β_T is called the *restriction* functor. Intuitively, the models of $Mod'(\Phi(T))$ contain those of $Mod(T)$ as a subclass, and can be "restricted" to that subclass via β.

We will use the notation $(\Phi, \alpha, \beta) : \mathcal{I} \rightarrowtail \mathcal{I}'$ to denote an extension map of institutions. It is easy to show that extension maps are closed under composition, and that they contain embeddings as a subcategory.

I am indebted to Till Mossakowski for pointing out that the above notion of extension map is closely related to his notion of *categorical retractive simulation*, which is a map of institutions (Φ, α, β) with a right-inverse left adjoint for each β_T [19]. In fact, categorical retractive simulations are a special case of the more general notion of an extension map.

The following proposition shows that if an institution $\mathcal{I}'$ has an entailment system making it into a complete logic, then we can use an extension map $\mathcal{I} \rightarrowtail \mathcal{I}'$ to "borrow" such an entailment system by the general method of [6] to make $\mathcal{I}$ a complete logic. We will later use this result with $\mathcal{I}' = MEqtl$ to endow other logics with sound and complete rules of deduction.

Proposition 16. *Let $(\Phi, \alpha, \beta) : \mathcal{I} \rightarrowtail \mathcal{I}'$ be an extension map such that isomorphic models in $\mathcal{I}$ satisfy the same sentences, and such that $\vdash'$ is a complete entailment system for $\mathcal{I}'$. Then, defining*

$$\Gamma \vdash_\Sigma \varphi \iff \textit{for all } \psi \in \alpha_\Sigma(\varphi),\ \emptyset'_\Sigma \cup \bigcup \alpha_\Sigma(\Gamma) \vdash'_{\Sigma'} \psi$$

yields a complete entailment system for $\mathcal{I}$.

Proof. We need to show that $\Gamma \vdash_\Sigma \varphi$ iff for all $M \in Mod(\Sigma, \Gamma), M \models_\Sigma \varphi$. By (Φ, α, β) being an extension map and isomorphic models in $\mathcal{I}$ satisfying the same sentences we have

[11] In the obvious sense of extension.

for all $M \in Mod(\Sigma, \Gamma), M \models_\Sigma \varphi$
iff
for all $M' \in Mod'(\Phi(\Sigma, \Gamma))$, $\beta(M') \models_\Sigma \varphi$
iff (by condition (iii))
for all $M' \in Mod'(\Phi(\Sigma, \Gamma))$, for all $\psi \in \alpha_\Sigma(\varphi)$, $M' \models_{\Sigma'} \psi$
iff (by $\vdash'$ a complete entailment system and condition (i))
for all $\psi \in \alpha_\Sigma(\varphi)$, $\emptyset'_\Sigma \cup \bigcup \alpha_\Sigma(\Gamma) \vdash'_{\Sigma'} \psi$. □

Theorem 17. *Let $(\Phi, \alpha, \beta) : \mathcal{I} \rightarrowtail \mathcal{I}'$ be an extension map of institutions with $\mathcal{I}$ and $\mathcal{I}'$ liberal. Then, free constructions are preserved by both extension and restriction in the sense that in the following diagram of adjoints associated to a theory morphism $H : T \to T'$*

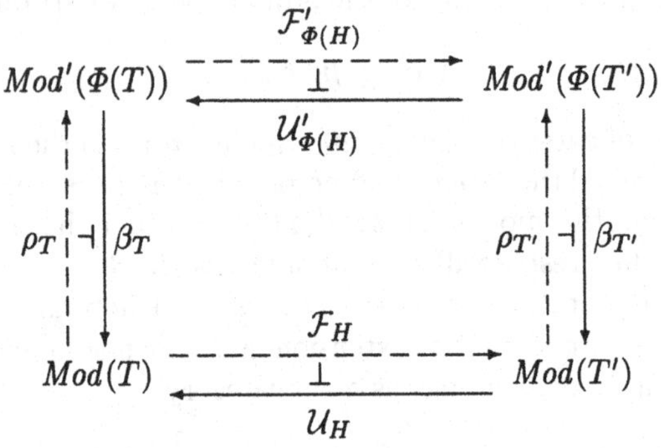

we have for each $M \in Mod(T)$

$$(i) \quad \mathcal{F}'_{\Phi(H)}(\rho_T(M)) \simeq \rho_{T'}(\mathcal{F}_H(M))$$
$$(ii) \quad \mathcal{F}_H(M) \simeq \beta_{T'}(\mathcal{F}'_{\Phi(H)}(\rho_T(M))) \ .$$

Proof. (i) follows immediately from the commutativity of the square of solid arrows by naturality of β, plus composition of adjoints being an adjoint, and adjoints being unique up to isomorphism. To see (ii), notice that

$$\beta_{T'}(\mathcal{F}'_{\Phi(H)}(\rho_T(M))) \simeq \beta_{T'}(\rho_{T'}(\mathcal{F}_H(M))) \simeq \mathcal{F}_H(M)$$

using first (i) and then that (Φ, α, β) is an extension. □

We will apply the notion of extension map and its properties to define maps into *MEqtl* in two important cases, namely order-sorted algebra (in Sections 11 and 12), and partial membership algebra (in Sections 13 and 14).

11 Order-Sorted Algebra Revisited

We should map to *MEqtl* the "best possible version" of order-sorted algebra (OSA). There are good reasons to believe that the versions now used like OSA^{GM} of Goguen and Meseguer [16] and OSA^{P} of Poigné [31] are unsatisfactory. I propose here a new version, OSA^{R}, that combines all the advantages and does not seem to have any drawbacks.

The following table summarizes some key desirable properties and how they are met, or fail to be met, by each variant of order-sorted algebra. The yes/no answers would have to be qualified in some cases, but this will have to wait for a longer discussion. Note that OSA^{GM} and OSA^{P} are Boolean complements of each other with respect to these properties.

	OSA^{GM}	OSA^{P}	OSA^{R}
Contains MSA as special case	Yes	No	Yes
Term algebras initial with minimal requirements	No	Yes	Yes
Pushouts of theories exist in general	No	Yes	Yes
Ad-hoc overloading possible	Yes	No	Yes
Terms of different sorts allowed in equations	Yes	No	Yes
Natural conservative extension map to *MEqtl*	No	Yes	Yes

Definition 18. An *order-sorted signature* is a triple $\Sigma = (S, \leq, \Sigma)$ with $S = (S, \leq)$ a poset and (S, Σ) a many-sorted signature.

We denote by $\hat{S} = S/{\equiv_{\leq}}$ the set of *connected components* of $(S, \leq)$, which is the quotient of S under the equivalence relation $\equiv_{\leq}$ generated by $\leq$. The equivalence $\equiv_{\leq}$ can be extended to sequences in the usual way.

The notion of a sensible signature is a minimal syntactic requirement to avoid excessive ambiguity. However, it permits fully flexible ad-hoc and subsort overloading of function symbols.

Definition 19. An order-sorted signature is called *sensible* if for any two operators $f : w \to s, f : w' \to s'$, with w and w' of same length, we have $w \equiv_{\leq} w' \;\Leftrightarrow\; s \equiv_{\leq} s'$.

Any order-sorted signature can be transformed into an isomorphic signature that is sensible and still has the desired ad-hoc and subsort overloadings.[12]

Notation. For connected components $k_1, \ldots, k_n, k \in \hat{S}$

$$f_k^{k_1 \ldots k_n} = \{f : s_1 \ldots s_n \to s \mid s_i \in k_i \;\; 1 \leq i \leq n, \; s \in k\}$$

denotes the family of "subsort polymorphic" operators with name f for those components.

Using the projection to the quotient $S \twoheadrightarrow \hat{S}$ we can define a many-sorted signature $(\hat{\Sigma}, \hat{S})$ with $f : k_1 \ldots k_n \to k$ iff $f_k^{k_1 \ldots k_n} \neq \emptyset$ in Σ.

---

[12] One can decorate each $f : s_1 \ldots s_n \longrightarrow s$ with the string of connected components $[s_1] \ldots [s_n][s]$ to get operators $f_{[s_1]\ldots[s_n][s]} : s_1 \ldots s_n \longrightarrow s$. Ad-hoc overloading can then be recovered by making suitable identifications between decorated function symbols without violating sensibility.

Definition 20. *Signature morphisms* $H : (S, \leq, \Sigma) \to (S', \leq', \Sigma')$ are many-sorted signature morphisms $H : (S, \Sigma) \to (S', \Sigma')$ with $H : (S, \leq) \to (S', \leq')$ monotonic and such that they preserve subsort overloading, that is, such that the following diagram commutes:

$$\begin{array}{ccc} \Sigma & \xrightarrow{\quad H \quad} & \Sigma' \\ \downarrow & & \downarrow \\ \hat{\Sigma} & \dashrightarrow & \hat{\Sigma}' \end{array}$$

Theorem 21. *The category* **OSSign** *of order-sorted signatures and homomorphisms is cocomplete.*

Definition 22. For $\Sigma = (S, \leq, \Sigma)$ an order-sorted signature, an *order-sorted Σ-algebra* A is a many-sorted (S, Σ)-algebra A such that

- whenever $s \leq s'$, then we have $A_s \subseteq A_{s'}$, and
- whenever $f : w \to s, f : w' \to s'$ in $f_k^{k_1 \ldots k_n}$ and $\overline{a} \in A^w \cap A^{w'}$, then we have $A_{f:w \to s}(\overline{a}) = A_{f:w' \to s'}(\overline{a})$.

An *order-sorted Σ-homomorphism* $h : A \to B$ is a many-sorted (S, Σ)-homomorphism such that whenever $s, s' \in k$ and $a \in A_s \cap A_{s'}$, then we have $h_s(a) = h_{s'}(a)$.

This defines a category $\mathbf{OSAlg}_\Sigma$.

Notice that, given an order-sorted Σ-algebra A, if we denote by $A_k = \bigcup_{s \in k} A_s$ for each $k \in \hat{S}$, then each $f_k^{k_1 \ldots k_n}$ determines a single *partial* function

$$A_{f_k^{k_1 \ldots k_n}} : A_{k_1} \times \cdots \times A_{k_n} \rightharpoonup A_k \ .$$

Similarly, an order-sorted Σ-homomorphism $h : A \to B$ determines an $\hat{S}$-family of functions

$$\{h_k : A_k \to B_k\}_{k \in \hat{S}} \ .$$

Theorem 23. *The category* $\mathbf{OSAlg}_\Sigma$ *has an initial algebra. Furthermore, if Σ is sensible, then the term algebra $\mathcal{T}_\Sigma$ with*

- *if $a : \lambda \to s$ then $a \in \mathcal{T}_{\Sigma,s}$,*
- *if $t \in \mathcal{T}_{\Sigma,s}$ and $s \leq s'$ then $t \in \mathcal{T}_{\Sigma,s'}$,*
- *if $f : s_1 \ldots s_n \to s$ and $t_i \in \mathcal{T}_{\Sigma,s_i}$ $1 \leq i \leq n$, then $f(t_1, \ldots, t_n) \in \mathcal{T}_{\Sigma,s}$,*

is initial.

Variable declarations have the form

$$\bar{X} = \{x_1 : s_1^1, \ldots, s_{l_1}^1; \ldots; x_m : s_1^m, \ldots, s_{l_m}^m\}$$

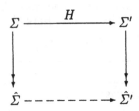

with $s^i_1, \ldots, s^i_{l_i} \in k_i \;\; 1 \leq i \leq m$, and then we can consider terms in the free algebra $\mathcal{T}_\Sigma(\tilde{X})$, which can be constructed as the initial algebra for the signature $\Sigma(\tilde{X})$ in which the variables x_i are added disjointly to Σ as constants with sorts s^i_j.

Conditional order-sorted Σ-equations are of the form

$$(\forall \tilde{X})\; t = t' \Leftarrow u_1 = v_1 \wedge \ldots \wedge u_n = v_n$$

with $t, t' \in \mathcal{T}_\Sigma(\tilde{X})_k$, and $u_i, v_i \in \mathcal{T}_\Sigma(\tilde{X})_{k_i} \;\; 1 \leq i \leq n$.

Satisfaction of such a conditional equation by an order-sorted Σ-algebra A can then be defined in the usual way using sort-preserving assignments $a : \tilde{X} \to A$.

An *order-sorted theory* is then a pair (Σ, Γ) with Γ a set of conditional Σ-equations. The category $\mathbf{OSAlg}_{\Sigma,\Gamma}$ of algebras for such a theory is the full subcategory of $\mathbf{OSAlg}_\Sigma$ defined by those algebras that satisfy the axioms in Γ.

Given a sensible order-sorted signature $(S, \leq, \Sigma)$ and a set Γ of conditional Σ-equations, we say that an unconditional equation $(\forall \tilde{X})\; t = t'$ is derivable from Γ iff it can be obtained by finite application of the following rules:

1. **Reflexivity**. For each variable declaration $\tilde{X}$ with $t \in \mathcal{T}_\Sigma(\tilde{X})$,

$$\frac{}{(\forall \tilde{X})\; t = t}$$

2. **Symmetry**.

$$\frac{(\forall \tilde{X})\; t = t'}{(\forall \tilde{X})\; t' = t}$$

3. **Transitivity**.

$$\frac{(\forall \tilde{X})\; t = t' \qquad (\forall \tilde{X})\; t' = t''}{(\forall \tilde{X})\; t = t''}$$

4. **Congruence**. For $f : s_1 \ldots s_n \to s$, $f : s'_1 \ldots s'_n \to s'$ in Σ with $s_1 \ldots s_n s \equiv_\leq s'_1 \ldots s'_n s'$, and for $t_i \in \mathcal{T}_\Sigma(\tilde{X})_{s_i}$, $t'_i \in \mathcal{T}_\Sigma(\tilde{X})_{s'_i}$, $1 \leq i \leq n$,

$$\frac{(\forall \tilde{X})\; t_1 = t'_1 \quad \ldots \quad (\forall \tilde{X})\; t_n = t'_n}{(\forall \tilde{X})\; f(t_1, \ldots, t_n) = f(t'_1, \ldots, t'_n)}$$

5. **Modus ponens**. For variable declarations $\tilde{X}, \tilde{Y}$, an assignment $\theta : \tilde{X} \to \mathcal{T}_\Sigma(\tilde{Y})$, and a conditional equation in Γ

$$(\forall \tilde{X})\; t = t' \Leftarrow u_1 = v_1 \wedge \ldots \wedge u_n = v_n$$

$$\frac{(\forall \tilde{Y})\; \overline{\theta}(u_1) = \overline{\theta}(v_1) \quad \ldots \quad (\forall \tilde{Y})\; \overline{\theta}(u_n) = \overline{\theta}(v_n)}{(\forall \tilde{Y})\; \overline{\theta}(t) = \overline{\theta}(t')}$$

Theorem 24. *Given an order-sorted theory (Σ, Γ), the above rules of deduction are sound and complete for deriving all the equations that are logical consequences of the axioms Γ.*

A signature morphism $H : (S, \leq, \Sigma) \to (S', \leq', \Sigma')$ induces a forgetful functor

$$\mathcal{U}_H : \mathbf{OSAlg}_{\Sigma'} \to \mathbf{OSAlg}_{\Sigma} \ ,$$

with $\mathcal{U}_H(A')_s = A'_{H(s)}$ and $\mathcal{U}_H(A')_f = A'_{H(f)}$ for each algebra $A' \in \mathbf{OSAlg}_{\Sigma'}$.

Definition 25. Given two order-sorted theories (Σ, Γ) and (Σ', Γ'), a *theory morphism* $H : (\Sigma, \Gamma) \to (\Sigma', \Gamma')$ is a signature morphism $H : \Sigma \to \Sigma'$ such that

$$\mathcal{U}_H(\mathbf{OSAlg}_{\Sigma',\Gamma'}) \subseteq \mathbf{OSAlg}_{\Sigma,\Gamma} \ ,$$

so that $\mathcal{U}_H$ restricts to a forgetful functor $\mathcal{U}_H : \mathbf{OSAlg}_{\Sigma',\Gamma'} \to \mathbf{OSAlg}_{\Sigma,\Gamma}$.

The institution $OSEqtl^{\mathrm{R}}$ is *liberal*, that is, for any theory morphism $H : (S, \leq, \Sigma, \Gamma) \to (S, \leq', \Sigma', \Gamma')$ we have an adjunction

$$\mathbf{OSAlg}_{\Sigma,\Gamma} \underset{\mathcal{U}_H}{\overset{\mathcal{F}_H}{\underset{\longleftarrow}{\overset{\dashrightarrow}{\bot}}}} \mathbf{OSAlg}_{\Sigma',\Gamma'}$$

12 Extending Order-Sorted Algebra to Membership Algebra

We can define an extension of institutions $(\Phi, \alpha, \beta) : OSEqtl^{\mathrm{R}} \rightarrowtail MEqtl$ as follows.

The functor Φ maps each order-sorted signature $\Sigma = (S, \leq, \Sigma)$ to the membership equational theory with signature $\Omega = (\hat{S}, \hat{\Sigma}, \pi : S \twoheadrightarrow \hat{S})$, where $\pi : S \twoheadrightarrow \hat{S}$ is the projection into $\equiv_{\leq}$-equivalence classes, and axioms

1. $(\forall x)\ x : s' \Leftarrow x : s$ for each $s \leq s'$ in $(S, \leq)$,
2. $(\forall x_1 \ldots x_n)\ f(x_1, \ldots, x_n) : s \Leftarrow x_1 : s_1 \wedge \ldots \wedge x_n : s_n$
   for each $f : s_1 \ldots s_n \to s$ in Σ.

The natural transformation α maps each order-sorted conditional equation

$$(\forall \tilde{X})\ t = t' \Leftarrow u_1 = v_1 \wedge \ldots \wedge u_n = v_n$$

to the sentence

$$(\forall x_1 \ldots x_m)\ t = t' \Leftarrow u_1 = v_1 \wedge \ldots \wedge u_n = v_n \wedge \tilde{X}$$

where, by abuse of language, if $\tilde{X} = x_1 : s_1^1, \ldots, s_{l_1}^1; \ldots; x_m : s_1^m, \ldots, s_{l_m}^m$ then in the above sentence $\tilde{X}$ denotes the conjunction

$$x_1 : s_1^1 \wedge \ldots \wedge x_1 : s_{l_1}^1 \wedge \ldots \wedge x_m : s_1^m \wedge \ldots \wedge x_m : s_{l_m}^m \ .$$

In general, Φ maps an order-sorted theory (Σ, Γ) to the theory obtained by adding to $\Phi(\Sigma)$ the axioms $\alpha(\Gamma)$; and maps an order-sorted theory morphism

$H : (\Sigma, \Gamma) \longrightarrow (\Sigma', \Gamma')$ to the theory morphism with many-kinded signature morphism the induced $\hat{H} : \hat{\Sigma} \longrightarrow \hat{\Sigma}'$, and with sort map $H : S \longrightarrow S'$.

The natural transformation β is given by the functor

$$(\_)^{\circ} : \mathbf{Alg}_{\Phi(\Sigma,\Gamma)} \to \mathbf{OSAlg}_{\Sigma,\Gamma}$$

sending each $A \in \mathbf{Alg}_{\Phi(\Sigma,\Gamma)}$ to the order-sorted algebra with carrier sets $\{A_s\}_{s \in S}$ that satisfies $A_s \subseteq A_{s'}$ whenever $s \leq s'$, and has operations $A_{f:s_1 \dots s_n \to s} : A_{s_1} \times \dots \times A_{s_n} \to A_s$ thanks to the axioms in $\Phi(\Sigma)$, and that satisfies the axioms Γ thanks to the axioms $\alpha(\Gamma)$.

Theorem 26. *The functor* $(\_)^{\circ} : \mathbf{Alg}_{\Phi(\Sigma,\Gamma)} \to \mathbf{OSAlg}_{\Sigma,\Gamma}$ *has a left adjoint that is a full coreflection*

$$(\_)^{\bullet} : \mathbf{OSAlg}_{\Sigma,\Gamma} \to \mathbf{Alg}_{\Phi(\Sigma,\Gamma)} \ ,$$

where, for each $A \in \mathbf{OSAlg}_{\Sigma,\Gamma}$, *we generate* $A^{\bullet}$ *as the smallest* $\Phi(\Sigma,\Gamma)$*-algebra with* $A_k \subseteq A^{\bullet}_k$ *for each* $k \in \hat{S}$, $A^{\bullet}_s = A_s$ *for each* $s \in S$, *and defining for each* $f : k_1 \dots k_n \to k$ *and* $(a_1, \dots, a_n) \in A^{\bullet}_{k_1} \times \dots \times A^{\bullet}_{k_n}$

$$A^{\bullet}_f(a_1,\dots,a_n) = \begin{cases} A_f(a_1,\dots,a_n) \in A_k & \textit{if } (a_1,\dots,a_n) \in dom(A_{f_k^{k_1 \dots k_n}}) \\ \text{the term } f(a_1,\dots,a_n) & \textit{otherwise} \ . \end{cases}$$

This defines our desired extension map of institutions $(\Phi, \alpha, \beta) : OSEqtl^{\mathrm{R}} \rightarrowtail MEqtl$. This map is particularly nice, in that it automatically endows an order-sorted algebra A with *error elements* (the new terms that have a kind but not a sort) and allows a runtime error recovery discipline in the corresponding algebra $A^{\bullet}$ that is simpler and more satisfactory than that of retracts [16]. In a sense, it provides within the membership algebra framework the analogue of what was called an "error-supersort" recovery discipline in OBJ, but here error terms have a kind, not a sort, a feature that avoids many complications.

Corollary 27. *The extension map of institutions* $(\Phi, \alpha, \beta) : OSEqtl^{\mathrm{R}} \rightarrowtail MEqtl$ *is also a conservative map of logics [24] when we take into account the corresponding entailment systems of each of the logics as defined by the inference rules in Sections 11 and 4.*

Corollary 28. *Let* $H : (\Sigma, \Gamma) \to (\Sigma', \Gamma')$ *be a morphism of order-sorted theories. Then free constructions are preserved "in both directions" by restriction and extension, in the sense that, denoting by* $\mathcal{F}_H$ *the left adjoint to the forgetful functor*

$$\mathcal{U}_H : \mathbf{OSAlg}_{\Sigma',\Gamma'} \to \mathbf{OSAlg}_{\Sigma,\Gamma} \ ,$$

and by $\mathcal{F}_{\Phi(H)}$ *the left adjoint to the forgetful functor*

$$\mathcal{U}_{\Phi(H)} : \mathbf{Alg}_{\Phi(\Sigma',\Gamma')} \to \mathbf{Alg}_{\Phi(\Sigma,\Gamma)} \ ,$$

we have for each $A \in \mathbf{OSAlg}_{\Sigma,\Gamma}$

1. $\mathcal{F}_{\Phi(H)}(A^\bullet) \simeq (\mathcal{F}_H(A))^\bullet$
2. $\mathcal{F}_H(A) \simeq (\mathcal{F}_{\Phi(H)}(A^\bullet))^\circ$.

In particular, given a signature $\Sigma = (S, \leq, \Sigma)$, we can consider the signature $\Delta = (S, \leq, \emptyset)$ (whose models are $(S, \leq)$-sets) and the theory inclusion $(\Delta, \emptyset) \hookrightarrow (\Sigma, \Gamma)$, whose left adjoint is the free order-sorted algebra construction on an $(S, \leq)$-set of variables $\tilde{X}$, so that we get the following isomorphisms for the free and initial order-sorted algebras:

$$\mathcal{T}_{\Sigma,\Gamma}(\tilde{X}) \simeq (\mathcal{T}_{\Phi(\Sigma,\Gamma)}(\tilde{X}^\bullet))^\circ$$
$$\mathcal{T}_{\Sigma,\Gamma} \simeq \mathcal{T}^\circ_{\Phi(\Sigma,\Gamma)} \ .$$

Therefore, for all practical purposes we can perform all the usual order-sorted algebra constructions within the membership algebra framework.

13 Partial Membership Algebra

The partial analogue of membership algebra is *partial membership algebra.* It has a similar expressiveness to deal with sorts and subsorts and with both equations and sort predicates, and generalizes several well-known partial specification formalisms.

The notion of signature is slightly different from that for the total case. As in Example 1, we specify a partial order $(S, \leq)$ on sorts, with each $s \leq s'$ being syntactic sugar for an axiom $(\forall x)\ x : s' \Leftarrow x : s$. Furthermore, we require that each connected component k of $(S, \leq)$ should have a top element $\top_k$. This is because, without such a top, definedness of a term t in the kind k would then have a *disjunctive* expression by a family of membership assertions, instead of the single assertion $t : \top_k$. Kinds are then defined as connected components of the poset of sorts. As before, given a poset $S = (S, \leq)$, we denote by $\hat{S} = S/{\equiv_\leq}$ the set of its connected components. Now, we will call the connected components in $\hat{S}$ the *kinds* of the poset S, and will denote them either by k or by $[s]$.

Definition 29. A *signature* in partial membership equational logic is a triple $\Omega = (S, \leq, \Sigma)$ where $(S, \leq)$ is a poset such that each of its kinds k has a top element $\top_k$, and Σ is an $\hat{S}$-kinded signature. We call $(S, \leq)$ the poset of *sorts* of Ω.

To avoid unnecessary ambiguity we will assume that if $f : w \to k$ and $f : w' \to k'$ are operator declarations in Σ with w and w' words of the same length, then $w = w' \ \Leftrightarrow\ k = k'$.

Definition 30. Given a signature $\Omega = (S, \leq, \Sigma)$, a *partial Ω-algebra* A assigns:

1. to each $s \in S$ a set A_s, in such a way that whenever $s \leq s'$ then we have $A_s \subseteq A_{s'}$, and
2. to each $f : k_1 \dots k_n \to k$ in Σ, a *partial* function

$$A_f : A_{\top_{k_1}} \times \cdots \times A_{\top_{k_n}} \rightharpoonup A_{\top_k} \ .$$

Given two partial Ω-algebras A and B, an *Ω-homomorphism* $h : A \to B$ is an S-indexed family of (total) functions $\{h_s : A_s \to B_s\}_{s \in S}$ such that:

1. If $a \in A_s$ and $s \leq s'$, then $h_s(a) = h_{s'}(a)$.
2. For each $f : k_1 \dots k_n \to k$ in Σ, and $(a_1, \dots, a_n) \in A_{\top_{k_1}} \times \dots \times A_{\top_{k_n}}$, if $A_f(a_1, \dots, a_n)$ is defined, then $B_f(h_{\top_{k_1}}(a_1), \dots, h_{\top_{k_n}}(a_n))$ is also defined and equal to $h_{\top_k}(A_f(a_1, \dots, a_n))$.

This determines a category $\mathbf{PAlg}_\Omega$.

Let $\Omega = (S, \leq, \Sigma)$ be a signature. Given a set of variables $X = \{x_1, \dots, x_m\}$, a *variable declaration* is a sequence

$$\tilde{X} = x_1 : \overline{s_1}; \dots ; x_m : \overline{s_m}$$

where for each $i = 1, \dots, m$, $\overline{s_i}$ is a set of sorts $s_1^i, \dots, s_{l_i}^i$ having all the same kind.

Given a variable declaration $\tilde{X}$, by slight abuse of language we also denote by X the $\hat{S}$-indexed family of subsets of X associated to $\tilde{X}$ in the obvious way.

Atomic Ω-formulas are either equations of the form $t = t'$ where $t, t' \in T_\Sigma(X)_k$ for some $k \in \hat{S}$ (where $T_\Sigma(X)$ denotes as usual the free Σ-algebra on the variables X), or membership assertions of the form $t : s$, where $s \in S$ and $t \in T_\Sigma(X)_{[s]}$. General *Ω-sentences* are then Horn clauses of the form:

$$(i) \ \ (\forall \tilde{X})\ t = t' \Leftarrow u_1 = v_1 \wedge \dots \wedge u_n = v_n \wedge w_1 : s_1 \wedge \dots \wedge w_m : s_m$$
$$(ii) \ \ (\forall \tilde{X})\ t : s \Leftarrow u_1 = v_1 \wedge \dots \wedge u_n = v_n \wedge w_1 : s_1 \wedge \dots \wedge w_m : s_m$$

where t, t', u_i, v_i, and w_j are all terms in $T_\Sigma(X)$.

Given a partial Ω-algebra A and a variable declaration $\tilde{X}$, we can define assignments $a : \tilde{X} \to A$ in the obvious way (if $x : \overline{s}$ and s appears in $\overline{s}$, then we must have $a(x) \in A_s$), and then we can define by induction an $\hat{S}$-indexed family of partial functions $\overline{a} : T_\Sigma(X) \rightharpoonup A_\top$, where $A_\top = \{A_{\top_k}\}_{k \in \hat{S}}$, extending a in the usual way.

For atomic sentences we then define satisfaction by $A, a \models_\Omega t = t'$ meaning that $\overline{a}(t)$ and $\overline{a}(t')$ are both defined and $\overline{a}(t) = \overline{a}(t')$ (that is, we take an *existence equation* interpretation), and by $A, a \models_\Omega t : s$ meaning that $\overline{a}(t)$ is defined and $\overline{a}(t) \in A_s$. Satisfaction of Horn sentences is then defined in the obvious way.

Given a set Γ of Ω-sentences, we define $\mathbf{PAlg}_{\Omega,\Gamma}$ as the full subcategory of $\mathbf{PAlg}_\Omega$ determined by those partial Ω-algebras that satisfy all the sentences in Γ. In other words, the pair $T = (\Omega, \Gamma)$ is a *theory*, and $\mathbf{PAlg}_T = \mathbf{PAlg}_{\Omega,\Gamma}$ is the category of its models.

Example 2. The theory of categories T_{Cat} is a theory in partial membership equational logic. Its poset of sorts has sorts *Object* and *Arrow*, with *Object* $\leq$ *Arrow*; therefore, there is only one kind. The one-kinded signature has two unary domain and codomain operations d, c, and a binary composition operation $\_;\_$. The theory T_{Cat} is presented by the following axioms:

$$
\begin{aligned}
(\forall f : \mathit{Arrow})\ & d(f) : \mathit{Object} \\
(\forall f : \mathit{Arrow})\ & c(f) : \mathit{Object} \\
(\forall a : \mathit{Object})\ & d(a) = a \\
(\forall a : \mathit{Object})\ & c(a) = a \\
(\forall f, g : \mathit{Arrow})\ & f; g : \mathit{Arrow} \Leftarrow c(f) = d(g) \\
(\forall f, g : \mathit{Arrow})\ & c(f) = d(g) \Leftarrow f; g : \mathit{Arrow} \\
(\forall f, g : \mathit{Arrow})\ & d(f; g) = d(f) \Leftarrow c(f) = d(g) \\
(\forall f, g : \mathit{Arrow})\ & c(f; g) = c(g) \Leftarrow c(f) = d(g) \\
(\forall a : \mathit{Object}, f : \mathit{Arrow})\ & f; a = f \Leftarrow c(f) = a \\
(\forall a : \mathit{Object}, f : \mathit{Arrow})\ & a; f = f \Leftarrow d(f) = a \\
(\forall f, g, h : \mathit{Arrow})\ & (f; g); h = f; (g; h) \Leftarrow c(f) = d(g) \wedge c(g) = d(h)
\end{aligned}
$$

It is easy to check that a model of T_{Cat} is exactly a category (in which objects coincide with identity arrows), and that a T_{Cat}-homomorphism is exactly a functor. Note that this example has some similarities with Example 1 (paths in a graph). A key difference is of course that the models in Example 1 are total, with error terms corresponding to undefined elements, where the models here are partial. However, one could easily define a partial variant $\mathtt{PATH}_p$ `[G :: GRAPH]` of the path parameterized module. The obvious theory morphism $T_{Cat} \longrightarrow \mathtt{PATH}_p$`[G :: GRAPH]` would then state that each free path category is indeed a category.

13.1 Partial Membership Algebra and Finitely Locally Presentable Categories

The expressive power of partial membership equational logic exactly corresponds to a logic whose categories of models are, up to equivalence of categories, the finitely locally presentable categories [12, 1].

Theorem 31. *Let (Ω, Γ) be a theory in partial membership equational logic. Then, $\mathbf{PAlg}_{\Omega,\Gamma}$ is a finitely locally presentable category.*

Proof sketch. First note that $\mathbf{PAlg}_{\Omega,\Gamma}$ is a full subcategory of $\mathbf{PAlg}_{\Omega}$.

Since (i) all categories of models definable in many-sorted Horn logic with equality are finitely locally presentable [1, pages 209–210 and Theorem 1.39], and (ii) a full reflective subcategory of a finitely locally presentable category that is closed under directed colimits is also finitely locally presentable [1, Theorem 1.39], it suffices to show:

1. $\mathbf{PAlg}_{\Omega}$ is definable in many-sorted Horn logic with equality, and
2. $\mathbf{PAlg}_{\Omega,\Gamma}$ is a full reflective subcategory of $\mathbf{PAlg}_{\Omega}$ closed under directed colimits.

To see 1 it is enough to check that, for $\Omega = (S, \leq, \Sigma)$, $\mathbf{PAlg}_\Omega$ is equivalent to the category of models of the many-kinded Horn theory with kinds $\hat{S}$, relation symbols

- for each $f : k_1 \ldots k_n \to k$ in Σ, a symbol $G(f) : k_1 \ldots k_n k$,
- for each sort $s \in S$ a unary predicate symbol $\_ : s$ of kind $[s]$,

and axioms

$$
\begin{array}{lll}
(i) & (\forall \overline{x}, y, y')\ y = y' \Leftarrow G(f)(\overline{x}, y) \wedge G(f)(\overline{x}, y'), & \text{for each } f \in \Sigma \\
(ii) & (\forall x)\ x : s' \Leftarrow x : s, \quad \text{for each } s \leq s' \text{ in } S & \\
(iii) & (\forall x : k)\ x : \top_k\ . &
\end{array}
$$

The idea is that we can regard each partial Ω-algebra as a relational structure by viewing each partial operation f as a relation $G(f)$ satisfying (i), and can express the subsort structure by the axioms (ii) – (iii) in such a way that Ω-homomorphisms exactly correspond to homomorphisms of the corresponding relational structure.

To see 2, by [1, Theorem 2.48], it is enough to show that $\mathbf{PAlg}_{\Omega,\Gamma}$ is closed in $\mathbf{PAlg}_\Omega$ under limits and directed colimits. Since the sorts of such limits and colimits are the set-theoretic limits and colimits of the corresponding diagrams of sets for each sort, and the operations are then induced in a universal way, it is somewhat tedious but not hard to check this property. □

Theorem 32. *Any finitely locally presentable category is equivalent to one of the form* $\mathbf{PAlg}_{\Omega,\Gamma}$ *for some theory* (Ω, Γ) *in partial membership equational logic.*

Proof sketch. It is enough to show that any finitely locally presentable category is equivalent to one of the form $\mathbf{PAlg}_{\Omega,\Gamma}$ in the special case in which (Ω, Γ) is an *essentially algebraic theory*, which, after a simple translation, is nothing but a theory in partial membership equational logic satisfying additional restrictions. Adámek and Rosický have already shown in [1, Theorem 3.36] that, indeed, any finitely locally presentable category is, up to equivalence, the category of algebras of an essentially algebraic theory.

Their notion of essentially algebraic theory is a special case of Reichel's *hierarchical equationally partial (hep)* theories [32] which can be regarded as a more general formulation of essentially algebraic theories as originally envisioned by Lawvere and Freyd [11]. Both the general hep notion and the more restricted one proposed by Adámek and Rosický have straightforward translations into partial membership equational logic. In fact, Reichel [32] makes explicit the embedding *HEP* $\hookrightarrow$ *PMSEqtl* into partial many-sorted equational logic with (conditional) existence equations, which is precisely the sublogic of *PMEqtl* obtained by requiring the poset of sorts $(S, \leq)$ to be discrete and restricting the sentences to Horn clauses involving only equations.

We give below the translation for the Adámek-Rosický formulation, in which a theory consists of a many-sorted signature Σ containing a subsignature Σ_t of total functions together with, for each $f \in \Sigma - \Sigma_t, f : w \to s$, a finite set $Def(f)$

of Σ_t-equations in the standard variables associated to the arity w, and finally a set E of Σ-equations.

An algebra for this theory is a partial many-sorted Σ-algebra A such that the operations in Σ_t are total, it satisfies the equations E as existence equations, and such that for each $f : w \to s$ in $\Sigma - \Sigma_t$, A_f is defined on $(a_1, \ldots, a_n) \in A^w$ iff $A, (a_1, \ldots, a_n) \models \mathit{Def}(f)$.

This is of course exactly an algebra in membership equational logic with discrete poset of sorts those of Σ, operations Σ, and axioms E together with the axioms

- $f(\overline{x}) = f(\overline{x})$, for each $f \in \Sigma_t$,
- $f(\overline{x}) = f(\overline{x}) \iff e_1 \wedge \ldots \wedge e_n$
  for each $f \in \Sigma - \Sigma_t$ with $\mathit{Def}(f) = \{e_1, \ldots, e_n\}$. □

13.2 The Institution of Partial Membership Algebra

Definition 33. Given two signatures $\Omega = (S, \leq, \Sigma)$ and $\Omega' = (S', \leq', \Sigma')$ in partial membership equational logic, a *signature morphism* $H : \Omega \to \Omega'$ is given by

1. a monotonic function $H_1 : (S, \leq) \to (S', \leq')$, and
2. an $\hat{S}^* \times \hat{S}$-indexed family of functions

$$H_2 = \{H_{2,w,k} : \Sigma_{w,k} \to \Sigma'_{\hat{H}_1(w),\hat{H}_1(k)}\}_{(w,k)\in\hat{S}^*\times\hat{S}}$$

where $\hat{H}_1$ is the induced function on connected components extended also homomorphically to strings. By abuse of notation, we will drop the subscripts in H_1, H_2 and write H throughout.

Such a signature morphism induces a forgetful functor

$$\mathcal{U}^p_H : \mathbf{PAlg}_{\Omega'} \to \mathbf{PAlg}_{\Omega} \ ,$$

where for each algebra $A' \in \mathbf{PAlg}_{\Omega'}$ we have

- $\mathcal{U}^p_H(A')_s = A'_{H(s)}$ for every $s \in S$,
- for each $f : k_1 \ldots k_n \to k$ in Σ,

$$\mathcal{U}^p_H(A')_f = A'_{H_{k_1\ldots k_n,k}(f)} \cap (A'_{H(\top_{k_1})} \times \cdots \times A'_{H(\top_{k_n})} \times A'_{H(\top_k)}) \ .$$

Definition 34. Given two theories (Ω, Γ) and (Ω', Γ'), a *theory morphism* $H : (\Omega, \Gamma) \to (\Omega', \Gamma')$ is a signature morphism $H : \Omega \to \Omega'$ such that

$$\mathcal{U}^p_H(\mathbf{PAlg}_{\Omega',\Gamma'}) \subseteq \mathbf{PAlg}_{\Omega,\Gamma} \ ,$$

so that $\mathcal{U}^p_H$ restricts to a forgetful functor $\mathcal{U}^p_H : \mathbf{PAlg}_{\Omega',\Gamma'} \to \mathbf{PAlg}_{\Omega,\Gamma}$.

The above definitions specify the category of signatures $\mathbf{Sign}_p$, and the model functor $\mathbf{PAlg} : \mathbf{Sign}_p^{op} \to \mathbf{Cat}$. We now need to define sentences. However, because of the useful generality of allowing signature morphisms with monotonic $H : (S, \leq) \to (S', \leq')$ that may not preserve the tops[13] of connected components, the definition of the sentence functor $sen_p : \mathbf{Sign}_p \to \mathbf{Set}$ is somewhat subtle. The problem is that the meaning of a sentence can be drastically altered by a small change in its "ambient" signature. Consider for example a signature Ω with one sort s and with a unary operation f and a binary operation g, and let $\Omega' \supseteq \Omega$ be the signature extending Ω by adding a new sort s' with $s \leq s'$. Then, the meaning of the Ω-sentence

$$(\dagger) \qquad (\forall x : s)\ f(x) = g(x, x)$$

is destroyed if it is mapped to itself by $sen(J)$, for $J : \Omega \hookrightarrow \Omega'$ the inclusion of signatures. This is because an Ω'-algebra A' can satisfy the sentence (†) while $\mathcal{U}_J^p(A')$ may not, since in A' for an assignment $a : \{x\} \to A'_s$ we can have $\overline{a}(f(x)), \overline{a}(g(x, x)) \in A'_{s'} - A'_s$. The proper translation of (†) as an Ω'-sentence is

$$(\forall x : s)\ f(x) = g(x, x) \wedge f(x) : s \wedge g(x, x) : s\ .$$

To deal with this problem we introduce the notion of ***Ω-envelope*** of a sentence. This is a semantically equivalent sentence that, when translated, captures the intended meaning.

Definition 35. Let Ω be a signature in partial membership equational logic. Then, for an equation $t = t'$ we define

$$env_\Omega(t = t') \;=\; t = t' \wedge \bigwedge_{u \in NVST(\{t, t'\})} u : \top_{k(u)}\ ,$$

where $NVST(\{t, t'\})$ is the set of nonvariable subterms of either t or t', and where for each such subterm u, $k(u)$ denotes its corresponding kind.

Similarly, for a membership assertion $t : s$ we define

$$env_\Omega(t : s) \;=\; t : s \wedge \bigwedge_{u \in NVST(t) - \{t\}} u : \top_{k(u)}\ .$$

Then, for a Horn sentence φ

$$(\forall \tilde{X})\ at \Leftarrow u_1 = v_1 \wedge \ldots \wedge u_n = v_n \wedge w_1 : s_1 \wedge \ldots \wedge w_m : s_m$$

with at an equation or membership assertion, we define $env_\Omega(\varphi)$ as the sentence

$$\begin{aligned}(\forall \tilde{X})\ env_\Omega(at) \Leftarrow\ & env_\Omega(u_1 = v_1) \wedge \ldots \wedge env_\Omega(u_n = v_n)\\ \wedge\ & env_\Omega(w_1 : s_1) \wedge \ldots \wedge env_\Omega(w_m : s_m)\end{aligned}$$

where now the consequent is in fact a conjunction.

---

[13] This generality is actually quite important in practice. Several examples of non-top-preserving theory morphisms can be found in [26], which provides a good case study of the usefulness and expressiveness of partial membership equational logic in the area of 2-categorical and double-categorical models for rewriting logic and tile logic, their relationships, and their concurrency applications.

Extending the satisfaction relation in the obvious way to universally quantified implications with conjunctions of atomic formulas in their antecedent and consequent parts, it is easy to check the following

Lemma 36. *For $A \in \mathbf{PAlg}_\Omega$ and φ a Horn sentence, $A \models_\Omega \varphi$ iff $A \models_\Omega env_\Omega(\varphi)$.*

This suggests extending the notion of sentence in *PMEqtl* from Horn clauses to universally quantified implications between conjunctions of atomic formulas. This of course does not change the classes of algebras that can be defined, since any such implication is equivalent to the set of Horn clauses obtained by implying each one of the atomic formulas in the consequent from the given set of premises, but will allow us to give a good description of the functor of sentences.

In fact, let $H : \Omega \to \Omega'$ be a signature morphism. Then, we have the following systematic relationship between sentences in Ω and sentences in Ω' with respect to satisfaction.

Lemma 37 (Satisfaction Lemma). *For $H : \Omega \to \Omega'$ a signature morphism, φ an implication between conjunctions of atomic formulas in the sentences of Ω, and $A' \in \mathbf{PAlg}_{\Omega'}$, we have*

$$\mathcal{U}^p_H(A') \models_\Omega \varphi \iff A' \models_{\Omega'} H(env_\Omega(\varphi))$$

where H denotes also the obvious translation of Ω-terms into Ω'-terms and of S-sorted variables into S'-sorted variables induced by the signature morphism H.

Proof sketch. We deal with the case of atomic sentences; the extension to implications between conjunctions of atomic formulas is straightforward.

Consider an Ω-equation $(\forall \tilde{X})\ t = t'$. We have $\mathcal{U}^p_H(A') \models_\Omega (\forall \tilde{X})\ t = t'$ iff (by previous lemma) $\mathcal{U}^p_H(A') \models_\Omega (\forall \tilde{X})\ env_\Omega(t = t')$.

Defining $H(\tilde{X})$ as the Ω'-declaration such that $x : s_1 \ldots s_n$ is in $\tilde{X}$ iff $x : H(s_1) \ldots H(s_n)$ is in $H(\tilde{X})$ for each $x \in X$, and noticing that, by construction, $\mathcal{U}^p_H(A')_s = A'_{H(s)}$, it is easy to show by reasoning on the corresponding assignments that $\mathcal{U}^p_H(A') \models_\Omega (\forall \tilde{X})\ env_\Omega(t = t')$ implies $A' \models_{\Omega'} (\forall H(\tilde{X}))\ env_{\Omega'}(H(t) = H(t'))$.

In a similar way, for a membership assertion $(\forall \tilde{X})\ t : s$, we have that $\mathcal{U}^p_H(A') \models_\Omega (\forall \tilde{X})\ env_\Omega(t : s)$ implies $A' \models_{\Omega'} (\forall H(\tilde{X}))\ env_{\Omega'}(H(t) : H(s))$.

In this way we can show the implication

$$\mathcal{U}^p_H(A') \models_\Omega \varphi \;\Rightarrow\; A' \models_{\Omega'} H(env_\Omega(\varphi)) \ .$$

The key idea for the converse implication can be captured by considering φ an atomic sentence of the form $(\forall \tilde{X})\ t : s$. Then $\mathcal{U}^p_H(A') \models_\Omega \varphi$ means that for any assignment $a : \tilde{X} \to \mathcal{U}^p_H(A')$, $\overline{a}(t)$ is defined and $\overline{a}(t) \in \mathcal{U}^p_H(A')_s = A'_{H(s)}$. This of course implies that for each nonvariable subterm u of t, $\overline{a}(u)$ is defined and $\overline{a}(u) \in \mathcal{U}^p_H(A')_{\top_{k(u)}} = A'_{H(\top_{k(u)})}$. Since each assignment $a : \tilde{X} \to \mathcal{U}^p_H(A')$ can be regarded as an assignment $a' : H(\tilde{X}) \to A'$ and viceversa, this is equivalent to requiring that for each $a' : H(\tilde{X}) \to A'$, $\overline{a'}(H(u))$ is defined and $\overline{a'}(H(u)) \in A'_{H(\mathsf{T}_{k(u)})}$, that is, $A' \models_{\Omega'} (\forall H(\tilde{X}))\ H(u) : H(\mathsf{T}_{k(u)})$. Therefore, we have the equivalence

$$\mathcal{U}^p_H(A') \models_\Omega (\forall \tilde{X})\ t : s \iff A' \models_{\Omega'} (\forall H(\tilde{X}))\ H(env_\Omega(t : s))$$

as desired. □

The obvious candidate for $sen_p(H)$ would then be $sen_p(H)(\varphi) = H(env_\Omega(\varphi))$ but this definition requires some qualifications in order to be functorial. Think for example of the case $H = 1_\Omega$, where $sen_p(H)$ has to be the identity!

This functoriality problem can be resolved by giving to sentences a somewhat more subtle structure. Notice that $sen_p(\Omega)$ is the set of sentences of the form

$$(\forall \tilde{X})\ at_1 \wedge \ldots \wedge at_n \Leftarrow at'_1 \wedge \ldots \wedge at'_m$$

with the at_i, at'_j Ω-atoms, which can be viewed as an equationally defined abstract data type in which $\_ = \_$ is commutative, $\_ \Leftarrow \_$ is a constructor, $(\forall \tilde{X})\_$ is another constructor subjected to a membership condition (the variables in the atomic formulas must appear in X), and $\_\wedge\_$ is an associative, commutative, and idempotent operator with neutral element *true*. The additional structure solving the functoriality problem is obtained by imposing the following additional equations on $sen_p(\Omega)$:

- $f(x_1, \ldots, x_i, \ldots, x_n) : s \wedge x_i : \mathsf{T}_{k_i} = f(x_1, \ldots, x_i, \ldots, x_n) : s$ for each $f : k_1 \ldots k_n \to k$ in Σ and $s \in k$,
- $(x : \mathsf{T}_k \wedge x = y) = (x = y)$ for all $k \in \hat{S}$,
- $(x : s \wedge x : s') = (x : s)$ for all $s \leq s'$ in S,

which of course always yield semantically equivalent sentences with respect to the satisfaction relation $\models_\Omega$. It is also easy to check that the above equations are Church-Rosser and terminating *modulo* the associativity, commutativity, and identity of the conjunction operator $\_\wedge\_$. We then define $sen_p(\Omega)$ as the abstract data type obtained by imposing the above additional equations, which can be explicitly described by considering the *canonical forms* of expressions (modulo associativity, commutativity, and identity) that are obtained using the above equations as rewrite rules.

Lemma 38. *With $sen_p(\Omega)$ defined as above, $sen_p(H)(\varphi) = H(env_\Omega(\varphi))$ defines a functor $sen_p : \mathbf{Sign}_p \to \mathbf{Set}$.*

Proof. First of all, sen_p preserves identities, because, by repeated application of the first two equations, we can show $sen_p(1_\Omega)(\varphi) = env_\Omega(\varphi) = \varphi$ for any sentence $\varphi \in sen_p(\Omega)$.

To see that sen_p preserves composition, let $\Omega \overset{H}{\to} \Omega' \overset{G}{\to} \Omega''$ be signature morphisms. We have to show that $GH(env_\Omega(\varphi)) = G(env_{\Omega'}(H(env_\Omega(\varphi))))$ in

$sen_p(\Omega'')$. It is enough to check this for atomic sentences, since the extension to implications is quite obvious. We will do the case of an equation. We have

$$\begin{aligned}
&G(env_{\Omega'}(H(env_{\Omega}(t = t')))) = \\
&G(env_{\Omega'}(H(t = t' \wedge \bigwedge_{u \in NVST(\{t,t'\})} u : \top_{k(u)}))) = \\
&G(env_{\Omega'}(H(t) = H(t') \wedge \bigwedge_{H(u) \in NVST(\{H(t),H(t')\})} H(u) : H(\top_{k(u)}))) = \\
&G(H(t) = H(t') \wedge \bigwedge_{H(u) \in NVST(\{H(t),H(t')\})} H(u) : \top_{k'(H(u))} \wedge \\
&\bigwedge_{H(u) \in NVST(\{H(t),H(t')\})} (H(u) : H(\top_{k(u)}) \wedge \bigwedge_{v \in NVST(H(u))} v : \top_{k'(v)})) = (*) \\
&G(H(t) = H(t') \wedge \bigwedge_{H(u) \in NVST(\{H(t),H(t')\})} H(u) : H(\top_{k(u)})) = \\
&GH(env_{\Omega}(t = t'))
\end{aligned}$$

where the step $(*)$ is justified by the fact that $H(\top_{k(u)}) \leq \top_{k'(H(u))}$ and that each nonvariable subterm of each $H(u)$ is also a nonvariable subterm of either $H(t)$ or $H(t')$, plus repeated application of the third equation. □

This lemma and the Satisfaction Lemma 37 show that indeed *PMEqtl* is an institution. We now proceed to show that it is a *liberal* institution.

Theorem 39. *PMEqtl is a liberal institution.*

Proof sketch. We need to show that if $H : (\Omega, \Gamma) \to (\Omega', \Gamma')$ is a theory morphism, then the functor $\mathcal{U}^p_H : \mathbf{PAlg}_{\Omega',\Gamma'} \to \mathbf{PAlg}_{\Omega,\Gamma}$ has a left adjoint. Since we have already shown that $\mathbf{PAlg}_{\Omega',\Gamma'}$ and $\mathbf{PAlg}_{\Omega,\Gamma}$ are finitely locally presentable categories in Theorem 31, by the Adjoint Functor Theorem for locally presentable categories [1, Theorem 1.66], it is enough to show that $\mathcal{U}^p_H$ preserves limits and directed colimits. This is somewhat tedious but not very hard to check, using the explicit set-theoretic descriptions of such limits and colimits. □

The above theorem relies on quite abstract principles and constructions, and does not provide an explicit, proof-theoretic, construction of the left adjoint. However, thanks to the general results about extension maps of liberal institutions in Section 10 (see Theorem 17) and our explicit construction of the left adjoint to the functor $\mathcal{U}_H$ for H a theory morphism in *MEqtl* given in Section 7 (see Theorem 10), we will derive in the next section a very explicit construction of the left adjoints to forgetful functors $\mathcal{U}^p_H$ for the above case of theory morphisms in *PMEqtl*.

14 Extending Partial Membership Algebra to Membership Algebra

We are now ready to define an extension map of institutions. Using Proposition 14, the corresponding generalized map of institutions can be characterized by a triple

$$(\Psi_0, \gamma, (\_)^{\circ}) : PMEqtl \rightarrowtail MEqtl \ .$$

The functor Ψ_0 sends signatures in *PMEqtl* to theories in *MEqtl*. A signature $\Omega = (S, \leq, \Sigma)$ is mapped to the theory $\Psi_0(\Omega)$ with signature $(\hat{S}, \hat{\Sigma}, \pi : S \twoheadrightarrow \hat{S})$, where $\pi : S \twoheadrightarrow \hat{S}$ is the projection into equivalence classes, and with axioms $(\forall x)\ x : s' \Leftarrow x : s$ for each $s \leq s'$ in $(S, \leq)$.

Signature morphisms are mapped in the obvious way, namely $\Psi_0(H_1, H_2) = (\hat{H}_1, H_1, H_2)$.

Therefore, there is a functor $\Psi^\circ : \mathbf{Sign}_p \to \mathbf{Sign}$ such that $\Psi^\circ = sign \circ \Psi_0$.

The family of functions

$$\{\gamma_\Omega : sen_p(\Omega) \to \mathcal{P}(sen(\Psi^\circ(\Omega)))\}_{\Omega \in \mathbf{Sign}_p}$$

maps each *canonical* form $(\forall \tilde{X})\ at_1 \wedge \ldots \wedge at_n \Leftarrow at'_1 \wedge \ldots \wedge at'_m$ in $sen_p(\Omega)$ to the set of sentences

$$\{(\forall X)\ at \Leftarrow env_\Omega(at'_1) \wedge \ldots \wedge env_\Omega(at'_m) \wedge \tilde{X} \mid at \text{ is an atomic formula in the conjunction } env_\Omega(at_1) \wedge \ldots \wedge env_\Omega(at_n)\}$$

where, by abuse of language, if $\tilde{X} = x_1 : s^1_1, \ldots, s^1_{l_1}; \ldots; x_m : s^m_1, \ldots, s^m_{l_m}$, then in the above sentences $\tilde{X}$ denotes the conjunction

$$x_1 : s^1_1 \wedge \ldots \wedge x_1 : s^1_{l_1} \wedge \ldots \wedge x_m : s^m_1 \wedge \ldots \wedge x_m : s^m_{l_m}\ .$$

The natural transformation $(\_)^\circ : \mathbf{Alg} \circ \Psi_0^{op} \Rightarrow \mathbf{PAlg}$ is the functor sending each algebra $A \in \mathbf{Alg}_{\Psi_0(\Omega)}$ to the partial Ω-algebra $A^\circ \in \mathbf{PAlg}_\Omega$ obtained by restricting each operation $A_f : A_{k_1} \times \cdots \times A_{k_n} \to A_k$ to the partial function

$$A^\circ_f = A_f \cap (A_{\top_{k_1}} \times \cdots \times A_{\top_{k_n}} \times A_{\top_k})\ .$$

We first need to check that this defines a generalized map of institutions by means of Proposition 14.

For each $H : \Omega \to \Omega'$ in $\mathbf{Sign}_p$, denoting $\Psi_0(\Omega') = (\Psi^\circ(\Omega'), Ax_{\Omega'})$, we need to show that for each $\varphi \in sen_p(\Omega)$ we have

$$\text{(ii)} \qquad \mathbf{Alg}_{\Psi^\circ(\Omega'), Ax_{\Omega'} \cup \Psi^\circ(H)(\gamma_\Omega(\varphi))} = \mathbf{Alg}_{\Psi^\circ(\Omega'), Ax_{\Omega'} \cup \gamma_{\Omega'}(H(\varphi))}$$

where we have shortened $sen_p(H)$ to H and $sen(\Psi^\circ(H))$ to $\Psi^\circ(H)$.

That is, we have to show that for any algebra $A \in \mathbf{Alg}_{\Psi_0(\Omega')}$ we have

$$A \models_{\Psi^\circ(\Omega')} \Psi^\circ(H)(\gamma_\Omega(\varphi)) \iff A \models_{\Psi^\circ(\Omega')} \gamma_{\Omega'}(H(\varphi))\ .$$

We will prove it for φ an equation of the form $(\forall \tilde{X})\ t = t'$, and will leave the remaining cases to the interested reader. On the one hand,

$$\begin{aligned}
&\Psi^\circ(H)(\gamma_\Omega((\forall \tilde{X})\ t = t')) = \\
&\Psi^\circ(H)(\{(\forall X)\ t = t' \Leftarrow \tilde{X}\} \cup \{(\forall X)\ u : \top_{k(u)} \Leftarrow \tilde{X} \mid u \in NVST(\{t, t'\})\}) = \\
&\{(\forall \hat{H}(X))\ H(t) = H(t') \Leftarrow H(\tilde{X})\} \cup \\
&\qquad \{(\forall \hat{H}(X))\ H(u) : H(\top_{k(u)}) \Leftarrow H(\tilde{X}) \mid u \in NVST(\{t, t'\})\}\ .
\end{aligned}$$

On the other hand, assuming $H(\top_{k(u)}) \neq \top_{k'(H(u))}$ (the other case being easier),

$$\gamma_{\Omega'}(H((\forall \tilde{X})\; t = t')) =$$
$$\gamma_{\Omega'}((\forall H(\tilde{X}))\; H(t) = H(t') \wedge \bigwedge_{u \in NVST(\{t,t'\})} H(u) : H(\top_{k(u)})) =$$
$$\{(\forall \hat{H}(X))\; H(t) = H(t') \Leftarrow H(\tilde{X})\} \cup$$
$$\{(\forall \hat{H}(X))\; H(u) : \top_{k'(H(u))} \Leftarrow H(\tilde{X}) \mid u \in NVST(\{t,t'\})\} \cup$$
$$\bigcup_{u \in NVST(\{t,t'\})} (\{(\forall \hat{H}(X))\; H(u) : H(\top_{k(u)}) \Leftarrow H(\tilde{X})\} \cup$$
$$\{(\forall \hat{H}(X))\; v : \top_{k'(v)} \Leftarrow H(\tilde{X}) \mid v \in NVST(H(u)) - \{H(u)\}\}) \;,$$

which is clearly semantically equivalent to the previous set of sentences under the axioms $(\forall x)\; x : s' \Leftarrow x : s$ for each $s \leq s'$ in $(S, \leq)$.

We have also to prove that for each $\Omega \in \mathbf{Sign}_p, \varphi \in sen_p(\Omega)$, and $A \in \mathbf{Alg}_{\Psi_0(\Omega)}$ we have

$$\text{(iii)} \qquad A \models_{\Psi^\circ(\Omega)} \gamma_\Omega(\varphi) \iff A^\circ \models_\Omega \varphi \;.$$

Again, let us treat the case $\varphi = (\forall \tilde{X})\; t = t'$, and let us leave the remaining cases to the interested reader. Then, $\gamma_\Omega((\forall \tilde{X})\; t = t')$ is the set of axioms

$$\{(\forall X)\; t = t' \Leftarrow \tilde{X}\} \cup \{(\forall X)\; u : \top_{k(u)} \Leftarrow \tilde{X} \mid u \in NVST(\{t,t'\})\} \;.$$

First note that an assignment $a : X \to A$ satisfying $\tilde{X}$ is exactly an assignment $a : \tilde{X} \to A^\circ$. Then note that $\overline{a}(t)$ and $\overline{a}(t')$ are defined in A° for such an assignment iff

$$A, a \models_{\Psi^\circ(\Omega)} (\forall X)\; u : \top_{k(u)} \Leftarrow \tilde{X}$$

for each $u \in NVST(\{t,t'\})$, because A satisfies the axioms $(\forall x)\; x : s' \Leftarrow x : s$ for each $s \leq s'$ in $(S, \leq)$.

And, assuming that both sides are defined in A°, then

$$A^\circ, a \models_\Omega (\forall \tilde{X})\; t = t' \quad \text{iff} \quad A, a \models_{\Psi^\circ(\Omega)} (\forall X)\; t = t' \Leftarrow \tilde{X} \;.$$

Finally, as in Proposition 14, Ψ_0 is extended to a signature-preserving functor $\Psi : \mathbf{Th}_p \to \mathbf{Th}$ sending each theory (Ω, Γ) to the theory $(\Psi^\circ(\Omega), Ax_\Omega \cup \bigcup \gamma_\Omega(\Gamma))$.

In order to prove that this generalized map of institutions is an extension map, we have to show now that for any $(\Omega, \Gamma) \in \mathbf{Th}_p$ the forgetful functor

$$(\_)^\circ : \mathbf{Alg}_{\Psi^\circ(\Omega), Ax_\Omega \cup \bigcup \gamma_\Omega(\Gamma)} \to \mathbf{PAlg}_{\Omega,\Gamma} \;,$$

which is just the restriction of the forgetful functor

$$(\_)^\circ : \mathbf{Alg}_{\Psi^\circ(\Omega), Ax_\Omega} \to \mathbf{PAlg}_\Omega \;,$$

has a left adjoint

$$(\_)^\bullet : \mathbf{PAlg}_{\Omega,\Gamma} \to \mathbf{Alg}_{\Psi^\circ(\Omega), Ax_\Omega \cup \bigcup \gamma_\Omega(\Gamma)}$$

such that the unit of the adjunction is an isomorphism.

But, since the restriction of $(\_)^\circ$ is a restriction to full subcategories closed under isomorphism, if we show that the original functor has a left adjoint $(\_)^\bullet$ whose unit is an isomorphism, and also show that the image under $(\_)^\bullet$ of $\mathbf{PAlg}_{\Omega,\Gamma}$ is included in $\mathbf{Alg}_{\Psi(\Omega,\Gamma)}$, that is, that $(\_)^\bullet$ also has a restriction, the result will clearly hold for any $(\Omega,\Gamma) \in \mathbf{Th}_p$ by taking such restriction.

Indeed, given $A \in \mathbf{PAlg}_\Omega$, define $A^\bullet \in \mathbf{Alg}_{\Psi_0(\Omega)}$ as the smallest algebra with $A_k \subseteq A^\bullet_k$ for each $k \in \hat{S}$, $A^\bullet_s = A_s$ for each $s \in S$, and defining for each $f : k_1 \dots k_n \to k$ and $(a_1, \dots, a_n) \in A^\bullet_{k_1} \times \dots \times A^\bullet_{k_n}$

$$A^\bullet_f(a_1,\dots,a_n) = \begin{cases} A_f(a_1,\dots,a_n) \in A_k & \text{if } (a_1,\dots,a_n) \in dom(A_f) \\ \text{the } \mathit{term}\ f(a_1,\dots,a_n) & \text{otherwise} \end{cases} .$$

Since the subsort structure is left unchanged, $A^\bullet$ clearly satisfies the subsort axioms in $\Psi_0(\Omega)$. Also, by construction we have $A = (A^\bullet)^\circ$ for any $A \in \mathbf{PAlg}_\Omega$. Taking such an identity map as the unit map, and assuming an algebra $B \in \mathbf{Alg}_{\Psi_0(\Omega)}$ and an Ω-homomorphism $h : A \to B^\circ$, we can inductively check that there is a unique $\Psi_0(\Omega)$-homomorphism $\overline{h} : A^\bullet \to B$ such that $\overline{h}^\circ = h$. The base case is $A \subseteq A^\bullet$ for which $\overline{h}$ must coincide with h. Then, reasoning inductively on the depth of terms in $A^\bullet$, if $\overline{h}$ is uniquely defined for terms of depth $\leq n$ and $f(a_1,\dots,a_n)$ is a term of depth $n+1$, we must necessarily have $\overline{h}(f(a_1,\dots,a_n)) = B_f(\overline{h}(a_1),\dots,\overline{h}(a_n))$, which uniquely defines $\overline{h}$ for terms of depth $n+1$ and ensures the homomorphic property.

We finally have to check that if $A \in \mathbf{PAlg}_{\Omega,\Gamma}$, then $A^\bullet \in \mathbf{Alg}_{\Psi(\Omega,\Gamma)}$. Since $A^\bullet$ clearly satisfies each of the axioms in $\Psi_0(\Omega)$, what we need to show is that if

$$(\forall \tilde{X})\ at \Leftarrow at_1 \wedge \dots \wedge at_n$$

is an axiom in Γ (it is enough to consider Horn clauses), then for each atomic formula at' in $env_\Omega(at)$ we have

$$A^\bullet \models (\forall X)\ at' \Leftarrow env_\Omega(at_1) \wedge \dots \wedge env_\Omega(at_n) \wedge \tilde{X}\ ,$$

which follows trivially from the already checked condition (ii) on $(\Psi_0, \gamma, (\_)^\circ)$ being a generalized map of institutions, since $(A^\bullet)^\circ = A$ and $A \models_\Omega (\forall \tilde{X})\ at \Leftarrow at_1 \wedge \dots \wedge at_n$ by hypothesis.

Thus, we have proved

Theorem 40. $(\Psi, \gamma, (\_)^\circ) : \mathit{PMEqtl} \rightarrowtail \mathit{MEqtl}$ *is an extension map of institutions.*

From this theorem, using the results in Section 10, we can derive the following useful corollaries.

Corollary 41. *We can "borrow" the entailment system of MEqtl to endow the institution PMEqtl with the following entailment system: Given a theory* $(\Omega,\Gamma) \in \mathbf{Th}_p$ *and a sentence* $\varphi \in sen_p(\Omega)$, *we define*

$$\Gamma \vdash^p_\Omega \varphi \iff Ax_\Omega \cup \bigcup \gamma_\Omega(\Gamma) \vdash_{\Psi^\circ(\Omega)} \psi \quad \text{for all } \psi \in \gamma_\Omega(\varphi)\ .$$

This entailment system makes PMEqtl into a complete *logic, and makes the extension map PMEqtl* $\rightarrowtail$ *MEqtl into a* conservative *generalized map of logics.*

Corollary 42. *A signature morphism* $H : \Omega \to \Omega'$ *is a theory morphism* $H : (\Omega, \Gamma) \to (\Omega', \Gamma')$ *in* $\mathbf{Th}_p$ *iff for each* $\varphi \in \Gamma$ *we have* $\Gamma' \vdash^{p}_{\Omega'} \varphi$.

Corollary 43. *Let* $H : (\Omega, \Gamma) \to (\Omega', \Gamma')$ *be a theory morphism in* $\mathbf{Th}_p$. *Then free constructions are preserved "in both directions" by restriction and extension, in the sense that, denoting by* $\mathcal{F}^p_H$ *the left adjoint to the forgetful functor*

$$\mathcal{U}^p_H : \mathbf{PAlg}_{\Omega',\Gamma'} \to \mathbf{PAlg}_{\Omega,\Gamma} \ ,$$

and by $\mathcal{F}_{\Psi(H)}$ *the left adjoint to the forgetful functor*

$$\mathcal{U}_{\Psi(H)} : \mathbf{Alg}_{\Psi(\Omega',\Gamma')} \to \mathbf{Alg}_{\Psi(\Omega,\Gamma)} \ ,$$

we have for each $A \in \mathbf{PAlg}_{\Omega,\Gamma}$

1. $\mathcal{F}_{\Psi(H)}(A^\bullet) \simeq (\mathcal{F}^p_H(A))^\bullet$
2. $\mathcal{F}^p_H(A) \simeq (\mathcal{F}_{\Psi(H)}(A^\bullet))^\circ$.

In particular, given a signature $\Omega = (S, \leq, \Sigma)$, *we can consider the signature* $\Delta = (S, \leq, \emptyset)$ *(whose models are* $(S, \leq)$*-sets) and the theory inclusion* $(\Delta, \emptyset) \hookrightarrow (\Omega, \Gamma)$, *whose left adjoint is the free algebra construction on an* $(S, \leq)$*-set of variables* $\tilde{X}$, *so that we get the following explicit construction of the free* $(\mathcal{T}^p_{\Omega,\Gamma}(\tilde{X}))$ *and initial* $(\mathcal{T}^p_{\Omega,\Gamma})$ *algebras in* $\mathbf{PAlg}_{\Omega,\Gamma}$:

$$\mathcal{T}^p_{\Omega,\Gamma}(\tilde{X}) = (\mathcal{T}_{\Psi(\Omega,\Gamma)}(\tilde{X}^\bullet))^\circ$$
$$\mathcal{T}^p_{\Omega,\Gamma} = \mathcal{T}^\circ_{\Psi(\Omega,\Gamma)} \ .$$

Therefore, both in terms of logical deduction and of free, relatively free, and initial algebras, we can perform all our constructions and reasoning for partial membership algebra within the logical framework of total membership algebra.

15 Membership Algebra as a Logical Framework for Equational Specification

Partial membership equational logic is a quite expressive formalism for specifying partial algebras, in a way entirely analogous to how membership equational logic is a flexible specification formalism. In particular, many other equational or Horn logics for partial specification can be mapped by embeddings or by weak embeddings into *PMEqtl*.

We have already discussed the embeddings

$$\mathit{EssAlg}^{\mathrm{AR}} \hookrightarrow \mathit{HEP} \hookrightarrow \mathit{PMSEqtl} \hookrightarrow \mathit{PMEqtl}$$

in the proof of Theorem 32, where the essentially algebraic theories in the sense of Adámek and Rosický [1] are a special case of hierarchical equationally partial theories in the sense of Reichel [32], and can then be naturally expressed as theories in partial many-sorted equational logic with (conditional) existence equations that, in turn, is a sublogic of partial membership equational logic, namely, the sublogic obtained by requiring that the posets of sorts are discrete and by allowing only equations as atoms.

We can in addition connect those embeddings with "Mossakowski's web" [27] of embeddings and weak embeddings of institutions showing the essential equivalence between many different formalisms for partial specification. Mossakowski has shown that the categories of models defined by each of the logics in his web are all the same, namely, the finitely locally presentable categories. We refer to [27] for the details of this web and briefly summarize below his notation for the different additional institutions:

- *LESketch* is the institution of models specifiable as limit-preserving functors into **Set** for left-exact sketches [2].
- *Limit* is the institution of models for limit theories in the sense of Coste [10].
- *PStrong* is the institution of partial many-sorted equational logic with *strong* equations [5].
- *PECE* is the sublogic of *PMSEqtl* where all equations in a clause's conditions are "definedness" existence equations of the form $t = t$.
- *COSASC* is the institution of conditional order-sorted algebras with sort constraints in the sense of [14, 25].

In addition, we consider the institution $\mathit{PMSHorn}^{=}$ of partial many-sorted Horn logic with equality, where the models are partial many-sorted algebras together with predicates, and the sentences are Horn clauses whose atomic formulas are either existence equations or predicates, with $M, a \models P(t_1, \ldots, t_n)$ meaning that $\overline{a}(t_i)$ is defined for $1 \leq i \leq n$, and $(\overline{a}(t_1), \ldots, \overline{a}(t_n)) \in P_M$. In a way entirely analogous to the embedding $\mathit{MSHorn}^{=} \hookrightarrow \mathit{MEqtl}$ studied in Section 3, we also have an embedding[14] $\mathit{PMSHorn}^{=} \hookrightarrow \mathit{PMEqtl}$. Furthermore, Mossakowski has also constructed mutual embeddings between *PMSEqtl* and an institution *RPCEL* essentially equivalent to $\mathit{PMSHorn}^{=}$.

The point is that, following such a web, all these institutions can be embedded, or weakly embedded, in *PMEqtl*, which in turn we have shown can be mapped by an extension map into *MEqtl*. We can summarize the entire situation, including our previous discussion of total equational specification formalisms in Section 8, in the following diagram, which justifies our claim that membership equational logic is indeed a suitable and simple logical framework for equational specification.

[14] To get an embedding in the other direction, one should restrict the signature morphisms in $\mathbf{Sign}_p$ by requiring that they preserve tops of connected components.

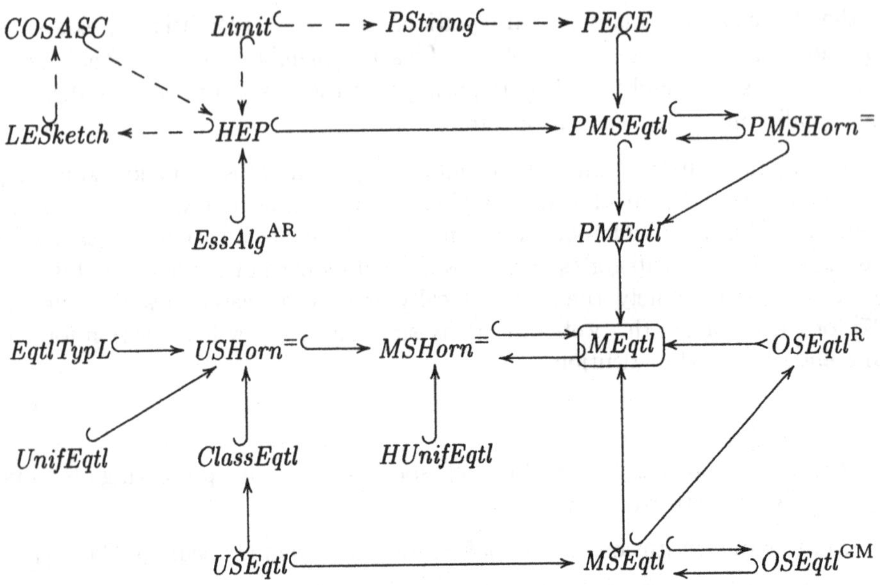

16 Concluding Remarks

I discuss below a number of future developments that seem particularly useful to exploit the results of this work in practice.

A matter that can be improved is the somewhat verbose translation of theories in the extension map $(\Psi, \gamma, (\_)^{\circ}) : \mathit{PMEqtl} \rightarrowtail \mathit{MEqtl}$, due to the replacement of atomic formulas by their corresponding envelopes. In practice many theories $T \in \mathbf{Th}_p$ are *strict*, in the precise sense that $\Psi(T)$ defines the same class of algebras as the theory obtained by the straightforward translation that maps a sentence $(\forall \tilde{X})\ at \Leftarrow at_1 \wedge \ldots \wedge at_n$ to the sentence $(\forall X)\ at \Leftarrow at_1 \wedge \ldots \wedge at_n \wedge \tilde{X}$. Sufficient conditions and algorithms to guarantee strictness will be discussed in future work.

The summary of logics that can be represented in *MEqtl* given in Section 15 suggests a natural path for mapping the Horn fragment *CASLHorn* of the CASL logic into membership equational logic. Theories in this fragment can be embedded very naturally into partial many-sorted Horn logic with equality $\mathit{PMSHorn}^{=}$. We then get the desired representation as the composition

$$\mathit{CASLHorn} \hookrightarrow \mathit{PMSHorn}^{=} \hookrightarrow \mathit{PMEqtl} \rightarrowtail \mathit{MEqtl}\ .$$

A somewhat less general subset of *CASLHorn* with subsorts understood as inclusions instead of as coercions, and with no extra predicates would of course have a straightforward embedding in *PMEqtl*, and would yield a very simple mapping into *MEqtl* if strictness checks of the kind mentioned above are used.

Besides having a logic with good properties, one is also interested in having theorem-proving techniques to verify properties of specifications in such a logic, including inductive properties of specifications with an initial semantics and of parameterized specifications. In this regard the prospects look quite promising. In joint work with Adel Bouhoula and Jean-Pierre Jouannaud [4, 3], we are developing a systematic study of such theorem-proving techniques, including Knuth-Bendix completion, tests for protecting and sufficiently complete theory extensions, and inductive theorem-proving techniques that extend to membership equational logic the rich body of results developed for many-sorted and order-sorted equational logics. Using the reflective capabilities of Maude we are beginning to build at SRI some theorem-proving tools written in Maude that implement some of these techniques and that will be used in the Cafe language and also in Maude itself.

The reflective character of rewriting logic and Maude [8] can also be very useful for implementing the actual representation map $(\Phi, \alpha, \beta) : \mathcal{L} \rightarrow \mathit{MEqtl}$ of a logic $\mathcal{L}$ into *MEqtl*. Although such a map is clearly a metalevel entity, it can be brought down to the object level by observing that, by reflection, there is an equationally defined data type $\mathtt{ADT}_{\mathit{MEqtl}}$ of theories in *MEqtl*, and that we can then implement the map of theories inside *MEqtl* by defining another data type $\mathtt{ADT}_{\mathcal{L}}$ for theories in $\mathcal{L}$ and an equationally defined function $\overline{\Phi} : \mathtt{ADT}_{\mathcal{L}} \rightarrow \mathtt{ADT}_{\mathit{MEqtl}}$ representing Φ. This technique, illustrated in [22], together with the reflective features of Maude, that can compile and execute the theories in $\mathtt{ADT}_{\mathit{MEqtl}}$, can then be used to translate a logic into *MEqtl* and to execute the translated theory.

Acknowledgements

Without the great kindness and generosity of Narciso Martí-Oliet, who has helped me in countless ways during the weeks near and beyond the deadline in the preparation, polishing, and criticism of the manuscript, this paper would never have made it to the WADT'97 proceedings. Thank you, Narciso, a friend in need and in deed.

As mentioned in the paper, this work is part of a broader research project. I am very thankful to Adel Bouhoula and Jean-Pierre Jouannaud for our fruitful collaboration on the proof-theoretic and theorem-proving aspects of membership algebra, and to Manuel Clavel, Steven Eker, and Patrick Lincoln for our joint work implementing both membership equational logic and rewriting logic in the Maude system.

I am also very thankful for the interest and the very helpful comments, at different stages of this work, that I have received from Maura Cerioli, Hans-Dieter Ehrich, Joseph Goguen, Narciso Martí-Oliet, Ugo Montanari, Till Mossakowski, Peter Mosses, Carolyn Talcott, and Martin Wirsing and Uwe Wolter. Finally, a detailed and thoughtful referee report prompted a number of improvements and clarifications in the exposition.

References

1. J. Adámek and J. Rosický. *Locally Presentable and Accessible Categories.* Cambridge University Press, 1994.
2. M. Barr and C. Wells. *Toposes, Triples and Theories.* Springer-Verlag, 1985.
3. A. Bouhoula, J.-P. Jouannaud, and J. Meseguer. Specification and proof in membership equational logic. Manuscript, SRI International, August 1996.
4. A. Bouhoula, J.-P. Jouannaud, and J. Meseguer. Specification and proof in membership equational logic. In M. Bidoit and M. Dauchet, editors, *Proceedings TAPSOFT'97*, volume 1214 of *Lecture Notes in Computer Science.* Springer-Verlag, 1997.
5. P. Burmeister. Partial algebras—Survey of a unifying approach towards a two-valued model theory for partial algebras. *Algebra Universalis*, 15:306–358, 1982.
6. M. Cerioli and J. Meseguer. May I borrow your logic? (Transporting logical structure along maps). *Theoretical Computer Science*, 173:311–347, 1997.
7. M. G. Clavel, S. Eker, P. Lincoln, and J. Meseguer. Principles of Maude. In J. Meseguer, editor, *Proc. First Intl. Workshop on Rewriting Logic and its Applications*, volume 4 of *Electronic Notes in Theoretical Computer Science.* Elsevier, 1996. `http://www1.elsevier.nl/mcs/tcs/pc/volume4.htm`.
8. M. G. Clavel and J. Meseguer. Reflection and strategies in rewriting logic. In J. Meseguer, editor, *Proc. First Intl. Workshop on Rewriting Logic and its Applications*, volume 4 of *Electronic Notes in Theoretical Computer Science.* Elsevier, 1996. `http://www1.elsevier.nl/mcs/tcs/pc/volume4.htm`.
9. CoFI Task Group on Semantics. CASL—The CoFI algebraic specification language, version 0.97, Semantics. `http://www.brics.dk/Projects/CoFI`, July 1997.
10. M. Coste. Localisation, spectra and sheaf representation. In M. P. Fourman, C. J. Mulvey, and D. S. Scott, editors, *Applications of Sheaves*, volume 753 of *Lecture Notes in Mathematics*, pages 212–238. Springer-Verlag, 1979.
11. P. Freyd. Aspects of topoi. *Bull. Austral. Math. Soc.*, 7:1–76, 1972.
12. P. Gabriel and F. Ulmer. *Lokal präsentierbare Kategorien.* Springer Lecture Notes in Mathematics No. 221, 1971.
13. J. Goguen and R. Burstall. Institutions: Abstract model theory for specification and programming. *Journal of the ACM*, 39(1):95–146, 1992.
14. J. Goguen, J.-P. Jouannaud, and J. Meseguer. Operational semantics of order-sorted algebra. In W. Brauer, editor, *Proceedings, 1985 International Conference on Automata, Languages and Programming*, volume 194 of *Lecture Notes in Computer Science*, pages 221–231. Springer-Verlag, 1985.
15. J. Goguen and J. Meseguer. Models and equality for logical programming. In H. Ehrig, G. Levi, R. Kowalski, and U. Montanari, editors, *Proceedings TAPSOFT'87*, volume 250 of *Lecture Notes in Computer Science*, pages 1–22. Springer-Verlag, 1987.
16. J. Goguen and J. Meseguer. Order-sorted algebra I: Equational deduction for multiple inheritance, overloading, exceptions and partial operations. *Theoretical Computer Science*, 105:217–273, 1992.
17. C. Hintermeier, C. Kirchner, and H. Kirchner. Dynamically-typed computations for order-sorted equational presentations. In *Proc. ICALP'94*, pages 60–75. Springer LNCS 510, 1994.
18. C. Hintermeier, C. Kirchner, and H. Kirchner. Dynamically-typed computations for order-sorted equational presentations. Technical Report 2208, INRIA, March 1994.

19. H.-J. Kreowski and T. Mossakowski. Equivalence and difference between institutions: simulating Horn Clause Logic with based algebras. *Math. Struct. in Comp. Sci.*, 5:189–215, 1995.
20. S. MacLane. *Categories for the Working Mathematician*. Springer-Verlag, 1971.
21. V. Manca, A. Salibra, and G. Scollo. Equational type logic. *Theoretical Computer Science*, 77:131–159, 1990.
22. N. Martí-Oliet and J. Meseguer. Rewriting logic as a logical and semantic framework. In J. Meseguer, editor, *Proc. First Intl. Workshop on Rewriting Logic and its Applications*, volume 4 of *Electronic Notes in Theoretical Computer Science*. Elsevier, 1996. `http://www1.elsevier.nl/mcs/tcs/pc/volume4.htm`.
23. J. Meseguer. Membership algebra. Lecture and abstract at the Dagstuhl Seminar on "Specification and Semantics," July 9, 1996.
24. J. Meseguer. General logics. In H.-D. E. et al., editor, *Logic Colloquium'87*, pages 275–329. North-Holland, 1989.
25. J. Meseguer and J. Goguen. Order-sorted algebra solves the constructor-selector, multiple representation and coercion problems. *Information and Computation*, 103(1):114–158, 1993.
26. J. Meseguer and U. Montanari. Mapping tile logic into rewriting logic. In F. Parisi-Presicce, ed., Proc. WADT'97, Springer LNCS, this volume, 1998.
27. T. Mossakowski. Equivalences among various logical frameworks of partial algebras. In H. K. Büning, editor, *Computer Science Logic, Paderborn, Germany, September 1995, Selected Papers*, volume 1092 of *Lecture Notes in Computer Science*, pages 403–433. Springer-Verlag, 1996.
28. P. D. Mosses. Unified algebras and institutions. In *Proc. Fourth Annual IEEE Symp. on Logic in Computer Science*, pages 304–312, Asilomar, California, June 1989.
29. P. D. Mosses. The use of sorts in algebraic specifications. In M. Bidoit and C. Choppy, editors, *Recent trends in Data Type Specification, 8th WADT, August 1991*, volume 655 of *Lecture Notes in Computer Science*, pages 66–91. Springer-Verlag, 1993.
30. F. Parisi-Presicce and S. Veglioni. Heterogeneous unified algebras. In *Proceedings of MFCS'93, 18th International Symposium on Mathematical Foundations of Computer Science*, pages 618–628. Springer LNCS 711, 1993.
31. A. Poigné. Parametrization for order-sorted algebraic specification. *Journal of Computer and System Sciences*, 40(2):229–268, 1990.
32. H. Reichel. *Initial Computability, Algebraic Specifications, and Partial Algebras*. Oxford University Press, 1987.
33. G. F. Stuart and W. W. Wadge. Classified model abstract data type specification. Manuscript, University of Victoria, 1991.
34. W. W. Wadge. Classified algebras. Technical report, University of Warwick, 1982.

Mapping Tile Logic into Rewriting Logic*

José Meseguer[1] and Ugo Montanari[2]

[1] Computer Science Laboratory, SRI International, Menlo Park, meseguer@csl.sri.com

[2] Dipartimento di Informatica, Università di Pisa, ugo@di.unipi.it

Abstract. *Rewriting logic* extends to concurrent systems with state changes the body of theory developed within the algebraic semantics approach. It is both a foundational tool and the kernel language of several implementation efforts (Cafe, ELAN, Maude). *Tile logic* extends (unconditional) rewriting logic since it takes into account state changes with side effects and synchronization. It is especially useful for defining compositional models of computation of reactive systems, coordination languages, mobile calculi, and causal and located concurrent systems. In this paper, the two logics are defined and compared using a recently developed algebraic specification methodology, *membership equational logic.* Given a theory T, the rewriting logic of T is the free monoidal *2*-category, and the tile logic of T is the free monoidal *double* category, both generated by T. An extended version of monoidal 2-categories, called *2VH-categories*, is also defined, able to include in an appropriate sense the structure of monoidal double categories. We show that 2VH-categories correspond to an extended version of rewriting logic, which is able to embed tile logic, and which can be implemented in the basic version of rewriting logic using suitable *internal strategies.* These strategies can be significantly simpler when the theory is *uniform.* A uniform theory is provided in the paper for CCS, and it is conjectured that uniform theories exist for most process algebras.

1 Introduction

Rewriting logic [27, 28, 31] extends to concurrent systems with state changes the body of theory developed within the algebraic semantics approach. It can also be

* Research supported by Office of Naval Research Contracts N00014-95-C-0225 and N00014-96-C-0114, by National Science Foundation Grant CCR-9633363, and by the Information Technology Promotion Agency, Japan, as part of the Industrial Science and Technology Frontier Program "New Models for Software Architechture" sponsored by NEDO (New Energy and Industrial Technology Development Organization). Also research supported in part by U.S. Army contract DABT63-96-C-0096 (DARPA); CNR Integrated Project *Metodi e Strumenti per la Progettazione e la Verifica di Sistemi Eterogenei Connessi mediante Reti di Comunicazione*; and Esprit Working Groups *CONFER2* and *COORDINA*. Research carried out while the second author was on leave at Computer Science Laboratory, SRI International, and visiting scholar at Stanford University.

considered as a semantic framework for concurrency. Its aim as a foundational tool is to faithfully express a wide range of models of computation. As shown in [27, 31], a rewriting theory $\mathcal{R}$ yields a cartesian 2-category [22] $\mathcal{L}(\mathcal{R})$ which does for $\mathcal{R}$ what a Lawvere theory $\mathcal{L}(T)$ does for an equational theory T. Moreover, the models of a rewrite theory $\mathcal{R}$ can be considered as 2-product-preserving 2-functors $M : \mathcal{L}_{\mathcal{R}} \to \mathbf{Cat}$, that is, as structures based on the 2-category **Cat** rather than on the category **Set**.

Rewriting logic is not only interesting as a theoretical framework. Several language implementation efforts (Cafe [18], ELAN [3], Maude [29, 10]) in various countries have adopted rewriting logic as their semantic basis, and support either executable specifications or parallel programming in rewriting logic. In particular the object-oriented language Maude developed at SRI International is based on rewriting logic and is efficiently implemented. Maude is also equipped with important extra features, like internal execution strategies and reflective logic, which allow it to easily embed a variety of other logics. A workshop has been recently dedicated to all the aspects of rewriting logic [32].

The tile model [19, 20, 21] relies on certain rewrite rules with side effects, called *tiles*, reminiscent of SOS rules [38] and SOS *contexts* [24]. A related model is also ***structured transition systems*** [13]. The tile model has been conceived with similar aims and similar algebraic structure as rewriting logic, and it extends rewriting logic (in the unconditional case), since it takes into account state changes with side effects and syncronization.

Tile systems can be seen as monoidal double categories [14] and tiles themselves as double cells. However, the tile model (as rewriting logic) can be presented in a purely logical form [21], tiles being just special sequents subject to certain inference rules and to certain proof normalization axioms. The fact that tile logic extends rewriting logic is reflected at the categorical level by the observation that 2-categories are a special case of double categories.

We now briefly introduce the tile model. A tile has the form:

$$s \xrightarrow[b]{a} s'$$

and states that the *initial configuration* s of the system evolves to the *final configuration* s' producing an *effect* b. However s is in general *open* (not closed) and the rewrite step producing the effect b is actually possible only if the subcomponents of s also evolve producing the *trigger* a. Both trigger a and effect b are called *observations*, and model the interaction, during a computation, of the system being described with its environment. More precisely, both system configurations are equipped with an *input* and an *output interface*, and the trigger just describes the evolution of the input interface from its initial to its final configuration. Similarly for the effect. It is convenient to visualize a tile as a two-dimensional structure (see Fig. 1), where the horizontal dimension corresponds to the extension of the system, while the vertical dimension corresponds to the extension of the computation. Actually, we should also imagine a third dimension (the thickness of the tile), which models parallelism: configurations, observations, interfaces and tiles themselves are all supposed to consist of several components in parallel.

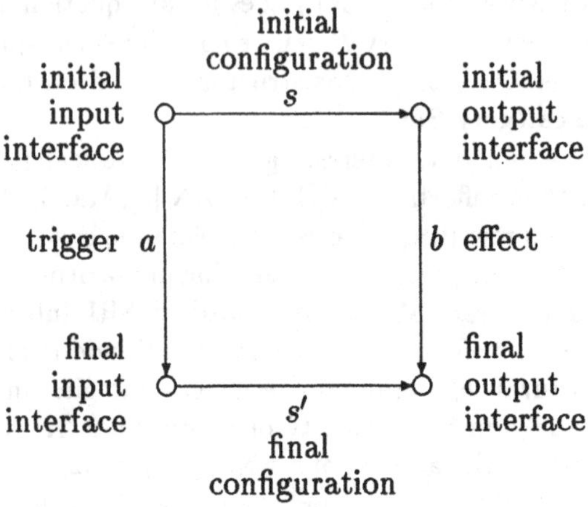

Fig. 1. A tile.

Both configurations and observations are assumed to be equipped with operations of parallel and sequential composition. Similarly, tiles themselves possess three operations of composition: parallel ($\_\otimes\_$), horizontal ($\_*\_$), and vertical ($\_\cdot\_$) composition.

The operation of parallel composition is self explanatory. Vertical composition models sequential composition of transitions and computations. Horizontal composition corresponds to synchronization: the effect of the first tile acts as trigger of the second tile, and the resulting tile expresses the synchronized behavior of both.

Computing in *tile logic* consists of starting from a set of basic tiles called *rewrite rules* and of applying the composition rules in all possible ways. In general the structure of a tile obtained in this way is specified by a *proof*, built up from the basic tiles used in the derivation and from the composition operations performed on them, up to certain normalizing axioms (which are those of monoidal double categories).

A tile reduces to a rewriting logic rule when trigger and effect are identities, i.e., when input and output interfaces stay idle during the step. In rewriting logic, the rewriting process is fully asynchronous, in the sense that matching rules can be independently applied, except when their redexes overlap. A tile with nontrivial trigger or effect, instead, can be applied only when it matches the whole state, or when it can be composed (usually horizontally or in parallel) with other tiles to build such a global tile.

Abstract semantics can be defined on tiles in the obvious way, i.e., by considering triggers and effects as external observations. Equivalences and congruences of configurations can then be defined in terms of traces, or via bisimilarity, or in other ways known for process algebras[3]. Logics of properties, and verification methods can also be defined in the usual way. We do not consider abstract semantics in this paper. Definitions and some important semantic properties (e.g., conditions for ensuring that bisimilarity is a congruence) are presented in [21].

Tiles have been used with success for modeling in detail several classes of applications. The most obvious is *coordination languages* [9]. In this research area, a common distinction is drawn between coordinators and basic (software) agents. Basic agents are considered as black boxes (to ensure interoperability) which interact with the coordinators via suitable protocols. Coordinators can be hierarchically nested and can compete and/or cooperate to achieve certain global goals, which can be fine grain, as transactions, or coarse grain, as work plans in computer-supported collaborative work (CSCW). Tiles are suitable for modeling coordination languages and systems, since triggers and effects naturally represent coordination protocols, and tiles themselves define the behavior of coordinators. In [36], a simple coordination model based on graph rewriting and synchronization is presented, and its behavior is directly described by a particular but important class of tiles (algebraic tiles). The hard computational problem of tile synchronization (which in general is NP-complete) is reduced to a distributed version of constraint solving, for which effective approaches exist. In [7, 8] the simplest possible interpretation is taken for system configurations, triggers and effects: markings of P/T Petri nets. However, horizontal composition of tiles yields *transition synchronization*, an important feature missing in ordinary nets (where only *token* synchronization is provided), and usually achieved only through complex constructions.

A main advantage of tiles is to offer an SOS-like compositional framework where data structures are not restricted to be syntactic terms. A small variation over ordinary terms which has proved very expressive is term graphs [1, 12]. They are analogous to terms, but the sharing of subterms can be specified also for closed[4] (i.e., without variables) term graphs. A first-class notion of sharing is essential for several important applications. One of them is *mobile processes.* Here the shared entities are names, which can be free or bound (i.e., global or local). Names usually are links to communication channels, or to objects, or to remote resources in general. Names have no meaning in themselves, except for specifying sharing, and in fact can be α-converted freely. Models of computation based on free and bound names are widespread (e.g., logic programming, λ-calculus, or process algebras with restriction or hiding), but the most general case is represented by models able to communicate names, and thus such that, when sending a private name to the external world, must make global a previously local name (extrusion). The most studied model of this kind is the π-calculus [35]. Name extrusion and handling makes π-calculus infinite branching,

[3] In tile rewrite systems modeling process algebras, the tiles with a closed agent as initial configuration have empty trigger, i.e., they can be considered as transitions of a labelled transition system, the label being the effect.

[4] Terms can share variables, but shared subterms of closed terms can be freely copied, always yielding an equivalent term.

and requires special notions of bisimulation. Tiles equipped with term graphs handle names as wires, i.e., every sharing connection must be explicitly defined. As a consequence, extrusion does not present any problem and can be specified using finite branching only. A presentation based on tiles of the ordinary abstract semantics of asynchronous π-calculus can be found in [15]. It employs the ordinary notion of tile bisimilarity.

Models for *concurrent and distributed systems* are the last application area we are describing. When concurrency is a primitive notion, models of process calculi usually include commuting transition diamonds[5] and observations of abstract locations or of causality links between events. Term graphs are again the most convenient structure. Sharing is used within configurations for modeling the operator $\_ \mid \_$ of process algebras, which in this context means sharing *the same location*. Within observations, sharing is used to express the fact that two events share *the same cause*, or, equivalently, that the same location has two different sublocations. In [16] it is shown that both operational and abstract semantics of CCS [34] with locations [4] can be conveniently rephrased within the tile model. Again, infinite branching is avoided and the ordinary notion of bisimilarity is employed.

The aim of this paper is to build a bridge from tile logic to rewriting logic. More generally, we think that it is conceptually useful to embed both models into a common semantic framework which makes clear not only their connections, but also their general role. For this purpose, a recent specification methodology, *membership equational logic* [30, 5, 33] and more specifically *partial membership equational logic* [33] (*PMEqtl*), is employed. *PMEqtl* theories consist of a partial ordering of sorts, a signature of operations, and Horn clauses whose atomic formulas are either equations $t = t'$, or assertions $t : s$ stating the membership of a term in a certain sort. *PMEqtl* extends order-sorted equational logic and supports partiality, subsorts, operator overloading and error specification. The key aspect of *PMEqtl* relevant to this paper is the ability to specify partial algebras with overloading, subsorts, and operations with equationally specified domains of definition. In fact, categories, and *a fortiori* 2-categories and double categories, are partial algebras of this kind, where the operation of arrow and cell composition is defined only if certain conditions are satisfied. A self-contained description of a simplified version of *PMEqtl*, *one-kinded partial membership equational logic*, is included. In particular, for this version of *PMEqtl* the notion of *tensor product* of theories is developed.

Tensor product (see for instance [25, 37]) is a well-known construction for ordinary algebraic (Lawvere) theories. Its importance can be understood by observing that the algebraic structures of a theory T can be defined not only on sets, the standard case $\mathbf{Alg}_T(\mathbf{Set})$, but also on any category $\mathbf{C}$ with suitable products or limits, to yield a category $\mathbf{Alg}_T(\mathbf{C})$. In particular, given two theories T and T', we can consider T'-algebras on the category of T-algebras, that

[5] Commuting transition diamonds define *concurrent* pairs of events, i.e., pairs of events which can occur in any order.

is the category $\mathbf{Alg}_{T'}(\mathbf{Alg}_T(\mathbf{C}))$, or instead T-algebras on the category of T'-algebras ($\mathbf{Alg}_T(\mathbf{Alg}_{T'}(\mathbf{C}))$). Regardless of the order, we obtain the same result up to isomorphism, namely the category of algebras for the *tensor product* $T \otimes T'$ of both theories. That is, we have isomorphisms:

$$\mathbf{Alg}_{T'}(\mathbf{Alg}_T(\mathbf{Set})) \simeq \mathbf{Alg}_T(\mathbf{Alg}_{T'}(\mathbf{Set})) \simeq \mathbf{Alg}_{T \otimes T'}(\mathbf{Set}).$$

The tensor product of theories is a very robust notion, rooted in the foundations of algebraic semantics, and very useful too, since the tensor theory $T \otimes T'$ can be effectively constructed. The construction for *PMEqtl* presented in this paper is somewhat more complex than, but analogous to, the construction for algebraic theories.

The relevance of the tensor product construction for tiles is that the theory of monoidal double categories can be obtained as the tensor product of the theory of categories (twice) with the theory of monoids. Thus, one can argue that if the desired model of computation must have operations of parallel and horizontal composition (to build more parallel and larger systems) and of vertical composition (to build longer computations), then monoidal double categories are the most natural answer. Of course it is always possible to add further structure restricting the models of interest to suitable subcategories.

Besides its use in specifying different models of computation by the tensor pruduct of theories, *PMEqtl* is also used in the paper for defining theory morphisms, which yield (as for Lawvere theories) corresponding adjunctions between categories of models. Free constructions of this kind are employed to define the algebraic semantics of tile rewrite systems.

The actual connection between rewriting logic and tile logic is obtained by defining in *PMEqtl* an extended version of 2-categories, called 2VH-categories, able to include in an appropriate sense the structure of double categories. In 2VH-categories, the operations of horizontal and vertical composition between tiles are derived operations based on those between 2-cells. It is possible to see that 2VH-categories correspond to an *extended* rewriting logic (*ERWL*), which is able to embed tile logic, and which can be implemented in the basic version of rewriting logic using suitable *internal strategies*. These strategies can be significantly simpler when the theory is *uniform*. In this case the formal translation presented in the paper yields an efficient implementation procedure. Interestingly enough, this is the case for CCS [34] and, we conjecture, for most process algebras. In the paper the case study of (finite) CCS is carried out in detail, and the corresponding rewriting theory is extracted and applied to a simple example.

The paper is organized as follows. Section 2 contains the background on partial membership equational logic, with double categories as an example of application of the tensor product construction. Section 3 introduces 2-categories as a special case of double categories, 2VH-categories, and computads[6] as models of rewrite systems. Section 4 defines tile rewrite systems, tile logic and extended

[6] Computads in this paper are essentially impoverished versions of monoidal double categories, being equipped with sets of double cells where no operation of composition is defined.

rewriting logic, and proves the adequacy of the latter. Also a tile rewrite system for CCS is introduced, which is uniform. An example of a nonuniform rewrite system is also provided.

2 Partial Membership Equational Logic

This section defines the basic notions of *partial membership equational logic* [33] (*PMEqtl*). This is a logic of partial algebras with subsorts and subsort polymorphism whose sentences are Horn clauses on equations $t = t'$ and membership assertions $t : s$. We treat here the *one kinded* case, in which the poset of sorts has a single connected component. A more detailed exposition for the many-kinded case can be found in [33].

2.1 Partial Algebras and Membership Equational Theories

Definition 1. *(signature)* A *signature* is a triple $\Omega = (S, \leq, \Sigma)$, with $(S, \leq)$ a poset with a top element $\top$, and $\Sigma = \{\Sigma_k\}_{k \in N}$ a family of sets indexed by natural numbers. Poset $(S, \leq)$ is called the poset of *sorts* of Ω. □

Definition 2. *(partial Ω-algebra)* Given a signature $\Omega = (S, \leq, \Sigma)$, a *partial Ω-algebra* A assigns:

(i) to each $s \in S$ a set A_s, in such a way that whenever $s \leq s'$, we have $A_s \subseteq A_{s'}$;
(ii) to each $f \in \Sigma_k$, $k \geq 0$, a partial function $A_f : A_\top^k \twoheadrightarrow A_\top$.

Given two partial Ω-algebras A and B, an *Ω-homomorphism* is a function $h : A_\top \to B_\top$ such that:

(i) for each $s \in S$, $h(A_s) \subseteq B_s$;
(ii) for each $f \in \Sigma_k$, $k \geq 0$, and $\mathbf{a} \in A_\top^k$, if $A_f(\mathbf{a})$ is defined, then $B_f(h^k(\mathbf{a}))$ is also defined and equal to $h(A_f(\mathbf{a}))$.

Notice that, because of condition **(i)**, for each $s \in S$ the function h restricts to a function $h|_s : A_s \to B_s$.

This determines a category $\mathbf{PAlg}_\Omega$.

□

Definition 3. *(declaration, formula and sentence)*

Let $\Omega = (S, \leq, \Sigma)$ be a signature. Given a set of variables $X = \{x_1, \ldots, x_m\}$, a *variable declaration* $\tilde{X}$ is a sequence

$$x_1 : \overline{s_1}, \ldots, x_m : \overline{s_m}$$

where for each $i = 1, \ldots, m$, $\overline{s_i}$ is a set of sorts $\{s_{i1}, \ldots, s_{il_i}\}$.

Atomic Ω-formulas are either equations

$$t = t'$$

where $t, t' \in T_\Sigma(X)$ (with $T_\Sigma(X)$ the usual free Σ-algebra on variables X) or membership assertions of the form

$$t : s$$

where $t \in T_\Sigma(X)$.

General Ω-sentences are then Horn clauses of the form

$$\forall \tilde{X}\ t = t' \Leftarrow u_1 = v_1\ \wedge\ \ldots\ \wedge\ u_n = v_n\ \wedge\ w_1 : s_1\ \wedge\ \ldots\ \wedge\ w_m : s_m$$

or of the form

$$\forall \tilde{X}\ t : s \Leftarrow u_1 = v_1\ \wedge\ \ldots\ \wedge\ u_n = v_n\ \wedge\ w_1 : s_1\ \wedge\ \ldots\ \wedge\ w_m : s_m$$

where the t, t', u_i, v_i and w_j are all terms in $T_\Sigma(X)$.

Definition 4. *(theory, model)*

Given a partial Ω-algebra A and a variable declaration $\tilde{X}$, we can define assignments $a : \tilde{X} \to A$ in the obvious way (if $x : \overline{s}$, and $s \in \overline{s}$, then we must have $a(x) \in A_s$) and then we can define a partial function $\overline{a} : T_\Sigma(X) \rightharpoonup A_\top$, extending a in the obvious way. For atomic sentences we then define satisfaction by

$$A, a \models t = t'$$

meaning that $\overline{a}(t)$ and $\overline{a}(t')$ are both defined and $\overline{a}(t) = \overline{a}(t)$ (that is, we take an *existence equation* interpretation) and by

$$A, a \models t : s$$

meaning that $\overline{a}(t)$ is defined and $\overline{a}(t) \in A_s$.

Satisfaction of Horn clauses is then defined in the obvious way. Given a set Γ of Ω-sentences, we then define $\mathbf{PAlg}_{\Omega,\Gamma}$ as the full subcategory of $\mathbf{PAlg}_\Omega$ determined by those partial Ω-algebras that satisfy all the sentences in Γ. In other words, the pair $T = (\Omega, \Gamma)$ is a *theory*, and $\mathbf{PAlg}_T = \mathbf{PAlg}_{\Omega,\Gamma}$ is the category of its models.

Example 1. (category)

The theory of categories T_{Cat} is a theory in partial membership equational logic. Its poset of sorts has sorts *Object* and *Arrow* with *Object* $\leq$ *Arrow*. There are two unary *domain* and *codomain* operations d and c, and a binary composition operation _;_. The theory is presented by the following axioms, for which we use a self-explaining Maude-like notation [10]. In the following, to avoid extra levels of indexing, we will denote theories by their Maude name, e.g., CAT or CAT instead of T_{Cat}.

```
fth CAT is
  sorts Object Arrow .
  subsorts Object < Arrow .
  ops d(_) c(_) _;_ .
  vars f g h : Arrow .
```

```
 var a : Object .
 mbs d(f) , c(f) : Object .
 eq d(a) = a .
 eq c(a) = a .
 cmb f;g : Arrow iff c(f) = d(g) .
 ceq d(f;g) = d(f) if c(f)=d(g) .
 ceq c(f;g) = c(g) if c(f)=d(g) .
 ceq a;f = f if d(f) = a .
 ceq f;a = f if c(f) = a .
 ceq (f;g);h = f;(g;h) if c(f) = d(g) and c(g) = d(h) .
endfth
```

It is easy to check that a model of CAT is exactly a category (in which objects coincide with identity arrows), and that a CAT-homomorphism is exactly a functor. □

Definition 5. *(signature and theory morphism)*
Given two signatures $\Omega = (S, \leq, \Sigma)$ and $\Omega' = (S', \leq', \Sigma')$, a *signature morphism* $H : \Omega \to \Omega'$ is given by:

1. a monotonic function $H : (S, \leq) \to (S', \leq')$, and
2. an N-indexed family of functions $\{H_k : \Sigma_k \to \Sigma'_k\}_{k \in N}$.

Such a signature morphism induces a forgetful functor $U_H : \mathbf{PAlg}_{\Omega'} \to \mathbf{PAlg}_{\Omega}$, where for each $A' \in \mathbf{PAlg}_{\Omega'}$ we have:

1. for each $s \in S$, $U_H(A')_s = A'_{H(s)}$;
2. for each[7] $f \in \Sigma_k$, $U_H(A')_f = A'_{H(f)} \cap (\underbrace{A'_{H(\top)} \times \ldots \times A'_{H(\top)}}_{k} \times A'_{H(\top)})$;
3. for each Ω'-homomorphism $h' : A' \to B'$, $U_H(h') = h'|_{H(\top)} : A'_{H(\top)} \to B'_{H(\top)}$, which is well-defined as a restriction of h' because h' is sort-preserving.

Given theories (Ω, Γ) and (Ω', Γ'), a *theory morphism* $H : (\Omega, \Gamma) \to (\Omega', \Gamma')$ is a signature morphism $H : \Omega \to \Omega'$ such that $U_H(\mathbf{PAlg}_{\Omega',\Gamma'}) \subseteq \mathbf{PAlg}_{\Omega,\Gamma}$, so that U_H restricts to a forgetful functor $U_H : \mathbf{PAlg}_{\Omega',\Gamma'} \to \mathbf{PAlg}_{\Omega,\Gamma}$. □

The reader is referred to [33] for proof-theoretical conditions on Γ and Γ' to ensure that a signature morphism $H : \Omega \to \Omega'$ is a theory morphism $H : (\Omega, \Gamma) \to (\Omega', \Gamma')$.

Proposition 6. (free construction associated to a theory morphism [33])
Given a theory morphism $H : (\Omega, \Gamma) \to (\Omega', \Gamma')$, its associated forgetful functor $U_H : \mathbf{PAlg}_{\Omega',\Gamma'} \to \mathbf{PAlg}_{\Omega,\Gamma}$ has a left adjoint $F_H : \mathbf{PAlg}_{\Omega,\Gamma} \to \mathbf{PAlg}_{\Omega',\Gamma'}$. □

[7] Notice that $U_H(A')_f = A'_{H(f)}$ would not be correct in general, since $U_H(A')_\top = A'_{H(\top)}$, where $H(\top)$ is not necessarily $\top$.

Definition 7. *(conservative, complete and persistent morphism)*
A theory morphism $H : (\Omega, \Gamma) \to (\Omega', \Gamma')$ is *conservative* (resp. *complete, persistent*) *w.r.t. sort s* if, for each algebra $A \in \mathbf{PAlg}_{\Omega,\Gamma}$, the component $(\eta_A)_s : A_s \to (U_H(F_H(A))_s$ corresponding to s of the unit of the adjunction associated to H is injective (resp. surjective, bijective). Morphism H is *conservative* (resp. *complete, persistent*) if it is conservative (resp. complete, persistent) w.r.t. all sorts $s \in S$. □

Definition 8. *(subalgebra)*
Given a signature $\Omega = (S, \leq, \Sigma)$ and a partial Ω-algebra A, an *Ω-subalgebra* B of A is an S-sorted family of subsets $\{B_s \subseteq A_s\}_{s \in S}$ such that:

(i) it is closed under the operations of Σ, that is, for each $f \in \Sigma_k$, and for each tuple $(b_1, \ldots, b_k) \in B_\top^k$, if $A_f(b_1, \ldots, b_k)$ is defined, then $A_f(b_1, \ldots, b_k) \in B_\top$;
(ii) it is closed under subsorts, in the sense that for each sort $s \in S$ we have $B_s = A_s \cap B_\top$.

It is clear that B with such operations and sorts is itself an Ω-algebra, and that the inclusion function $B \subseteq A$ is an Ω-homomorphism. □

It is also easy to prove the following.

Lemma 9. *For any set Γ of Horn sentences in partial membership equational logic, the category $\mathbf{PAlg}_{\Omega,\Gamma}$ is closed under Ω-subalgebras, i.e., if $A \in \mathbf{PAlg}_{\Omega,\Gamma}$ and B is an Ω-subalgebra of A, then $B \in \mathbf{PAlg}_{\Omega,\Gamma}$.* □

Example 2. (subcategory)
For the theory $(\Omega, \Gamma) = CAT$ of Example 1, the subalgebras of a category $\mathbf{C}$ are exactly its subcategories. □

Note that the notion of Ω-subalgebra just defined is strictly stronger than that of Ω-monomorphism. It is easy to check that, in the category $\mathbf{PAlg}_\Omega$, $m : C \to A$ is a monomorphism iff the associated function $m : C_\top \to A_\top$ is injective. Of course, by taking the smallest image of m, any such monomorphism always factors through an isomorphism and an inclusion $C \xrightarrow{\sim} B \hookrightarrow A$, where B is a *weak* subalgebra, as defined below.

Definition 10. *(weak subalgebra)*
Given a signature $\Omega = (S, \leq, \Sigma)$ and a partial Ω-algebra A, a *weak Ω-subalgebra* of A is a partial Ω-algebra B such that $B_\top \subseteq A_\top$ and such that the inclusion map $B \hookrightarrow A$ is an Ω-homomorphism. □

In general, given a set Γ of Horn sentences in partial membership equational logic, and a partial algebra $A \in \mathbf{PAlg}_{\Omega,\Gamma}$, a weak subalgebra B of A need not satisfy the sentences Γ. For example, given a nonempty category $\mathbf{C}$, the weak subalgebra with same arrows and objects as $\mathbf{C}$, but with all operations everywhere undefined is not a category. However, for $(\Omega, \Gamma) = CAT$, the following relationship happens to hold between subalgebras and weak subalgebras.

Example 3. Given a category **C**, if $\mathbf{D} \subseteq \mathbf{C}$ is a weak subalgebra and **D** itself is a category, then $\mathbf{D} \subseteq \mathbf{C}$ is a subalgebra, that is, a subcategory.

2.2 The Tensor Product Construction

Following lines similar to those in [39], and using also the categorical axiomatization of canonical inclusions in a category **C** as a poset category of special monos $\mathbf{I} \subseteq \mathbf{C}$ satisfying suitable axioms suggested in [26], one can, given a signature $\Omega = (S, \leq, \Sigma)$, define partial algebras in a category **C** with finite limits and a suitable poset of canonical inclusions **I**. Each sort s has an associated object A_s, and if $s \leq s'$ there is a canonical inclusion $A_s \hookrightarrow A'_s$ in **I**. Given $f \in \Sigma_k$ we associate to it an arrow $Dom(A_f) \to A_\top$ in **C**, where $Dom(A_f)$ is a subobject with a canonical inclusion $Dom(A_f) \hookrightarrow A^k_\top$. In this way we can define categories $\mathbf{PAlg}_\Omega(\mathbf{C})$ and $\mathbf{PAlg}_{\Omega,\Gamma}(\mathbf{C})$ so that our categories $\mathbf{PAlg}_\Omega$ and $\mathbf{PAlg}_{\Omega,\Gamma}$ are the special case $\mathbf{PAlg}_\Omega(\mathbf{Set})$ and $\mathbf{PAlg}_{\Omega,\Gamma}(\mathbf{Set})$. It is not hard to check that $\mathbf{PAlg}_\Omega$ and $\mathbf{PAlg}_{\Omega,\Gamma}$ are categories with limits, and that Ω-subalgebra inclusions $A \subseteq B$ constitute a poset category of canonical inclusions. Therefore, given two theories $T = (\Omega, \Gamma)$ and $T' = (\Omega', \Gamma')$, we can consider the category $\mathbf{PAlg}_T(\mathbf{PAlg}_{T'})$. For example, for T the theory of monoids and T' the theory of categories, $\mathbf{PAlg}_T(\mathbf{PAlg}_{T'})$ is the category of strict monoidal categories (see Section 3.2).

Since the poset $\mathbf{I} \subseteq \mathbf{PAlg}_{T'}$ of canonical inclusions that we have chosen is that of subalgebra inclusions, given a theory $T = (\Omega, \Gamma)$, each subsort relation $s \leq s'$ in Ω will be interpreted as a T'-subalgebra inclusion $A_s \hookrightarrow A'_s$, and each partial k-ary operation f in Ω will be interpreted as a T'-subalgebra inclusion $Dom(A_f) \hookrightarrow A^k_\top$ together with a T'-homomorphism $Dom(A_f) \to A_\top$. Furthermore, in order for these interpretations to yield a T-algebra structure in $\mathbf{PAlg}_{T'}$, the axioms Γ must be satisfied.

In a way analogous to algebraic theories [25, 17], to lim theories [40] and to sketches [23], there is then a theory $T \otimes T'$ in partial membership equational logic such that

$$\mathbf{PAlg}_{T\otimes T'} \simeq \mathbf{PAlg}_T(\mathbf{PAlg}_{T'}) \simeq \mathbf{PAlg}_{T'}(\mathbf{PAlg}_T).$$

Notice that we could have chosen a bigger poset $\mathbf{I}' \subseteq \mathbf{PAlg}_{T'}$ of subalgebra inclusions, yielding a looser definition of $\mathbf{PAlg}_T(\mathbf{PAlg}_{T'})$. A natural choice for $\mathbf{I}'$ would have been the set of weak subalgebra inclusions. This would yield a notion of tensor product of theories equivalent to the tensor product of their corresponding sketches. However, as we have already pointed out, the notion of weak subalgebra is too loose, giving rise in general to somewhat unintuitive models; for this reason we favor instead the notion of tensor product associated to subalgebras. Nevertheless, in the special case $T' = CAT$, because of the property mentioned in Example 3, the definition of $\mathbf{PAlg}_T(\mathbf{PAlg}_{CAT})$ is the same whether we choose subalgebras or instead weak subalgebras as canonical inclusions in $\mathbf{PAlg}_{CAT}$.

The explicit definition of $T \otimes T'$ is as follows.

Definition 11. *(tensor product)*
Let $T = (\Omega, \Gamma)$ and $T' = (\Omega', \Gamma')$ be theories in partial membership equational logic, with $\Omega = (S, \leq, \Sigma)$ and $\Omega' = (S', \leq', \Sigma')$. Then their *tensor product* $T \otimes T'$ is the theory with signature $\Omega \otimes \Omega'$ having:

(i) poset of sorts $(S, \leq) \times (S', \leq')$;
(ii) signature $\Sigma \otimes \Sigma'$, with an operator $f^l \in (\Sigma \otimes \Sigma')_n$ for each $f \in \Sigma_n$, and with an operator $g^r \in (\Sigma \otimes \Sigma')_m$ for each $g \in \Sigma'_m$. In particular, for f a constant in Σ_0 we get a constant f^l in $(\Sigma \otimes \Sigma')_0$.

The axioms of $T \otimes T'$ are the following:

A. Inherited Axioms.
For each axiom in Γ

$$\alpha = \forall(x_1 : \overline{s_1}, \ldots, x_m : \overline{s_m})\ \varphi(\mathbf{x}) \Leftarrow c(\mathbf{x})$$

with $\overline{s_i} = \{s_{i1}, \ldots, s_{il_i}\}$, $1 \leq i \leq m$, we introduce an axiom

$$\alpha^l = \forall(x_1 : \overline{s_1^l}, \ldots, x_m : \overline{s_m^l})\ \varphi^l(\mathbf{x}) \Leftarrow c^l(\mathbf{x})$$

with $\overline{s_i^l} = \{(s_{i1}, \top'), \ldots, (s_{il_i}, \top')\}$, $1 \leq i \leq m$, and with φ^l, c^l the obvious translations of φ, c obtained by replacing each $f \in \Sigma$ by its corresponding f^l.
Similarly, we define for each axiom $\beta \in \Gamma'$ the axiom β^r and impose all these axioms.

B. Subalgebra Axioms.
(i) For each $f \in \Sigma_n$ and each $s' \in S', s' \neq \top'$, we introduce the axiom:

$$\forall(x_1 : (\top, s'), \ldots, x_n : (\top, s'))$$
$$f^l(x_1, \ldots, x_n) : (\top, s') \Leftarrow f^l(x_1, \ldots, x_n) : (\top, \top').$$

(ii) For each $g_m \in \Sigma'_m$ and each $s \in S, s \neq \top$, we introduce the axiom:

$$\forall(x_1 : (s, \top'), \ldots, x_m : (s, \top'))$$
$$g^r(x_1, \ldots, x_m) : (s, \top') \Leftarrow g^r(x_1, \ldots, x_m) : (\top, \top').$$

(iii) For each $(s, s') \in S \times S'$ with $s \neq \top$ and $s' \neq \top'$, we have the axiom:

$$\forall x : (\top, \top')\ \ x : (s, s') \Leftarrow x : (\top, s') \wedge x : (s, \top').$$

C. Homomorphism Axioms.

For each $f \in \Sigma_n, g \in \Sigma'_m, n + m \geq 0$, we introduce the axiom:

$$\forall \mathbf{x}\ \ f^l(g^r(\mathbf{x_{1\cdot}}), \ldots, g^r(\mathbf{x_{n\cdot}})) = g^r(f^l(\mathbf{x_{\cdot 1}}), \ldots, f^l(\mathbf{x_{\cdot m}})) \Leftarrow$$
$$\Leftarrow \textstyle\bigwedge_{1 \leq i \leq n} g^r(\mathbf{x_{i\cdot}}) : (\top, \top')\ \wedge\ \bigwedge_{1 \leq j \leq m} f^l(\mathbf{x_{\cdot j}}) : (\top, \top').$$

where

$$
\begin{aligned}
\mathbf{x} \; &= \{x_{ij} : (\mathsf{T}, \mathsf{T}')\}^{1\leq i\leq n}_{1\leq j\leq m}, \\
\mathbf{x_{i.}} &= \{x_{ij} : (\mathsf{T}, \mathsf{T}')\}_{1\leq j\leq m}, \quad 1 \leq i \leq n \\
\mathbf{x_{.j}} &= \{x_{ij} : (\mathsf{T}, \mathsf{T}')\}_{1\leq i\leq n}, \quad 1 \leq j \leq m.
\end{aligned}
$$

□

The essential property of $T \otimes T'$ is expressed in the following theorem, whose proof will be given elsewhere.

Theorem 12. (models of the tensor product)
Let T, T' be theories in partial membership equational logic. Then we have the following isomorphisms of categories:

$$\mathbf{PAlg}_{T\otimes T'} \simeq \mathbf{PAlg}_T(\mathbf{PAlg}_{T'}) \simeq \mathbf{PAlg}_{T'}(\mathbf{PAlg}_T).$$

□

A useful property of the tensor product of theories that we will make use of without proof is its *functoriality* in the category of theories. Therefore, if $H : T_1 \to T_2$ and $G : T_1' \to T_2'$ are theory morphisms, we have an associated theory morphism:

$$H \otimes G : T_1 \otimes T_1' \to T_2 \otimes T_2'.$$

It can be shown that the tensor product of theories is associative and commutative up to isomorphism, that is, that we have natural isomorphisms of theories $T \otimes T' \simeq T' \otimes T$ and $T \otimes (T' \otimes T'') \simeq (T \otimes T') \otimes T''$ giving a symmetric monoidal category structure to the category of theories.

Example 4. (double category)
A *double category* has been defined [14] as a category structure on **Cat**, the category of categories, that is, an object of $\mathbf{PAlg}_{CAT}(\mathbf{PAlg}_{CAT}) = \mathbf{DCat}$. The theory $CAT \otimes CAT$ then axiomatizes double categories in partial membership equational logic.

Spelling out the specification of $T \otimes T'$ for the case of $T = T' = CAT$ we get the following poset of sorts, where *Square* is the top:

$$
\begin{aligned}
&(Object, Object) = Object, && (Arrow, Arrow) = Square, \\
&(Arrow, Object) = Harrow, && (Object, Arrow) = Varrow, \\
&Object \leq Harrow \leq Square, && Object \leq Varrow \leq Square.
\end{aligned}
$$

For the operations in $\Omega \otimes \Omega'$ we adopt the following N-E-W-S notation:

$$d^l = w,\; c^l = e,\; d^r = n,\; c^r = s,\; (\_;\_)^l = \_ * \_,\; (\_;\_)^r = \_ \cdot \_.$$

The presentation of double categories in Maude-like notation is thus as follows.

```
fth DCAT is
  sorts Object Harrow Varrow Square .
  subsorts Object < Harrow Varrow < Square .
  ops n(_) e(_) w(_) s(_) _*_ _·_ .
  vars f h : Harrow .
  vars u v : Varrow .
  vars A B C D : Square .
  *** Inherited Axioms: Horizontal
  mbs w(A) , e(A) : Varrow .
  eq w(v) = v .
  eq e(v) = v .
  cmb A*B : Square iff e(A) = w(B) .
  ceq w(A*B) = w(A) if e(A) = w(B) .
  ceq e(A*B) = e(B) if e(A) = w(B) .
  ceq v*A = A if w(A) = v .
  ceq A*v = v if e(A) = v .
  ceq (A*B)*C = A*(B*C) if e(A) = w(B) and e(B) = w(C) .
  *** Inherited Axioms: Vertical
  mbs n(A) , s(A) : Harrow .
  eq n(h) = h .
  eq s(h) = h .
  cmb A·B : Square iff s(A) = n(B) .
  ceq n(A·B) = n(A) if s(A) = n(B) .
  ceq s(A·B) = s(B) if s(A) = n(B) .
  ceq h·A = A if n(A) = h .
  ceq A·h = h if s(A) = h .
  ceq (A·B)·C = A·(B·C) if s(A) = n(B) and s(B)= n(C) .
  *** Subalgebra Axioms
  cmb A : Object if A : Harrow and A : Varrow .
  mbs w(h) , e(h) , n(v) , s(v) : Object .
  mb f*h : Harrow .
  mb u·v : Varrow .
  *** Homomorphism Axioms
  eq n(w(A)) = w(n(A)) .
  eq n(e(A)) = e(n(A)) .
  eq s(w(A)) = w(s(A)) .
  eq s(e(A)) = e(s(A)) .
  ceq w(A·B) = w(A)·w(B) if s(A) = n(B) .
  ceq e(A·B) = e(A)·e(B) if s(A) = n(B) .
  ceq n(A*B) = n(A)*n(B) if e(A) = w(B) .
  ceq s(A*B) = s(A)*s(B) if e(A) = w(B) .
  ceq (A*B)·(C*D) = (A·C)*(B·D) if e(A) = w(B)
      and e(C) = w(D) and s(A) = n(C) and s(B) = n(D) .
endfth
```

Notice that in the above axiomatization we do not present the literal instances of the axioms, but equivalent forms. For example, we get `w(h) : Object` from $w(h) : Varrow$ (by inherited axioms), plus $w(h) : Harrow$ (by the subalgebra axiom properly speaking), plus the subalgebra axiom forcing $Harrow \cap Varrow = Object$. □

In the following, we enrich our Maude-like notation with the tensor product construction. The presentation of double categories thus becomes much simpler:

```
fth DCAT is CAT ⊗ CAT renamed by (
  sorts (Object,Object) to Object . (Arrow,Arrow) to Square .
  sorts (Arrow,Object) to Harrow . (Object,Arrow) to Varrow .
  ops d left to w . c left to e . d right to n . c right to s .
  ops _;_ left to _*_ . _;_ right to _·_ )
endfth
```

3 Relating 2-Categories and Double Categories

We define 2-categories and 2VH-categories as models of suitable *PMEqtl* theories. We then enrich such structures to their monoidal version by tensoring with the theory of monoids, define computads and study the relatively free constructions associated to these categories of models.

3.1 2-Categories and 2VH-Categories

We now consider 2-categories [22]. They are probably the best known kind of enriched category. In particular, they yield models of rewriting logic in a very natural way [27]. Here, however, they can be considered as special cases of double categories, where vertical arrows are identified with objects. In 2-categories, squares are called cells, and horizontal arrows are called arrows. Moreover, north and south source and target are called d and c, while west and east source and target are l and r. Also, horizontal composition is denoted $\_;\_$ and vertical composition is denoted $\_\circ\_$. We can therefore specify 2-categories as follows:

```
fth 2CAT is including DCAT renamed by (
  sort Square to Cell  Harrow to Arrow  Varrow to Object .
  ops w to l . e to r . n to d . s to c .
  ops _*_ to _;_ . _·_ to _o_ ) .
endfth
```

We now introduce an extended version of 2-categories, called 2VH-categories, able to include in an appropriate sense the structure of double categories. In the following Maude-like presentation, the theory 2CAT is imported as such, without any renaming. In addition, new sorts *Harrow*, *Varrow* and *Square* are introduced, which correspond to the homonymous sorts of double categories. The poset of sorts is shown in Figure 2.

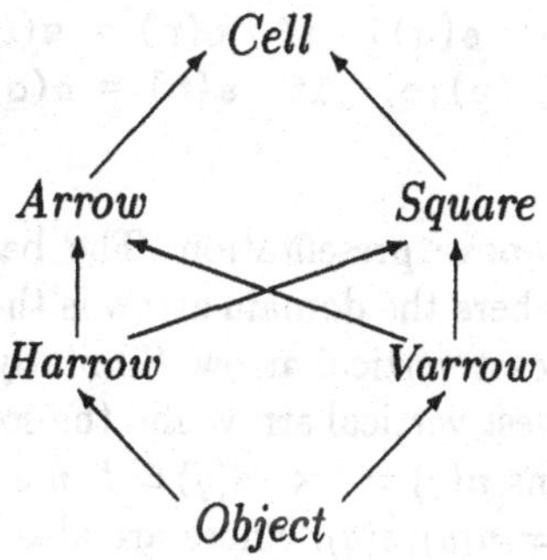

Fig. 2. The sort poset for 2VH-categories.

The basic intuition is that, if we are given a 2-category with subcategories *Harrow* and *Varrow* of *Arrow* such that they are disjoint except for objects, and such that the horizontal and vertical components can be recovered from their composition, then we can form a double category by considering squares with horizontal and vertical sides, and we can define their horizontal and vertical composition by using the already existing cell composition of the 2-category.

```
fth 2VHCAT is including 2CAT
  sorts Harrow Varrow Square .
  subsorts Object < Harrow Varrow < Arrow Square < Cell .
  ops n(_) e(_) w(_) s(_) _*_ _·_ .
  vars f h : Harrow .
  vars u v : Varrow .
  var t : Arrow .
  vars p q : Square .
  vars A B : Cell .
  cmb t : Object  if  t : Harrow and t : Varrow .
  cmb n(A) : Cell iff A : Square and n(A) : Harrow .
  cmb e(A) : Cell iff A : Square and e(A) : Varrow .
  cmb w(A) : Cell iff A : Square and w(A) : Varrow .
  cmb s(A) : Cell iff A : Square and s(A) : Harrow .
  ceq n(q) = h if d(q) = h;v .
  ceq e(q) = v if d(q) = h;v .
  ceq w(q) = v if c(q) = v;h .
  ceq s(q) = h if c(q) = v;h .
  eq d(q) = n(q);e(q) .
  eq c(q) = w(q);s(q) .
  cmb f;h : Harrow  iff r(f) = l(h) .
  cmb u;v : Varrow  iff r(u) = l(v) .
  cmb p*q : Square iff e(p) = w(q) .
  cmb A*B : Cell iff A , B : Square and e(A) = w(B) .
  cmb p·q : Square iff s(p) = n(q) .
  cmb A·B : Cell iff A , B : Square and s(A) = n(B) .
```

```
  ceq p*q = (n(p);q)∘(p;s(q))  if  e(p) = w(q) .
  ceq p·q = (p;e(q))∘(w(p);q)  if  s(p) = n(q) .
endfth
```

We comment on the above presentation. The basic idea is to represent a square q as a special cell where the domain arrow is the composition of the north horizontal arrow and the east vertical arrow. Similarly, the codomain arrow of q is the composition of the west vertical arrow and the south horizontal arrow. This is the meaning of equations $n(q) = h \Leftarrow d(q) = h; v$ and $e(q) = v \Leftarrow d(q) = h; v$, and of the equation $d(q) = n(q); e(q)$. There are also three analogous equations for $c(q)$.

To make this idea precise it is necessary to make both horizontal arrows and vertical arrows into subcategories of the category of arrows of the 2-category. This is expressed by the subsort ordering $Harrow, Varrow < Arrow$ and by clauses $f; g : Harrow \Leftrightarrow r(f) = l(g)$ and $u; v : Varrow \Leftrightarrow r(u) = l(v)$. Furthermore, the clause $t : Object \Leftarrow t : Harrow \wedge t : Varrow$ states that horizontal arrows and vertical arrows are disjoint except for objects.

Notice that the operations $\_*\_$ and $\_\cdot\_$ of horizontal and vertical square composition are defined only on adjacent squares (and not on other cells) according to clauses $A * B : Cell \Leftrightarrow A, B : Square \wedge e(A) = w(B) \wedge A * B : Square$. Similarly, N-E-W-S operations are defined only on squares according to the clause $n(A) : Cell \Leftrightarrow A : Square \wedge n(A) = Harrow$, and the other three analogous clauses.

Finally, the operation of horizontal composition of squares is defined in terms of horizontal and vertical composition of cells according to equation $p * q = (n(p); q) \circ (p; s(q)) \Leftarrow e(p) = w(q)$. Similarly for vertical composition of squares.

We then have a natural theory morphism from DCAT to 2VHCAT.

Proposition 13. *Let V be the signature morphism from DCAT to 2VHCAT relating homonymous sorts and operators. Then V is a theory morphism.* □

The proof is not difficult, and corresponds to showing that all the axioms of double categories can be proved from the axioms of 2VH-categories. The theory morphism above can be trivially specified in Maude-like notation.

```
view V from DCAT to 2VHCAT is
endview
```

3.2 Monoidal Models and Computads

In this section we add a monoidal structure to the models previously presented. This is quite important for making our approach expressive, since the monoidal operation represents parallel composition. The extension is almost effortless thanks to the tensor product of theories introduced in Section 2.2. In fact it is enough to introduce a theory MON of monoids and then to apply the tensor construction. We directly show all our extensions in Maude-like notation. We also introduce computads, which are an impoverished version of monoidal double categories, able to model tile rewrite systems.

The theory of monoids is as follows, where the monoidal operation is $\_ \otimes \_$.

```
fth MON is
  sort Monoid .
  ops 1 _⊗_ .
  vars a b c : Monoid .
  mb a⊗b : Monoid .
  eq a⊗1 = a .
  eq 1⊗a = a .
  eq a⊗(b⊗c) = (a⊗b)⊗c .
endfth
```

The theory of (strict) monoidal categories can be specified as follows:

```
fth MONCAT is MON ⊗ CAT renamed by (
  sorts (Monoid,Object) to Object . (Monoid,Arrow) to Arrow .
  ops 1 left to 1 . ⊗ left to ⊗ .
  ops d right to d . c right to c . ; right to ; )
endfth
```

Notice that we overload the symbol $\otimes$ using it for both the tensor construction on theories and for the monoidal operation within a theory.

In the above construction we have two subalgebra axioms:

$$1 : \mathit{Object}, \quad a \otimes b : \mathit{Object}$$

while homomorphism axioms are the well-known axioms of monoidal product[8]:

$$d(f \otimes g) = d(f) \otimes d(g),\ c(f \otimes g) = c(f) \otimes c(g),$$
$$f_1 \otimes g_1 \ ; \ f_2 \otimes g_2 = (f_1; f_2) \otimes (g_1; g_2).$$

We can now derive the theory of monoidal double categories.

```
fth MONDCAT is MON ⊗ DCAT renamed by (
  sorts (Monoid,Object) to Object . (Monoid,Square) to Square .
  sorts (Monoid,Harrow) to Harrow . (Monoid,Varrow) to Varrow .
  ops 1 left to 1 . ⊗ left to ⊗ .
  ops n right to n . e right to e . s right to s .
  ops w right to w . * right to *  . · right to · )
endfth
```

[8] Homomorphism axioms between the 1 operation of the monoid and the operations of the category:

$$d(1) = 1, \quad c(1) = 1, \quad 1\,;1 = 1$$

are already subsumed by other existing axioms.

Subalgebra axioms and homomorphism axioms are as follows:

$$\begin{gathered}1\colon Object,\ a\otimes b\colon Object,\ f\otimes g\colon Harrow,\ u\otimes v\colon Varrow,\\ n(A\otimes B)=n(A)\otimes n(B),\ e(A\otimes B)=e(A)\otimes e(B),\\ w(A\otimes B)=w(A)\otimes w(B),\ s(A\otimes B)=s(A)\otimes s(B),\\ A_1\otimes B_1\ ;\ A_2\otimes B_2=(A_1;A_2)\otimes(B_1;B_2),\\ A_1\otimes B_1\ *\ A_2\otimes B_2=(A_1*A_2)\otimes(B_1*B_2).\end{gathered}$$

Notice that we could have obtained the same theory MONDCAT by applying the tensor product construction to MONCAT and CAT, or to CAT and MONCAT (but not to MONCAT and MONCAT!). We call **MonDCat** = **PAlg**$_{MONDCAT}$ the category of monoidal double categories.

We now introduce the monoidal version of 2VH-categories and of the theory morphism V.

```
fth MON2VHCAT is MON ⊗ 2VHCAT renamed by (
  sorts (Monoid,Object) to Object . (Monoid,Arrow) to Arrow .
  sorts (Monoid,Cell) to Cell . (Monoid,Harrow) to Harrow .
  sorts (Monoid,Varrow) to Varrow . (Monoid,Square) to Square .
  ops 1 left to 1 . ⊗ left to ⊗ .
  ops d right to d . c right to c .
  ops l right to l . r right to r . ; right to ; .
  ops ∘ right to ∘ . n right to n . e right to e .
  ops w right to w . s right to s .
  ops * right to * . · right to · )
endfth
```

We call **Mon2VHCat** = **PAlg**$_{MON2VHCAT}$ the category of monoidal 2VH-categories.

```
view W from MONDCAT to MON2VHCAT is id ⊗ V .
endview
```

We now present the main result of the paper.

Theorem 14. *The theory morphism W from MONDCAT to MON2VHCAT is:*

(i) *complete;*
(ii) *persistent w.r.t. sorts Object, Harrow and Varrow.*

□

We will be interested in the following property:

Definition 15. *(uniform monoidal double category)*
A monoidal double category **D** is *uniform* if the monoidal 2VH-category $F_W(\mathbf{D})$ satisfies the following additional conditional membership axiom:

$$\forall(A\colon Cell,\ \ f,h\colon Harrow,\ \ u,v\colon Varrow)\qquad A\colon Square \Leftarrow d(A)=f;v \wedge c(A)=u;h.$$

□

Whenever this axiom is satisfied, it is easy to check whether a cell is a square, because it is enough to check whether the condition in the right member of the clause is satisfied.

We now give, as usual in Maude-like form, the presentation of *computads*, which (when provided with suitable generators) correspond to tile rewrite systems. Computads are reminiscent of monoidal double categories in that they also have a monoidal category of horizontal arrows and a monoidal category of vertical arrows sharing the objects. However, instead of squares they have *rules*, for which no operation is defined, unless they happen to be horizontal or vertical arrows. For horizontal composition this is expressed by the rule:

$$A * B : Rule \Leftrightarrow A, B : Harrow \wedge e(A) = w(B) \wedge A * B : Harrow$$

which states that horizontal composition $A*B$ of two elements (of the top sort) is defined iff they are adjacent horizontal arrows, and in this case it is an horizontal arrow.

```
fth CTD is
  including MONCAT renamed by (
  sorts Object to Object . Arrow to Harrow .
  ops 1 to 1 . ⊗ to ⊗ .
  ops d to w . c to e . ; to * ) .
  including MONCAT renamed by (
  sorts Object to Object . Arrow to Varrow .
  ops 1 to 1 . ⊗ to ⊕ .
  ops d to n . c to s . ; to ∘ ) .
  sort Rule .
  subsorts Object < Harrow Varrow < Rule .
  vars a b : Object .
  vars f g : Harrow .
  vars u v : Varrow .
  vars A B : Rule .
  mb f⊗g: Harrow .
  mb u⊕v: Varrow .
  eq a⊗b = a⊕b .
  cmb A : Harrow if A⊗B : Rule .
  cmb B : Harrow if A⊗B : Rule .
  cmb A : Varrow if A⊕B : Rule .
  cmb B : Varrow if A⊕B : Rule .
  cmb f*g : Harrow iff e(f) = w(g) .
  cmb A*B : Rule iff A , B : Harrow and e(A) = w(B) .
  cmb u·v : Varrow iff s(u) = n(v) .
  cmb A·B : Rule iff A , B : Varrow and s(A) = n(B) .
  cmb A : Object if A : Harrow and A : Varrow .
  mbs w(h) , e(h) , n(v) , s(v) : Object .
```

```
  mbs w(A) , e(A) : Varrow .
  mbs n(A) , s(A) : Harrow .
  eq n(w(A)) = w(n(A)) .
  eq n(e(A)) = e(n(A)) .
  eq s(w(A)) = w(s(A)) .
  eq s(e(A)) = e(s(A)) .
endfth
```

Notice that the structure of monoidal categories is imported for both horizontal and vertical arrows. *A priori* the two monoidal operators $\otimes$ and $\oplus$ on vertical and horizontal arrows are different (but they coincide on objects). However, they will be both mapped to the same operator when defining the theory morphism from computads to monoidal double categories.

We call $\mathbf{Ctd} = \mathbf{PAlg}_{CTD}$ the category of computads. Computads are essentially an impoverished version of monoidal double categories and can be used to present a monoidal double category as the free extension of a given computad. Therefore one can easily show the existence of a theory morphism from CTD to MONDCAT.

Proposition 16. (morphism from CTD to MONDCAT)
Let S be the signature morphism from CTD to MONDCAT mapping sort Rule to sort Square, operator $\oplus$ to operator $\otimes$ and for the rest relating homonymous sorts and operators. Then S is a theory morphism. □

The above property can be easily proved, since the axioms which must be satisfied by A are the same which must be satisfied by $U_S(A)$. The only exception is when a square $\alpha \in A_{Square}$ is not a horizontal arrow. Then, as an element of $U_S(A)_{Rule}$, horizontal composition is not defined on α, and thus no related axioms must be satisfied. Similarly, if $\alpha \in A_{Square}$ is not a vertical arrow, in $U_S(A)_{Rule}$, vertical composition is not defined on it. Finally, if $\alpha \in A_{Square}$ is neither a horizontal nor a vertical arrow, in $U_S(A)_{Rule}$, no monoidal operation is defined on it.

The signature morphism above can be specified in Maude-like notation.

```
view S from CTD to MONDCAT is
  sort Rule to Square .
  op ⊕ to ⊗ .
endview
```

We can now compose W and S to obtain a morphism Z from CTD to MON2VHCAT and apply the result of Proposition 6 in a couple of interesting cases.

```
view Z from CTD to MON2VHCAT is S;W .
endview
```

Proposition 17. (adjunctions from **Ctd** to **MonDCat** and to **Mon2VHCat**)
The forgetful functor U_S : **MonDCat** $\rightarrow$ **Ctd** *has a left adjoint* F_S : **Ctd** $\rightarrow$ **MonDCat**. *Similarly, the forgetful functor* U_Z : **Mon2VHCat** $\rightarrow$ **Ctd** *has a left adjoint* F_Z : **Ctd** $\rightarrow$ **Mon2VHCat**. *Furthermore,* F_Z *can be obtained by composing* F_S *with the left adjoint* F_W *to the forgetful functor* U_W : **Mon2VHCat** $\rightarrow$ **MonDCat**. □

4 Tile Logic and Enriched Rewriting Logic

We now formally introduce (the monoidal version of) tile rewrite systems and tile logic.

Definition 18. *(many-sorted hyper-signature)*
Given a set S of sorts, a *(many-sorted, hyper) signature* is an $S^* \times S^*$-indexed family of sets $\Sigma = \{\Sigma_{v,w}\}_{(v,w)\in S^*\times S^*}$, where S^* denotes the free monoid on set S. Each $f \in \Sigma_{v,w}$ is denoted $f : v \rightarrow w$. □

Definition 19. *(strict monoidal category freely generated by a signature)*
Given a signature Σ, $\mathbf{M}(\Sigma)$ is the strict monoidal category freely generated by Σ. □

Notice that the above well known free construction can be easily seen to be the left adjoint functor construction associated to a theory morphism. It is enough to define an (ordinary, total, algebraic) theory MONGRPH of *monoidal graphs* consisting of nodes (equipped with a monoidal operation) and of hyperarcs (equipped with source and target operations), and to consider the theory morphism G: MONGRPH $\rightarrow$ MONCAT mapping nodes into objects and hyperarcs into arrows; then $\mathbf{M} = F_G$. Since Σ is a model of MONGRPH, $\mathbf{M}(\Sigma)$ is then a model of MONCAT, i.e., a strict monoidal category.

Definition 20. *(tile rewrite system, tile logic)*
Let Σ_h and Σ_v be two disjoint signatures, called the *horizontal* and the *vertical* signature respectively. They must share the same set of sorts S.
A *tile rewrite system* $\mathcal{R} = C(S, \Sigma_h, \Sigma_v, R)$ is the computad where the set of objects is the free monoid on S, the horizontal (resp. vertical) arrows are the arrows of $\mathbf{M}(\Sigma_h)$ (resp. $\mathbf{M}(\Sigma_v)$), and the set of rules is R.
The *tile logic* of $\mathcal{R}$ is the monoidal double category $F_S(\mathcal{R})$ freely generated from $\mathcal{R}$ by the left adjoint functor F_S described by Proposition 17.
In the following, for α either a rule in $\mathcal{R}$ or a square $\alpha \in F_S(\mathcal{R})_{Square}$ we write $\alpha : f \xrightarrow[v]{u} g$ if $n(\alpha) = f$, $e(\alpha) = v$, $w(\alpha) = u$ and $s(\alpha) = g$, and we also write $\mathcal{R} \vdash_{tile} f \xrightarrow[v]{u} g$. □

As shown in [27], the Lawvere 2-theory $\mathcal{L}_\mathcal{R}$ of an unconditional rewrite theory $\mathcal{R}$ is the free 2-category with finite 2-products generated by $\mathcal{R}$ when viewed as an appropriate computad. Therefore, in 2-categorical terms rewriting logic is the logic associated to the models of the theory $P2CAT$ of 2-categories with finite 2-products. Regarding the cartesian 2-product $\_\times\_$ as a monoidal operation $\_\otimes\_$ exactly corresponds to defining a theory morphism $Y : MON2CAT \to P2CAT$, where $MON2CAT$ is the tensor product $MON \otimes 2CAT$. The forgetful functor U_Y : **P2Cat** $\to$ **Mon2Cat** has then a left adjoint F_Y, which allows us to consider the notion of a *monoidal rewrite theory* as an appropriate computad $\mathcal{R}$ that has first a free extension to a monoidal Lawvere theory $\mathcal{L}_{\mathcal{R}}^{\otimes} \in$ **Mon2Cat**, and whose standard 2-Lawvere theory is $\mathcal{L}_{\mathcal{R}} = F_Y(\mathcal{L}_{\mathcal{R}}^{\otimes})$.

The desired link between tile logic and rewriting logic can be obtained by considering monoidal rewrite theories that generate monoidal Lawvere theories not just in **Mon2Cat**, but in the richer category **Mon2VHCat**. We shall call such rewrite theories *enriched rewrite theories.* As computads they coincide with tile rewrite systems $\mathcal{R}$, that is, they can be regarded as computads $\mathcal{R} \in$ **Ctd**. Their logic is then captured by their monoidal 2VH Lawvere theory, namely $F_Z(\mathcal{R}) \in$ **Mon2VHCat**.

Definition 21. *(enriched rewriting logic)*
Given a tile rewrite system $\mathcal{R} = C(S, \Sigma_h, \Sigma_v, R)$, the *enriched rewriting logic* ($ERWL$) of $\mathcal{R}$ is the monoidal 2VH-category $F_Z(\mathcal{R})$ freely generated from $\mathcal{R}$ by the left adjoint functor F_Z described by Proposition 17.
We write $\mathcal{R} \vdash_{ERWL} f \xrightarrow[v]{u} g$ iff there is a square $\alpha : f \xrightarrow[v]{u} g \in F_Z(\mathcal{R})_{Square}$. □

Tile logic and enriched rewriting logic are then systematically related by the left adjoint F_W to the forgetful functor U_W : **Mon2VHCat** $\to$ **MonDCat**, that for any tile rewrite system $\mathcal{R}$ gives us a unit map

$$\eta : F_S(\mathcal{R}) \to U_W(F_Z(\mathcal{R}))$$

mapping tile logic into enriched rewriting logic, since (by Proposition 17) we have $F_Z(\mathcal{R}) = F_W(F_S(\mathcal{R}))$. Since by Theorem 14 we know that η is surjective on squares and bijective on arrows, we then obtain the following adequacy property for this mapping.

Corollary 22. (adequacy of enriched rewriting logic)
Given a tile rewrite system $\mathcal{R} = C(S, \Sigma_h, \Sigma_v, R)$ *we have*

$$\mathcal{R} \vdash_{tile} f \xrightarrow[v]{u} g \iff \mathcal{R} \vdash_{\mathrm{ERWL}} f \xrightarrow[v]{u} g$$

□

The practical importance of this result is that we can use an implementation of rewriting logic such as those currently available (Cafe [18], ELAN [3], Maude [10]) to perform deductions in tile logic. Notice that the above equivalence depends crucially on the rewrite proof α being a *square* $\alpha : f; v \to u; h$, since in $F_Z(\mathcal{R})$ we can have many rewrite proofs α that are not squares and therefore are *not* valid tile proofs. The constraint that the rewrite proof α must be a square can be imposed and implemented by means of the notion of an *internal strategy* [11] that restricts the rewrite proofs by requiring the overall proof to be an appropriate composition of squares yielding a square. Of course, if the tile rewrite system is uniform in the sense of the following definition, then the strategy imposing the constraint that proofs are squares can be defined much more simply, since one only needs to check the sides of the overall proof.

Definition 23. *(uniform tile rewrite system)*
A tile rewrite system $\mathcal{R}$ is *uniform* if its associated monoidal double category $F_S(\mathcal{R})$ is uniform (see Definition 15). □

As an interesting example of a uniform system, we now present CCS, Milner's Calculus for Communicating Systems [34]. For the tile presentation, we follow [21].

Example 5. (tile and rewriting logic for CCS)

Syntax of CCS. Let Δ be the alphabet for basic actions (which is ranged over by α) and $\overline{\Delta}$ the alphabet of complementary actions ($\Delta = \overline{\overline{\Delta}}$ and $\Delta \cap \overline{\Delta} = \emptyset$); the set $\Lambda = \Delta \cup \overline{\Delta}$ will be ranged over by λ. Let $\tau \notin \Lambda$ be a distinguished action, and let $\Lambda \cup \{\tau\}$ (ranged over by μ) be the set of CCS actions.

The syntax of finite CCS agents is defined by the following grammar:

$$P ::= \ nil \ \Big| \ \mu.P \ \Big| \ P\backslash\alpha \ \Big| \ P+P \ \Big| \ P \mid P.$$

Operational Semantics of CCS. Given a process P, its dynamic behaviour can be described by a suitable transition system, along the lines of the SOS approach, where the transition relation is freely generated from the following set of inference rules.

$$\frac{}{\mu.P \xrightarrow{\mu} P} \qquad \frac{P \xrightarrow{\mu} Q \ \ \mu \notin \{\alpha, \overline{\alpha}\}}{P\backslash\alpha \xrightarrow{\mu} Q\backslash\alpha}$$

$$\frac{P \xrightarrow{\mu} Q}{P+R \xrightarrow{\mu} Q} \qquad \frac{P \xrightarrow{\mu} Q}{R+P \xrightarrow{\mu} Q}$$

$$\frac{P \xrightarrow{\mu} Q}{P|R \xrightarrow{\mu} Q|R} \qquad \frac{P \xrightarrow{\alpha} Q, \ P' \xrightarrow{\overline{\alpha}} Q'}{P|P' \xrightarrow{\tau} Q|Q'} \qquad \frac{P \xrightarrow{\mu} Q}{R|P \xrightarrow{\mu} R|Q}.$$

Given the transition

$$((a.nil + b.nil) \mid \overline{a}.nil)\backslash a \xrightarrow{\tau} (nil \mid nil)\backslash a,$$

its proof is as follows:

$$\cfrac{\cfrac{\cfrac{\cfrac{}{a.nil \xrightarrow{a} nil}}{a.nil + b.nil \xrightarrow{a} nil} \qquad \cfrac{}{\overline{a}.nil \xrightarrow{\overline{a}} nil}}{(a.nil + b.nil) \mid \overline{a}nil \xrightarrow{\tau} nil \mid nil}}{((a.nil + b.nil) \mid \overline{a}nil)\backslash a \xrightarrow{\tau} (nil \mid nil)\backslash a}$$

We now present the tile rewrite system for CCS.

Signatures of the CCS Rewrite System. There is only one sort $\underline{1}$. The free monoid generated by it is represented by underlined natural numbers with $\underline{n} \otimes \underline{m} = \underline{m+n}$. For the horizontal signature the operators are: $nil \in (\Sigma_h)_{\underline{0},\underline{1}}$; $\mu._{-}$ and $_{-}\backslash\alpha \in (\Sigma_h)_{\underline{1},\underline{1}}$; $_{-}+_{-}$ and $_{-} \mid _{-} \in (\Sigma_h)_{\underline{2},\underline{1}}$; and $!(_{-}) \in (\Sigma_h)_{\underline{1},\underline{0}}$. The latter constructor, called *eraser*, is needed to discard the rejected alternative after a choice step. For the vertical signature the operators are $\underline{\mu}._{-} \in (\Sigma_v)_{\underline{1},\underline{1}}$.

Rules of the CCS Rewrite System.

$$\begin{array}{lll}
\mathtt{Pref}_\mu : \mu \xrightarrow[\underline{\mu}]{\underline{1}} \underline{1} & \mathtt{Res}_\mu : \backslash\alpha \xrightarrow[\underline{\mu}]{\underline{\mu}} \backslash\alpha & \text{for} \quad \mu \notin \{\alpha, \overline{\alpha}\} \\
\mathtt{Suml}_\mu : + \xrightarrow[\underline{\mu}]{\underline{\mu}\otimes\underline{1}} \underline{1}\otimes ! & \mathtt{Sumr}_\mu : + \xrightarrow[\underline{\mu}]{\underline{1}\otimes\underline{\mu}} ! \otimes \underline{1} & \\
\mathtt{Compl}_\mu : \mid \xrightarrow[\underline{\mu}]{\underline{\mu}\otimes\underline{1}} \mid & \mathtt{Compr}_\mu : \mid \xrightarrow[\underline{\mu}]{\underline{1}\otimes\underline{\mu}} \mid & \mathtt{Synch}_\lambda : \mid \xrightarrow[\underline{\tau}]{\underline{\lambda}\otimes\underline{\overline{\lambda}}} \mid .
\end{array}$$

We associate a name to every rule of the tile rewrite system in order to refer to them later. The rules closely correspond to the ordinary SOS rules. For instance rule $\mathtt{Pref}_\mu$ states, as its SOS counterpart, that constructor μ can be deleted, i.e., it can be replaced by the identity $\underline{1}$. Furthermore, the trigger is also the identity, and thus the corresponding SOS rule is an axiom. Finally, the effect is $\underline{\mu}$ and this corresponds to the label of the transition in the SOS case. As another example, rule $\mathtt{Suml}_\mu$ defines left choice. The initial configuration is the constructor $+ : \underline{2} \to \underline{1}$, while the final configuration is $\underline{1}\otimes! : \underline{2} \to \underline{1}$, which states that the first component is preserved and the second component is discarded. The trigger states that in the first component we must have an action $\underline{\mu}$ while no action (i.e., identity action) is required on the second component. Action $\underline{\mu}$ is then transferred to the effect. We call **CCS** the computad defined in this way.

The tile corresponding to the previous example is obtained as follows:

$$\alpha = (((nil * \mathtt{Prefix_a}) \otimes (nil * b) * \mathtt{Suml_a}) \otimes (nil * \mathtt{Prefix_{\overline{a}}})) * \mathtt{Synch_a} * \mathtt{Res}_\tau .$$

CCS Rules and Tiles as Cells. According to Definition 21, we can now employ the functor F_Z to derive the enriched rewriting logic semantics of our CCS rewrite system. For this purpose, we first specify the cell version of our rewrite rules and then apply *ERWL* rewriting.

$$\begin{array}{ll}
\mathtt{Pref}_\mu : \underline{\mu}[\mu(p)] \to p & \mathtt{Res}_\mu : \underline{\mu}[p\backslash\alpha] \to \underline{\mu}[p]\backslash\alpha \quad \text{for } \mu \notin \{\alpha, \overline{\alpha}\} \\
\mathtt{Suml}_\mu : \underline{\mu}[p+q] \to \underline{\mu}[p]\otimes!(q) & \mathtt{Sumr}_\mu : \underline{\mu}[p+q] \to \underline{\mu}[q]\otimes!(p) \\
\mathtt{Compl}_\mu : \underline{\mu}[p \mid q] \to \underline{\mu}[p] \mid q & \mathtt{Compr}_\mu : \underline{\mu}[p \mid q] \to p \mid \underline{\mu}[q] \\
\mathtt{Synch}_\lambda : \underline{\tau}[p \mid q] \to \underline{\lambda}[p] \mid \underline{\overline{\lambda}}[q]. &
\end{array}$$

We use standard term notation for left and right hand sides of rewriting rules, i.e., for the domain and codomain arrows of cells. This is possible since almost all our horizontal and vertical constructors are term constructors, i.e., return one value. The only exception is discharger, which returns no value. To accommodate for the discharger, we extend the ordinary term notation introducing the monoidal operation. Notice that the parallel composition $t\otimes!(t')$ still returns one value, and thus can be safely used in a subterm position. It is easy to see that, thanks to the homomorphism axioms of monoidal categories, the following properties hold, where t and t' are terms without variables:

$$t \otimes !(t') = !(t') \otimes t \qquad !(t) \otimes !(t') = !(t') \otimes !(t) \qquad t''(t \otimes !(t')) = t''(t) \otimes !(t').$$

To distinguish between horizontal and vertical arrows, when composed to form an arrow, we use square brackets. For instance in the left hand side $\underline{\tau}[p \mid q]$ of $\mathbf{Synch}_\lambda$, the horizontal part is inside the brackets and the vertical part is outside, while in the right hand side $\underline{\lambda}[p] \mid \underline{\overline{\lambda}}[q]$ the vertical part is inside the brackets and the horizontal part is outside.

We can now show the derivation in *ERWL* style for our running example.

$$\begin{array}{lll}
\underline{\tau}[((a.nil + b.nil) \mid \overline{a}.nil)\backslash a] & \rightarrow [\underline{\tau}[(a.nil + b.nil) \mid \overline{a}.nil]]\backslash a & \rightarrow \\
([\underline{a}[a.nil + b.nil]] \mid [\underline{\overline{a}}[\overline{a}.nil]])\backslash a & \rightarrow ([\underline{a}[a.nil]] \otimes !(b.nil) \mid [\underline{\overline{a}}[\overline{a}.nil]])\backslash a & \rightarrow \\
([\underline{a}[a.nil]] \otimes !(b.nil) \mid nil)\backslash a & \rightarrow (nil \otimes !(b.nil) \mid nil)\backslash a & = \\
(nil \mid nil)\backslash a \otimes !(b.nil). & &
\end{array}$$

□

It is natural to proceed top down in the proof, as in a goal-oriented evaluation of SOS inference rules. We thus start with the initial configuration and the expected action. Notice that the presence of more than one level of square brackets in all the intermediate terms of our *ERWL* derivation makes clear that all its sub-derivations are cells but not squares. Only the entire derivation is a square, and actually it is easy to see that it can be made equal, employing the axioms of MON2VHCAT, to the square α defined above.

In general it is not enough to check that the domain (resp. codomain) arrow of a cell is the composition of a horizontal and a vertical arrow (resp. a vertical and a horizontal arrow) to make sure that the cell is a square. The following well-known "four bricks" computad provides a conterexample.

Example 6. (four bricks counterexample)

Let S consist of the underlined natural numbers, and let all constructors, both horizontal and vertical, have $\underline{1}$ as source and target. The set of rules is as follows:

$$\alpha = f_1 \xrightarrow[v_1']{v_1} g_4' * h_1 \qquad \beta = g_2 \xrightarrow[u_2]{v_1' \cdot w_2} g_2' \qquad \epsilon = h_1 \xrightarrow[w_2]{w_4} h_3$$

$$\gamma = h_3 * g_2' \xrightarrow[v_3]{v_3'} f_3 \qquad \delta = g_4' \xrightarrow[w_4 \cdot v_3']{u_4} g_4.$$

It is easy to see that the cell

$$f_1; \beta; v_3 \ \circ \ \alpha; w_2; g_2', v_3 \ \circ \ v_1; g_4'; \epsilon; g_2'; v_3 \ \circ \ v_1; g_4'; w_4; \gamma \ \circ \ v_1; \delta; f_3$$

is not a square. In fact, even if domain and codomain arrows are of the right form, this cell cannot be obtained from the rules just employing horizontal and vertical compositions $*$ and $\cdot$.

□

However, counterexamples as shown above cannot be found for the CCS rewrite system because of the following property:

Proposition 24. *The tile rewrite system* **CCS** *of Example 5 is uniform.*

The above result guarantees that a very simple implementation exists of the tile logic for CCS into rewriting logic. In fact there is no need to keep the proof terms while rewriting, since to show that an *ERWL* derivation corresponds to a tile derivation it is enough to check the domain and codomain arrows of the resulting cell.

5 Conclusion

In this paper we have employed partial membership equational logic to define (monoidal) double categories, 2-categories and their generalization, 2VH-categories. The aim was to build a tight connection between two very general models of computation, tile logic and rewriting logic. Tile logic is especially useful for defining compositional models of computation of reactive systems, coordination languages, mobile calculi, and causal and located concurrent systems. Rewriting logic extends the algebraic semantics approach to concurrent systems with state changes and is the basis of several existing languages. In particular, the language Maude developed at SRI is efficiently implemented and is equipped with important extra features, like execution strategies and reflective logic, which allow it to easily embed a variety of other logics. We have shown that rewriting logic derivation (of squares) within 2VH-categories coincides with tile logic derivation within double categories, when starting from the same computad. Thus if an implementation of 2VH-categories within Maude is provided using internal strategies, the models of computation specified via tile rewrite sistems can be actually run in Maude. Also, the notion of uniform tile rewrite system was introduced, for which the internal strategy is particularly simple, and a tile rewrite system for CCS was presented which is actually uniform.

Several lines of future research are possible and promising at this point. One consists of actually building in Maude the internal strategies above and of carrying out experiments about the flexibility and efficiency of execution in Maude of tile specifications. In particular, it will be very important to define specific formats of tile rewrite systems which guarantee uniformity. We are confident that most process algebras will fall in these classes. Results in this direction would make it easy to experiment with Maude about new calculi and about verification techniques for reactive systems. Finally, the definition of more refined theories can be tackled for particular versions of the tile model. For example, most of the applications described in the introduction require a version of tiles (and thus of double categories) which is *symmetric* monoidal rather than simply monoidal, and some of them require *cartesian* double categories. Definitions in the general case (where the structure should appear on both dimensions) are far from trivial [2], since natural transformations on double categories are defined in terms of four functors from a double to a four-fold category. Upcoming work by Roberto Bruni and the authors [6] proposes suitable notions of symmetric monoidal and cartesian double categories, and shows that the relation between monoidal double categories and monoidal 2VH-categories exploited in this paper can be smoothly extended to the symmetric monoidal and cartesian case. Also, preliminary results show for instance the possibility of programming and executing in Maude the tile systems recently proposed in the literature [16] for located and mobile calculi.

6 Acknowledgments

We would like to thank Narciso Martí-Oliet and Roberto Bruni for their comments.

References

1. H.P. Barendregt, M.C.J.D. van Eekelrn, J.R.W. Glauert, J.R. Kennaway, M.J. Plasmeijer, M.R. Sleep, *Term Graph Reduction*, Proc. PARLE, Springer LNCS 259, 141–158, 1987.
2. A. Bastiani, C. Ehresmann *Multiple Functors I: Limits Relative to Double Categories*, Cahiers de Topologie ed Géométrie Différentielle **15** (3), 1974, pp. 545-621.
3. P. Borovanský, C. Kirchner, H. Kirchner, P.-E. Moreau, and M. Vittek. ELAN: A logical framework based on computational systems. In *Proc. 1st Intl. Workshop on Rewriting Logic and its Applications*, ENTCS, North Holland, 1996.
4. G. Boudol, I. Castellani, M. Hennessy, A. Kiehn, *Observing Localities*, Theoretical Computer Science, 114: 31–61, 1993.
5. Adel Bouhoula, Jean-Pierre Jouannaud, and José Meseguer. Specification and proof in membership equational logic. In M. Bidoit and M. Dauchet, editors, *Proceedings of TAPSOFT'97*. Springer LNCS 1214, 1997.
6. R. Bruni, J. Meseguer, and U. Montanari. *Process and Term Tile Logic*. Technical Report, SRI International, in preparation.
7. R. Bruni, U. Montanari, *Zero-Safe Nets, or Transition Synchronization Made Simple*. In: Catuscia Palamidessi, Joachim Parrow, Eds, EXPRESS'97, ENTCS, Vol. 7, 1997.
8. R. Bruni, U. Montanari, *Zero-Safe Nets: The Individual Token Approach*. In: Francesco Parisi-Presicce, Ed., Proc. 12th WADT Workshop on Algebraic Development Techniques, Spinger LNCS, 1998, this volume.
9. P. Ciancarini, C. Hankin, Eds., *Coordination Languages and Models*, LNCS 1061, 1996.
10. Manuel G. Clavel, Steven Eker, Patrick Lincoln, and José Meseguer. Principles of Maude. In: J. Meseguer, Guest Ed., First International Workshop on Rewriting Logic and its Applications, ENTCS 4 (1996).
11. M. Clavel, J. Meseguer, *Internal Strategies in a Reflective Logic*. In: J. Meseguer, Guest Ed., First International Workshop on Rewriting Logic and its Applications, ENTCS 4 (1996).
12. A. Corradini, F. Gadducci, *A 2-Categorical Presentation of Term Graph Rewriting*. In: Eugenio Moggi, Giuseppe Rosolini, Eds., Category Theory and Computer Science 1997, Springer LNCS 1290, 1997, pp.87-105.

13. A. Corradini, U. Montanari, *An Algebraic Semantics for Structured Transition Systems and its Application to Logic Programs*, Theoretical Computer Science **103**, 1992, pp.51-106.
14. C. Ehresmann, *Catégories Structurées*: I and II, Ann. Éc. Norm. Sup. 80, Paris (1963), 349-426; III, Topo. et Géo. diff. V, Paris (1963).
15. G. Ferrari, U. Montanari, *A Tile-Based Coordination View of the Asynchronous π-calculus*. In: Igor Prvara, Peter Ruzicka, Eds., Mathematical Foundations of Computer Science 1997, Springer LNCS 1295, 1997, pp. 52-70.
16. G. Ferrari, U. Montanari, *Tiles for Concurrent and Located Calculi*. In: Catuscia Palamidessi, Joachim Parrow, Eds, EXPRESS'97, ENTCS, Vol. 7.
17. P. Freyd. Algebra valued functors in general and tensor products in particular. *Coll. Math.*, 14:89–106, 1966.
18. K. Futatsugi and T. Sawada. Cafe as an extensible specification environment. In *Proc. of the Kunming International CASE Symposium, Kunming, China, November*, 1994.
19. F. Gadducci, *On the Algebraic Approach to Concurrent Term Rewriting*, PhD Thesis, Università di Pisa, Pisa. Technical Report TD-96-02, Department of Computer Science, University of Pisa, 1996.
20. F. Gadducci, U. Montanari, *The Tile Model*. In: Gordon Plotkin, Colin Stirling, and Mads Tofte, Eds., Proof, Language and Interaction: Essays in Honour of Robin Milner, MIT Press, to appear. Also Technical Report TR-96-27, Department of Computer Science, University of Pisa, 1996
21. F. Gadducci, U. Montanari, *Tiles, Rewriting Rules and CCS*. In: J. Meseguer, Guest Ed., 1st Int. Workshop on Rewriting Logic and its Applications, ENTCS 4 (1996).
22. G.M. Kelly, R.H. Street, *Review of the Elements of 2-categories*, Lecture Notes in Mathematics 420, 1974, pp. 75-103.
23. C. Lair, Etude Générale de la Catégorie des esquisses, Esquisses Math. 24 (1974).
24. K.G. Larsen, L. Xinxin, *Compositionality Through an Operational Semantics of Contexts*, in Proc. ICALP'90, LNCS 443, 1990, pp. 526-539.
25. F.W. Lawvere. Some algebraic problems in the context of functorial semantics of algebraic theories. In *Proc. Midwest Category Seminar II*, pages 41–61. Springer Lecture Notes in Mathematics No. 61, 1968.
26. N. Martí-Oliet, J. Meseguer, Inclusion and Subtypes I: First-order Case, *J. Logic Computat.*, Vol.6 No.3, pp.409-438, 1996.
27. J. Meseguer, *Rewriting as a Unified Model of Concurrency*, SRI Technical Report, CSL-93-02R, 1990. See the appendix on *Functorial Semantics of Rewrite Systems*.
28. J. Meseguer, *Conditional Rewriting Logic as a Unified Model of Concurrency*, Theoretical Computer Science **96**, 1992, pp. 73-155.
29. J. Meseguer. A logical theory of concurrent objects and its realization in the Maude language. In Gul Agha, Peter Wegner, and Akinori Yonezawa, Eds., *Research Directions in Concurrent Object-Oriented Programming*, pp. 314–390. MIT Press, 1993.
30. J. Meseguer. Membership algebra. Lecture and abstract at the Dagstuhl Seminar on "Specification and Semantics," July 9, 1996.
31. J. Meseguer, *Rewriting Logic as a Semantic Framework for Concurrency: A Progress Report*, in: U. Montanari and V. Sassone, Eds., *CONCUR'96: Concurrency Theory*, Springer LNCS 1119, 1996, 331-372.
32. J. Meseguer, Ed., *Procs. Rewriting Logic and Applications*, First International Workshop, ENTCS 4 (1996).

33. J. Meseguer, *Membership Equational Logic as a Logical Framework for Equational Specification.* In: Francesco Parisi-Presicce, Ed., Proc. 12th WADT Workshop on Algebraic Development Techniques, Spinger LNCS, 1998, this volume.
34. R. Milner, *Communication and Concurrency,* Prentice-Hall, 1989.
35. R. Milner, J. Parrow, D. Walker, *A Calculus of Mobile Processes* (parts I and II), Information and Computation, 100:1–77, 1992.
36. U. Montanari, F. Rossi, *Graph Rewriting and Constraint Solving for Modelling Distributed Systems with Synchronization,* in: Paolo Ciancarini and Chris Hankin, Eds., Coordination Languages and Models, LNCS 1061, 1996, pp. 12-27. Full paper submitted for publication.
37. B. Pareigis, *Categories and Functors,* Academic Press, 1970.
38. G. Plotkin, *A Structural Approach to Operational Semantics,* Technical Report DAIMI FN-19, Computer Science Department, Aarhus University, 1981.
39. A. Poigné. Algebra categoricaly. In D. Pitt et al., editor, *Category Theory and Computer Programming,* volume 240 of *LNCS,* pages 76–102. Springer-Verlag, 1985.
40. P. Gabriel and F. Ulmer. *Lokal präsentierbare Kategorien.* Springer Lecture Notes in Mathematics No. 221, 1971.

An Algebra of Mixin Modules*

Davide Ancona and Elena Zucca

DISI - Università di Genova
Via Dodecaneso, 35, 16146 Genova (Italy)
email: {davide,zucca}@disi.unige.it

Abstract. Mixins are modules which may contain deferred components, i.e. components not defined in the module itself, and allow definitions to be overridden. We give an axiomatic definition of a set of operations for mixin combination, corresponding to a variety of constructs existing in programming languages (merge, hiding, overriding, functional composition, ...). In particular, we show that they can all be expressed in terms of three primitive operations (namely, sum, reduct and freeze), which are characterized by a small set of axioms. We show that the given axiomatization is sound w.r.t. to a model provided in some preceding work. Finally, we prove the existence of a normal form for mixin expressions.

Introduction

The notion of *mixin*, firstly introduced in the context of object oriented programming [10], has recently become the subject of increasing interest in many respects and with many slight variations in the intended meaning [9, 11, 7, 15, 22]. Some preceding work of us [4, 6] has been devoted to a rigorous formulation of the notion, covering and making precise the various ways in which the word is used in the literature; we refer to this formulation in the discussion below.

Mixins (or *mixin modules*) are a generalization of usual modules in programming languages, which are collections of definitions of heterogeneous components, e.g. types, functions, procedures, exceptions and so on (typical examples are Modula-2 or Standard ML). The generalization consists in two main features.

First, some of the components can be only declared in the module, without having an associated definition; we say that these components are *deferred* (the terminology comes from object oriented languages). A mixin with deferred components cannot be used in isolation; a binary *merge* operation allows to combine two mixins, say M_1, M_2, in such a way that a deferred component in M_1 can be *concreted* by a definition provided in M_2, and conversely. This operation is commutative, associative and no conflicting definitions are allowed (i.e. a component defined in both arguments). In the resulting mixin $M_1 \oplus M_2$ there can still be deferred components (if some deferred component in one argument has not been matched in the other), or we can get a *concrete module*, i.e. a module with no deferred components, which can be effectively used. An important remark is that

* This work has been partially supported by Murst 40% - Modelli della computazione e dei linguaggi di programmazione and CNR - Formalismi per la specifica e la descrizione di sistemi ad oggetti.

the symmetry of the merge operator allows recursive definitions to span module boundaries, with a great benefit for modularity, as illustrated e.g. in [15, 6].

The second extension w.r.t. usual modules, again inspired by the object oriented approach, is the possibility that, assembling modules together, some definition in a module is replaced by a new definition provided in another module. This feature is called *overriding* and is typical of inheritance in object oriented languages: anyway, the concept turns out to be completely orthogonal to the object oriented nature of the language and can be formulated independently from the notions of object, class and subtyping hierarchies.

Formally, overriding can be seen as another binary operation s.t. $M_1 \Leftarrow M_2$ is the same of $M_1 \oplus M_2$, except that there can be conflicting definitions and in this case the definitions in M_2 take the precedence. Overriding can be seen as the composition of two different operations: a *restrict* operation whose effect is to "cancel" some definitions in a module, and the merge operation (this view of overriding is originally due to [9]).

Since definitions of components can refer to each other, redefining a component, say m, can actually change the behavior of other components, e.g. a component m' defined by $m' = \ldots\ m\ \ldots$. This is not always the case: some languages allows the user to explicitly specify whether, in case of redefinition of m, m' should refer to the new or to the old version. We will say that m is *virtual* in the first case, *frozen* in the second (cfr. virtual and non virtual methods in C++, while the term *frozen* has been introduced in [9]).

In [4, 6] we have proposed a formal model for mixin modules. The basic idea is to see a mixin as a function from input to output components, where output components are those defined in the module, while input components are those which definitions in the module can depend on (hence deferred and virtual components). Moreover, we have defined a kernel language of mixin modules, i.e. a set of operators for composing mixins corresponding to a variety of constructs existing in programming languages (including merge, restrict, inheritance/overriding, hiding, functional composition). An important point is that we have not fixed an underlying *core* language (following the ML terminology), but provided a language of modules which can be instantiated on top of a variety of different languages. This point of view (the module language is a small language of its own, with its typing rules, as much independent as possible from the core language) was a design goal of the SML module system [23] and has been recently recognized as fundamental both from the type theoretic [20, 18] and the software engineering [11, 7, 22] point of view.

In this paper, we take a different approach to the formal definition of operators for composing mixin modules, i.e. we give an *axiomatic characterization*, in the spirit of the seminal paper [8]. In particular, we show that all the mixin operators can be expressed in terms of three primitive operations (namely, sum, reduct and freeze) which are characterized by a small set of axioms. We show that the given axiomatization is sound w.r.t. the previously presented model. Finally, we prove the existence of a normal form for mixin expressions.

The paper is organized as follows: Sect.1 is a summary of our preceding work in [4, 6] providing the notions needed in the sequel; in Sect.2 we give the axiomatic definition of the primitive (Sect.2.1) and derived (Sect.2.2) operations;

in Sect.3 we state our main technical results, which are: the soundness of the axiomatization w.r.t. to the model (Sect.3.1), the normal form theorem (Sect.3.2) and some algebraic laws holding for derived operations (Sect.3.3). Finally, in Sect.4 we summarize the contribution of the paper and outline further work.

1 Mixins and Their Composition Operators

In this section, we briefly present our formal model for mixin modules and a set of operations for composing mixins with their interpretation in the model, by giving a summary of some preceding work [4, 6] which is needed in the following; anyway, here we have to fix some extra requirement (see Def.6 below) at the level of the core language for guaranteeing the validity of some expected properties of mixin composition operators.

While the paper is self-contained on the technical side, for more examples and discussions about mixins we refer to our previous work in [4, 6, 5].

1.1 A Formal Model for Mixins

On the semantic side, a quite natural view of a module with deferred components is as a function F which, for any given assignment of values to these components, gives values to the defined components. Since we want to abstract w.r.t. the nature of components in a specific language, we take the approach of *institutions* [17] and model a collection of (names and types of) components by a *signature* Σ and all the possible assignments of values to the components as the class of the *models* over Σ, denoted $Mod(\Sigma)$. The concrete definitions of signatures and models will depend on the (semantic framework modeling) the underlying core language (*core framework*). Hence, $F\colon Mod(\Sigma^{in}) \to Mod(\Sigma^{out})$, where $\Sigma^{in}, \Sigma^{out}$ are the signatures representing the input (deferred) and output (defined) components, respectively.

In order to model overriding, one needs to refine the above intuition. Indeed, redefining a virtual component can change the behavior of other components. Formally, it is necessary to keep track of the dependency from virtual components, hence virtual components (which are part of the output signature as defined components) must be present in the input signature, too. In summary, Σ^{in} represents all the components which definitions in the module may depend on, i.e. deferred and virtual components; $\Sigma^{in} \setminus \Sigma^{out}, \Sigma^{in} \cap \Sigma^{out}$ and $\Sigma^{out} \setminus \Sigma^{in}$ represent the deferred, virtual and frozen components, respectively. This model of overriding is the adaptation, in a different context, of the semantics of inheritance developed by W. Cook [13] and U. S. Reddy [24].

As an example of mixin module, consider the following definition:

```
mixin M =
 deferred leq:int*int→bool
 frozen eq(i1,i2:int):bool=leq(i1,i2) and leq(i2,i1)
 frozen lth(i1,i2:int):bool=not leq(i2,i1)
 lleq(i1,i2,j1,j2:int):bool=lth(i1,j1) or (eq(i1,j1) and leq(i2,j2))
 frozen llth(i1,i2,j1,j2:int):bool=not lleq(j1,j2,i1,i2)
end
```

The mixin `M` is defined on top of a very simple functional language, supporting only basic types; its input signature Σ^{in} is {`leq`, `lleq`}, whereas its output signature Σ^{out} is {`eq`, `lth`, `lleq`, `llth`}; hence, $\Sigma^{in} \setminus \Sigma^{out}$ = {`leq`}, $\Sigma^{in} \cap \Sigma^{out}$ = {`lleq`} and $\Sigma^{out} \setminus \Sigma^{in}$ = {`eq`, `lth`, `llth`}.

The semantics of the functions defined in `M` depends on the deferred component `leq` whose definition must be provided from the outside. The functions `eq` and `lth` are frozen, therefore any redefinition of them has no effect on the semantics of `lleq`. The function `lleq` is virtual (assuming that functions are virtual by default), hence any redefinition of it is propagated also to `llth`.

1.2 Operators on Mixin Modules

In this subsection we give the formal definition of a set of operators for composing mixin modules, i.e. a kernel language of mixins. In this language, any expression M has a type, modeling the interface of the module, which is a pair of signatures in the underlying core framework; we write $M: \Sigma^{in} \to \Sigma^{out}$ and the intended meaning is that $\Sigma^{in} \setminus \Sigma^{out}$, $\Sigma^{in} \cup \Sigma^{out}$ and $\Sigma^{out} \setminus \Sigma^{in}$ are the deferred, virtual and frozen components, respectively. If $M: \Sigma^{in} \to \Sigma^{out}$, then the semantics of M is a function from $Mod(\Sigma^{in})$ into $Mod(\Sigma^{out})$.

For each operator, we first give a typing rule specifying compatibility conditions between the types of the arguments, and the resulting type of the result; then we give the semantic interpretation of the operator.

Typing and semantic rules are given assuming that there is a fixed *(flat) core framework* modeling the underlying core language. The precise definition is quite technical and provided in the next subsection; for an intuitive understanding of the rules it should be enough the explanation below.

A (flat) core framework consists of three ingredients: a category of signatures **Sig**, a model functor Mod: $\mathbf{Sig}^{op} \to \mathbf{Set}$ and a *fix* operator (a family of functions, one for each signature). Signatures and models have the usual meaning they have in institutions [17]; anyway, we require some additional properties. Signatures are required to form a *boolean signature category*, i.e. we assume *inclusions* between signatures with related operations of union, intersection and difference. The model functor is required to preserve all finite colimits, in order to guarantee the existence of the *amalgamated sum* of models (see e.g. [16]). The family of *fix* functions, instead, is completely new w.r.t. to the standard notions in institutions theory. For each signature Σ, fix_Σ is a function

$fix_\Sigma: (Mod(\Sigma) \to Mod(\Sigma)) \to Mod(\Sigma)$

and the expected meaning is that $fix(F)$ returns "the least fixed point" of F, in a sense to be made precise depending on the core framework.

The fix operator is needed for modeling operators which transform some virtual component into frozen.

For instance, we can transform the function `lleq` of the previously defined mixin `M` into a frozen function, obtaining a new mixin where the definition of `llth` is now equivalent to `not (lth(j1,i1) or (eq(j1,i1) and leq(j2,i2)))`.

Intuitively we have replaced the invocation of `lleq` in the body of `llth` with its current definition (whose evaluation requires in the general case a least fixed point operator since definitions can be mutually recursive).

The typing and semantic rules for the operators are given in Fig.1 and Fig.2, respectively. We give here below some intuitive explanation.

Merge This operator allows to "sum" two mixin modules, say M_1 and M_2, obtaining a new module where the deferred components of M_1 are concreted by the definitions given in M_2, if any, and conversely. Two mixin modules can be merged together only if no components are defined in both. The semantic clause expresses the fact that the definitions in $M_1 \oplus M_2$ are obtained taking the union of the definitions of M_1 and M_2; moreover, it is necessary to apply the *fix* operator in order to eliminate from the input signature the deferred components of one argument concreted by frozen components of the other.

Freeze This operator allows to make a module independent from the redefinition of some components, say Σ^{fr}; hence these components, if they were virtual, become frozen, i.e. disappear from the input signature. The semantics is given by means of the *fix* operator; the intuition is that all the components will refer from now on to the values of the Σ^{fr}-components as they are determined by the current definitions.

Restrict This operation allows to "throw away" some definitions in a module, i.e. to cancel the corresponding components from the output signature. Restrict is different from hiding, since a virtual component whose definition is thrown away remains in the interface of the module as deferred and can be redefined later, while an hidden component becomes not visible from the outside. The semantic clause expresses the fact that some definitions are forgotten (as formally expressed by the reduct functor).

Hiding This operation allows to hide some defined component from the outside. Hiding deferred components makes no sense since definitions of other components could depend on them. Hiding virtual components requires first to apply the *fix* operator, in such a way that all the other definitions will refer from now on to their current definitions.

Overriding This operator allows to merge two mixins with conflicting defined components, by overriding the definitions of M_1 by the corresponding definitions of M_2. Note that the overriding operator coincides with the merge operator when there are no components defined in both the mixins; indeed, it can be expressed as a combination of the merge and the restrict operation (see Sect.2).

Functional Composition This operator is a generalization of the application of parameterized modules; here the formal parameters are the deferred components $\Sigma_2^{in} \setminus \Sigma_2^{out}$ of M_2, whereas the actual parameters are the defined components Σ_1^{out} of M_1. Again, the semantics is expressed in terms of the *fix* operator, in order to correctly handle the virtual components of M_1.

1.3 Flat Core Frameworks

This is a technical subsection providing the formal definition of (flat) core frameworks. The keyword "flat" refers to the fact that in [4, 6] we define also a different version of core frameworks, called *sorted core frameworks*, where signatures have an explicit notion of *types* or *sorts*. Sorted core frameworks are more adequate for modeling languages with type definitions; anyway, w.r.t. to the main aim of this paper, which is to study algebraic laws holding between operators, this notion

$$\text{(M-ty)}\ \frac{M_i\colon \Sigma_i^{in} \to \Sigma_i^{out}, i = 1,2}{M_1 \oplus M_2\colon (\Sigma_1^{in} \cup \Sigma_2^{in}) \setminus (\Sigma_1^{fr} \cup \Sigma_2^{fr}) \to \Sigma_1^{out} \cup \Sigma_2^{out}}\ \begin{array}{l}\Sigma_1^{out} \cap \Sigma_2^{out} = \emptyset \\ \Sigma_i^{fr} = \Sigma_i^{out} \setminus \Sigma_i^{in}, i = 1,2\end{array}$$

$$\text{(F-ty)}\ \frac{M\colon \Sigma^{in} \to \Sigma^{out}}{\textbf{freeze } \Sigma^{fr} \textbf{ in } M\colon \Sigma^{in} \setminus \Sigma^{fr} \to \Sigma^{out}}\ \Sigma^{fr} \subseteq \Sigma^{out}$$

$$\text{(R-ty)}\ \frac{M\colon \Sigma^{in} \to \Sigma^{out}}{\textbf{restrict } \Sigma^{rs} \textbf{ in } M\colon \Sigma^{in} \to \Sigma^{out} \setminus \Sigma^{rs}}\ \Sigma^{rs} \subseteq \Sigma^{out}$$

$$\text{(H-ty)}\ \frac{M\colon \Sigma^{in} \to \Sigma^{out}}{\textbf{hide } \Sigma^{hd} \textbf{ in } M\colon \Sigma^{in} \setminus \Sigma^{hd} \to \Sigma^{out} \setminus \Sigma^{hd}}\ \Sigma^{hd} \subseteq \Sigma^{out}$$

$$\text{(O-ty)}\ \frac{M_i\colon \Sigma_i^{in} \to \Sigma_i^{out} \quad i = 1,2}{M_1 \Leftarrow M_2\colon (\Sigma_1^{in} \cup \Sigma_2^{in}) \setminus (\Sigma_1^{fr} \cup \Sigma_2^{fr}) \to \Sigma_1^{out} \cup \Sigma_2^{out}}\ \begin{array}{l}\Sigma_1^{fr} = (\Sigma_1^{out} \setminus \Sigma_2^{out}) \setminus \Sigma_1^{in} \\ \Sigma_2^{fr} = \Sigma_2^{out} \setminus \Sigma_2^{in}\end{array}$$

$$\text{(FC-ty)}\ \frac{M_i\colon \Sigma_i^{in} \to \Sigma_i^{out} \quad i = 1,2}{M_2 \circ M_1\colon (\Sigma_1^{in} \setminus \Sigma_1^{out}) \cup (\Sigma_2^{in} \cap \Sigma_2^{out}) \to \Sigma_2^{out}}\ \begin{array}{l}\Sigma_2^{in} \setminus \Sigma_2^{out} = \Sigma_1^{out} \\ (\Sigma_1^{in} \setminus \Sigma_1^{out}) \cap \Sigma_2^{out} = \emptyset\end{array}$$

Fig. 1. Typing rules

would have implied a more involved technical treatment, without a significant change in the properties of the operators.
Notation. If **C** is a category, then $|\mathbf{C}|$ denotes the class of its objects.

Definition 1. A *signature category* is a category **Sig** with all finite colimits whose objects are called *signatures*.

If Σ_1 and Σ_2 are two signatures, then we denote by $\Sigma_1 + \Sigma_2$ and $\emptyset$ the unique (up to isomorphism) coproduct of Σ_1 and Σ_2 and initial object in **Sig**, respectively.

Definition 2. A *boolean signature category* is a pair <**Sig**, $\mathcal{I}$> where **Sig** is a signature category and $\mathcal{I}$ is a subcategory of **Sig** with $|\mathcal{I}| = |\mathbf{Sig}|$ and s.t.

- $\mathcal{I}$ is a distributive lattice with bottom element (denoted by $\emptyset$); we call the morphisms in $\mathcal{I}$ *inclusions*, use the notation $\Sigma_1 \subseteq \Sigma_2$ if there is an inclusion from Σ_1 into Σ_2, and denote this (unique) inclusion by i_{Σ_1,Σ_2}. We call *union* (denoted by $\Sigma_1 \cup \Sigma_2$) and *intersection* (denoted by $\Sigma_1 \cap \Sigma_2$), respectively, the join and the meet of Σ_1 and Σ_2 in $\mathcal{I}$. For any morphism $\sigma\colon \Sigma_1 \to \Sigma_2$ and signature $\Sigma_1' \subseteq \Sigma_1$ we write $\sigma_{|\Sigma_1'}$ for the composition $\sigma \circ i_{\Sigma_1',\Sigma_1}$;
- for any $\Sigma_1, \Sigma_2 \in |\mathbf{Sig}|$ there exists a signature Σ s.t. $\Sigma \cup \Sigma_1 = \Sigma_2 \cup \Sigma_1$ and $\Sigma \cap \Sigma_1 = \emptyset$. It easy to show that such a signature (denoted by $\Sigma_2 \setminus \Sigma_1$) is unique;

$$\text{(M-sem)}\ \frac{[\![M_i]\!] = F_i, i = 1,2}{[\![M_1 \div M_2]\!] = \lambda A.\mathit{fix}(\lambda B.F_1((A+B_{|\Sigma^{fr}})_{|\Sigma_1^{in}}) + F_2((A+B_{|\Sigma^{fr}})_{|\Sigma_2^{in}}))}\ \Sigma^{fr} = \Sigma_1^{fr} \cup \Sigma_2^{fr}$$

$$\text{(F-sem)}\ \frac{[\![M]\!] = F}{[\![\textbf{freeze } \Sigma^{fr} \textbf{ in } M]\!] = \lambda A.\mathit{fix}(\lambda B.F(A + B_{|\Sigma^{fr} \cap \Sigma^{in}}))}$$

$$\text{(R-sem)}\ \frac{[\![M]\!] = F}{[\![\textbf{restrict } \Sigma^{rs} \textbf{ in } M]\!] = \lambda A.(F(A))_{|\Sigma^{out} \setminus \Sigma^{rs}}}$$

$$\text{(H-sem)}\ \frac{[\![M]\!] = F}{[\![\textbf{hide } \Sigma^{hd} \textbf{ in } M]\!] = \lambda A.(\mathit{fix}(\lambda B.F(A + B_{|\Sigma^{hd} \cap \Sigma^{in}})))_{|\Sigma^{out} \setminus \Sigma^{hd}}}$$

$$\text{(O-sem)}\ \frac{[\![M_i]\!] = F_i, i = 1,2}{[\![M_1 \Leftarrow M_2]\!] = \lambda A.\mathit{fix}(\lambda B.(F_1((A+B_{|\Sigma^{fr}})_{|\Sigma_1^{in}}))_{|\Sigma_1^{out} \setminus \Sigma_2^{out}} + F_2((A+B_{|\Sigma^{fr}})_{|\Sigma_2^{in}}))}$$

with $\Sigma^{fr} = \Sigma_1^{fr} \cup \Sigma_2^{fr}$

$$\text{(FC-sem)}\ \frac{[\![M_i]\!] = F_i, i = 1,2}{[\![M_2 \circ M_1]\!] = \lambda A.F_2(\mathit{fix}(\lambda B.F_1(A_{|\Sigma_1^{in} \setminus \Sigma_1^{out}} + B_{|\Sigma_1^{in} \cap \Sigma_1^{out}})) + A_{|\Sigma_2^{in} \cap \Sigma_2^{out}})}$$

Fig. 2. Semantic rules

- $\emptyset$ is initial in **Sig** and for any $\Sigma_1, \Sigma_2 \in |\mathbf{Sig}|$, $\Sigma_1 \hookrightarrow \Sigma_1 \cup \Sigma_2 \hookleftarrow \Sigma_2$ is a pushout for $\Sigma_1 \hookleftarrow \Sigma_1 \cap \Sigma_2 \hookrightarrow \Sigma_2$.

Definition 3. A *model part* is a pair <**Sig**, *Mod*> where

- **Sig** is a signature category;
- *Mod* is a functor, $Mod\colon \mathbf{Sig}^{op} \to \mathbf{Set}$; for any signature Σ, objects in $Mod(\Sigma)$ are called *models* over Σ or *Σ-models*; for any signature morphism $\sigma\colon \Sigma \to \Sigma'$, $Mod(\sigma)$ is called the *reduct* functor and denoted by $-_{|\sigma}$; we write $-_{|\Sigma_1}$ for $-_{|i_{\Sigma_1,\Sigma_2}}$ if $\Sigma_1 \subseteq \Sigma_2$.

A model part <**Sig**, *Mod*> is *regular* iff *Mod* preserves all finite colimits.

Definition 4. A model part <**Sig**, *Mod*> satisfies the amalgamation property iff for any pushout diagram

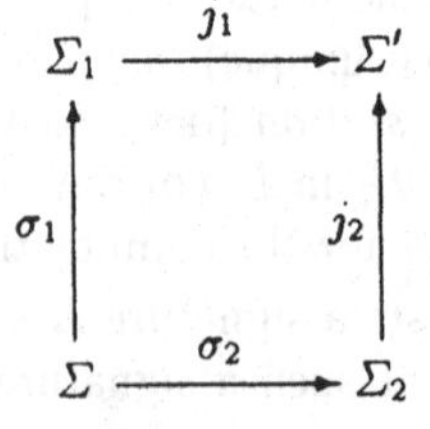

and for every pair of models $A_i \in Mod(\Sigma_i)$, $i = 1, 2$, s.t. $A_{1|\sigma_1} = A_{2|\sigma_2}$, there exists a unique model in $Mod(\Sigma')$, denoted by $A_1 + A_2$, s.t. $(A_1 + A_2)_{|j_i} = A_i$, $i = 1, 2$. We call $A_1 + A_2$ the *amalgamated sum* of A_1 and A_2.

Fact 5 *Every regular model part satisfies the amalgamation property.*

In particular, note that for any regular model part $<\mathbf{Sig}, Mod>$, if $\Sigma_1 \xrightarrow{j_1} \Sigma_1 + \Sigma_2 \xleftarrow{j_2} \Sigma_2$ is a coproduct and $A_i \in Mod(\Sigma_i)$, $i = 1, 2$, then there exists a unique model $A_1 + A_2$ in $Mod(\Sigma_1 + \Sigma_2)$ s.t. $(A_1 + A_2)_{|j_i} = A_i$, $i = 1, 2$. Indeed $\Sigma_1 \xrightarrow{j_1} \Sigma_1 + \Sigma_2 \xleftarrow{j_2} \Sigma_2$ is a pushout for $\Sigma_1 \leftarrow \emptyset \rightarrow \Sigma_2$ and $Mod(\emptyset)$ is a singleton.

Definition 6. A *flat core framework* is a triple $<\mathbf{Sig}, Mod, fix>$ where

- $<\mathbf{Sig}, Mod>$ is a regular model part with **Sig** a boolean signature category;
- fix is a family of (total) functions $fix_\Sigma : (Mod(\Sigma) \to Mod(\Sigma)) \to Mod(\Sigma)$, indexed over the signatures Σ in **Sig** and satisfying the following properties:
  1. (fix-point) for any $F: Mod(\Sigma) \to Mod(\Sigma)$,
     $F(fix_\Sigma(F)) = fix_\Sigma(F)$;
  2. (uniformity) for any $F: Mod(\Sigma) \to Mod(\Sigma)$, $F': Mod(\Sigma') \to Mod(\Sigma')$ and $\sigma: \Sigma' \to \Sigma$, if $(F(A))_{|\sigma} = F'(A_{|\sigma})$ for any $A \in Mod(\Sigma)$, then
     $(fix_\Sigma(F))_{|\sigma} = fix_{\Sigma'}(F')$;
  3. (currying) for any $F: Mod(\Sigma) \times Mod(\Sigma) \to Mod(\Sigma)$,
     $fix_\Sigma(fix_\Sigma \circ \overline{F}) = fix_\Sigma(F \circ <id, id>)$, where $\overline{F}$ and $<id, id>$ are the functions defined by $\overline{F}(A)(B) = F(A, B)$, $<id, id>(A) = <A, A>$, for any $A, B \in Mod(\Sigma)$.

Property 1 reflects the intuition that each fix_Σ is actually a fixed-point operator. This is essential for proving that functional composition can be expressed by means of the merge operator.

Property 2 (together with property 1) states that each fix_Σ is a *least* fixed point operator and ensures that the freeze operator well-behaves when composed with the merge operator.

Finally, property 3 is needed for proving that iterative applications of the freeze operator can be always reduced to a single application.

2 An Axiomatic Definition

In this section we give an axiomatic definition of the operators on mixin modules. More precisely, the axiomatization is given in two steps: first, we define three primitive operations characterized by a small set of axioms; then, we give for each operator of the previous section an axiom stating that the operator can be expressed in terms of these three primitive operations.

2.1 Primitive Operations

The primitive operations we consider are sum, reduct and freeze. Their typing rules are given in Fig.3.

$$\text{(Sum-ty)}\ \frac{M_1\colon \Sigma^{in} \to \Sigma_1^{out} \qquad M_2\colon \Sigma^{in} \to \Sigma_2^{out}}{M_1 + M_2\colon \Sigma^{in} \to \Sigma_1^{out} + \Sigma_2^{out}}$$

$$\text{(Reduct-ty)}\ \frac{M\colon \Sigma^{in} \to \Sigma^{out}}{{}_{\sigma^{in}|}M_{|\sigma^{out}}\colon \Sigma'^{in} \to \Sigma'^{out}} \quad \begin{array}{l}\sigma^{in}\colon \Sigma^{in} \to \Sigma'^{in}\\ \sigma^{out}\colon \Sigma'^{out} \to \Sigma^{out}\end{array}$$

$$\text{(Freeze-ty)}\ \frac{M\colon \Sigma^{in} + \Sigma^{fr} \to \Sigma^{out}}{\mathit{freeze}_{\sigma^{fr}}(M)\colon \Sigma^{in} \to \Sigma^{out}} \ \sigma^{fr}\colon \Sigma^{fr} \to \Sigma^{out}$$

Fig. 3. Typing rules

The sum $M_1 + M_2$ is obtained from M_1, M_2 sharing their input components (which must be the same) and taking the disjoint union of the definitions. The reduct ${}_{\sigma^{in}|}M_{|\sigma^{out}}$ is obtained from M by renaming in an independent way input and output components. Note that the signature morphism σ^{in} goes from the old to the new input signature, while the signature morphism σ^{out} goes in the opposite direction; that allows in particular to add dummy input components and to forget output components. The freeze operator allows to bind some input components (Σ^{fr}) to some definitions in the module, in the way specified by the σ^{fr} signature morphism.

Note that for defining primitive operations we do not need to assume that **Sig** is a boolean signature category, but just a signature category.

Notation. We use the following abbreviations for the reduct: if $j^{in}\colon \Sigma'^{in} \to \Sigma^{in}$ and $j^{out}\colon \Sigma^{out} \to \Sigma'^{out}$ are injections of coproducts, then we write ${}_{\Sigma^{in}|}M_{|\Sigma^{out}}$ instead of ${}_{j^{in}|}M_{|j^{out}}$. Note that, since $\Sigma \xrightarrow{id} \Sigma \longleftarrow \emptyset$ is a coproduct, if $M\colon \Sigma^{in} \to \Sigma^{out}$, then we write ${}_{\Sigma^{in}|}M_{|\Sigma^{out}}$ instead of ${}_{id_{\Sigma^{in}}|}M_{|id_{\Sigma^{out}}}$. Finally, if $M\colon\Sigma^{in} \to \Sigma^{out}$, then we write ${}_{\sigma^{in}|}M$ instead of ${}_{\sigma^{in}|}M_{|\Sigma^{out}}$, $M_{|\sigma^{out}}$ instead of ${}_{\Sigma^{in}|}M_{|\sigma^{out}}$.

If $\sigma_1\colon \Sigma_1 \to \Sigma$ and $\sigma_2\colon \Sigma_2 \to \Sigma$ are two signature morphisms, then we denote by $[\sigma_1, \sigma_2]$ the unique morphism from $\Sigma_1 + \Sigma_2$ to Σ making commute the following diagram:

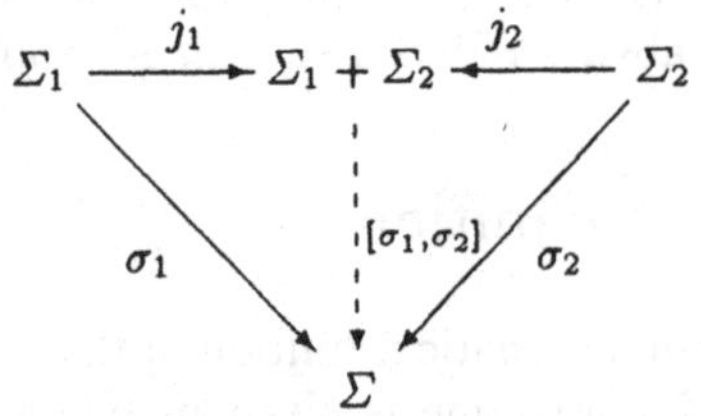

Moreover, if $\sigma_1\colon \Sigma_1 \to \Sigma'_1$ and $\sigma_2\colon \Sigma_2 \to \Sigma'_2$ are two signature morphisms, then we denote by $\sigma_1 + \sigma_2$ the morphism $[j'_1 \circ \sigma_1, j'_2 \circ \sigma_2]\colon \Sigma_1 + \Sigma_2 \to \Sigma'_1 + \Sigma'_2$, where j'_1 and j'_2 are the injections of the coproduct $\Sigma'_1 + \Sigma'_2$.

The axiomatic characterization of the three primitive operators is given in Fig.4.

Types of variables and functionalities of signature morphisms are omitted in Fig.4 and listed below.

$$(M_1 + M_2)_{|\Sigma_1^{out}} = M_1 \quad (1)$$

$$(M_1 + M_2)_{|\Sigma_2^{out}} = M_2 \quad (2)$$

$$M = M_1 + M_2 \text{ if } M_{|\Sigma_i^{out}} = M_i, i = 1, 2 \quad (3)$$

$${}_{\Sigma^{in}|}M_{|\Sigma^{out}} = M \quad (4)$$

$${}_{\sigma_1^{in} \circ \sigma_2^{in}|}M_{|\sigma_2^{out} \circ \sigma_1^{out}} = {}_{\sigma_1^{in}|}{}_{\sigma_2^{in}|}M_{|\sigma_2^{out}|\sigma_1^{out}} \quad (5)$$

$$\mathit{freeze}_{\sigma_2^{fr}}(\mathit{freeze}_{\sigma_1^{fr}}(M)) = \mathit{freeze}_{[\sigma_1^{fr}, \sigma_2^{fr}]}(M) \quad (6)$$

$$\mathit{freeze}_{\sigma_0^{fr}}(M) = M \quad (7)$$

$$\mathit{freeze}_{\sigma^{fr}}(M_{|\sigma^{out}}) = (\mathit{freeze}_{\sigma^{out} \circ \sigma^{fr}}(M))_{|\sigma^{out}} \quad (8)$$

$${}_{\sigma^{in}|}(\mathit{freeze}_{\sigma^{fr}}(M)) = \mathit{freeze}_{\sigma'^{fr}}({}_{\sigma^{in} + \sigma'^{in}|}M) \quad (9)$$

$$\mathit{freeze}_{\sigma_1^{fr}}(M_1) + \mathit{freeze}_{\sigma_2^{fr}}(M_2) = \mathit{freeze}_{\sigma_1^{fr} + \sigma_2^{fr}}({}_{\Sigma'^{in}|}M_1 + {}_{\Sigma'^{in}|}M_2) \quad (10)$$

Fig. 4. Axioms for the primitive operations

Axioms (1), (2) and (3) ensure that the sum operator enjoys the amalgamation property; they hold for $M_i\colon \Sigma^{in} \to \Sigma_i^{out}$, $i = 1, 2$, $M\colon \Sigma^{in} \to \Sigma_1^{out} + \Sigma_2^{out}$.

Axioms (4) and (5) express the functoriality of the reduct operator; they hold for $M\colon \Sigma^{in} \to \Sigma^{out}$ and $\sigma_1^{in}\colon \Sigma'^{in} \to \Sigma''^{in}$, $\sigma_2^{in}\colon \Sigma^{in} \to \Sigma'^{in}$, $\sigma_1^{out}\colon \Sigma''^{out} \to \Sigma'^{out}$, $\sigma_2^{out}\colon \Sigma'^{out} \to \Sigma^{out}$.

Axiom (6) shows how the composition of freeze works; it is easy to prove that such a composition is both commutative and associative. The axiom holds for $M\colon \Sigma_1^{fr} + \Sigma_2^{fr} + \Sigma^{in} \to \Sigma^{out}$ and $\sigma_i^{fr}\colon \Sigma_i^{fr} \to \Sigma^{out}$, $i = 1, 2$.

Axiom (7) states that freeze is the identity whenever we consider the (unique) morphism σ_0^{fr} from the initial object $\emptyset$ to Σ^{out} (assuming $M\colon \Sigma^{in} \to \Sigma^{out}$).

Axioms (8) and (9) describe how the freeze operator behaves w.r.t. the reduct operator; these axioms hold for $M\colon \Sigma^{fr} + \Sigma^{in} \to \Sigma^{out}$ and $\sigma^{fr}\colon \Sigma^{fr} \to \Sigma^{out}$, $\sigma'^{fr}\colon \Sigma'^{fr} \to \Sigma^{out}$, $\sigma^{in}\colon \Sigma^{in} \to \Sigma'^{in}$, $\sigma'^{in}\colon \Sigma^{fr} \to \Sigma'^{fr}$, $\sigma^{out}\colon \Sigma'^{out} \to \Sigma^{out}$ s.t. $\sigma'^{fr} \circ \sigma'^{in} = \sigma^{fr}$.

Finally, axiom (10) describes how the freeze operator behaves w.r.t. the sum operator; the axiom holds for $M_i\colon \Sigma_i^{fr} + \Sigma^{in} \to \Sigma_i^{out}$ and $\sigma_i^{fr}\colon \Sigma_i^{fr} \to \Sigma_i^{out}$, $i = 1, 2$. We have used Σ'^{in} as an abbreviation for $\Sigma_1^{fr} + \Sigma_2^{fr} + \Sigma^{in}$.

From axioms (1) to (10) several useful laws can be deduced (see Fig.5): distributivity of the reduct w.r.t. the sum (ax.(i), (ii) and (iii)), commutativity (ax. (iv)), associativity (ax.(v)) and existence and unicity (w.r.t. a fixed input signature) of the neutral element of sum (ax.(vi) and (vii)) (we have omitted the straightforward proofs for reasons of space).

2.2 Derived Operations

In Fig.6 we give a set of axioms which state that each high-level operator defined in Sect.1.2 can be derived from the primitive operations. Note that in order to

$$
\begin{aligned}
M_{1|\sigma_1^{out}} + M_{2|\sigma_2^{out}} &= (M_1 + M_2)_{|\sigma_1^{out}+\sigma_2^{out}} && \text{(i)}\\
M_{|\sigma_1^{out}} + M_{|\sigma_2^{out}} &= M_{|[\sigma_1^{out},\sigma_2^{out}]} && \text{(ii)}\\
{}_{\sigma^{in}|}(M_1 + M_2) &= {}_{\sigma^{in}|}M_1 + {}_{\sigma^{in}|}M_2 && \text{(iii)}\\
M_1 + M_2 &= M_2 + M_1 && \text{(iv)}\\
(M_1 + M_2) + M_2 &= M_1 + (M_2 + M_3) && \text{(v)}\\
M_1 + M_{2|\emptyset} &= M_1 && \text{(vi)}\\
M_{1|\emptyset} &= M_{2|\emptyset} && \text{(vii)}
\end{aligned}
$$

Fig. 5. Derived axioms for the primitive operations

$$
\begin{aligned}
M_1 \oplus M_2 &= \mathit{freeze}_{\Sigma^{fr}\cap\Sigma^{in}}({}_{\Sigma^{in}|}M_1 + {}_{\Sigma^{in}|}M_2) && (11)\\
\textbf{freeze } \Sigma^{fr} \textbf{ in } M &= \mathit{freeze}_{\Sigma^{fr}\cap\Sigma^{in}}(M) && (12)\\
\textbf{restrict } \Sigma^{rs} \textbf{ in } M &= M_{|\Sigma^{out}\setminus\Sigma^{rs}} && (13)\\
\textbf{hide } \Sigma^{hd} \textbf{ in } M &= (\mathit{freeze}_{\Sigma^{hd}\cap\Sigma^{in}}(M))_{|\Sigma^{out}\setminus\Sigma^{hd}} && (14)\\
M_1 \Leftarrow M_2 &= (\textbf{restrict } \Sigma_1^{out} \cap \Sigma_2^{out} \textbf{ in } M_1) \oplus M_2 && (15)\\
M_2 \circ M_1 &= \textbf{hide } \Sigma_1^{out} \textbf{ in } M_1 \oplus M_2 && (16)
\end{aligned}
$$

Fig. 6. Axioms for the high-level operations

deal with high-level operations, we need to consider again the more specific notion of boolean signature category defined in Sect.1.3, supporting signature inclusion, intersection, union and difference. However, recall that by our assumptions on boolean signature categories, if $\Sigma_1 \cap \Sigma_2 = \emptyset$, then $\Sigma_1 \hookrightarrow \Sigma_1 \cup \Sigma_2 \hookleftarrow \Sigma_2$ is a coproduct. In particular, this holds when $\Sigma_2 = \Sigma \setminus \Sigma_1$. Therefore, if $M: \Sigma^{in} \rightarrow \Sigma^{out}$ and $\Sigma^{fr} \subseteq \Sigma^{in} \cap \Sigma^{out}$, then $\mathit{freeze}_{\Sigma^{fr}}(M)$ is equivalent to $\mathit{freeze}_{\sigma^{fr}}(M)$, where σ^{fr} is the inclusion of Σ^{fr} into Σ^{out} (note that $\Sigma^{in} = \Sigma^{fr} + (\Sigma^{in} \setminus \Sigma^{fr})$). Analogously, if $\Sigma^{in} \subseteq \Sigma'^{in}$ and $\Sigma'^{out} \subseteq \Sigma^{out}$, then ${}_{\Sigma'^{in}|}M_{|\Sigma'^{out}}$ is equivalent to ${}_{\sigma^{in}|}M_{|\sigma^{out}}$, where $\sigma^{in} = i_{\Sigma^{in},\Sigma'^{in}}$ and $\sigma^{out} = i_{\Sigma'^{out},\Sigma^{out}}$.

3 Technical Results

In this section we present our main technical results. First, we prove (Sect.3.1) that the given set of axioms is sound w.r.t. to the model presented in Sect.1, enriched by a semantic interpretation of the three primitive operations. Then we give a normal form theorem for mixin expressions (Sect.3.2); finally we state a number of derived axioms for high-level operations.

3.1 Soundness

In order to show that the model presented in Sect.1 actually satisfies the axiomatization given in Sect.2, we have to provide a semantic interpretation of the three primitive operations. That is done in Fig.7.

$$\text{(Sum-sem)}\ \frac{[\![M_i]\!] = F_i, i = 1, 2}{[\![M_1 + M_2]\!] = \lambda A.F_1(A) + F_2(A)}$$

$$\text{(Reduct-sem)}\ \frac{[\![M]\!] = F}{[\![{}_{\sigma^{in}|}M_{|\sigma^{out}}]\!] = \lambda A.(F(A_{|\sigma^{in}}))_{|\sigma^{out}}}$$

$$\text{(Freeze-sem)}\ \frac{[\![M]\!] = F}{[\![\mathit{freeze}_{\sigma^{fr}}(M)]\!] = \lambda A.\mathit{freeze}(\lambda B.F(A + B_{|\sigma^{fr}}))}$$

Fig. 7. Semantic rules

Proposition 7. *The axioms of Fig.4 and Fig.6 hold when we interpret mixin operators as defined in Fig.2 and Fig.7.*

Proof. See [2].

3.2 Normal Form

Definition 8. Let X be a non-empty set of typed mixin variables. Then ME_X denotes the set of the mixin expressions over X, including X and inductively defined by the rules in Fig.1 and Fig.3.

Notation. Let I be a finite set of indexes and $\{M_i : \Sigma^{in} \to \Sigma_i^{out}\}_{i \in I}$ be an I-indexed set of mixin expressions in ME_X. Then $\sum_{i \in I} M_i$ is defined as follows:

$$\sum_{i \in I} M_i = M_j \text{ if } I = \{j\}; \qquad \sum_{i \in I} M_i = M_j + \sum_{i \in J} M_i \text{ if } I = J + \{j\}.$$

Note that the notation is consistent since the sum operation is both commutative and associative (see ax.(iv) and (v) of Fig.5). Moreover the axioms (i), (ii) and (iii) can be easily generalized to this notation.

Proposition 9. *Let M be a mixin expression in ME_X. Then there exist variables x_i in X and morphisms σ_i^{in}, σ^{out} and σ^{fr}, $i \in I$ (I finite), s.t.*
$M = (\mathit{freeze}_{\sigma^{fr}}(\sum_{i \in I} {}_{\sigma_i^{in}|}x_i))_{|\sigma^{out}}$

Proof. See [2] □

3.3 Properties of High-Level Operations

In Fig.8 we state some (intuitively expected) properties of the high-level operators. They all can be derived from ax.(1) to ax.(16)

Let us start by proving that the merge operator is commutative (ax.(viii)). Indeed, by ax.(11) and ax.(iv), $M_1 \oplus M_2 = \mathit{freeze}_{\Sigma^{fr} \cap \Sigma^{in}}({}_{\Sigma^{in}|}M_1 + {}_{\Sigma^{in}|}M_2) = \mathit{freeze}_{\Sigma^{fr} \cap \Sigma^{in}}({}_{\Sigma^{in}|}M_2 + {}_{\Sigma^{in}|}M_1) = M_2 \oplus M_1$.

A less trivial prove is the associativity of merge (ax.(ix)).

$$\begin{aligned}
M_1 \oplus M_2 &= M_2 \oplus M_1 && \text{(viii)}\\
(M_1 \oplus M_2) \oplus M_3 &= M_1 \oplus (M_2 \oplus M_3) && \text{(ix)}\\
M_1 \oplus \textbf{restrict } \Sigma_2^{out} \textbf{ in } M_2 &= M_1 \text{ if } \Sigma_2^{in} \subseteq \Sigma_1^{in} && \text{(x)}\\
M_1 \Leftarrow M_2 &= M_1 \oplus M_2 \text{ if } \Sigma_1^{out} \cap \Sigma_2^{out} = \emptyset && \text{(xi)}\\
M_1 \Leftarrow M_2 &= M_2 \text{ if } \Sigma_1^{in} \subseteq \Sigma_2^{in}, \Sigma_1^{out} \subseteq \Sigma_2^{out} && \text{(xii)}\\
M_1 \Leftarrow \textbf{restrict } \Sigma_2^{out} \textbf{ in } M_2 &= M_1 \text{ if } \Sigma_2^{in} \subseteq \Sigma_1^{in} && \text{(xiii)}
\end{aligned}$$

Fig. 8. Some properties of high-level operations

Lemma 10. *Let* $M_i: \Sigma^{in} \to \Sigma_i^{out}$, $i = 1, 2$, *be two mixins, with* $\Sigma_1^{out} \cap \Sigma_2^{out} = \emptyset$ *and let* $\Sigma^{fr}, \Sigma'^{fr}$ *be two signatures s.t.* $\Sigma^{fr} \subseteq \Sigma'^{fr}$, $\Sigma^{fr} \subseteq \Sigma_1^{out} \cap \Sigma^{in}$, $\Sigma'^{fr} \subseteq (\Sigma_1^{out} \cup \Sigma_2^{out}) \cap \Sigma^{in}$. *Then,*
$freeze_{\Sigma'^{fr}}({}_{\Sigma^{in}}|freeze_{\Sigma^{fr}}(M_1) + M_2) = freeze_{\Sigma'^{fr}}(M_1 + M_2)$.

Lemma 11. *Let* $M: \Sigma^{in} \to \Sigma^{out}$ *be a mixin and let* $\Sigma^{fr}, \Sigma'^{in}$ *be two signatures s.t.* $\Sigma^{fr} \subseteq \Sigma^{out} \cap \Sigma^{in}$, $\Sigma^{in} \subseteq \Sigma'^{in}$. *Then,*
$freeze_{\Sigma'^{fr}}({}_{\Sigma'^{in}}|M) = freeze_{\Sigma'^{fr} \setminus (\Sigma'^{in} \setminus \Sigma^{in})}({}_{(\Sigma'^{in} \setminus (\Sigma^{in} \cup \Sigma^{fr})) \cup \Sigma^{in}}|M)$.

From lemma 10 and 11 (see [2] for the proofs) we can derive:
$(M_1 \oplus M_2) \oplus M_3 = freeze_{\Sigma^{fr} \cap \Sigma^{in}}({}_{\Sigma^{in}}|M_1 + {}_{\Sigma^{in}}|M_2 + {}_{\Sigma^{in}}|M_3) = M_1 \oplus (M_2 \oplus M_3)$, with $\Sigma^{fr} = \Sigma_1^{fr} \cup \Sigma_2^{fr} \cup \Sigma_3^{fr}$, $\Sigma_i^{fr} = \Sigma_i^{out} \setminus \Sigma_i^{in}$, $i = 1, 2, 3$, $\Sigma^{in} = \Sigma_1^{in} \cup \Sigma_2^{in} \cup \Sigma_3^{in}$. Note that from this proof we can easily derive the following law (for a finite non empty set of indexes I): $\oplus_{i \in I} M_i = freeze_{\Sigma^{fr} \cap \Sigma^{in}}(\sum_{i \in I} {}_{\Sigma^{in}}|M_i)$, where $M_i: \Sigma_i^{in} \to \Sigma_i^{out}$, $i \in I$, $\Sigma^{fr} = \bigcup_{i \in I}(\Sigma_i^{out} \setminus \Sigma_i^{in})$, $\Sigma^{in} = \bigcup_{i \in I} \Sigma_i^{in}$, $\Sigma_i^{out} \cap \Sigma_j^{out} = \emptyset$, for any $i, j \in I$, $i \neq j$.

The existence of the neutral element (ax.(x)) derives from ax.(4), (7) and (vi):
$M_1 \oplus M_{2|\emptyset} = freeze_{(\Sigma_1^{out} \setminus \Sigma_1^{in}) \cap \Sigma_1^{in}}({}_{\Sigma_1^{in}}|M_1 + {}_{\Sigma_1^{in}}|M_{2|\emptyset}) = freeze_{\emptyset}(M_1) = M_1$.

The overriding operator reduces to merge (ax.(xi)) when the output signatures are disjoint: $M_1 \Leftarrow M_2 = M_{1|\Sigma_1^{out} \setminus \Sigma_2^{out}} \oplus M_2 = M_{1|\Sigma_1^{out}} \oplus M_2 = M_1 \oplus M_2$.

The existence of the left neutral element for overriding is ensured by ax.(xii)): $M_1 \Leftarrow M_2 = M_{1|\Sigma_1^{out} \setminus \Sigma_2^{out}} \oplus M_2 = M_{1|\emptyset} \oplus M_2 = M_2$. From ax.(xii) one can easily deduce idempotency of overriding: $M \Leftarrow M = M$.

The existence of the right neutral element for overriding is stated in ax.(xiii): $M_1 \Leftarrow M_{2|\emptyset} = M_{1|\Sigma_1^{out} \setminus \emptyset} \oplus M_{2|\emptyset} = M_1$.

4 Conclusion

We have given an axiomatic characterization of the notion of mixin module, following the spirit of the seminal paper [8]. We have proved that such axiomatization is sound w.r.t. the intended model, previously defined in [4, 6], and that each mixin expression is provably equal to a simple normal form where the freeze operator is applied only once. Moreover, many other expected properties of the mixin operators can be deduced from the axioms. A more comprehensive treatment of the topics developed here and in [4, 6] can be found in [2].

Even though inspired by [8], our work is different w.r.t. the well-established algebraic treatment of module composition. Here we address the problem of combining together programming rather than specification modules; therefore, we have to consider, together with classical operators (like export and renaming), new operators (like freeze) related with the notions of module modification and extension which hardly make sense in the context of specifications.

Several recent papers [12, 18, 19, 20, 21] have pointed out the importance of modularity mechanisms independent of the underlying core language and supporting the notions of separate compilation and linking. At our knowledge, our proposal of axiomatization for mixins is the first which supports these principles.

What is missing up to now is, on the one hand, a complete calculus for mixins equipped with a reduction semantics and, on the other, a more strict integration of our framework with the notions of type system and type checking.

We plan to develop a calculus for mixins (see [3]), analogously to what has been done in [14]; the calculus, again, has to be independent of the core language and sound w.r.t. the model. Note that, by virtue of the results of this paper, it suffices to define a calculus involving only the primitive operators; the high-level operators come for free.

On the side of type checking we are studying the possibility of defining model parts of institutions parameterized by a type system (see [1]), in order to obtain module systems with type sharing (at the level of both deferred and defined type components), separate type checking and linking built on top of advanced typed languages (e.g., supporting polymorphism and subtyping).

References

1. D. Ancona. An algebraic framework for separate compilation. Technical Report DISI-TR-97-10, Dipartimento di Informatica e Scienze dell'Informazione, Università di Genova, 1997. Submitted for publication.
2. D. Ancona. *Modular Formal Frameworks for Module Systems*. PhD thesis, Dipartimento di Informatica, Università di Pisa, 1998. To appear.
3. D. Ancona and E. Zucca. A theory of mixin modules: algebraic laws and reduction semantics. In preparation.
4. D. Ancona and E. Zucca. An algebraic approach to mixins and modularity. In M. Hanus and M. Rodríguez Artalejo, editors, *ALP '96 - 5th Intl. Conf. on Algebraic and Logic Programming*, number 1139 in Lecture Notes in Computer Science, pages 179–193, Berlin, 1996. Springer Verlag.
5. D. Ancona and E. Zucca. Overriding operators in a mixin-based framework. In H. Glaser, P. Hartel, and H. Kuchen, editors, *Proc. PLILP '97 - 9th International Symposium on Programming Languages, Implementations, Logics, and Programs*, number 1292 in Lecture Notes in Computer Science, pages 47–61, Berlin, September 1997. Springer Verlag.
6. D. Ancona and E. Zucca. A theory of mixin modules: basic and derived operators. *Mathematical Structures in Computer Science*, 1998. To appear.
7. G. Banavar and G. Lindstrom. An application framework for module composition tools. In *ECOOP '96*, number 1098 in Lecture Notes in Computer Science, pages 91–113. Springer Verlag, July 1996.
8. J.A. Bergstra, J. Heering, and P. Klint. Module algebra. *Journ. ACM*, 37(2):335–372, 1990.

9. G. Bracha. *The Programming Language JIGSAW: Mixins, Modularity and Multiple Inheritance.* PhD thesis, Dept. Comp. Sci., Univ. Utah, 1992.
10. G. Bracha and W. Cook. Mixin-based inheritance. In *ACM Symp. on Object-Oriented Programming: Systems, Languages and Applications 1990*, pages 303–311. ACM Press, October 1990. SIGPLAN Notices, volume 25, number 10.
11. G. Bracha and G. Lindstrom. Modularity meets inheritance. In *Proc. International Conference on Computer Languages*, pages 282–290, San Francisco, April 1992. IEEE Computer Society.
12. L. Cardelli. Program fragments, linking, and modularization. In *Proc. 24th ACM Symp. on Principles of Programming Languages*, pages 266–277. ACM Press, January 1997.
13. W. Cook. *A Denotational Semantics of Inheritance.* PhD thesis, Dept. Comp. Sci. Brown University, 1989.
14. J. Courant. A module calculus enjoying the subject-reduction property. In *TAPSOFT '97: Theory and Practice of Software Development*, number 1214 in Lecture Notes in Computer Science. Springer Verlag, April 1997.
15. D. Duggan and C. Sourelis. Mixin modules. In *Intl. Conf. on Functional Programming*, Philadelphia, May 1996. ACM Press.
16. H. Ehrig and B. Mahr. *Fundamentals of Algebraic Specification 1. Equations and Initial Semantics*, volume 6 of *EATCS Monograph in Computer Science.* Springer Verlag, 1985.
17. J.A. Goguen and R.M. Burstall. Institutions: Abstract model theory for computer science. *Journ. ACM*, 39:95–146, 1992.
18. R. Harper and M. Lillibridge. A type theoretic approach to higher-order modules with sharing. In *ACM Symp. on Principles of Programming Languages 1994*, pages 127–137. ACM Press, 1994.
19. M. P. Jones. Using parameterized signatures to express modular structure. In *Proc. 23rd ACM Symp. on Principles of Programming Languages*, pages 68–78, St. Petersburg Beach, Florida, Jan 1996. ACM Press.
20. X. Leroy. Manifest types, modules and separate compilation. In *ACM Symp. on Principles of Programming Languages 1994*, pages 109–122. ACM Press, 1994.
21. X. Leroy. A modular module system. Technical Report 2866, Institute National de Recherche en Informatique et Automatique, April 1996.
22. M. Van Limberghen and T. Mens. Encapsulation and composition as orthogonal operators on mixins: A solution to multiple inheritance problems. *Object Oriented Systems*, 3(1):1–30, 1996.
23. R. Milner, M. Tofte, and R. Harper. *The Definition of Standard ML.* The MIT Press, Cambridge, Massachussetts, 1990.
24. U. S. Reddy. Objects as closures: Abstract semantics of object-oriented languages. In *Proc. ACM Conf. on Lisp and Functional Programming*, pages 289–297, 1988.

Completeness of a Logical System for Structured Specifications

Tomasz Borzyszkowski*

University of Gdańsk, Institute of Mathematics,
Division of Computer Science

Abstract. The main aim of this paper is to present a completeness proof of a formal system for reasoning about logical consequences of structured specifications. The system is based on the proof rules for structural specifications build in an arbitrary institution as presented in [ST 88]. The proof of its completeness is inspired by the proof due to M. V. Cengarle (see [Cen 94]) for specifications in first-order logic and the logical system for reasoning about them presented also in [Wir 91].

1 Introduction

In a number of papers on algebraic specifications (see [Cen 94, Far 92, ST 88, SST 92, Tar 86, Wir 91]) the main goal was to build:

- a specification formalism flexible enough to allow one to cope with various problems of software engineering;
- a sound and complete logic for reasoning about such specifications.

We follow these goals when the first and a part of the second aim (soundness) is achieved by using structured specifications and the logical system built over an arbitrary logic (formalized as an institution) in a way similar to [ST 88]. The rest (completeness) was inspired by the completeness proof of the Π_s system presented in [Cen 94].

In the preliminary sections we recall the concept of an institution and introduce various properties of the underlying logical system formalized as an institution (see [GB 90]). In particular, we define the closure of an institution under conjunction and negation in a way that ensures the deduction property and introduce the interpolation and amalgamation properties for an arbitrary institution. These properties are then used to study some crucial properties of a specification formalism, which we introduce following [ST 88]. One such a property is existence of an equivalent normal form for any specification in our formalism. Most importantly, we show that a natural proof system for deriving logical consequences of structural specifications is sound and complete for any institution satisfying the basic closure and interpolation properties. This generalizes to an arbitrary institution the results of Cengarle [Cen 94] on completeness of a similar system for specifications in the first-order logic.

* e-mail: mattb@univ.gda.pl, http://monika.univ.gda.pl/~mattb

2 Institutions

While developing a specification system independently of the underlying logical system, it is necessary to formalize in a sufficiently abstract way the informal mathematical concept of a logical system. Our choice of an abstract formalization depends on what we mean by a logical system. Following [GB 90], in the model–theoretic tradition of logic:

> "One of the most essential elements of a logical system is its relationship of *satisfaction* between its *syntax* (i.e. its sentences) and its *semantics* (i.e. models)..."

Based on this principle, the notion of a logical system is formalized in [GB 90] as a mathematical object called *institution*.

An institution consists of a collection of signatures, together with a set of Σ-sentences, a collection of Σ-models and a satisfaction relation between Σ-models and Σ-sentences, for each signature Σ. The only requirement is that when we change signatures (by signature morphisms) the induced translations of sentences and models preserve the satisfaction relation.

Definition 1 [GB 90]. An institution **INS** consists of:

- a category $\mathbf{Sign_{INS}}$ of signatures;
- a functor $\mathbf{Sen_{INS}} : \mathbf{Sign_{INS}} \to \mathbf{Set}$, giving a set $\mathbf{Sen_{INS}}(\Sigma)$ of Σ-sentences for each signature $\Sigma \in |\mathbf{Sign_{INS}}|$; and
- a functor $\mathbf{Mod_{INS}} : \mathbf{Sign}^{op}_{\mathbf{INS}} \to \mathbf{Cat}$, giving a category $\mathbf{Mod_{INS}}(\Sigma)$ of Σ-models for each signature $\Sigma \in |\mathbf{Sign_{INS}}|$; and
- for each $\Sigma \in |\mathbf{Sign_{INS}}|$, a satisfaction relation $\models^{\mathbf{INS}}_{\Sigma} \subseteq |\mathbf{Mod_{INS}}(\Sigma)| \times \mathbf{Sen_{INS}}(\Sigma)$ such that for any signature morphism $\sigma : \Sigma \to \Sigma'$, Σ-sentence $\varphi \in \mathbf{Sen_{INS}}(\Sigma)$ and Σ'-model $M' \in |\mathbf{Mod_{INS}}(\Sigma')|$ the following *satisfaction condition* holds:

$$M' \models^{\mathbf{INS}}_{\Sigma'} \mathbf{Sen_{INS}}(\sigma)(\varphi) \quad \text{iff} \quad \mathbf{Mod_{INS}}(\sigma)(M') \models^{\mathbf{INS}}_{\Sigma} \varphi.$$

□

Examples of various logical systems formalized as institutions can be found in [GB 90]. In the next two definitions we define what it means that an institution "has conjunction" and "has negation".

Definition 2. We say that an institution **INS** *has conjunction* if for every signature $\Sigma \in |\mathbf{Sign_{INS}}|$ and finite set of Σ-sentences $\{\varphi_i\}_{i \in \mathcal{I}} \subseteq \mathbf{Sen_{INS}}(\Sigma)$ there exists a Σ-sentence, which we denote by $\bigwedge_{i \in \mathcal{I}} \varphi_i$, such that for every Σ-model $M \in |\mathbf{Mod_{INS}}(\Sigma)|$:

$$M \models^{\mathbf{INS}}_{\Sigma} \bigwedge_{i \in \mathcal{I}} \varphi_i \quad \text{iff for every } i \in \mathcal{I} \quad M \models^{\mathbf{INS}}_{\Sigma} \varphi_i.$$

□

We can similarly define what it means that an institution "has infinite conjunction":

Definition 3. We say that an institution **INS** *has infinite conjunction* if for every signature $\Sigma \in |\mathbf{Sign_{INS}}|$ and set of Σ-sentences $\{\varphi_i\}_{i\in\mathcal{I}} \subseteq \mathbf{Sen_{INS}}(\Sigma)$, where $\mathcal{I}$ is a (possibly infinite) set of indices, there exists a Σ-sentence, which we denote by $\bigwedge_{i\in\mathcal{I}} \varphi_i$, such that for every Σ-model $M \in |\mathbf{Mod_{INS}}(\Sigma)|$:

$$M \models^{\mathbf{INS}}_{\Sigma} \bigwedge_{i\in\mathcal{I}} \varphi_i \quad \text{iff for every } i \in \mathcal{I} \quad M \models^{\mathbf{INS}}_{\Sigma} \varphi_i.$$

□

Obviously, if an institution has infinite conjunction, then it has conjunction as well.

Definition 4. We say that an institution **INS** *has negation*, if for every signature $\Sigma \in |\mathbf{Sign_{INS}}|$ and Σ-sentence $\varphi \in \mathbf{Sen_{INS}}(\Sigma)$ there exists a Σ-sentence, which we denote by $\neg\varphi$, such that for every Σ-model $M \in |\mathbf{Mod_{INS}}(\Sigma)|$:

$$M \models^{\mathbf{INS}}_{\Sigma} \neg\varphi \quad \text{iff it is not true that:} \quad M \models^{\mathbf{INS}}_{\Sigma} \varphi.$$

□

In the rest of the paper the following abbreviations are used:

- for any set of sentences $\Phi \subseteq \mathbf{Sen_{INS}}(\Sigma)$ and $M \in \mathbf{Mod_{INS}}(\Sigma)$ we define $M \models^{\mathbf{INS}}_{\Sigma} \Phi$ as an abbreviation for *for every sentence* $\varphi \in \Phi$: $M \models^{\mathbf{INS}}_{\Sigma} \varphi$, and similarly for every class of models $\mathcal{M} \subseteq \mathbf{Mod_{INS}}(\Sigma)$ and sentence $\varphi \in \mathbf{Sen_{INS}}(\Sigma)$ we define $\mathcal{M} \models^{\mathbf{INS}}_{\Sigma} \varphi$ as an abbreviation for *for every model* $M \in \mathcal{M}$: $M \models^{\mathbf{INS}}_{\Sigma} \varphi$;
- for any sentences $\varphi, \psi \in \mathbf{Sen_{INS}}(\Sigma)$ we define $\varphi \models^{\mathbf{INS}}_{\Sigma} \psi$ as an abbreviation for *for every model* $M \in \mathbf{Mod_{INS}}(\Sigma)$, $M \models^{\mathbf{INS}}_{\Sigma} \psi$ *whenever* $M \models^{\mathbf{INS}}_{\Sigma} \varphi$, and similarly $\Phi \models^{\mathbf{INS}}_{\Sigma} \varphi$, for any set of sentences $\Phi \subseteq \mathbf{Sen_{INS}}(\Sigma)$, as an abbreviation for *for every model* $M \in \mathbf{Mod_{INS}}(\Sigma)$, $M \models^{\mathbf{INS}}_{\Sigma} \varphi$ *whenever* $M \models^{\mathbf{INS}}_{\Sigma} \Phi$;
- if an institution **INS** has conjunction and negation, then for any sentences $\varphi_1, \varphi_2 \in \mathbf{Sen_{INS}}(\Sigma)$ we define $\varphi_1 \wedge \varphi_2$ as an abbreviation for the sentence $\bigwedge_{i\in\{1,2\}} \varphi_i$ and $\varphi_1 \Rightarrow \varphi_2$ as an abbreviation for the sentence: $\neg(\varphi_1 \wedge \neg\varphi_2)$;
- the following abbreviations will be used: $d\ \varphi$ for $\mathbf{Sen_{INS}}(d)(\varphi)$, $M|_\sigma$ for $\mathbf{Mod_{INS}}(\sigma)(M)$ and $\models$ for $\models^{\mathbf{INS}}_{\Sigma}$ when it is clear what they mean;
- for any set of sentences $\Phi \subseteq \mathbf{Sen_{INS}}(\Sigma)$ we write $\bigwedge \Phi$ as an abbreviation for $\bigwedge_{i\in\mathcal{I}} \varphi_i$ where $\Phi = \{\varphi_i \mid i \in \mathcal{I}\}$, similarly we write $\bigwedge d\Phi$ for $\bigwedge_{i\in\mathcal{I}} d\,\varphi_i$.

Theorem 5 Deduction. *For any institution* **INS** *that has conjunction and negation,* $\Sigma \in |\mathbf{Sign_{INS}}|$ *and any sentences* $\varphi_1, \varphi_2, \varphi_3 \in \mathbf{Sen_{INS}}(\Sigma)$ *we have:*

$$\varphi_1 \wedge \varphi_2 \models \varphi_3 \quad \textit{iff} \quad \varphi_1 \models \varphi_2 \Rightarrow \varphi_3.$$

Proof. Directly from the definition. □

The above theorem shows that "semantic" deduction is a property of all institutions having conjunction and negation.

3 $(\mathcal{D},\mathcal{T})$-institutions

In specification formalisms (see [Cen 94, Far 92, ST 88, SST 92, Wir 91]) signature morphisms are used in at least two ways: the first is to hide some symbols in the signature of the specification and the second is to add and/or rename some symbols in the signature. According to this observation, in each institution we distinguish two classes of signature morphisms: $\mathcal{D}$ – a class of signature morphisms proper for hiding symbols and a class $\mathcal{T}$ for the second purpose.

Definition 6 $(\mathcal{D},\mathcal{T})$-institution. Let $\mathcal{D},\mathcal{T} \subseteq \mathbf{Sign_{INS}}$ be classes of signature morphisms in an institution **INS**. We say that the institution **INS** is $(\mathcal{D},\mathcal{T})$-institution iff:

- classes $\mathcal{D}$ and $\mathcal{T}$ are closed under composition and include all isomorphisms;
- for every $(d : \Sigma \to \Sigma_1) \in \mathcal{D}$ and $(t : \Sigma \to \Sigma_2) \in \mathcal{T}$ there exist $(t' : \Sigma_1 \to \Sigma') \in \mathcal{T}$ and $(d' : \Sigma_2 \to \Sigma') \in \mathcal{D}$ such that they form the following pushout in **Sign**:

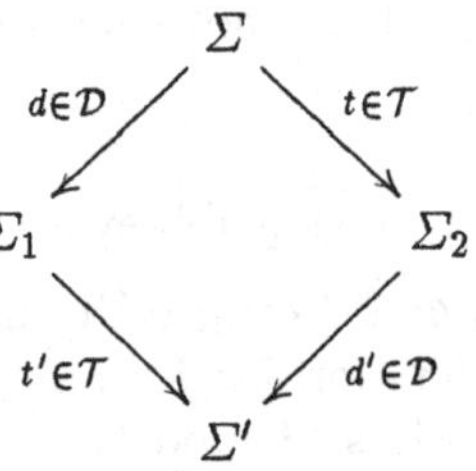

□

The pushout requirement from the above definition obviously does not hold for every classes $\mathcal{D}$ and $\mathcal{T}$. For example:

Example 1. Let us consider any institution **INS** where $\mathbf{Sign_{INS}}$ is the category of algebraic signatures with derived morphisms $\mathbf{AlgSig}^{der}$ (see [SB 83]) and let all the derived morphisms from $\mathbf{AlgSig}^{der}$ be in both classes of morphisms $\mathcal{D}$ and $\mathcal{T}$. Then the pushout from Definition 6 does not exist in general because the category $\mathbf{AlgSig}^{der}$ does not have all pushouts. On the other hand, when for instance $\mathcal{D}$ is the class of inclusions and $\mathcal{T}$ is the class of all derived morphisms then the pushout exists.
A positive example could be any institution **INS** with (finitely) cocomplete category of signatures $\mathbf{Sign_{INS}}$, e.g. the category of algebraic signatures **AlgSig** is such a category.

Another property of the logical system, used in the completeness theorem (see Section 5 and [Cen 94]), is the interpolation property. The definition of the $(\mathcal{D},\mathcal{T})$-interpolation property is inspired by the formalization of the Craig Interpolation Theorem presented in [Tar 86].

Definition 7 $(\mathcal{D},\mathcal{T})$-interpolation property. A $(\mathcal{D},\mathcal{T})$-institution **INS** satisfies $(\mathcal{D},\mathcal{T})$-interpolation property iff for any $d, d' \in \mathcal{D}$ and $t, t' \in \mathcal{T}$ that form a pushout in $\mathbf{Sign_{INS}}$ (as in Definition 6) and $\varphi_i \in \mathbf{Sen_{INS}}(\Sigma_i)$ for $i = 1, 2$, if $\mathbf{Sen_{INS}}(t')(\varphi_1) \models^{\mathbf{INS}}_{\Sigma'} \mathbf{Sen_{INS}}(d')(\varphi_2)$ then there exists $\varphi \in \mathbf{Sen_{INS}}(\Sigma)$ such that:

$$\varphi_1 \models^{\mathbf{INS}}_{\Sigma_1} \mathbf{Sen_{INS}}(d)(\varphi) \quad \text{and} \quad \mathbf{Sen_{INS}}(t)(\varphi) \models^{\mathbf{INS}}_{\Sigma_2} \varphi_2.$$

□

Of course, the interpolation property does not hold in every institution. For instance, it does not hold for institution **EQ** of equational logic (see [BHK 90]). However, it does hold for some richer institutions like e.g. institution **FOLEQ** of the first order logic with equality (see [BM 86]).

Definition 8 $(\mathcal{D}, \mathcal{T})$-amalgamation property. A $(\mathcal{D}, \mathcal{T})$-institution **INS** satisfies $(\mathcal{D}, \mathcal{T})$-amalgamation property iff for any $d, d' \in \mathcal{D}$ and $t, t' \in \mathcal{T}$ that form a pushout in $\mathbf{Sign_{INS}}$ (as in Definition 6) and for any $M_1 \in \mathbf{Mod_{INS}}(\Sigma_1)$ and $M_2 \in \mathbf{Mod_{INS}}(\Sigma_2)$, if $M_1|_d = M_2|_t$ then there exists a unique model $M' \in \mathbf{Mod_{INS}}(\Sigma')$ such that $M'|_{t'} = M_1$ and $M'|_{d'} = M_2$. □

Assumption 9. In the rest of the paper we will work with an arbitrary but fixed $(\mathcal{D}, \mathcal{T})$-institution that has conjunction and negation, and for which $\mathcal{D} \subseteq \mathcal{T}$.

4 Specifications

In the rest of the paper we will work with specifications similar to specifications defined in [ST 88]. As in [ST 88], we assume that software systems, described by specifications, are adequately represented as models of the underlying institution. This means that ultimately a specification describes a signature and a class of models over this signature, called the *models of the specification.* Specifications presented in this section (and also these in [ST 88]) determine a class of specifications **Spec** with the semantics that yields for any specification $SP \in$ **Spec** its signature $\mathbf{Sig}[SP] \in |\mathbf{Sign_{INS}}|$ and the collection of its models $\mathbf{Mod}[SP] \subseteq |\mathbf{Mod}(\mathbf{Sig}[SP])|$. If $\mathbf{Sig}[SP] = \Sigma$, we will call SP a Σ-specification; the collection of all Σ-specifications is denoted by $\mathbf{Spec}_\Sigma$.

There are two differences between specifications presented here and those from [ST 88]: we work only with a subset of SBOs[2] presented in [ST 88], and specifications presented here are built over an arbitrary $(\mathcal{D}, \mathcal{T})$-institution **INS** satisfying $(\mathcal{D}, \mathcal{T})$-amalgamation, whereas in [ST 88] they are built over any institution.

Definition 10 Specifications. Specifications over a $(\mathcal{D}, \mathcal{T})$-institution **INS** satisfying the $(\mathcal{D}, \mathcal{T})$-amalgamation property and their semantics are defined inductively as follows:

[2] SBO is the abbreviation for Specification-Building Operation.

1. Any presentation $\langle \Sigma, \Phi \rangle$, where $\Sigma \in \mathbf{Sign_{INS}}$ and $\Phi \subseteq \mathbf{Sen_{INS}}(\Sigma)$, is a specification with the following semantics:
   $\mathbf{Sig}[\langle \Sigma, \Phi \rangle] = \Sigma$;
   $\mathbf{Mod}[\langle \Sigma, \Phi \rangle] = \{M \in |\mathbf{Mod_{INS}}(\Sigma)| \mid M \models_{\Sigma}^{\mathbf{INS}} \Phi\}$.
2. For any signature Σ and Σ-specifications SP_1 and SP_2, their union $SP_1 \cup SP_2$ is a specification with the following semantics:
   $\mathbf{Sig}[SP_1 \cup SP_2] = \Sigma$;
   $\mathbf{Mod}[SP_1 \cup SP_2] = \mathbf{Mod}[SP_1] \cap \mathbf{Mod}[SP_2]$.
3. For any morphism $(t : \Sigma \to \Sigma') \in \mathcal{T}$ and Σ-specification SP, **translate** SP **by** t is a specification with the following semantics:
   $\mathbf{Sig}[\mathbf{translate}\ SP\ \mathbf{by}\ t] = \Sigma'$;
   $\mathbf{Mod}[\mathbf{translate}\ SP\ \mathbf{by}\ t] = \{M' \in |\mathbf{Mod_{INS}}(\Sigma')| \mid M'|_t \in \mathbf{Mod}[SP]\}$.
4. For any morphism $(d : \Sigma \to \Sigma') \in \mathcal{D}$ and Σ'-specification SP', **derive** SP' **by** d is a specification with the following semantics:
   $\mathbf{Sig}[\mathbf{derive}\ SP'\ \mathbf{by}\ d] = \Sigma$;
   $\mathbf{Mod}[\mathbf{derive}\ SP'\ \mathbf{by}\ d] = \{M'|_d \mid M' \in \mathbf{Mod}[SP']\}$.

□

Definition 11 Semantic consequence. We say that a Σ-sentence φ is a *semantic consequence* of a Σ-specification SP (written $SP \models_{\Sigma} \varphi$) if $\mathbf{Mod}[SP] \models_{\Sigma}^{\mathbf{INS}} \varphi$. □

Next we define an entailment relation $\vdash_{\Sigma} \subseteq \mathbf{Spec}_{\Sigma} \times \mathbf{Sen}(\Sigma)$, parametrized by the proof system $\vdash_{\Sigma}^{\mathbf{INS}}$ for the underlying $(\mathcal{D}, \mathcal{T})$-institution **INS**, which soundly and completely approximates the semantic consequence relation $\models_{\Sigma} \subseteq \mathbf{Spec}_{\Sigma} \times \mathbf{Sen}(\Sigma)$. Such a relation, inspired by a similar relation from [ST 88], could be presented by the following set of rules:

$$\text{(CR)}\ \frac{\{SP \vdash_{\Sigma} \varphi_i\}_{i \in \mathcal{I}} \quad \{\varphi_i\}_{i \in \mathcal{I}} \vdash_{\Sigma}^{\mathbf{INS}} \varphi}{SP \vdash_{\Sigma} \varphi} \qquad \text{(basic)}\ \frac{\varphi \in \Phi}{\langle \Sigma, \Phi \rangle \vdash_{\Sigma} \varphi}$$

$$\text{(sum1)}\ \frac{SP_1 \vdash_{\Sigma} \varphi}{SP_1 \cup SP_2 \vdash_{\Sigma} \varphi} \qquad \text{(sum2)}\ \frac{SP_2 \vdash_{\Sigma} \varphi}{SP_1 \cup SP_2 \vdash_{\Sigma} \varphi}$$

$$\text{(trans)}\ \frac{SP \vdash_{\Sigma} \varphi}{\mathbf{translate}\ SP\ \mathbf{by}\ t \vdash_{\Sigma'} t\varphi} \qquad \text{(derive)}\ \frac{SP' \vdash_{\Sigma'} d\varphi}{\mathbf{derive}\ SP'\ \mathbf{by}\ d \vdash_{\Sigma} \varphi}$$

where $(t : \Sigma \to \Sigma') \in \mathcal{T}$ and $(d : \Sigma \to \Sigma') \in \mathcal{D}$.

The rule (CR) incorporates the proof system for the underlying $(\mathcal{D}, \mathcal{T})$-institution which, we assume, is sound. The main difference between the above set of rules and those presented in [ST 88] are rules (trans) and (derive). In [ST 88] morphisms occurring in the (trans) and (derive) rules (and in corresponding SBOs) could be any signature morphisms whereas in the rules presented above morphisms are restricted to fixed classes of morphisms: $\mathcal{T}$ for the rule (trans) and $\mathcal{D}$ for the rule (derive). This is an obvious consequence of the difference between Definition 10 and the definition of specifications presented in [ST 88].

Let us notice that all the SBOs presented in [Cen 94] can be expressed by the generic SBOs presented in this section. Moreover, the proof rules presented in [Cen 94] can be derived from the rules presented above.

The above entailment relation is sound wrt the semantical consequence relation. The proof follows directly from the proof of soundness presented in [ST 88]. Now, to prove its completeness we need some more notions. Similar definitions were presented in [Cen 94] (cf. also [BHK 90]).

Definition 12 Normal form. We say that the specification SP is in the normal form if it has a form: **derive** $\langle \Sigma, \Phi \rangle$ **by** d, where $(d : \mathbf{Sig}[SP] \to \Sigma) \in \mathcal{D}$ and $\Phi \subseteq \mathbf{Sen}_{\mathbf{INS}}(\Sigma)$. □

Definition 13. Specifications SP_1 and SP_2 are equivalent (written $SP_1 \cong SP_2$) if $\mathbf{Sig}[SP_1] = \mathbf{Sig}[SP_2]$ and $\mathbf{Mod}[SP_1] = \mathbf{Mod}[SP_2]$. □

The following definition introduces operation **nf** which for every specification SP gives a specification $\mathbf{nf}(SP)$ which is in the normal form and such that $\mathbf{nf}(SP)$ is equivalent to SP in the sense of Definition 13.

Definition 14 nf operation. **nf** operation on Σ-specification SP is defined as follows:

1. If SP is a specification expression of the form $\langle \Sigma, \Phi \rangle$, then
   $\mathbf{nf}(SP) = \mathbf{derive}\ \langle \Sigma, \Phi \rangle\ \mathbf{by}\ id_\Sigma$;
2. If SP is a specification expression of the form $SP_1 \cup SP_2$, then
   $\mathbf{nf}(SP) = \mathbf{derive}\ \langle \Sigma', t_1' \Phi_2 \cup d_2' \Phi_1 \rangle\ \mathbf{by}\ d$, where
   $\mathbf{nf}(SP_i) = \mathbf{derive}\ \langle \Sigma_i, \Phi_i \rangle\ \mathbf{by}\ d_i$ for $i = 1, 2$, $d = d_1; d_2'$ and $\Sigma', t_1' \in \mathcal{T}$ and $d_2' \in \mathcal{D}$ are given by a pushout in **Sign**:

$$
\begin{array}{ccc}
\Sigma & \xrightarrow{d_1} & \Sigma_1 \\
\downarrow{\scriptstyle d_2} & \searrow{\scriptstyle d} & \downarrow{\scriptstyle d_2'} \\
\Sigma_2 & \xrightarrow{t_1'} & \Sigma'
\end{array}
$$

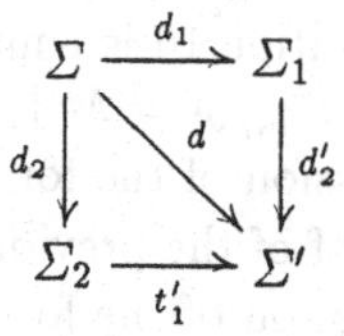

3. If SP is a specification expression of the form **translate** SP_1 **by** t, then
   $\mathbf{nf}(SP) = \mathbf{derive}\ \langle \Sigma', t' \Phi_1 \rangle\ \mathbf{by}\ d_1'$, where $\mathbf{nf}(SP_1) = \mathbf{derive}\ \langle \Sigma_1, \Phi_1 \rangle\ \mathbf{by}\ d_1$ and $\Sigma', t' \in \mathcal{T}$ and $d_1' \in \mathcal{D}$ are given by a pushout in **Sign**:

$$
\begin{array}{ccc}
\mathbf{Sig}[SP_1] & \xrightarrow{d_1} & \Sigma_1 \\
\downarrow{\scriptstyle t} & & \downarrow{\scriptstyle t'} \\
\mathbf{Sig}[SP] & \xrightarrow{d_1'} & \Sigma'
\end{array}
$$

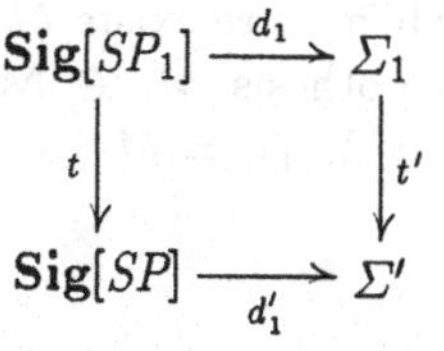

4. If SP is a specification expression of the form **derive** SP_1 **by** d, then **nf**(SP) = **derive** $\langle \Sigma_1, \Phi_1 \rangle$ **by** $(d; d_1)$, where **nf**(SP_1) = **derive** $\langle \Sigma_1, \Phi_1 \rangle$ **by** d_1.

□

The following theorem expresses the main property of **nf** operation (see also [BHK 90, Cen 94] for a similar result):

Theorem 15. *For any specification SP, we have* **nf**(*SP*) $\cong$ *SP*.

Proof. By induction on the structure of *SP*. The signature part of the equivalence, **Sig**[**nf**(SP)] = **Sig**[SP], follows directly from Definition 14.
Proof of the model part, **Mod**[**nf**(SP)] = **Mod**[SP]:

1. SP is a specification expression of the form $\langle \Sigma, \Phi \rangle$.
   This case is obvious, since the reduct along identity is the identity.
2. SP is a specification expression of the form $SP_1 \cup SP_2$.
   $\subseteq$: Let $M \in$ **Mod**[**nf**(SP)]. Then there exists a model $M' \in$ **Mod**[$\langle \Sigma', t'_1\Phi_2 \cup d'_2\Phi_1 \rangle$] such that $M'|_d = M$. It means that $M' \models t'_1\Phi_2$ and $M' \models d'_2\Phi_1$, which by the satisfaction condition is equivalent to $M'|_{t'_1} \models \Phi_2$ and $M'|_{d'_2} \models \Phi_1$. By Definitions 10 and 14, we have $M'|_{d_2;t'_1} \in$ **Mod**[**nf**(SP_2)] and $M'|_{d_1;d'_2} \in$ **Mod**[**nf**(SP_1)]. By the induction hypothesis and because $M'|_{d_2;t'_1} = M'|_d = M$ and similarly $M'|_{d_1;d'_2} = M'|_d = M$: $M \in$ **Mod**[SP_1] $\cap$ **Mod**[SP_2] = **Mod**[SP].
   $\supseteq$: Let $M \in$ **Mod**[SP]. Then $M \in$ **Mod**[SP_1] and $M \in$ **Mod**[SP_2], which by the induction hypothesis yields $M \in$ **Mod**[**nf**(SP_1)] and $M \in$ **Mod**[**nf**(SP_2)]. By the definitions, there exist models $M_i \in$ **Mod**[$\langle \Sigma_i, \Phi_i \rangle$] for $i = 1, 2$ such that $M_i|_{d_i} = M$ and $M_i \models \Phi_i$ for $i = 1, 2$. By the $(\mathcal{D}, \mathcal{T})$-amalgamation property there exists a unique model $M' \in$ **Mod**(Σ') such that $M'|_{d'_2} = M_1$ and $M'|_{t'_1} = M_2$. Now we have $M'|_{d'_2} \models \Phi_1$ and $M'|_{t'_1} \models \Phi_2$, which by the satisfaction condition and Definition 10 is equivalent to $M' \in$ **Mod**[$\langle \Sigma', d'_2\Phi_1 \cup t'_1\Phi_2 \rangle$], and because $M = M'|_{d_1;d'_2} = M'|_d$, we have $M \in$ **Mod**[**nf**(SP)].
3. SP is a specification expression of the form **translate** SP_1 **by** t.
   Proof is similar to the proof of the previous case.
4. SP is a specification expression of the form **derive** SP_1 **by** d.
   $\subseteq$: Let $M \in$ **Mod**[**nf**(SP)]. Then by definitions there exists a model $M_1 \in$ |**Mod**(Σ_1)| such that $M_1 \models \Phi_1$ and $M_1|_{d;d_1} = M$. Now, by the induction hypothesis $M_1|_{d_1} \in$ **Mod**[SP_1], which by Definition 10 means that $M \in$ **Mod**[SP].
   $\supseteq$: Let $M \in$ **Mod**[SP]. Then there exists $M_1 \in$ **Mod**[SP_1] such that $M_1|_d = M$. By the induction hypothesis $M_1 \in$ **Mod**[**nf**(SP_1)]. Now, there exists $M_2 \in$ **Mod**(Σ_1) such that $M_2|_{d_1} = M_1$, and because $M_2|_{d;d_1} = M$ we have $M \in$ **Mod**[**nf**(SP)].

□

The above theorem is very important from our point of view and its proof is crucial for understanding the rules presented above and the proof of their completeness. It allows us to replace any specification by its appropriate normal form, for which some basic properties are more transparent.

5 Completeness

In this section we present a proof of completeness of entailment relation $\vdash_\Sigma$ wrt semantical consequence relation $\models_\Sigma$. Both relations were defined in Section 4.

Theorem 16 Completeness. *Let* **INS** *be an* $(\mathcal{D},\mathcal{T})$*-institution that has infinite conjunction and negation. If:*

1. *Institution* **INS** *satisfies* $(\mathcal{D},\mathcal{T})$*-interpolation and* $(\mathcal{D},\mathcal{T})$*-amalgamation properties, and*
2. *Entailment relation* $\vdash_\Sigma^{\mathbf{INS}}$ *used in rule (CR) is complete for institution* **INS**

then for any Σ*-specification SP over institution* **INS** *and any* Σ*-sentence* φ,

$$SP \models_\Sigma \varphi \quad \textit{implies} \quad SP \vdash_\Sigma \varphi.$$

Proof. By induction on the structure of *SP*:

1. If *SP* is a specification expression of the form $\langle\Sigma,\Phi\rangle$, then

   $$\langle\Sigma,\Phi\rangle \models_\Sigma \varphi \;\text{ iff }\; \mathbf{Mod}[\langle\Sigma,\Phi\rangle] \models_\Sigma^{\mathbf{INS}} \varphi \;\text{ iff }\; \Phi \models_\Sigma^{\mathbf{INS}} \varphi$$

   and this, by assumption 2, is equivalent to $\Phi \vdash_\Sigma^{\mathbf{INS}} \varphi$. Now, if $\varphi \in \Phi$ then the rule (basic) completes the proof. If $\varphi \notin \Phi$, then (CR) and (basic) rules must be used to complete the proof.
2. Let *SP* be a specification expression of the form $SP_1 \cup SP_2$ and let $\mathbf{nf}(SP_i) = \mathbf{derive}\ \langle\Sigma_i,\Phi_i\rangle\ \mathbf{by}\ (d_i : \Sigma \to \Sigma_i)$ for $i = 1,2$.
   Then $\mathbf{nf}(SP) = \mathbf{derive}\ \langle\Sigma', t_2'\Phi_1 \cup d_1'\Phi_2\rangle\ \mathbf{by}\ (d : \Sigma \to \Sigma')$, where $d_1' \in \mathcal{D}$, $t_2' \in \mathcal{T}$ and Σ' are given by the following pushout in **Sign**:

   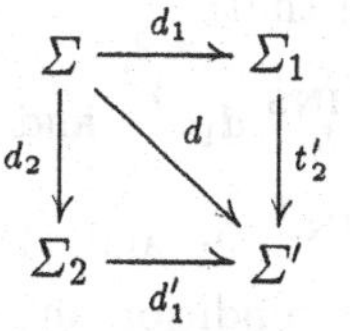

   From Theorem 15 we have $\mathbf{Mod}[\mathbf{nf}(SP)] \models_\Sigma^{\mathbf{INS}} \varphi$. Therefore by the satisfaction condition $t_2'\Phi_1 \cup d_1'\Phi_2 \models_{\Sigma'}^{\mathbf{INS}} d\varphi$, which is equivalent to $t_2' \bigwedge \Phi_1 \models_{\Sigma'}^{\mathbf{INS}} \bigwedge d_1'\Phi_2 \Rightarrow d\varphi$ in **INS**. Since $d = d_2; d_1'$, this is equivalent to $t_2' \bigwedge \Phi_1 \models_{\Sigma'}^{\mathbf{INS}} d_1'(\bigwedge \Phi_2 \Rightarrow d_2\varphi)$. By $(\mathcal{D},\mathcal{T})$-interpolation property for **INS**, we have that there exists a Σ-sentence φ_3 such that:

   $$(1) \quad \bigwedge \Phi_1 \models_{\Sigma_1}^{\mathbf{INS}} d_1\varphi_3 \quad \text{and} \quad (2) \quad d_2\varphi_3 \models_{\Sigma_2}^{\mathbf{INS}} \bigwedge \Phi_2 \Rightarrow d_2\varphi.$$

$\mathbf{Mod}[\langle\Sigma_1,\Phi_1\rangle] \models^{\mathbf{INS}}_{\Sigma_1} \Phi_1$ and (1) implies $\mathbf{Mod}[\langle\Sigma_1,\Phi_1\rangle] \models^{\mathbf{INS}}_{\Sigma_1} d_1\varphi_3$, which by the satisfaction condition is equivalent to $\mathbf{Mod}[\mathbf{nf}(SP_1)] \models^{\mathbf{INS}}_{\Sigma} \varphi_3$ and so, by Theorem 15 to $\mathbf{Mod}[SP_1] \models^{\mathbf{INS}}_{\Sigma} \varphi_3$. Now, by the induction hypothesis we obtain $SP_1 \vdash_\Sigma \varphi_3$.

(2) by Theorem 5 is equivalent to $\bigwedge\Phi_2 \models^{\mathbf{INS}}_{\Sigma_2} d_2(\varphi_3 \Rightarrow \varphi_1)$, which from $(\mathcal{D},\mathcal{T})$-interpolation property implies that there exists a Σ-sentence φ_4 such that:

$$(3)\quad \bigwedge\Phi_2 \models^{\mathbf{INS}}_{\Sigma_2} d_2\varphi_4 \quad\text{and}\quad (4)\quad \varphi_4 \models^{\mathbf{INS}}_{\Sigma} \varphi_3 \Rightarrow \varphi.$$

From $\mathbf{Mod}[\langle\Sigma_2,\Phi_2\rangle] \models^{\mathbf{INS}}_{\Sigma_2} \Phi_2$ and (3), we have $\mathbf{Mod}[\langle\Sigma_2,\Phi_2\rangle] \models^{\mathbf{INS}}_{\Sigma_2} d_2\varphi_4$ and so, by the satisfaction condition and Theorem 15 $\mathbf{Mod}[SP_2] \models^{\mathbf{INS}}_{\Sigma} \varphi_4$, which by the induction hypothesis gives: $SP_2 \vdash_\Sigma \varphi_4$. The following derivation completes the proof:

$$(\mathrm{CR})\ \frac{(\mathrm{sum1})\dfrac{SP_1 \vdash_\Sigma \varphi_3}{SP_1 \cup SP_2 \vdash_\Sigma \varphi_3} \quad (\mathrm{sum2})\dfrac{SP_2 \vdash_\Sigma \varphi_4}{SP_1 \cup SP_2 \vdash_\Sigma \varphi_4} \quad \{\varphi_4,\varphi_3\} \vdash^{\mathbf{INS}}_{\Sigma} \varphi}{SP_1 \cup SP_2 \vdash_\Sigma \varphi}$$

where $\{\varphi_4,\varphi_3\} \vdash^{\mathbf{INS}}_{\Sigma} \varphi$ follows from (4) by Theorem 5 and because $\vdash^{\mathbf{INS}}_{\Sigma}$ completely approximate $\models^{\mathbf{INS}}_{\Sigma}$ (assumption 2).

3. If SP is a specification expression of the form **translate** SP' **by** $(t : \Sigma' \to \Sigma)$, then let $\mathbf{nf}(SP') =$ **derive** $\langle\Sigma_1,\Phi_1\rangle$ **by** $(d_1 : \Sigma' \to \Sigma_1)$ and $\mathbf{nf}(SP) =$ **derive** $\langle\Sigma_1',t'\Phi_1\rangle$ **by** $(d_1' : \Sigma \to \Sigma_1')$, where t', d_1' and Σ_1' are given by a pushout diagram in **Sign**:

$$\begin{array}{ccc} \Sigma' & \xrightarrow{\;t\;} & \Sigma \\ {\scriptstyle d_1}\downarrow & & \downarrow{\scriptstyle d_1'} \\ \Sigma_1 & \xrightarrow[\;t'\;]{} & \Sigma_1' \end{array}$$

Now, similarly to case 2 $SP \models_\Sigma \varphi$ iff $\mathbf{Mod}[\mathbf{nf}(SP)] \models^{\mathbf{INS}}_{\Sigma} \varphi$.

By the satisfaction condition, we obtain $\mathbf{Mod}[\langle\Sigma_1',t'\Phi_1\rangle] \models^{\mathbf{INS}}_{\Sigma_1'} d_1'\varphi$, which is equivalent to $t'\bigwedge\Phi_1 \models^{\mathbf{INS}}_{\Sigma_1'} d_1'\varphi$. By the $(\mathcal{D},\mathcal{T})$-interpolation property, there exists a Σ'-sentence φ' such that:

$$(1)\quad \bigwedge\Phi_1 \models^{\mathbf{INS}}_{\Sigma_1} d_1\varphi' \quad\text{and}\quad (2)\quad t\varphi' \models^{\mathbf{INS}}_{\Sigma} \varphi.$$

Because $\mathbf{Mod}[\langle\Sigma_1,\Phi_1\rangle] \models^{\mathbf{INS}}_{\Sigma_1} \Phi_1$ and (1), we have $\mathbf{Mod}[\langle\Sigma_1,\Phi_1\rangle] \models^{\mathbf{INS}}_{\Sigma_1} d_1\varphi'$ and by the satisfaction condition and Theorem 15 $\mathbf{Mod}[SP'] \models^{\mathbf{INS}}_{\Sigma'} \varphi'$, which by the induction hypothesis is equivalent to $SP' \vdash_{\Sigma'} \varphi'$. The following derivation completes this case:

$$(\mathrm{CR})\ \frac{(\mathrm{trans})\dfrac{SP' \vdash_{\Sigma'} \varphi'}{\textbf{translate}\ SP'\ \textbf{by}\ t \vdash_\Sigma t\varphi'} \quad t\varphi' \vdash^{\mathbf{INS}}_{\Sigma} \varphi}{\textbf{translate}\ SP'\ \textbf{by}\ t \vdash_\Sigma \varphi}$$

where $t\varphi' \vdash^{\mathbf{INS}}_{\Sigma} \varphi$ follows from (2) by assumption 2.

4. If SP is a specification expression of the form **derive** SP' **by** d, where $d : \Sigma \to \Sigma'$, then $SP \models_\Sigma \varphi$ iff $(\mathbf{Mod}[SP'])|_d \models^{\mathbf{INS}}_\Sigma \varphi$. By the satisfaction condition, we have $\mathbf{Mod}[SP'] \models^{\mathbf{INS}}_{\Sigma'} d\varphi$, and by the induction hypothesis $SP' \vdash_{\Sigma'} d\varphi$. Application of the (derive) rule completes the proof.

□

Definition 17 Compactness. Institution **INS** is *compact* iff for any Σ-sentence $\varphi \in \mathbf{Sen}_{\mathbf{INS}}(\Sigma)$ and any set of Σ-sentences $\Phi \subseteq \mathbf{Sen}_{\mathbf{INS}}(\Sigma)$:
if $\Phi \models \varphi$, then there exists a finite set $\Psi \subseteq \Phi$ such that $\Psi \models \varphi$. □

Having compactness, we can modify Theorem 16 and we obtain:

Corollary 18. *Let* **INS** *be a compact* $(\mathcal{D}, \mathcal{T})$*-institution that has conjunction and negation. If* **INS** *satisfies assumptions 1 and 2 of Theorem 16, then for any* Σ*-specification* SP *over institution* **INS** *and any* Σ*-sentence* φ*,*

$$SP \models_\Sigma \varphi \qquad \textit{implies} \qquad SP \vdash_\Sigma \varphi.$$

Proof. An obvious modification of the proof of Theorem 16: in each case when from $\Phi \models \varphi$ we deduce $\bigwedge \Phi \models \varphi$, here we first have to choose a finite set $\Psi \subseteq \Phi$ such that $\Psi \models \varphi$, and then work with $\bigwedge \Psi \models \varphi$. □

The assumptions of the above corollary are satisfied for instance by any $(\mathcal{D}, \mathcal{T})$-institution **FOEQ**[3]. If we extend the institution **EQ**[4] to institution having infinite conjunction and negation (let us denote it by $\mathbf{EQ}^{\wedge\neg}$), then the interesting question is what is the smallest extension of $\mathbf{EQ}^{\wedge\neg}$ for which Corollary 18 holds.

Definition 19. We say that a specification in the normal form:
derive $\langle \Sigma, \Phi \rangle$ **by** d is *finite* iff the set of sentences Φ is finite.
We also say that specifications defined by Definition 10 are *finite* iff in point 1 of Definition 10 we additionally assume that the set Φ is finite. □

Theorem 20. *The normal form of a finite specification is finite.*

Proof. By induction on the structure of SP. □

Corollary 21. *Let* **INS** *be a* $(\mathcal{D}, \mathcal{T})$*-institution that has conjunction and negation. If* **INS** *satisfies assumptions 1 and 2 of Theorem 16, then for any finite* Σ*-specification* SP *over the institution* **INS** *and any* Σ*-sentence* φ*,*

$$SP \models_\Sigma \varphi \qquad \textit{implies} \qquad SP \vdash_\Sigma \varphi.$$

[3] **FOEQ** is the institution of the first order logic with equality (see [GB 90]).
[4] **EQ** is the institution of the many sorted logic with equality (see [GB 90]).

Proof. By inspection of the proof of Theorem 16. It is easy to check that all the sets of sentences used there are finite if *SP* is finite. □

The $(\mathcal{D},\mathcal{T})$-interpolation defined by Definition 7 is the strong version of interpolation. If we consider the weak version of $(\mathcal{D},\mathcal{T})$-interpolation (see [DGS 93]), i.e. where sentences φ_1 and φ from Definition 7 are replaced by sets of sentences, then:

Theorem 22. *$(\mathcal{D},\mathcal{T})$-institution* **INS** *that has infinite conjunction or has conjunction and is compact, satisfies the weak $(\mathcal{D},\mathcal{T})$-interpolation property iff it satisfies the $(\mathcal{D},\mathcal{T})$-interpolation property.*

Proof. Directly from the definition. □

Corollary 23. *Let* **INS** *be a $(\mathcal{D},\mathcal{T})$-institution that has infinite conjunction, or has conjunction and is compact, and has negation. If:*

1. *Institution* **INS** *satisfies weak $(\mathcal{D},\mathcal{T})$-interpolation and $(\mathcal{D},\mathcal{T})$-amalgamation properties, and*
2. *Entailment relation $\vdash_{\Sigma}^{\mathbf{INS}}$ used in rule (CR) is complete for the institution* **INS**

then for any Σ-specification SP over the institution **INS** *and any Σ-sentence φ,*

$$SP \models_{\Sigma} \varphi \qquad \textit{implies} \qquad SP \vdash_{\Sigma} \varphi.$$

Proof. Directly from Theorems 16 and 22. □

6 An example

In this section we present an example of a specification which satisfies assumptions of Corollary 18. It is a transformation of Example 5.3.3 from [Cen 94]. It shows how to use the system presented in this paper for putting specifications together.

Example 2. In this example we will work with specifications over $(\mathcal{D},\mathcal{T})$-institution **FOEQ** (see section 5).

First we define two specifications: the first ST specifying stacks and the second N specifying natural numbers. Then we put them together to obtain specification N-ST of stacks of natural numbers. Let us start with signatures:

SIG-ST = **sig**
sorts *El*; *St*
opns $empty : St$;
$push : El \times St \to St$;
$top : St \to El$;
$pop : St \to St$
rels $is\_empty \subseteq St$
end

SIG-N = **sig**
sorts *N*
opns $zero : N$;
$succ : N \to N$
rels $is\_zero \subseteq N$
end

In the next step we define specifications of stacks and natural numbers:

$$
\begin{array}{rl}
ST = \langle\ SIG\text{-}ST, & \{\forall_{e:El}.\forall_{x:St}.pop(push(e,x)) = x; \\
& \ \forall_{e:El}.\forall_{x:St}.top(push(e,x)) = e\ \}\ \rangle \\[1ex]
N = \langle\ SIG\text{-}N, & \{\forall_{m,n:Nat}.succ\ m = succ\ n \Rightarrow m = n; \\
& \ \forall_{m:Nat}.\neg succ\ m = zero; \\
& \ is\_zero\ zero; \\
& \ \forall_{m:Nat}.\neg is\_zero(succ\ m)\}\ \rangle
\end{array}
$$

Now we put above specifications together to obtain a specification of stacks of natural numbers. Let us consider the following pushout in **Sign**:

$$
\begin{array}{ccc}
SIG\text{-}EL & \xrightarrow{\ d\ } & SIG\text{-}ST \\
{\scriptstyle t}\downarrow & & \downarrow{\scriptstyle t'} \\
SIG\text{-}N & \xrightarrow[\ d'\ ]{} & SIG\text{-}N\text{-}ST
\end{array}
$$

where $SIG\text{-}EL$ = **sig sorts** *El* **end**, $t(El) = N$ and d is an inclusion. From the above we can define $N\text{-}ST$ = (**translate** N **by** d') $\cup$ (**translate** ST **by** t'), and prove several properties of the $N\text{-}ST$ specification, e.g.:
$N\text{-}ST \vdash_{SIG-N-ST} \forall_{x:St}.is\_zero(top(push(zero,x)))$. In the following proof we write $\vdash$ as an abbreviation for $\vdash_{SIG-N-ST}$:

$$
(\mathrm{CR})\ \dfrac{\dfrac{(3)}{N\text{-}ST \vdash is\_zero(zero)} \qquad \dfrac{(2)}{N\text{-}ST \vdash \forall_{n:N}.\forall_{x:St}.top(push(n,x)) = n}\ (1)}{N\text{-}ST \vdash \forall_{x:St}.is\_zero(top(push(zero,x)))}
$$

where (1) is a proof in **FOLEQ** of the following judgement:

$$
\left\{\begin{array}{l} is\_zero(zero), \\ \forall_{n:N}.\forall_{x:St}.top(push(n,x)) = n \end{array}\right\} \vdash^{\mathbf{FOLEQ}}_{SIG-N-ST} \forall_{x:St}.is\_zero(top(push(zero,x))),
$$

(2) is the following proof:

$$
(\mathrm{sum2})\ \dfrac{(\mathrm{trans})\ \dfrac{\dfrac{(\mathrm{basic})}{ST \vdash_{SIG-ST} \forall_{n:El}.\forall_{x:St}.top(push(n,x)) = n}}{\mathbf{translate}\ ST\ \mathbf{by}\ t' \vdash \forall_{n:N}.\forall_{x:St}.top(push(n,x)) = n}}{N\text{-}ST \vdash \forall_{n:N}.\forall_{x:St}.top(push(n,x)) = n}
$$

and finally, (3) is:

$$
(\mathrm{sum1})\ \dfrac{(\mathrm{trans})\ \dfrac{\dfrac{(\mathrm{basic})}{NAT \vdash_{SIG-N} is\_zero(zero)}}{\mathbf{translate}\ N\ \mathbf{by}\ d' \vdash is\_zero(zero)}}{N\text{-}ST \vdash is\_zero(zero)}
$$

□

7 Future work

In this paper we have studied a logical system for reasoning about a logical consequences of structured specifications in an arbitrary institution. We identify formal properties of the underlying institution that ensure its (soundness and) completeness. The next interesting step will be to define a logical system for reasoning about the refinement relation between specifications and try to obtain a correctness result for it. The refinement relation (see [Cen 94]) can be defined as a partial order relation that holds between a more "abstract" and more "concrete" specification.

Another problem is using specifications built over a logical system which is not rich enough to fulfill requirements of Theorem 16 (or its corollaries form Section 5). In my opinion, this problem can be resolved in at least two ways. The first is to find a conservative extension (possibly canonical for some classes of logics) for which Theorem 16 (or its corollaries from Section 5) holds. An alternative is to use an institution representation (see [Tar 95]) of the underlying logic in a richer logic or a logical framework like LF or HOL.

It seems also possible to obtain a similar result (soundness and correctness) for the structural part of the CASL language (see [Mos 97]) or at least for a reasonable part of the CASL language by instantiating and perhaps extending the results presented here.

8 Acknowledgments

I would like to thank Andrzej Tarlecki for his support, encouragement and helpful comments, Maria Victoria Cengarle and Martin Wirsing for their work on algebraic specifications and in particular on proof systems for structured specifications and also people from the Institute of Computer Science of the Polish Academy of Sciences for useful discussions and stimulating atmosphere. This research was partially supported by KBN grant 8T11C01811.

References

[BM 86] J. L. Bell, M. Machover. A course in mathematical logic. North-Holland, 1986.

[BHK 90] J. A. Bergstra, J. Heering, P. Klint. Module algebra. *Journal of the ACM*, 37(2):335-372, April 1990.

[Cen 94] M. V. Cengarle. Formal Specifications with High-Order Parameterization. *Ph.D. thesis*, Institut für Informatik, Ludwig-Maximilians-Universität Müenchen, 1994.

[DGS 93] R. Diaconescu, J. Goguen, P. Stefaneas. Logical Support for Modularization. In: G. Huet, G. Plotkin, editors *Logical Environments, Proceedings of a Workshop held in Edinburgh, Scotland, May 1991*, Cambridge University Press, pages 83–130, 1993.

[Far 92] J. Farrés Casals. Verification in ASL and Related Specification Languages. *Ph.D. thesis*, report CST-92-92, Dept. of Computer Science, University of Edinburgh, 1992.

[GB 90] J. A. Goguen, R. M. Burstall. Institutions: abstract model theory for specifications and programming. *Report ECS-LFCS-90-106*, University of Edinburgh, 1990.

[GM 85] J. A. Goguen, J.Meseguer. Completeness of many-sorted equational logic. *Houston Journal of Mathematics* , volume. 11(3), pages 307–334, 1985.

[Mos 97] P. D. Mosses. CoFI: The Common Framework Initiative for Algebraic Specification and Development. *Theory and Practice of Software Development*, volume 1214 of LNCS, pages 115–137, Springer-Verlag, 1997.

[SB 83] D. Sannella, R. Burstall. Structured Theories in LCF. *Proc. 8th Colloq. on Trees in Algebra and Programming*, L'Aquila, LNCS 159, pp. 377–391, Springer 1983.

[SST 92] D. Sannella, S. Sokołowski, A. Tarlecki. Towards formal development of programs from algebraic specification: parameterization revised. *Acta Informatica*, volume 29, pages 689–736, 1992.

[ST 88] D. Sannella, A. Tarlecki. Specifications in an Arbitrary Institution *Information and Computation*, volume 76, pages 165–210, 1988.

[ST 95] D. Sannella, A. Tarlecki. Essential concepts of algebraic specification and program development. *To appear in: Formal Aspects of Computing.*

[Tar 86] A. Tarlecki. Bits and pieces of the theory of institutions. *Proc. Workshop on Category Theory and Computer Programming*, Guildford. Springer LNCS 240, pages 334-363, 1986.

[Tar 95] A. Tarlecki. Moving between logical systems. *Recent Trends in Data Type Specifications. Selected Papers. 11th Workshop on Specification of Abstract Data Types ADT'95,* Olso, September 1995, eds. M. Haveraaen, O. J. Dahl, O. Owe, Springer LNCS 1130, pages 478–502, 1996.

[Wir 91] M. Wirsing. Structured Specifications: Syntax, Semantics and Proof Calculus. In F. L. Bauer, W. Brauer, and H. Schwichtenberg, editors, *Logic and Algebra of Specification*, volume 94 of *NATO ASI Series F: Computer and Systems Sciences*, pages 411–442. Springer Verlag, 1991.

Zero-Safe Nets: The Individual Token Approach*

Roberto Bruni and Ugo Montanari

Dipartimento di Informatica, Università di Pisa, Italia.
bruni,ugo@di.unipi.it.

Abstract. In this paper we provide both an operational and an abstract concurrent semantics for zero-safe nets under the *individual token* philosophy. The main feature of zero-safe nets is a primitive notion of *transition synchronization*. Besides ordinary places, called *stable* places, zero-safe nets come equipped with *zero* places, which are empty in any stable marking. *Connected transactions* represent *basic atomic computations* of the system between stable markings. They must satisfy two main requirements: 1) to model interacting activities which cannot be decomposed into disjoint sub-activities, and 2) not to consume stable tokens which were generated in the same transaction. Zero tokens acts as triggers for the firings of the transitions which compose the transaction. The abstract counterpart of a zero-safe net consists of a P/T net where each transition locates a distinguished transaction. In the second part of the paper, following the *Petri nets are monoids* approach, we make use of category theory to analyze and motivate our framework. More precisely, the operational semantics of zero-safe nets is characterized as an adjunction, and the derivation of abstract P/T nets as a coreflection.

1 Introduction

Petri nets [18] are one of the most attractive models of concurrency, which also offers a basic concurrent framework often used as a semantic foundation on which to interpret many concurrent languages [21, 10, 17, 6, 8, 1]. However, the basic net model does not have any synchronization mechanism among transitions, while this feature is essential to write modular and expressive programs. For

* Research supported by Office of Naval Research Contracts N00014-95-C-0225 and N00014-96-C-0114, National Science Foundation Grant CCR-9633363, and by the Information Technology Promotion Agency, Japan, as part of the Industrial Science and Technology Frontier Program "New Models for Software Architechture" sponsored by NEDO (New Energy and Industrial Technology Development Organization). Also research supported in part by U.S. Army contract DABT63-96-C-0096 (DARPA); CNR Integrated Project *Metodi e Strumenti per la Progettazione e la Verifica di Sistemi Eterogenei Connessi mediante Reti di Comunicazione*; and Esprit Working Groups *CONFER2* and *COORDINA*. Research carried on in part while the second author was on leave at Computer Science Laboratory, SRI International, Menlo Park, USA, and visiting scholar at Stanford University.

instance, all the above translations involve complex constructions for the net defining the synchronized composition of two programs.

Zero-safe nets (also *ZS nets*), introduced in [4], extend Petri nets along this direction, coming equipped with a very general notion of transition synchronization as a built-in feature. ZS nets are based on the notion of *zero* places. Tokens produced in a zero place act as triggers for the firing of transitions which are able to consume them. A distinguished set of *stable* places is also present. Stable markings (consisting only of stable tokens) describe the abstract-level markings, whilst non-stable markings (those involving zero tokens) define non-observable states of the refined model. A synchronized evolution of a ZS net starts at some stable marking, evolves through non-observable states and finally leads to a new observable state. A 'refined' ZS net and an 'abstract' Petri net are supposed to model the same given system. The latter offers the synchronized view and the former specifies how every transition of the latter is actually achieved as a different coordinated collection of firings, called *transactions*. However, the concurrent semantics of an operational model is usually defined by considering as equivalent all the computations where the same concurrent events are executed in different orders. Thus, we would like to identify those transactions which are equivalent from a *concurrent* viewpoint. The simplest approach, presented in [4], relies on the *collective token* philosophy [9], CTph for short. This school of thought identifies all the firing sequences obtained by repeatedly permuting pairs of concurrently enabled firings. The major drawback of this approach is that causal dependencies on zero tokens are lost. It follows that the class of computations captured by abstract nets may turn out to be far too generic for many applications. In this paper we present an alternative approach based on the *individual token* philosophy [9], ITph for short, where a wider class of aspects can be taken into account. In fact, we identify transactions which refer to isomorphic Goltz-Reisig processes [11]. The induced equivalence classes are called *connected transactions*.

The ZS net MS in Fig. 1 will be our running example. Net MS represents a *multicasting system*. As in a broadcasting system, an agent can simultaneously send the same message to an unlimited number of receivers, but here the receivers are not necessarily all the remaining agents, and thus several one-to-many communications can take place concurrently. Each token in place a is a different *active* (i.e., ready to communicate) agent. Transition *new* permits to create an unlimited number of agents. A firing of *send* opens a one-to-many communication: a message is put in the buffer z and the agent is suspended until the end of the transaction. A firing of *copy* adds a new copy of the message. A firing of *receive* synchronizes an active agent with a copy of the message and then suspends the agent. The transaction is completed when all the copies have been received. At the end of a session, all the suspended agents are moved into place b. Transition *reset* makes an agent active again. We call *copy policy* any strategy for making copies of the messages in the buffer. E.g., in the *sequential copying* policy, every time *copy* produces two copies of the message in the buffer, at most one of them is used to produce other copies. In the CTph, only transmis-

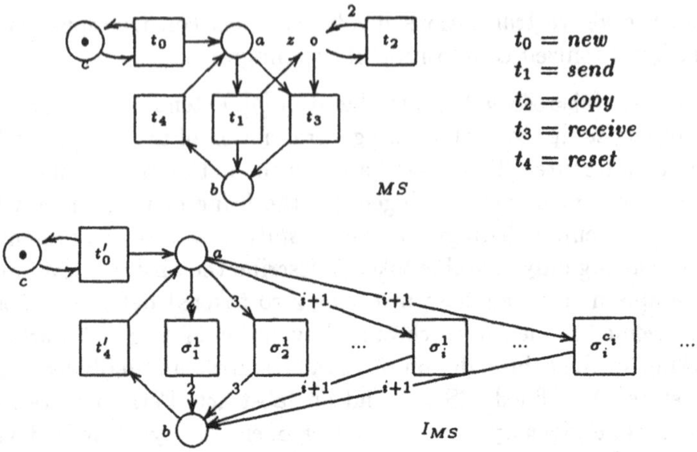

Fig. 1. A ZS net for a multicasting system, and its causal abstract net (we extend the "circles and boxes" representation for nets by drawing smaller circles for zero places).

sions differing for the number of involved agents can be distinguished. E.g., the sequential copy sc_4 = *send-copy-{receive,copy}-{receive,copy}-2receive* and the balanced copy bc_4 = *send-copy-2copy-4receive* are equivalent one-to-four transmissions [4]. On the contrary, the ITph distinguishes different copy policies. The resulting infinite *causal abstract* P/T net I_{MS} is displayed in Fig. 1. Net I_{MS} comes equipped with a *causal refinement morphism* ϵ_{MS} to the ZS net MS. Morphism ϵ_{MS} maps the places of I_{MS} into the homonymous stable places of MS, and each transition of I_{MS} into a *connected transaction* of MS (i.e., transition σ_n^k corresponds to the one-to-n transmission which follows the k-th codified copy policy). For each one-to-n transmission there are c_n copy policies (e.g., $c_4 = 2$, and the transitions σ_4^1 and σ_4^2 identify the equivalence classes of sc_4 and bc_4).

The paper is organized as follows: after briefly recalling some basic definitions of net theory, in Section 3 we present ZS nets and their operational and abstract ITph semantics. We introduce the notion of *causal firing sequences*, as enriched firing sequences. This allows a concise representation of concatenable processes, and has a suggestive implementation on a machine whose states are collections of token-stacks. Then, the evolution of a ZS net is defined in terms of equivalence classes of causal firing sequences (satisfying some additional requirement), and ordinary P/T nets are defined as the abstract counterparts of ZS nets. In Section 4, by employing some elementary category theory, we give evidence that our constructions are *natural*. Both the operational semantics of ZS nets and the derivation of the abstract P/T nets are characterized as two universal constructions (the former is an *adjunction* and the latter is a *coreflection*), following the *Petri nets are monoids* style. The universal properties of the two constructions state that they are the 'best' possible choices. Due to space limitation, some proofs are omitted and others just sketched.

2 Preliminaries

A *net* N is a triple $(S_N, T_N; F_N)$, where $S_N \neq \emptyset$ is the set of *places* $a, a', \ldots$, T_N is the set of *transitions* $t, t', \ldots$ (with $S_N \cap T_N = \emptyset$), and $F_N \subseteq (S_N \times T_N) \cup (T_N \times S_N)$ is called the *flow relation*. We will denote $S_N \cup T_N$ by N whenever no confusion arises. Subscripts will be omitted if they are obvious from the context. For $x \in N$, the set ${}^\bullet x = \{y \in N \mid yFx\}$ ($x^\bullet = \{y \in N \mid xFy\}$) is called the *pre-set* (*post-set*) of x. Let also ${}^\circ N = \{x \in N \mid {}^\bullet x = \emptyset\}$ and $N^\circ = \{x \in N \mid x^\bullet = \emptyset\}$ be the sets of *initial* and *final elements* of N, resp. A place a is said to be *isolated* iff ${}^\bullet a \cup a^\bullet = \emptyset$. We assume that for any transition t, ${}^\bullet t \neq \emptyset$.

A *P/T net* is a tuple $N = (S, T; F, W, u_{\text{in}})$ s.t. $(S, T; F)$ is a net, function $W : F \longrightarrow \mathbb{N}$ assigns a positive *weight* to each arc and multiset $u_{\text{in}} : S \longrightarrow \mathbb{N}$ is the *initial marking*. Relation F may be seen as a function $F : ((S \times T) \cup (T \times S)) \longrightarrow \{0, 1\}$, with $xFy \iff F(x, y) \neq 0$. Then, if we replace $\{0, 1\}$ with $\mathbb{N}$, F becomes a *multiset* relation and W is unnecessary.

A *marking* $u : S \longrightarrow \mathbb{N}$ is a finite multiset of places. It can be written either as $u = \{n_1 a_1, \ldots, n_k a_k\}$ where $n_i \in \mathbb{N}$, $n_i > 0$ (if $n_i = 0$ then the corresponding term $n_i a_i$ is safely omitted) dictates the number of occurrences (*tokens*) of the place a_i in u, i.e. $n_i = u(a_i)$, or as a formal sum $u = \bigoplus_{a_i \in S} n_i a_i$ (the order of summands is immaterial, and the addition is defined by taking $(\bigoplus_i n_i a_i) \oplus (\bigoplus_i m_i a_i) = (\bigoplus_i (n_i + m_i) a_i)$ and 0 as the neutral element). For any transition $t \in T$ let $pre(t)$ and $post(t)$ be the multisets over S such that $pre(t)(a) = F(a, t)$ and $post(t)(a) = F(t, a)$ $\forall a \in S$.

The interleaving behaviour of a net is usually described in terms of firing sequences. Given a P/T net N let u and u' be two markings of N. Then, a transition $t \in T_N$ is *enabled at* u iff $pre(t)(a) \leq u(a)$, $\forall a \in S_N$. Moreover, we say that u evolves to u' under the *firing* of t, written $u[t\rangle u'$, if and only if t is enabled at u and $u'(a) = u(a) - pre(t)(a) + post(t)(a)$, $\forall a \in S$. A *firing sequence* from u_0 to u_n is a sequence of markings and firings such that $u_0[t_1\rangle u_1 \ldots u_{n-1}[t_n\rangle u_n$. Given a marking u of N the set $[u\rangle$ of its *reachable markings* is the smallest set of markings such that $u \in [u\rangle$, and moreover $\forall u' \in [u\rangle$ such that $u'[t\rangle u''$ for some transition t, then $u'' \in [u\rangle$. Besides firings and firing sequences, *steps* and *steps sequences* are introduced. A step allows the simultaneous execution of several independent transitions. Eventually, we say that a net is *safe* if, for all reachable markings, a bound n can be given for the number of tokens in each place, i.e. $\forall u \in [u_{\text{in}}\rangle, \forall a \in S,\ u(a) \leq n$.

3 Zero Safe Computations

We augment P/T nets with special places called zero places. Their role is to coordinate the atomic execution of complex collections of transitions.

Definition 1 (ZS net). A *zero-safe* net (*ZS net* for short) is a 5-tuple $B = (S_B, T_B; F_B, u_B; Z_B)$ where $N_B = (S_B, T_B; F_B, u_B)$ is the *underlying* P/T net, and the set $Z_B \subseteq S_B$ is the set of *zero places*. The places in $S_B \setminus Z_B$ are called *stable places*. A *stable marking* is a multiset of stable places.

Stable markings describe *observable* states of the system. The presence of some zero tokens in a marking makes it unobservable (e.g., *non-stable*). State changes are given in terms of *connected steps.* A connected step may involve the synchronization of several transitions, but it can be applied only if the starting state contains enough stable tokens to enable all the transitions independently. No token can be left on zero places at the end of the step (neither can be found there at the beginning of the step). Thus, all the zero tokens which are produced are also consumed in the same step. *Connected transactions* are *atomic* connected steps which consume all the stable tokens of the starting state.

In the ITph, a marking may be seen as an indexed (over the places of the net) collection of ordered sequences of tokens, and the firing of a transition specify which tokens (of each ordered sequence) are consumed and also the correspondence between each token in the reached marking with either a produced token or an idle token of the original marking. Using multisets instead of ordered sequences would make it impossible to recognize which token was produced by which firing, as it happens for the CTph.

Example 1. In our running example, suppose that the current marking is $\{a, b\}$. If t_4 fires, then a new token is produced in place a. Then, a firing of t_1 consumes a token from place a. In the ITph approach, it makes a difference if t_1 gets the token produced by t_4 or the one already present in a (in the former case the firing of t_1 causally depends on that of t_4 while in the latter case the firings of t_1 and of t_4 are concurrent activities). In the CTph approach the two firings are always concurrent, since the initial marking enables both t_1 and t_4, i.e., the execution of t_4 does not modify the enabling condition of t_1.

The Stacks Based Approach. The approach we propose is very similar to the one adopted in [19]: we choose a canonical interpretation of the tokens that are to be consumed and produced in a firing and we introduce *permutation firings* with the task of rearranging the orderings of the indexed sequences of tokens. A marking becomes a collection of stacks, one for each place, that can be accessed by transitions through a firing to extract and to insert tokens. A permutation firing is just a re-organization of the current state (i.e., of the stacks). We will denote the token stack associated to a certain place a with the term a-stack.

Definition 2 (Causal firing, permutation firing). Let N be a P/T net, and $s = u[t\rangle u'$ be a firing of N for some marking u and transition t. We interpret firing s as a *causal firing* by assuming that s consumes the 'first' $pre(t)(a)$ tokens of the a-stack of u and produces the 'first' $post(t)(a)$ tokens of the a-stack of u', for each place a. Given a marking $u = \{n_a a\}_{a \in S_N}$ of N, a *symmetry* p on u is a vector of *permutations* $p = \langle \pi_a \rangle_{a \in S_N}$ with $\pi_a \in \Pi(n_a)$, $\forall a \in S_N$, i.e. each π_a is a permutation of n_a elements. We denote by $\Pi(u)$ the set of all symmetries on u. Each symmetry p on u induces a *permutation firing* $s = u[p\rangle u$ on the net. A *causal firing sequence* is a finite sequence $\omega = s_1 \cdots s_n$ of causal and permutation firings such that $s_i = u_{i-1}[X_i\rangle u_i$ with $X_i \in T_N \cup \Pi(u_{i-1})$ for $i = 1, \ldots, n$. We say that ω *starts at* u_0 (written $O(\omega) = u_0$) and *ends in* u_n (written $D(\omega) = u_n$).

Example 2. Let N_{MS} be the underlying net of the ZS net MS in Fig. 1. A causal firing sequence for N_{MS} is $\omega = \{b,c\}[t_0\rangle\{a,b,c\}\ [t_4\rangle\{2a,c\}\ [t_1\rangle\{a,b,c,z\}$ $[t_3\rangle\{2b,c\}$. At the beginning the stacks of places a and z are empty and the stacks of places b and c contain one token each. After the firing of t_0 the token in the c-stack is replaced by a new one and a token is also inserted in the a-stack. The firing of t_4 consumes the token in the b-stack and puts a new token on top of the a-stack. Transition t_1 consumes the token on top of the a-stack and inserts a token both in the b-stack and in the z-stack. The firing of t_3 consumes the unique token in the z-stack and also the token produced by t_0 in the a-stack, and it inserts a token on top of the b-stack. Since the sequence ω does not involve any symmetry, it follows that the latest tokens produced are the first to be consumed next. To represent the sequence where t_1 depends on t_0 (and t_3 depends on t_4) we have two possibilities. The first one is to execute t_0 after t_4 (they are concurrently enabled), thus obtaining the sequence $\omega' = \{b,c\}[t_4\rangle\{a,c\}[t_0\rangle\{2a,c\}[t_1\rangle\{a,b,c,z\}[t_3\rangle\{2b,c\}$. The second possibility is to reorganize the a-stack just before the execution of t_1. This can be done via a symmetry $p = \langle(1\ 2)_a\rangle \in \Pi(2a \oplus c)$, thus obtaining the sequence $\omega'' = \{b,c\}[t_0\rangle\{a,b,c\}[t_4\rangle\{2a,c\}[p\rangle\{2a,c\}[t_1\rangle\{a,b,c,z\}[t_3\rangle\{2b,c\}$.

Review of Concatenable Processes. Causal firing sequences define a correspondence among the tokens produced and consumed via firings. This is due to the implicit orders which are imposed on the markings and is strictly related to a *process* view of computations. *Concatenable processes* [5, 20] are obtained from processes by imposing a total ordering on the origins that are instances of the same place and, similarly, on the destinations.

A net K is a *deterministic occurrence net* iff $\forall a \in S_K,\ |{}^\bullet a| \leq 1 \wedge |a^\bullet| \leq 1$ and F_K^* is acyclic (F^* denotes the reflexive and transitive closure of relation F), i.e., $\forall x,y \in K,\ xF_K^*y \wedge yF_K^*x \implies x = y$). A (Goltz-Reisig) *process* for a P/T net N is a mapping $P : K \longrightarrow N$ from an occurrence net K to N such that $P(S_K) \subseteq S_N$, $P(T_K) \subseteq T_N$, ${}^\circ K \subseteq S_K$, and $\forall t \in T_K,\ \forall a \in S_N,\ F_N(a, P(t)) = \left|P^{-1}(a) \cap {}^\bullet t\right| \wedge F_N(P(t), a) = \left|P^{-1}(a) \cap t^\bullet\right|$. As usual we denote the set of *origins* (i.e., minimal or initial places) and *destinations* (i.e., final or maximal places) with $O(K) = {}^\circ K$ and $D(K) = K^\circ \cap S_K$, resp. Two processes P and P' of N are *isomorphic* and thus identified if there exists an isomorphism $\psi : K_P \longrightarrow K_{P'}$ such that $P' \circ \psi = P$.

Given a set S with a labelling function $l : S \longrightarrow S'$, a *label-indexed ordering function* for l is a family $\beta = \{\beta_a\}_{a \in S'}$ of bijections, where $\beta_a : l^{-1}(a) \longrightarrow \{1, \ldots, \left|l^{-1}(a)\right|\}$. A *concatenable process* for a P/T net N is a triple $C = (P, {}^\circ\ell, \ell^\circ)$ where $P : K \longrightarrow N$ is a process for N and ${}^\circ\ell, \ell^\circ$ are label-indexed ordering functions for the labelling function P restricted to $O(K)$ and $D(K)$, resp. Two concatenable processes C and C' are isomorphic if P_C and $P_{C'}$ are isomorphic via a mapping preserving all the orderings.

A partial binary operation $\_;\_$ (associative up to iso and with identities) of concatenation of concatenable processes (whence their names) can be easily defined: we take as source (target) the image through P of the initial (maximal) places of K_P; then the composition of $C = (P, {}^\circ\ell, \ell^\circ)$ and $C' = (P', {}^\circ\ell', \ell'^\circ)$

is realized by merging, when it is possible, the maximal places of K_P with the initial places of $K_{P'}$ according to their labelling and ordering functions so to match those places one-to-one. Concatenable processes admit also a monoidal *parallel* composition $\_ \otimes \_$, which can be represented by putting two processes side by side. Due to space limitation, we refer the interested reader to [5] for the formal definitions.

3.1 From Causal Sequences to Processes

It may be easily noticed that each causal firing sequence uniquely determines a concatenable process. Informally the construction associates an elementary (concatenable) process to each causal and permutation firing.

From causal firings to processes. Let N be a P/T net and $s = u[t\rangle u'$ be a causal firing, with $u = \{n_a a\}_{a \in S_N}$, $pre(t) = \{h_a a\}_{a \in S_N}$, $post(t) = \{k_a a\}_{a \in S_N}$. The associated concatenable process is $pr(s) = (P : K \longrightarrow N, {}^{\circ}\ell, \ell^{\circ})$, where:

- $T_K = \{\tilde{t}\}$, $P(\tilde{t}) = t$; $S_K = \{\tilde{a}_i \mid a \in S_N,\ 1 \le i \le n_a + k_a\}$, $P(\tilde{a}_i) = a$;
- ${}^{\bullet}\tilde{t} = \{\tilde{a}_i \mid a \in S_N,\ 1 \le i \le h_a\}$, $\tilde{t}^{\bullet} = \{\tilde{a}_i \mid a \in S_N,\ h_a + 1 \le i \le h_a + k_a\}$, thus $O(K) = \{\tilde{a}_i \in S_K \mid i \le h_a \vee i \ge h_a + k_a + 1\}$ and $D(K) = \{\tilde{a}_i \in S_K \mid i \ge h_a + 1\}$;
- $\forall \tilde{a}_i \in O(K),\ {}^{\circ}\ell_a(\tilde{a}_i) = \begin{cases} i & \text{if } 1 \le i \le h_a \\ i - k_a & \text{if } h_a + 1 + k_a \le i \le n_a + k_a \end{cases}$;
- $\forall \tilde{a}_i \in D(K),\ \ell^{\circ}_a(\tilde{a}_i) = i - h_a$.

A brief explanation is necessary. The occurence net K contains a unique transition $\tilde{t}$ (mapped onto transition t of N), and a place for each consumed, produced, and idle token of s. For each place $a \in S_N$ we need exactly $n_a + k_a$ different places in S_K. We denote the generic i-th place associated to place a with $\tilde{a}_i$. The set $\{\tilde{a}_i \mid a \in S_N,\ 1 \le i \le h_a\}$ represents the tokens which are consumed by the causal firing of t, i.e. the 'first' h_a tokens of each a-stack in the starting state. The set $\{\tilde{a}_i \mid a \in S_N,\ h_a + 1 \le i \le h_a + k_a\}$ represents the tokens which are produced by the causal firing of t, i.e. the 'first' k_a tokens of each a-stack in the ending state. The set $\{\tilde{a}_i \mid a \in S_N,\ h_a + k_a + 1 \le i \le n_a + k_a\}$ contains the remaining idle tokens, i.e. the 'last' $n_a - h_a$ token of each a-stack of both u and u'. Functions ${}^{\circ}\ell$ and ℓ° are defined accordingly with this assumptions.

From permutation firings to processes. Let N be a P/T net and $s = u[p\rangle u'$ be a permutation firing, with $u = \{n_a a\}_{a \in S_N}$ and $p = \langle \pi_a \rangle_{a \in S_N}$. The associated concatenable process $pr(s) = (P : K \longrightarrow N, {}^{\circ}\ell, \ell^{\circ})$ is defined as follows:

- $T_K = \emptyset$ (it follows that $O(K) = D(K) = S_K$);
- $S_K = \{\tilde{a}_i \mid a \in S_N,\ 1 \le i \le n_a\}$, $P(\tilde{a}_i) = a$;
- $\forall \tilde{a}_i \in O(K),\ {}^{\circ}\ell_a(\tilde{a}_i) = i$; $\forall \tilde{a}_i \in D(K),\ \ell^{\circ}_a(\tilde{a}_i) = \pi_a(i)$.

In this case the set of transitions is empty and all the tokens stay idle. The generic place $\tilde{a}_i$ of k denotes the instance of place a which corresponds to the i-th token (from the top) of the a-stack of the starting state. The re-organization induced by the permutation firing is provided by the functions ${}^{\circ}\ell$ and ℓ°.

The concatenable process associated to a (finite) causal firing sequence is given by the concatenation of the concatenable processes associated to each step of the given sequence. In what follows we denote with $pr(\omega)$ the concatenable process associated with the causal firing sequence ω (up to iso).

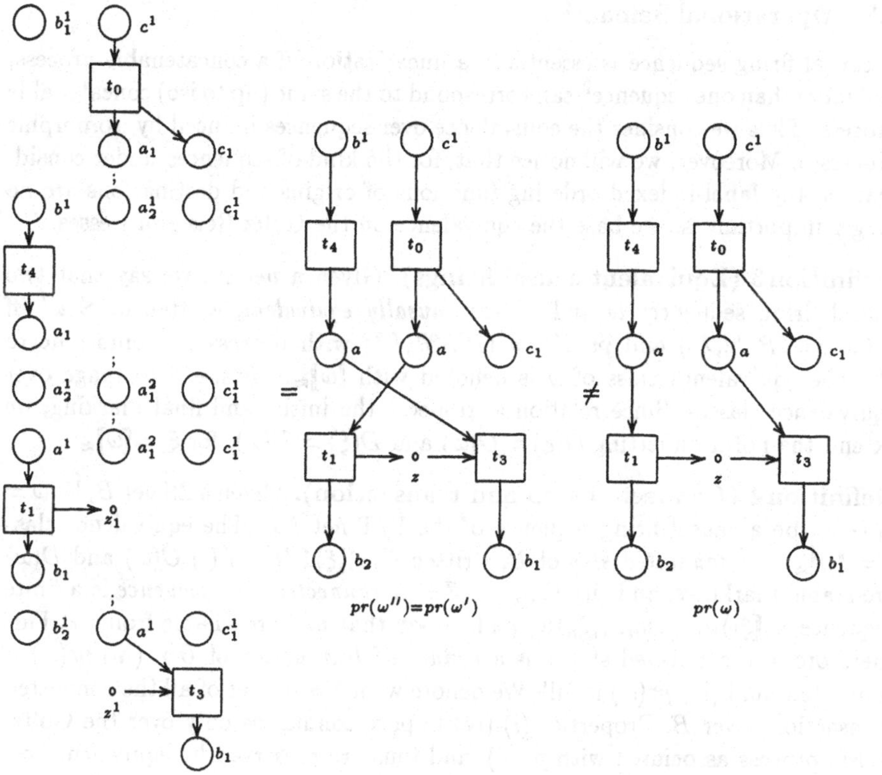

Fig. 2. The concatenable processes derived from sequences ω'', ω', and ω of Ex. 2.

Example 3. The concatenable processes derived from the sequences of Ex. 2 are presented in Fig. 2 (we use the standard notation that labels the places and transitions of the occurence net K with their images in N; a superscript for any initial place and a subscript for any final place denotes the value of ${}^{\circ}\ell$ and ℓ°, resp.), the construction of $pr(\omega'')$ being explained in details.

Terminology. Let us introduce some properties of processes that we will use extensively. A process is *active* iff it includes at least one transition, *inactive* otherwise. An active process is *decomposable into parallel activities* iff it is the parallel composition of two (or more) active processes. If such a decomposition does not exist, then the process is called *connected.* A connected process may involve idle places, but it does not admit disjoint activities. The resources which are first produced and then consumed (i.e., the 'inner' places) are called *evolution places.* More formally, a concatenable process $C = (P : K \longrightarrow N, {}^{\circ}\ell, \ell^{\circ})$ is *connected* iff the set of transitions of K is non-empty and moreover, for each pair (t, t') of transitions of K there exists an undirected path (through the arcs of the flow relation) connecting t and t'. Process C is *full* iff it does not contain idle (i.e., isolated) places (i.e., $\forall a \in S_K,\ |{}^{\bullet}a| + |a^{\bullet}| \geq 1$). Finally, the set of *evolution places* of process C is $E_C = \{P(a) \mid a \in K,\ |{}^{\bullet}a| = |a^{\bullet}| = 1\}$.

3.2 Operational Semantics

A causal firing sequence is essentially a linearization of a concatenable process, and more than one sequence[2] can correspond to the same (up to iso) concatenable process. Thus we consider the equivalence over sequences induced by isomorphic processes. Moreover, we will notice that, for the kind of sequences under consideration, the label-indexed ordering functions of origins and destinations are no longer important, so we base the equivalence on the Goltz-Reisig processes.

Definition 3 (Equivalent causal firings). Given a net N, we say that two causal firing sequences ω and ω' are *causally equivalent*, written $\omega \approx \omega'$ iff $pr(\omega) = (P, {}^{\circ}\ell, \ell^{\circ})$ and $pr(\omega') = (P', {}^{\circ}\ell', \ell'^{\circ})$ with process P isomorphic to P'. The equivalence class of ω is denoted with $[\![\omega]\!]_{\approx}$. We use ξ to range over equivalence classes. Since relation $\approx$ respects the initial and final marking, we extend the notation letting $O(\xi) = O(\omega)$ and $D(\xi) = D(\omega)$, for $\xi = [\![\omega]\!]_{\approx}$.

Definition 4 (Connected step and transaction). Given a ZS net B, let $\omega = s_1 \cdots s_n$ be a causal firing sequence of the P/T net N_B. The equivalence class $\xi = [\![\omega]\!]_{\approx}$ is a *connected step* of B, written $O(\xi)[\![\xi\rangle D(\xi)$, if (i) $O(\omega)$ and $D(\omega)$ are stable markings, and (ii) $E_{pr(\omega)} \subseteq Z_B$. A *connected step sequence* is a finite sequence $u_0[\![\xi_1\rangle u_1 \dots u_{n-1}[\![\xi_n\rangle u_n$ and we say that u_n is reachable from u_0. Furthermore, the connected step ξ is a *connected transaction* of B if (iii) $pr(\omega)$ is connected, and (iv) $pr(\omega)$ is full. We denote with Ξ_B the set of all the connected transactions over B. Properties (i)-(iv) impose conditions only over the Goltz-Reisig process associated with $pr(\omega)$, and thus are preserved by equivalence $\approx$.

We claim that connected transactions are a good definition for the basic behaviours of the systems. Our assertion is supported by the fact that connected transactions denote *atomic computations that cannot be extended further*. Atomicity follows immediately from the connectedness of the associated processes. The second argument deserves a more precise explanation. An atomic behaviour can be extended if there exists a broader atomic behaviour of which the former is a sub-part. From our viewpoint, the only interaction allowed in a ZS net is given by the flow of tokens through zero places. Since connected steps and transactions start and also end in stable markings, it is impossible to hook them in a wider atomic computation by means of zero tokens. This is very clear for transactions, because they consume all the needed resources. This is not the case of connected steps, since it could be possible for some resource to stay idle during the whole sequence of moves. However this kind of resources are stable and not connected to the rest of the step, thus, any other activity involving them is intrinsically concurrent w.r.t. the step under consideration. It follows that any wider behaviour extending a connected step is not atomic (i.e., it can be expressed in terms of concurrent components).

[2] Sequences differing in the order in which concurrent firings are executed or for the way in which equivalent symmetries are performed are identified.

Example 4. Let us consider the ZS net MS of Fig. 1. The equivalence class of the causal firing sequence $\{a\}[t_1\rangle\{b,z\}[t_4\rangle\{a,z\}[t_3\rangle\{b\}$ is not a connected step since the prop. (ii) is not satisfied. Class $[\![\{2a,c\}[t_1\rangle\{a,b,c,z\}[t_3\rangle\{2b,c\}[t_0\rangle\{a,2b,c\}]\!]_{\approx}$ is a connected step but not a connected transaction since the constraint (iii) is not satisfied. The class of $\{4a\}[t_1\rangle\{3a,b,z\}[t_2\rangle\{3a,b,2z\}[t_3\rangle\{2a,2b,z\}[t_3\rangle\{a,3b\}$ is a connected step but not a connected transaction since the prop. (iv) is not satisfied. The class of $\{5a\}[t_1\rangle u_1[t_2\rangle u_2[t_2\rangle u_3[t_2\rangle u_4[t_3\rangle u_5[t_3\rangle u_6[t_3\rangle u_7[t_3\rangle\{5b\}$, where intermediate markings are the obvious ones, is a connected transaction.

3.3 Abstract Semantics

Next, we define an abstract view of the system modelled by a ZS net. Since transactions rewrite multisets of stable tokens, it is natural to choose a net as a candidate for the abstraction. Since the ordering of tokens in the pre-set (post-set) of a transition is useless we should abstract from it. This is already done via the equivalence classes of causal firing sequences. When restricted to connected steps, this equivalence intuitively corresponds to limit the symmetries of permutation firings to be vectors of permutations over the zero places only, with the assumption that the stable tokens which are produced in a transaction are not reused during the same transaction. The last statement was also the basis for the CTph approach.

Example 5. Consider the ZS net MS in Fig. 1. Let $\omega = \{2a\}[t_1\rangle\{a,b,z\}[t_3\rangle\{2b\}$, $s = \{2a\}[p\rangle\{2a\}$ and $s' = \{2b\}[p'\rangle\{2b\}$, where p and p' are the symmetries which swap the two tokens in a and b, resp. The causal sequences ω, $s\omega$, and $\omega s'$ define the same connected transaction $\xi = [\![\omega]\!]_{\approx}$, but $pr(s\omega) \neq pr(\omega) \neq pr(\omega s')$. If we represent the connected transaction ξ as a transition t of a net, then its preset (as well as its postset) is an unordered multiset. This means that when t fires it is impossible to distinguish among the two tokens in b that it produces, and also among the two tokens in a that it consumes. We can conclude that it makes no sense to have many different transitions to represent behaviours that we cannot reproduce at the abstract level. Thus we are forced to identify $pr(s\omega)$, $pr(\omega)$, $pr(\omega s')$, and also $pr(s\omega s')$.

Definition 5 (Causal abstract net). Let $B = (S_B, T_B; F_B, u_B; Z_B)$ a ZS net. Net $I_B = (S_B \backslash Z_B, \Xi_B; F, u_B)$, with $F(a,\delta) = pre(\delta)(a)$ and $F(\delta,a) = post(\delta)(a)$, is the *causal abstract net* of B, where Ξ_B is the set of all the connected transaction of B, and $pre(\delta)$ and $post(\delta)$ denote the multisets $O(\delta)$ and $D(\delta)$, resp.

Example 6. We conclude this section by illustrating the causal abstract net of the multicasting system. The net I_{MS} is (partially!) depicted in Fig. 1. Transition t_0' creates a new communicating process and it corresponds to $[\![\omega_0]\!]_{\approx}$ with $\omega_0 = \{c\}[t_0\rangle\{a,c\}$. Similarly t_4' is the equivalence class of the the firing of t_4 in the marking $\{b\}$. Each σ_i^k describes a different one-to-i communication, where index k identifies the copy policy. A generic one-to-i communication can be essentially described as follows: a firing of t_1 initiates the communication, then the system executes as many firings of t_2 as the number of copies of the message needed (i.e., $i-1$ since a message is already present in the buffer), and finally i firings of t_3 synchronize the messages with different active processes. In the ITph we distinguish among tokens in a same marking, which were created by different firings. In this way we have a one-to-one correspondence among copy policies and the complete[3] binary trees with exactly i leaves (we don't distinguish between 'left' and 'right' children).

For any i, the total number of copy policies can be derived as follows. For $i = 1$ there is only one tree whose root is the unique leaf. If $i = 2h + 1$ for some integer $h > 0$, it follows that one of the subtrees rooted in a child of the root is a complete binary tree with $j \leq h$ leaves, while the subtree rooted in the other child is a complete binary tree with $i - j$ leaves; for any j, we know that there are $c_j \cdot c_{i-j}$ possible trees made in this way, thus $c_i = \sum_{j=1}^{h} c_j \cdot c_{i-j}$. If $i = 2h$ for some integer $h > 0$, we adopt an analogous reasoning to deduce that for any $j < h$ there are $c_j \cdot c_{i-j}$ possible complete binary trees such that exactly j leaves belongs to one of the subtrees rooted in the children of the root. The case $j = h$ requires more attention. In fact, if the two subtrees have the same number h of leaves then there are $\frac{c_h \cdot (c_h+1)}{2}$ possible ways for choosing them. It follows that $c_i = \frac{c_h \cdot (c_h+1)}{2} + \sum_{j=1}^{h-1} c_j \cdot c_{n-j}$. Since there are no transitions in MS requiring 2 or more zero token, there are no other transactions.

4 Universal Constructions

The aim of this section is to propose an algebraic characterization of the definitions and the constructions presented in the previous section. To this purpose, we make use of some elementary concepts of category theory. The first notion consists of the *category of models* itself: objects are models and arrows represent some notion of simulation. The choice of arrows is very informative, since they complement and in a sense redefine (e.g., isomorphic objects are often identified) the meaning of models. We define a suitable category **dZPetri** (where ZS nets are considered as programs) where net morphisms satisfy an important additional condition. Then we consider a construction that exhibits an *adjunction* from **dZPetri** to a category **ZSCGraph** consisting of some kind of machines, equipped with operations and transitions between states. It is proved that this adjunction is strictly related to the semantics of ZS nets defined in the previous section. Our second construction starts from a complex category **ZSC** of ZS nets (which is however strictly related to **ZSCGraph**), having the ordinary category **Petri** of P/T nets as a subcategory, and yields a *coreflection* corresponding exactly to the construction of the causal abstract net in Def. 5.

4.1 Review of 'Petri Nets are Monoids'

Petri net theory can be profitably developed within category theory [22, 13, 2]. We follow the approach initiated in [13] (other references are [14, 5, 15, 16]).

[3] We say that a binary tree is *complete* if any internal node has exactly two children.

A (place/transition) *Petri net* is a graph $(S^{\oplus}, T, \partial_0, \partial_1)$ where the set of nodes is the free commutative monoid $S^{\oplus}$ over the set of *places* S (functions $\partial_0, \partial_1 : T \longrightarrow V$ are called *source* and *target*, resp., and we write $t : u \longrightarrow v$, with obvious meaning, to shorten the notation). A *Petri net morphism* is a graph morphism $h = (f : T \longrightarrow T', g : S^{\oplus} \longrightarrow S'^{\oplus})$ (i.e., $g(\partial_i(u)) = \partial'_i(f(u))$ for $i = 0, 1$) where g is a monoid homomorphism (this defines the category **Petri**).

In [14, 5] it has been shown that it is possible to enrich the algebraic structure of transitions in order to capture some basic constructions on nets. As an example, the forgetful functor from **CMonRPetri** [14] to **Petri** has a left adjoint which associates to each Petri net N its *marking graph* $\mathcal{C}[N]$, which corresponds to the ordinary operational semantics of N (i.e., its arrows are the step sequences of N). The objects of **CMonRPetri** are *reflexive Petri commutative monoids* (i.e., Petri nets together with a function $id : S^{\oplus} \longrightarrow T$, where T is a commutative monoid $(T, \otimes, 0)$ and ∂_0, ∂_1 and id are monoid homomorphisms), and its arrows are Petri net morphisms preserving identities and the monoidal structures.

The algebraic structure of process is well captured in [20]. There it is shown how to associate a free symmetric strict monoidal category (see Appendix A) $\mathcal{F}[N]$ to each net N in such a way that, under two suitable axioms, it characterizes the concatenable processes of N. This is due to the existence of a left adjoint functor $\mathcal{F} : \mathbf{Petri} \longrightarrow \mathbf{SSMC}^{\oplus}$ to the forgetful functor $\mathcal{U} : \mathbf{SSMC}^{\oplus} \longrightarrow \mathbf{Petri}$. Given a net N the category $\mathcal{F}[N]$ has the elements of $S_N^{\oplus}$ as objects, while its arrows are generated by the following inference rules

$$\frac{u \in S_N^{\oplus}}{id_u : u \longrightarrow u \in \mathcal{F}[N]} \qquad \frac{t : u \longrightarrow v \in T_N}{t : u \longrightarrow v \in \mathcal{F}[N]} \qquad \frac{a, b \in S_N}{c_{a,b} : a \oplus b \longrightarrow b \oplus a \in \mathcal{F}[N]}$$

$$\frac{\alpha : u \longrightarrow v,\ \beta : u' \longrightarrow v' \in \mathcal{F}[N]}{\alpha \otimes \beta : u \oplus u' \longrightarrow v \oplus v' \in \mathcal{F}[N]} \qquad \frac{\alpha : u \longrightarrow v,\ \beta : v \longrightarrow w \in \mathcal{F}[N]}{\alpha ; \beta : u \longrightarrow w \in \mathcal{F}[N]}$$

modulo the axioms expressing that $\mathcal{F}[N]$ is a strict monoidal category, and the axioms stating that the collection[4] $\{c_{u,v}\}_{u,v \in S_N^{\oplus}}$ plays the role of the symmetry natural isomorphism which makes $\mathcal{F}[N]$ into a ssmc. This axiomatization will be useful to shorten the notation in the sketched proof of Th. 16.

Theorem 6. *Given a net N, the concatenable processes of N are isomorphic to the arrows of the category $\mathcal{P}[N]$, which is the monoidal quotient of the free ssmc on N ($\mathcal{F}[N]$) modulo the axioms*

$$c_{a,b} = id_{a \oplus b} \quad \textit{if } a \neq b \in S_N, \textit{ and} \tag{1}$$
$$s; t; s' = t \quad \textit{if } t \in T_N, \textit{ and } s, s' \textit{ are symmetries.} \tag{2}$$

The previous construction provides an algebraic view of net computations which is strictly related to a process understanding of the causal behaviour of a net, but is not functorial. The main problem is that there exist reasonable morphisms of nets which cannot be extended to a monoidal functor. We illustrate

[4] Symmetries $c_{u,v}$ for $u, v \in S_N^{\oplus}$ denote any term obtained from $c_{a,b}$ for $a, b \in S_N$ by applying recursive rules analogous to axioms (3) given in Th. 11.

below the example presented in [16]. We will show that this kind of morphisms can be avoided in the category of ZS nets, our choice being justified by the necessity to preserve atomic behaviours through morphisms.

Example 7. Consider the nets N and N' pictured below and the net morphism $f : N \longrightarrow N'$ s.t. $f(t_i) = t'_i$, $f(a_i) = a'$, $f(b) = b'$ and $f(c) = c'$ for $i = 0, 1$.

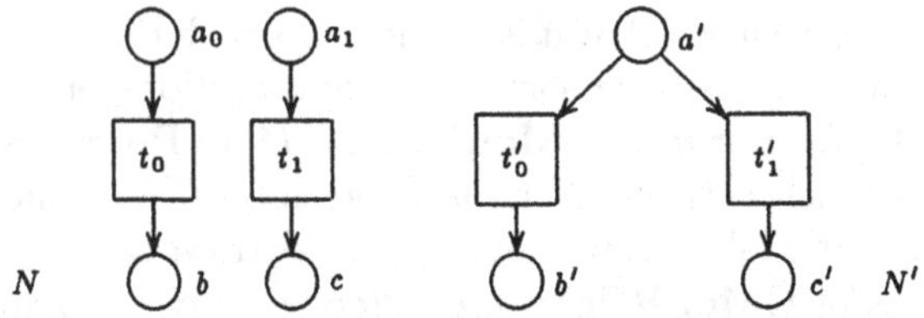

Morphism f cannot be extended to a functor $\mathcal{P}[f] : \mathcal{P}[N] \longrightarrow \mathcal{P}[N']$. In fact, supposing that such an extension F exists, then $F(t_0 \otimes t_1) = F(t_0) \otimes F(t_1) = t'_0 \otimes t'_1$ by the monoidality of F. Since $t_0 \otimes t_1 = t_1 \otimes t_0$ in $\mathcal{P}[N]$, then $t'_0 \otimes t'_1 = t'_1 \otimes t'_0$ which is impossible, as the two expressions denote different processes in $\mathcal{P}[N']$.

4.2 Operational Semantics as Adjunction

Definition 7 (Category dZPetri). A *ZS net* is a Petri net where the set of places $S = L \cup Z$ is partitioned into *stable* and *zero* places. A *ZS net morphism* is a Petri net morphism $(f, g) : N \longrightarrow N'$ where homomorphism g preserves partitioning of places (i.e., if $a \in Z$ then $g(a) \in Z'^{\oplus}$ and if $a \in S \setminus Z$ then $g(a) \in (S' \setminus Z')^{\oplus}$) and satisfies the additional condition of mapping zero places into pairwise disjoint (non-empty) zero markings (*disjoint image* property). We call *disjoint* any morphism of this kind. This defines the category **dZPetri**.

Since $S^{\oplus}$ is a free commutative monoid we may represent the set of nodes of a ZS net as $L^{\oplus} \times Z^{\oplus}$, and ZS net morphisms as triples (f, g_L, g_Z), where g_L and g_Z are monoid homomorphisms on the monoids of stable and zero places, resp.

Example 8. The graph corresponding to the ZS net MS defined in Fig. 1 has the following set of arcs: $T_{MS} = \{t_0 : (c, 0) \longrightarrow (a \oplus c, 0), t_1 : (a, 0) \longrightarrow (b, z), t_2 : (0, z) \longrightarrow (0, 2z), t_3 : (a, z) \longrightarrow (b, 0), t_4 : (b, 0) \longrightarrow (a, 0)\}$.

Disjoint morphisms play a very important role here. If we restrict to consider disjoint morphisms only, then we avoid the awkward situation arising from Ex. 7. Moreover, if we identified two different zero places via a (non-disjoint) morphism then the behaviour of the abstract model might dramatically change. Since we use zero places to specify a synchronization mechanism, it is important to ensure that this mechanism is always preserved.

The next definition introduces a category of more structured models, which is reminiscent of the constructions both of marking graphs and free ssmc's.

Definition 8 (Category ZSCGraph). A *ZS causal graph* $E = ((L \cup Z)^{\oplus}, (T, \otimes, 0, id, *), \partial_0, \partial_1)$ is both a ZS net and a reflexive Petri monoid. In addition, it comes equipped with a partial function $\_ * \_$ called *horizontal composition*:

$$\frac{\alpha : (u, x) \longrightarrow (v, y),\ \beta : (u', y) \longrightarrow (v', y')}{\alpha * \beta : (u \oplus u', x) \longrightarrow (v \oplus v', y')}$$

and a collection of *horizontal swappings* $\{e_{x,y} : (0, x \oplus y) \longrightarrow (0, y \oplus x)\}_{x,y \in Z^{\oplus}}$. Horizontal composition is associative and has identities $id_{(0,x)}$ for any $x \in Z^{\oplus}$. The monoidal operator $\_ \otimes \_$ is functorial w.r.t. horizontal composition, and the *horizontal naturality* axiom $e_{x,x'} * (\beta \otimes \alpha) = (\alpha \otimes \beta) * e_{y,y'}$ holds for any $\alpha : (u, x) \longrightarrow (v, y)$ and $\beta : (u', x') \longrightarrow (v', y')$. Moreover, the following *coherence* axioms are satisfied for any $x, y, y' \in Z^{\oplus}$: $e_{x,y} * e_{y,x} = id_{(0,x \oplus y)}$, and $e_{x,y \oplus y'} = (e_{x,y} \otimes id_{(0,y')}) * (id_{(0,y)} \otimes e_{x,y'})$. A morphism h between two ZS causal graphs E and E' is a monoidal disjoint morphism which in addition respects horizontal composition and swappings This defines the category **ZSCGraph**.

Horizontal composition is the key feature of our approach. It behaves like sequential composition on zero places and like the ordinary parallel composition on stable places. This is necessary to avoid the construction of steps which reuse stable tokens. Swappings are used to specify the causality relation among produced and consumed zero tokens.

Proposition 9. *If $\alpha : (u, 0) \longrightarrow (v, 0)$ and $\alpha' : (u', 0) \longrightarrow (v', 0)$ are two transitions of a ZS causal graph then $\alpha \otimes \alpha' = \alpha' \otimes \alpha$ and $\alpha * \alpha' = \alpha \otimes \alpha'$.*

Corollary 10. *The full subcategory of* **ZSCGraph** *whose objects are Petri nets (i.e., $Z = \emptyset$) is isomorphic to* **CMonRPetri**.

Theorem 11. *The obvious forgetful functor $\mathcal{U}$:* **ZSCGraph** $\longrightarrow$ **dZPetri** *has a left adjoint $\mathcal{CG}$:* **dZPetri** $\longrightarrow$ **ZSCGraph**, *which maps a ZS net B into the ZS causal graph $\mathcal{CG}[B]$, whose arrows are generated by the following inference rules*

$$\frac{t : (u, x) \longrightarrow (v, y) \in T_B}{t : (u, x) \longrightarrow (v, y) \in \mathcal{CG}[B]} \qquad \frac{\alpha : (u, x) \longrightarrow (v, y),\ \beta : (u', x') \longrightarrow (v', y') \in \mathcal{CG}[B]}{\alpha \otimes \beta : (u \oplus u', x \oplus x') \longrightarrow (v \oplus v', y \oplus y') \in \mathcal{CG}[B]}$$

$$\frac{(u, x) \in L_B^{\oplus} \times Z_B^{\oplus}}{id_{(u,x)} : (u, x) \longrightarrow (u, x) \in \mathcal{CG}[B]} \qquad \frac{z, z' \in Z_B}{d_{z,x} : (0, z \oplus x) \longrightarrow (0, x \oplus z) \in \mathcal{CG}[B]}$$

$$\frac{\alpha : (u, x) \longrightarrow (v, y),\ \beta : (u', y) \longrightarrow (v', z) \in \mathcal{CG}[B]}{\alpha * \beta : (u \oplus u', x) \longrightarrow (v \oplus v', z) \in \mathcal{CG}[B]}$$

modulo the axioms expressing that the arrows form a (strict) monoid with unit $id_{(0,0)}$, and that horizontal composition $\_ * \_$ is associative and has identities $id_{(0,x)}$, the functoriality axiom for the tensor product, and the axioms expressing that the collection of swappings $d_{x,y}$ plays the role of the 'horizontal' natural isomorphism: $d_{x,x'} * (\beta \otimes \alpha) = (\alpha \otimes \beta) * d_{y,y'}$, and $d_{z,z'} * d_{z',z} = id_{z \oplus z'}$, for any arrows $\alpha : (u, x) \longrightarrow (v, y), \beta : (u', x') \longrightarrow (v', y') \in \mathcal{CG}[B]$, and for any $z, z' \in Z_B$, where $d_{x,y}$ for $x, y \in Z_B^{\oplus}$ denotes any term obtained from the basic symmetries by applying recursively the rules:

$$\begin{aligned} d_{0,x} &= id_{(0,x)} = d_{x,0}, \\ d_{z \oplus x,y} &= (id_{(0,z)} \otimes d_{x,y}) * (d_{z,y} \otimes id_{(0,x)}), \text{ and} \\ d_{x,y \oplus z} &= (d_{x,y} \otimes id_{(0,z)}) * (id_{(0,y)} \otimes d_{x,z}). \end{aligned} \tag{3}$$

Proof. (Sketch). It follows immediately from the definition that $\mathcal{CG}[\mathcal{B}]$ is a ZS causal graph. We need to show that it is the *free* ZS causal graph on B. Let $\eta_B : B \longrightarrow \mathcal{U}[\mathcal{CG}[\mathcal{B}]]$ the disjoint ZS net morphism which is the identity on places and the obvious injection on transitions. We show that η_B is universal, i.e., for any ZS causal graph E and for any disjoint ZS net morphism $h = (f, g_L, g_Z) : B \longrightarrow \mathcal{U}[E]$, there exists a unique ZS causal graph morphism $k : \mathcal{CG}[\mathcal{B}] \longrightarrow E$ such that $h = \eta_B; \mathcal{U}[k]$ (in **dZPetri**). Thus, morphisms k and h must agree on the generators of $\mathcal{CG}[\mathcal{B}]$ and the extension of k to tensor and horizontal composition is uniquely determined by its definition on the generators. The proof can be completed just showing that k preserves the axioms which generate $\mathcal{CG}[B]$.

The notion of adjunction between a category with 'more structure' (**ZSCGraph** in our case), and a similar category but with 'less structure' (**dZPetri**) is useful to characterize natural constructions. In fact, the left-adjoint to the usually obvious forgetful functor which deletes the 'extra' structure is unique (up to iso) and represents the best possible way for adding this structure.

Theorem 12. *When restricted to P/T nets, functor $\mathcal{CG}$ coincides with $\mathcal{C}$.*

The previous theorem shows that the algebraic semantics of ZS nets is an extension of the ordinary semantics of P/T nets. Unfortunately, the ZS causal graph $\mathcal{CG}[B]$ is still too concrete w.r.t. the operational semantics of ZS nets. Thus, we need two more axioms (analogous to axioms (1) and (2) of Th. 6).

Definition 13. Given a ZS net B, let $\mathcal{CG}[B]/\Psi$ be the quotient of the free ZS causal graph $\mathcal{CG}[B]$ generated by B in **ZSCGraph** modulo the axioms

$$d_{z,z'} = id_{(0,z\oplus z')} \text{ if } z \neq z' \in Z_B, \text{ and} \tag{4}$$

$$d * t * d' = t \text{ if } t \in T_B, \text{ and } d, d' \text{ are swappings.} \tag{5}$$

The quotient $\mathcal{CG}[B]/\Psi$ is s.t. for any ZS causal graph morphism $k : \mathcal{CG}[B] \longrightarrow E$ respecting axioms (4) and (5) (i.e., $k(d_{z,z'}) = id_{(0,k(z)\oplus k(z'))}$, and $k(d*t*d') = k(t)$), there is a unique arrow k_Ψ such that $k = Q_\Psi; k_\Psi$ (in **ZSCGraph**), where $Q_\Psi : \mathcal{CG}[B] \longrightarrow \mathcal{CG}[B]/\Psi$ is the obvious morphism associated to the (least) congruence generated by the imposed axiomatization.

Proposition 14. *For any disjoint morphism $h : B \longrightarrow B'$ in* **dZPetri** *there exists a unique extension $\hat{h} : \mathcal{CG}[B]/\Psi \longrightarrow \mathcal{CG}[B']/\Psi$ of h in* **ZSCGraph**.

Proof. Take $k = Q'_\Psi \circ \mathcal{CG}[h] : \mathcal{CG}[B] \longrightarrow \mathcal{CG}[B']/\Psi$. Morphism k respects axioms (4) and (5) (because h is disjoint and maps transitions to transitions), thus k_Ψ is uniquely determined. Then take $\hat{h} = k_\Psi$.

Example 9. Let MS be the ZS net of our running example whose set of arcs is defined in Ex. 8. For instance the arrow $t_1 * t_3 \in \mathcal{CG}[MS]/\Psi$ has source $(2a, 0)$ and target $(2b, 0)$. Instead, notice that the arrow $(t_1 \otimes id_{(a,0)}) * (id_{(b,0)} \otimes t_3)$ goes from $(3a \oplus b, 0)$ to $(a \oplus 3b, 0)$. As another example, the following expressions are all identified in $\mathcal{CG}[MS]/\Psi$, i.e., they all denote the same arrow: $t_1 * t_2 * (t_2 \otimes t_3) * (t_3 \otimes t_3) = t_1 * t_2 * (t_2 \otimes id_{(0,z)}) * (t_3 \otimes t_3 \otimes t_3) = t_1 * t_2 * d_{z,z} * (t_2 \otimes id_{(0,z)}) * (t_3 \otimes t_3 \otimes t_3) = t_1 * t_2 * (id_{(0,z)} \otimes t_2) * (t_3 \otimes t_3 \otimes t_3) = t_1 * t_2 * (t_3 \otimes t_2) * (t_3 \otimes t_3)$.

Definition 15 (Prime Arrow). An arrow $\alpha : (u, 0) \longrightarrow (v, 0)$ of a ZS causal graph E is *prime* iff α cannot be expressed as the monoidal composition of non-trivial arrows (i.e., $\nexists\beta, \gamma \in H, \beta \neq id_{(0,0)} \neq \gamma$ such that $\alpha = \beta \otimes \gamma$).

Example 10. In our running example, some prime arrows of $\mathcal{CG}[MS]$ are $t_0, t_1 * t_3$, and $t_1 * t_2 * (t_2 \otimes t_2) * (t_3 \otimes t_2 \otimes t_3 \otimes t_3) * (t_3 \otimes t_3)$. As a counterexample, the arrow $(t_1 \otimes t_1) * d_{z,z} * (t_2 \otimes t_3) * (t_3 \otimes t_3)$ is not prime.

Theorem 16. *Given a ZS net B, there is a one-to-one correspondence between arrows $\alpha : (u, 0) \longrightarrow (v, 0) \in \mathcal{CG}[B]/\Psi$ and the connected steps of B. Moreover, if such an arrow is prime (and is not an identity) then the corresponding connected step is a connected transaction.*

Proof. (Sketch). For any arrow β of $\mathcal{CG}[B]/\Psi$ we define inductively on the structure of β a concatenable process $C(\beta)$ of N_B as follows: $C(t) = t$, $C(id_{(u,x)}) = id_u \otimes id_x$, $C(d_{z,z}) = c_{z,z}$, $C(\beta' \otimes \beta'') = C(\beta') \otimes C(\beta'')$ and $C(\beta' * \beta'') = (C(\beta') \otimes u'); (v \otimes C(\beta''))$ if $\beta' : (u, x) \longrightarrow (v, y)$ and $\beta'' : (u', x') \longrightarrow (v', y')$. It can be verified that any different expression denoting β yields the same result. Moreover, if $\alpha : (u, 0) \longrightarrow (v, 0) \in \mathcal{CG}[B]/\Psi$ then the process obtained from $C(\alpha)$ by forgetting the label-indexed ordering functions of origins and destinations denotes a connected step of B.

Conversely, let $\xi = [\![\omega]\!]_\approx$ (for some causal firing sequence ω) be a connected step. The concatenable process $pr(\omega)$ of N_B can be denoted algebraically as the sequential and parallel composition of transitions, identities and symmetries. Moreover, we can take an equivalent process C without stable symmetries, i.e., C can be expressed as $\alpha_1; \ldots; \alpha_n$ where $\alpha_i = \beta_i \otimes u_i \otimes x_i$ with $u_i \in L_B^\oplus$, $x_i \in Z_B^\oplus$ and $\beta_i \in T_B \cup \{c_{kz,z}\}_{z \in Z_B, k \in \mathbb{N}}$. Then take $\alpha' = \alpha_1' * \ldots * \alpha_n'$ where $\alpha_i' = \beta_i' \otimes x_i$ and $\beta_i' = \beta_i$ if $\beta_i \in T_B$ and $\beta_i' = d_{kz,z}$ if $\beta_i = c_{kz,z}$ for some zero place z, and integer k. Eventually, $\alpha = \alpha' \otimes u'$ where u' is the multiset of idle tokens.

This result states the correspondence among algebraic and operational semantics.

4.3 Abstract Semantics as Coreflection

Finally, we present the universal construction of the abstract semantics of ZS nets. We make use of a category **ZSC** whose objects are ZS nets and whose morphisms allow for the refinement of a transition into a connected transaction. This construction is somehow reminiscent of the construction of **ImplPetri** in [14].

Definition 17. Given ZS net B, a *causal abstract transition* of a $\mathcal{CG}[B]/\Psi$ is either a prime arrow of $\mathcal{CG}[B]/\Psi$ or a transition of B. Given two ZS net B and B', a *causal refinement morphism* $h : B \longrightarrow B'$ is a disjoint ZS net morphism $h = (f, g_L, g_Z)$ from B to (the image through the forgetful functor of) $\mathcal{CG}[B']/\Psi$ such that function f maps transitions into causal abstract transitions.

Since morphism h is disjoint, a transition can be refined into a transaction iff both its preset and its postset are stable. Transition involving zero places can only be mapped to transitions.

Lemma 18. *Given a causal refinement morphism $h : B \longrightarrow B'$, it uniquely extends to a morphisms $\hat{h} : \mathcal{CG}[B]/\Psi \longrightarrow \mathcal{CG}[B']/\Psi$ in* **ZSCGraph**, *which preserves prime arrows.*

Definition 19 (Category ZSC). The category **ZSC** has ZS nets as objects and causal refinement morphisms as arrows, their composition being defined through the extension in **ZSCGraph** given by Lemma 18.

Theorem 20. *Category* **Petri** *is embedded in* **ZSC** *fully and faithfully as a coreflective subcategory. Furthermore, the right adjoint of the coreflection $\mathcal{I}[\_]$ maps every ZS net B into its causal abstract net I_B (see Def. 5).*

Proof. (Sketch). The connected transactions (i.e. prime arrows, by Theorem 16) of a P/T net are all and only its transitions. Thus a causal refinement morphism $h : N \longrightarrow N'$ maps transitions into transitions. Next we want to prove that the obvious inclusion functor from **Petri** to **ZSC** has a right adjoint $\mathcal{I}[\_]$: **ZSC** $\longrightarrow$ **Petri** such that $\mathcal{I}[\_]$ maps each ZS net B into its causal abstract net I_B. We verify that $\mathcal{I}[\_]$ extends to a functor. Consider a causal refinement morphism $h = (f, g_L, g_Z) : B \longrightarrow B'$. Let $\hat{h} : \mathcal{CG}[B]/\Psi \longrightarrow \mathcal{CG}[B']/\Psi$ be the unique extension of h in **ZSCGraph**. Morphism $\hat{h}$ preserves prime arrows (by Lemma 18). Then we define $\mathcal{I}[h] = (f', g)$ with $f'(\xi) = \hat{h}(\xi)$ for any $\xi \in \Xi_B$ and $g(a) = g_L(a)$ for any $a \in L_B$. It follows that the unit component η_N of the adjunction is the identity and the counit component ϵ_B maps each transition of the abstract net into the appropriate connected transaction.

Category **ZSC** can be thought to represent the operational models, while **Petri** defines 'abstract' models. The functor $\mathcal{I}$: **ZSC** $\longrightarrow$ **Petri** that maps each ZS net B onto its abstract P/T net I_B, is the right adjoint to the inclusion functor. For every ZS net B there is a unique arrow $\epsilon_B : \mathcal{I}[B] \longrightarrow B$ with the universal property that, given any abstract model N in **Petri**, for every arrow $h : N \longrightarrow B$ there is a unique arrow $h' : N \longrightarrow \mathcal{I}[B]$ with $h = h'; \epsilon_B$. This situation is ideal from a semantic point of view. In fact $\mathcal{I}[B]$ can be understood as an abstraction of model B (e.g. its behaviour), with the additional advantage of being at the same time a model itself. The universal property above means that if we observe models from an abstract point of view (i.e. via morphisms originating from objects in **Petri**), then there is an isomorphism (via left composition with ϵ_B) between observations of B and observations of its abstract counterpart $\mathcal{I}[B]$. Thus in a sense, seen from **Petri**, B is the same as I_B.

5 Conclusion

We have proposed ZS nets as a model which offers the basis for a uniform approach to concurrent language translations. E.g., CCS-style languages may be easily modelled by representing the channels as zero places, in the style of our multicasting example. In this paper we have based our constructions on the so-called individual token philosophy [9]. Correspondingly, our categorical models rely on monoidal graphs equipped with an operation of horizontal composition together with a collection of special transitions called *swappings* to represent the permutations of tokens, along the style of [5]. An alternative and simpler approach corresponds to the so-called *collective token philosophy* as illustrated in [4]. We noticed that, whatever the adopted philosophy is, an identical restriction, called 'disjoint image property', must be required for the arrows of categories **ZSC** and **ZSN** (of [4]), which are used to define the abstract semantics. However, depending on the chosen approach – ITph vs CTph – two notion of transactions may be defined, each leading to different operational and abstract models. This should help to clarify the distinction between the two philosophies also from a pragmatic perspective rather than just from an academic viewpoint.

As a final remark, *symmetric, strict monoidal double categories* [3] seem to offer an alternative categorical characterization for the semantics of ZS nets. In this sense a ZS net could be viewed as a simple instance of a *tile rewrite system* [3, 7] where basic tiles are net transitions, and the horizontal composition of tiles corresponds to composition $\_ * \_$. The vertical composition of tiles would essentially build causal step sequences.

References

1. E. Best, R. Devillers and J. Hall. The Box Calculus: A New Causal Algebra with Multi-label Communication. In *Advances in Petri Nets '92, LNCS, n. 609*, 21–69. Springer-Verlag, 1992.
2. C. Brown and D. Gurr. A Categorical Linear Framework for Petri Nets. In *Proceedings of the 5th LICS Symposium*, 208–218, 1990.
3. R. Bruni, J. Meseguer, and U. Montanari. *Process and Term Tile Logic.* Technical Report, SRI International, to appear.
4. R. Bruni and U. Montanari. Zero-Safe Nets, or Transition Synchronization Made Simple. In *Proceedings of EXPRESS'97, ENTCS, Vol.7, 1997.*
5. P. Degano, J. Meseguer, and U. Montanari. Axiomatizing the Algebra of Net Computations and Processes. *Acta Informatica*, 33(7):641–667, October 1996.
6. P. Degano, R. De Nicola, and U. Montanari. A Distributed Operational Semantics for CCS based on Condition/Event Systems. *Acta Informatica*, 26:59–91, 1988.
7. F. Gadducci and U. Montanari. The Tile Model In: Gordon Plotkin, Colin Stirling, and Mads Tofte, Eds., *Proof, Language and Interaction: Essays in Honour of Robin Milner* MIT Press, to appear.
8. R. Gorrieri and U. Montanari. On the Implementation of Concurrent Calculi into Net Calculi: Two Case Studies *TCS 141, 1-2*, 1995, 195–252.
9. R.J. Van Glabbeek and G.D. Plotkin. Configuration Structures. In D. Kozen, editor, *Proceedings of the 10th LICS Symposium, IEEE*, pages 199–209, 1995.
10. R. Van Glabbeek and F. Vaandrager. Petri Net Models for Algebraic Theories of Concurrency. In *Proc. of PARLE, LNCS, n. 259*, 224–242. Springer-Verlag, 1987.
11. U. Goltz and W. Reisig. The Non-Sequential Behaviour of Petri Nets. *Information and Computation*, 57:125–147, 1983.

12. S. MacLane. *Categories for the Working Mathematician.* Springer-Verlag, 1971.
13. J. Meseguer and U. Montanari Petri Nets are Monoids: A New Algebraic Foundations for Net Theory. *Proc. 3rd LICS Symposium,* IEEE 1988:155–164.
14. J. Meseguer and U. Montanari. Petri Nets are Monoids. *Information and Computation,* 88(2):105–155, October 1990.
15. J. Meseguer, U. Montanari, and V. Sassone. Process versus Unfolding Semantics for Place/Transition Petri Nets. *TCS, Volume 153,* issue 1-2, (1996) pages 171-210.
16. J. Meseguer, U. Montanari, and V. Sassone. Representation Theorems for Petri Nets. Festschrift in honor of Prof. Wilfried Brauer to appear.
17. E.R. Olderog. Operational Petri Net Semantics for CCSP. In G. Rozenberg, editor, *Advances in Petri Nets '87, LNCS, n. 266,* 196–223. Springer-Verlag, 1987.
18. W. Reisig. *Petri Nets.* Springer-Verlag, 1985.
19. G. Ristori. *Modelling Systems with Shared Resources via Petri Nets .* PhD thesis TD 05/94, Department of Computer Science, University of Pisa, 1994.
20. V. Sassone. An Axiomatization of the Algebra of Petri Net Concatenable Processes. *Theoretical Computer Science,* vol. 170, n.1–2, pp 277–296, 1996.
21. G. Winskel. Event Structure Semantics of CCS and Related Languages. In *Proceedings of ICALP '82, LNCS, n. 140,* pages 561–567. Springer-Verlag, 1982.
22. G. Winskel. Petri Nets, Algebras, Morphisms and Compositionality. *Information and Computation,* 72:197–238, 1987.

A   Symmetric, Strict Monoidal Categories

A *symmetric, strict monoidal category* [12], *ssmc* for short, is a quadruple $(\mathcal{C}, \otimes, e, \gamma)$ where $\mathcal{C}$ is the underlying category (with composition $\_;\_$ and identity id_x for each object x), functor $\otimes : \mathcal{C} \times \mathcal{C} \longrightarrow \mathcal{C}$ is the *tensor product*, object e of $\mathcal{C}$ is called the *unit object*, the diagrams

$$
\begin{array}{ccc}
\mathcal{C} \times \mathcal{C} \times \mathcal{C} & \xrightarrow{\otimes \times 1} & \mathcal{C} \times \mathcal{C} \\
{\scriptstyle 1 \times \otimes} \downarrow & & \downarrow {\scriptstyle \otimes} \\
\mathcal{C} \times \mathcal{C} & \xrightarrow{\otimes} & \mathcal{C}
\end{array}
\qquad
\mathcal{C} \xrightarrow{\langle 1, e \rangle} \mathcal{C} \times \mathcal{C} \xleftarrow{\langle e, 1 \rangle} \mathcal{C}
$$

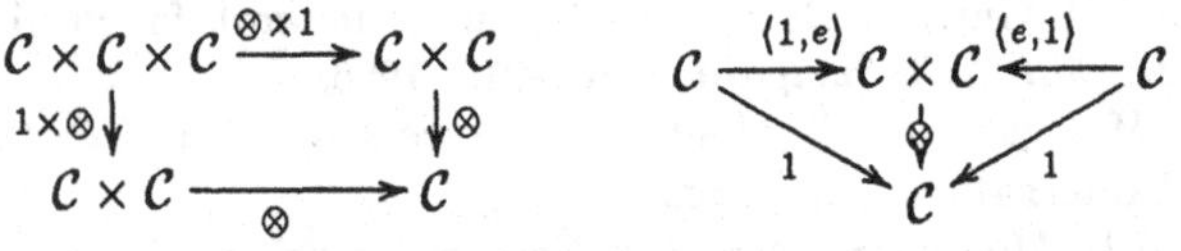

commute (where $\langle \_ , \_ \rangle$ denotes the pairing of functors induced by the cartesian product of categories), and natural transformation $\gamma : \_{-1} \otimes \_{-2} \Rightarrow \_{-2} \otimes \_{-1}$ is an isomorphism called *symmetry* satisfying the Kelly-MacLane *coherence axioms* $\gamma_{x \otimes y, z} = (id_x \otimes \gamma_{y,z}); (\gamma_{x,z} \otimes id_y)$, and $\gamma_{x,y}; \gamma_{y,x} = id_{x \otimes y}$ (for any objects x, y and z). A *symmetric strict monoidal functor* is a functor $F_\otimes : \mathcal{C} \longrightarrow \mathcal{C}'$ which preserves the monoidal structure and the symmetries.

Let **SSMC** be the category of symmetric strict monoidal categories and symmetric strict monoidal functors. We denote with **SSMC**$^\oplus$ the full subcategory of **SSMC** consisting of the monoidal categories whose objects form *free* commutative monoids.

Implementation of Derived Programs (Almost) for Free*

Maura Cerioli and Elena Zucca

DISI–Dipartimento di Informatica e Scienze dell'Informazione,
Università di Genova, Via Dodecaneso, 35, 16146 Genova, Italy,
e-mail: {cerioli,zucca}@disi.unige.it

Introduction

In the process of top-down software development, an implementation step can consist of two different kinds of refinement:

- "local", i.e. replacing a module A by a more specific module B which simulates the behavior of A;
- "global", i.e. passing from a more abstract specification or programming language, say $\mathcal{I}$, to a less abstract, say $\mathcal{I}'$.

A property that usually holds in practice is that these two kinds of refinement can be composed, i.e., if a module A is correctly implemented by B, then all the programs in $\mathcal{I}$ using A can be correctly transformed in programs in $\mathcal{I}'$ using B, provided that we are able to translate linguistic constructs from $\mathcal{I}$ to $\mathcal{I}'$ (in other words, we get "for free" the implementation of derived programs).

Our aim is to give a model for this situation independent from the particular formalisms $\mathcal{I}$ and $\mathcal{I}'$ which are involved, in the spirit of the theory of institutions [3]. Indeed, commonly used notions of refinement of formalisms (see [7] for further references, too) do not support this intuition, since the translation of expressions from $\mathcal{I}$ to $\mathcal{I}'$ is not given once and for all, but depends on (is parameterized by) the specific signature.

The framework we propose is partly similar to that of parchments, that are syntactic representations of institutions where expressions over a signature Σ are terms over an algebraic signature $Lang(\Sigma)$ (see e.g. [2]); but we take the stronger uniformity requirement that this algebraic signature is independent from the specific signature Σ, which only impacts on the choice of variables. Moreover, since our aim is to have a notion of formalism including both specification and programming languages, the expression we consider are not just boolean sentences, but are classified by their *types*. Putting the two things together, we get a notion of *typed uniform parchment* (Section 1), which on one side seems general enough for capturing most common institutions, on the other side allows to express notions closer to programming languages, in particular the factorization of an implementation step in a global and a local part mentioned above (Section 2).

* Partially supported by Murst 40% - Modelli della computazione e dei linguaggi di programmazione and CNR - Formalismi per la specifica e la descrizione di sistemi ad oggetti.

Finally, the notion of implementation presented here allows to relate individual values within the carriers of the involved models, generalizing the original approach by Hoare [4]. Since this *concrete data-type implementation* maps individual elements, it is possible to verify the correctness of an implementation w.r.t. a singled out function of the signature, while the available notions of implementation require to take into account whole models (or even specifications) at one time. This is, at our knowledge, the first attempt at rephrasing concrete data-type implementation in an institutional framework.

1 Typed Uniform Parchments

The institution theory has been originally developed in order to provide a flexible formalism to represent specification languages focusing on the validity relation between boolean sentences and models.

In order to represent not only axiomatic frameworks, but also a broader range of applications, like database query systems, knowledge representation systems and programming languages, institutions can be *generalized* (see e.g. [2]) replacing boolean sentences by "expressions" (we will adopt this term from now on) to be evaluated in some generic category of *values*.

Definition 1 [2]. A *generalized institution* is a tuple $(\mathbf{Sign}, \mathit{Exp}, \mathit{Mod}, [\![\_]\!])$, where:

- **Sign** is a a category,
- Exp: $\mathbf{Sign} \to \mathbf{Set}$ is a functor giving the *expressions* over a signature,
- Mod: $\mathbf{Sign}^{op} \to \mathbf{Cat}$ is a functor giving the *models* over a signature,
- $[\![\_]\!]$: $|\mathit{Mod}| \times \mathit{Exp} \stackrel{..}{\longrightarrow} \mathcal{V}$ is an extra-natural transformation, giving the *evaluation* of an expression in a model, with $\mathcal{V}$ the *universe of admissible values*; the application of $[\![\_]\!]_\Sigma$ to arguments $ex \in \mathit{Exp}(\Sigma)$ and $M \in |\mathit{Mod}(\Sigma)|$ will be denoted by $[\![ex]\!]_\Sigma^M$.

The extra-naturality of evaluation means that for each signature morphism $\sigma: \Sigma \to \Sigma'$, each Σ-expression ex and each Σ'-model M'

$$[\![\mathit{Exp}(\sigma)(ex)]\!]_{\Sigma'}^{M'} = [\![ex]\!]_\Sigma^{\mathit{Mod}(\sigma)(M)}.$$

In order to give a natural description of programming languages, we endow generalized institutions with a notion of *typing*. A similar notion is that of *concrete institutions* (see e.g. [1]), i.e. standard institutions with boolean sentences, where signatures have an underlying set of *sorts* or *types* and models over a signature Σ with types S have an underlying *carrier* which is an S-sorted set.

Definition 2. A *typed institution* $\mathcal{TGI}$ consists of a generalized institution $\mathcal{GI} = (\mathbf{Sign}, \mathit{Exp}, \mathit{Mod}, [\![\_]\!])$ and a *typing system* for $\mathcal{GI}$, i.e.:

- a functor Types: $\mathbf{Sign} \to \mathbf{Set}$, giving the *types* of a signature;
- a natural transformation τ: $\mathit{Exp} \stackrel{.}{\longrightarrow} \mathit{Types}$, *typing* the expressions;
- an extra-natural transformation $[\_]$: $|\mathit{Mod}| \times \mathit{Types} \stackrel{..}{\longrightarrow} \mathcal{T}$, with $\mathcal{T} \subseteq \wp(\mathcal{V})$ the *universe of admissible type values*, giving the *carrier* of a type in a model; the application of $[\_]_\Sigma$ to arguments $t \in \mathit{Types}(\Sigma)$ and $M \in |\mathit{Mod}(\Sigma)|$ will be denoted by $[t]_\Sigma^M$;

satisfying, for each $\Sigma \in |\mathbf{Sign}|$, $M \in |Mod(\Sigma)|$ and $ex \in Exp(\Sigma)$, the following *type preservation* condition:

$$[\![ex]\!]^M_\Sigma \in [\tau_\Sigma(ex)]^M_\Sigma .$$

Parchments have been introduced in [2] as syntactical presentations of institutions, allowing not only to avoid the (often tedious and/or difficult) check that the satisfaction condition holds, but also to combine institutions, presented by parchments, in a more convenient way (see e.g. [5]).

The intuition behind parchments is to define the set of the sentences over a signature Σ by initiality, i.e. as the set of the terms of a distinguished sort in an algebraic signature $Lang(\Sigma)$, where $Lang$ is a functor intuitively associating a *language* with any signature. Moreover, a model over Σ can be defined by giving a morphism $M: \Sigma \to \Gamma$ where Γ is a distinguished signature, intuitively the "universal language". The validity of sentences in a model is then derived from the evaluation of the term algebra $W_{Lang(\Gamma)}$ into a fixed algebra G^V, intuitively the "semantical universe".

The *typed uniform parchments* (shorty tu parchments from now on) we present in this section differ from parchments in two respects. First of all, as they describe typed institutions, tu parchments have two levels of language: the *type* language, describing for each signature the set of acceptable types, and the *value* language, describing the (typed) expressions denoting values. The two levels also apply to the semantic part; so we have two universal languages and two algebras giving the semantical universes.

Second, we require a much stronger uniformity in the way of associating a language with a given signature. Indeed, in most commonly used institutions and in programming languages too, there is a unique language fixed once and for all, and a specific signature only contributes in providing a family of symbols which play the role of variables. This is patent in programming languages, where correct programs (expressions) using a module are usually determined by a fixed grammar (the language description) and a set of identifiers, depending on the individual module. But this also applies to specification frameworks. Let us consider, for instance, the ground terms over an algebraic signature $\Sigma = \langle S, O\rangle$. They can be seen as the terms over another signature $AV\Sigma(S)$, with sorts $\{term(s) \mid s \in S\} \cup \{op(s_1 \ldots s_k \to s) \mid s_1, \ldots, s_k, s \in S\}$ and an operation $\_(\_, \ldots, \_): op(s_1 \ldots s_k \to s) \times term(s_1) \times \ldots \times term(s_k) \to term(s)$ for any $s_1, \ldots, s_k, s \in S$. Each operation $f \in O_{s_1 \ldots s_k, s}$ is seen as a variable of sort $op(s_1 \ldots s_k \to s)$. Moreover, the sorts of the signature $AV\Sigma(S)$ are, in turn, terms over a signature $AT\Sigma$, with operations $term(\_)$ and $op(\_ \ldots \_ \to \_)$, over the set S of type variables. Hence the specific signature Σ only determines the sets S and O of variables to be used respectively to build the *type* and the *value* expressions. Thus, in tu parchments the language components of both levels factorize into a functor, providing for each signature its *names*, and an algebraic signature, describing how these names are used to build the (type and value) expressions.

Notation 1: Indexed Sets. Let us denote by $SSet: \mathbf{Set}^{op} \to \mathbf{Cat}$ the *indexed set* functor, associating with each set S the category of S-sorted families of sets (and S-sorted families of functions as arrows) and with each function renaming the indexes the *reduct.* Moreover, **SSet** denotes the flattening of *SSet*, that is the category of sorted sets, where an object is a pair $\langle S, X\rangle$ with S a set and X an S-family of sets, and an arrow from $\langle S, X\rangle$ into $\langle S', X'\rangle$ is a pair $\langle f: S \to S', \{h_s: X_s \to X'_{f(s)}\}_{s\in S}\rangle$. Finally, *Sorts*: **SSet** $\to$ **Set** denotes the functor giving the sorts of a sorted set.

Here, we chose to use (many-sorted) algebraic signatures to describe the language components, for simplicity. However, the results we present are only based on the existence of the free algebra construction, and the commutativity of the reduct functor w.r.t. the carrier functor.

Notation 2: Algebraic Signatures. We denote by **AlgSign** the category of algebraic many-sorted signatures and by *Sorts*: **AlgSign** $\to$ **Set** the functor giving the sorts of an algebraic signature; when clear from the context, we will use σ for *Sorts*(σ).

The functor **ALG**: **AlgSign**$^{op} \to$ **Cat** associates with each $\Sigma \in |\mathbf{AlgSign}|$ the category of the many-sorted total algebras over Σ.

For each $\Sigma \in |\mathbf{AlgSign}|$ let W_Σ: *SSet*(*Sorts*(Σ)) $\to$ **ALG**(Σ) denote the left adjoint of the carrier functor $|\_|$: **ALG**(Σ) $\to$ *SSet*(*Sorts*(Σ)), with unit η^Σ and counit ϵ^Σ.

Moreover, for each $A \in |\mathbf{ALG}(\Sigma)|$ and $f: X \to |A|$, we denote by $eval_\Sigma^{A,f}: W_\Sigma(X) \to A$ the unique free extension of f, that is $eval_\Sigma^{A,f} = \epsilon_A^\Sigma \cdot W_\Sigma(f)$. Finally, if $A = W_{\Sigma'}(Y)_{|\sigma}$ for some morphism $\sigma: \Sigma \to \Sigma'$ in **AlgSign**, then $eval_\Sigma^{A,f}$ will be denoted by $W_\sigma(f)$.

Let us collect a few properties of $W_\sigma(f)$, that will be used in the next section.

Lemma 3. *For each* $\sigma: \Sigma \to \Sigma'$, $\sigma': \Sigma' \to \Sigma''$ *in* **AlgSign**, $f: X \to |W_{\Sigma'}(Y)_{|\sigma}|$, $f': Y \to |W_{\Sigma''}(Z)_{|\sigma'}|$ *and* $f'': X \to Y_{|\sigma}$ *in* **Set**:

1. $W_{\sigma'\cdot\sigma}(|W_{\sigma'}(f')_{|\sigma}| \cdot f) = W_{\sigma'}(f')_{|\sigma} \cdot W_\sigma(f)$
2. $|W_{\sigma'}(f')_{|\sigma}| \cdot (\eta_Y^{\Sigma'})_{|\sigma} = f'_{|\sigma}$
3. $W_{\sigma'\cdot\sigma}(f'_{|\sigma} \cdot f'') = W_{\sigma'}(f')_{|\sigma} \cdot W_\sigma((\eta_Y^{\Sigma'})_{|\sigma} \cdot f'')$

In the following a *type language* will be a signature $T\Sigma$ with fixed sorts n, e of the *name* and *expression* types, respectively. Intuitively, the $T\Sigma$-terms of sort e represent the types of expressions, while those of sort n represent auxiliary types, whose elements will be used for constructing the expressions. In most common cases, the auxiliary types collect function and procedures, like in the following example of many-sorted signatures

Example 1.

$AT\Sigma =$ **sorts** n, e
opns $\{op(\_\ldots\_ \to \_): e^k \times e \to n \mid k \geq 0\}$

Each type language implicitly defines a category of sets with *substitutions* as arrows, where type names are allowed to be expanded into type expressions. This is technically described in terms of the *Kleisli* category for the monad induced by the adjunction between the carrier functor and the free algebra construction.

Notation 3: Type Languages. We denote by **AlgSign***, the sub-category of **AlgSign** with fixed sorts n, e of the *name and expression types*, respectively, and signature morphisms preserving such fixed sorts. Signatures in **AlgSign*** will be called *type languages*.

The following notations are introduced for a given type language $T\Sigma$.

Adjunction Let us denote by $F^{T\Sigma} = \underset{T\Sigma}{W}(\_ : e)$ the left adjoint to $G^{T\Sigma} = |\_|_e$ with $\eta_e^{T\Sigma} = (\eta^{T\Sigma}_{\_:e})_e$ as unit and $\epsilon_e^{T\Sigma} = \epsilon^{T\Sigma} \cdot \underset{T\Sigma}{W}(E_{|\_|})$ as counit of the adjunction, obtained by composing the usual adjunction between indexed sets and algebras with the following adjunction $\langle \_ : e, \_{}_e, id, E \rangle$ between sets and indexed sets:

- $\_ : e$: **Set** $\rightarrow SSet(Sorts(\Sigma))$ associates any X with the family of sets that are all empty but for the index e, for which X is yielded; analogously for functions;
- $\_{}_e$: $SSet(Sorts(\Sigma)) \rightarrow$ **Set** is the projection over the index e;
- E is, for each family of sets X, the embedding into X of the family having all components empty but for the index e, for which X_e is yielded.

Substitution category Let us denote by $\mathbf{Set}_{T\Sigma}$ the *Kleisli*'s category over the monad induced by the adjunction between $F^{T\Sigma}$ and $G^{T\Sigma}$, having sets as objects, the functions from X into $|\underset{T\Sigma}{W}(Y : e)|_e$ as arrows from X to Y, $\eta_e^{T\Sigma}$ as identity, and composition defined by $g \overset{T\Sigma}{\circ} f = G^{T\Sigma}({\epsilon_e^{T\Sigma}}_{F^{T\Sigma}(Z)} \cdot F^{T\Sigma}(g)) \cdot f$ for each $f: X \rightarrow Y$ and $g: Y \rightarrow Z$.
Moreover, for each $f: X \rightarrow Y$ in **Set** we will denote by $f_{T\Sigma} = {\eta_e^{T\Sigma}}_Y \cdot f: X \rightarrow Y$ its corresponding arrow in $\mathbf{Set}_{T\Sigma}$.

Name and Expression Functors The functors $|\underset{T\Sigma}{W}(\_ : e)|_n : \mathbf{Set}_{T\Sigma} \rightarrow \mathbf{Set}$ and $|\underset{T\Sigma}{W}(\_ : e)|_e : \mathbf{Set}_{T\Sigma} \rightarrow \mathbf{Set}$, give the sets of *name* and *expression* types and their homomorphical translations, that are the composition of the projections over the n and e components of the carrier with the generalization[2] of $F^{T\Sigma}$. We will also denote by $|\underset{T\Sigma}{W}(\_ : e)|_{n \uplus e} : \mathbf{Set}_{T\Sigma} \rightarrow \mathbf{Set}$ the coproduct of such functors, yielding on each set the disjoint union of the nameable and expression types, with silent injections.

The intuition behind the choice of symbols in the following definition is that a T is used to decorate the parts regarding the *type level*, an N for the *auxiliary names* and their valuations and a V for the elements of the *value level* that are directly involved in building and evaluating value expression.

Definition 4. A *typed uniform parchment*, from now on tu parchment, is a tuple $\mathcal{P} = (\mathbf{Sign}, TN, T\Sigma, \mathcal{TN}, G^T, \nu^T, N, V\Sigma, \mathcal{N}, G^V, \nu^N)$ where

Signatures **Sign** is a category of *signatures*,

Type level

- TN: **Sign** $\rightarrow$ **Set** is a functor giving the *type names* of a signature,
- $T\Sigma \in |\mathbf{AlgSign}^*|$ is the *type language*,
- $\mathcal{TN}$ is a set giving the *universe of types*;
- G^T is a model of $T\Sigma$ term-generated by the valuation $\nu^T : \mathcal{TN} \rightarrow |G^T|_e$, s.t. $|G^T|_n, |G^T|_e \subseteq |\mathbf{Set}|$;

Value level

- N: **Sign** $\rightarrow$ **SSet** is a functor giving the *(value) names* of a signature s.t. $Sorts \cdot N = |\underset{T\Sigma}{W}(TN(\_) : e)|_n$;
- $V\Sigma$: $\mathbf{Set}_{T\Sigma} \rightarrow$ **AlgSign** is a functor giving the *(value) languages*, s.t. $Sorts \cdot V\Sigma = |\underset{T\Sigma}{W}(\_ : e)|_{n \uplus e}$ (*well-typedness*)
- $\mathcal{N}$ is a sorted set giving the *universe of values*, s.t. $Sorts(\mathcal{N}) = |\underset{T\Sigma}{W}(\mathcal{TN} : e)|_n$;

---

[2] If $\langle F, G, \eta, \epsilon \rangle : \mathbf{X} \rightarrow \mathbf{A}$ is an adjunction, then F may be generalized to a functor $F^+ : \mathbf{X}_{GF} \rightarrow \mathbf{A}$ by $F^+(f: X \rightarrow Y) = \epsilon_{F(Y)} \cdot F(f)$.

- G^V is a model of $V\Sigma(\mathcal{TN})$ term-generated by the valuation $\nu^N\colon \mathcal{N} \to |G^V|$ s.t. for any $s \in |W_{T\Sigma}(\mathcal{TN} : e)|_{n\uplus e}$, $|G^V|_s = eval_{T\Sigma}^{G^T,\nu^T}(s)$.

Since the sort component of the indexed set morphism given by the application of N or $V\Sigma$ to arrows is fixed, in the following we will omit it, provided that no ambiguity arises.

Note that the value languages are indexed on sets instead of signatures. This, together with the well-typedness condition, guarantees a strong uniformity. In particular, if $|\mathbf{Set}|$ is the power-set of some universe $\mathcal{U}$, intuitively representing all possible *type names*, then it is possible to prove that $V\Sigma(\mathcal{U})$ and the family of the $V\Sigma(f)$ for f an endomorphism of $\mathcal{U}$ describe $V\Sigma$ up to isomorphism, so that $V\Sigma$ can be equivalently presented by *one* signature and a bunch of morphisms.

Let us illustrate the definition on the example of a simple algebraic language. Since standard many-sorted algebras have a very poor language, with only function application, we have enriched them by a few simple constructs, in order to better show the features of tu parchments. The languages defined by $AT\Sigma$ and $AV\Sigma$ do not take into account static constraints (i.e. also non well-formed expressions are obtained by the λ-abstraction construct applied to terms on a larger set of variables, which are semantically evaluated to a special $\perp$ value). Static constraints could be easily enforced adding axioms, that is using a category of specifications instead of $\mathbf{AlgSign}^*$. Here we take this approach for sake of simplicity.

Example 2. Let us describe the components of the parchment $\mathcal{ALG}$. Let us fix a universe $\mathcal{X}$ of variables to be used in value expressions.

Signatures ASign = AlgSign,
Type Level

- $ATN =$ *Sorts*,
- $AT\Sigma$ is described in Example 1;
- $\mathcal{ATN}$ is some universe of sets; let $\mathcal{V}$ denote the universe of all the elements of sets in $\mathcal{ATN}$ and $Val = \{V \mid V\colon Dom(V) \to \mathcal{V}\}$ with $Dom(V) \subseteq \mathcal{X}$ denote the set of *environments*;
- $A\nu^T(A) = (Val \to A^\perp)$, where $A^\perp = A \cup \{\perp\}$, for each $A \in \mathcal{ATN}$
  $op^{AG^T}((Val \to A_1^\perp)\ldots(Val \to A_n^\perp) \to (Val \to A^\perp))$
  $= (A_1 \times \ldots \times A_n \to A^\perp)$

Value Level

- the functor AN on a signature $\langle S, O\rangle$ yields the indexed set having as component of index $op(s_1 \ldots s_n \to s)$ the set $O_{s_1\ldots s_n,s}$ and is analogously defined on morphisms;
- For each $S \in |\mathbf{Set}|$,
  $AV\Sigma(S) =$
  sorts $|W_{AT\Sigma}(S : e)|_{n\uplus e}$
  opns $\{\_(\_,\ldots,\_)\colon op(s_1\ldots s_k \to s) \times s_1 \times \ldots \times s_k \to s \mid k \geq 0, s, s_1, \ldots, s_k \in S\}$
  $\cup\{x\colon \to s \mid x \in \mathcal{X}, s \in S\}$
  $\cup\{\lambda x_1 \ldots x_k.\_\colon s \to op(s_1 \ldots s_k \to s) \mid k \geq 0, s, s_1, \ldots, s_k \in S\}$

The translation along some $f: S \to |W_{AT\Sigma}(S' : e)|_e$ is $|W_{AT\Sigma}(f : e)|_{n \uplus e}$ on sorts and the family of identities on operations.

- $\mathcal{AN}_{op(A_1 \ldots A_n \to A)} = \{f: A_1 \times \ldots \times A_n \to A^{\perp}\}$, for each $A, A_1, \ldots, A_n \in \mathcal{ATN}$
- $A\nu^N(f) = f$, for each $f \in \mathcal{AN}$
  for each $\_(\_, \ldots, \_): op(A_1 \ldots A_n \to A) \times A_1 \times \ldots \times A_n \to A$,

$$f(t_1, \ldots, t_n)^{AG^V} = \lambda V. f^{\perp}(t_1(V), \ldots, t_n(V))$$

  where $f: A_1^{\perp} \times \ldots \times A_n^{\perp} \to A^{\perp}$ is defined by $f^{\perp}(a_1, \ldots, a_n) = f(a_1, \ldots, a_n)$ if $a_i \in A_i$ for $i = 1 \ldots n$, $\perp$ otherwise
  $x^{AG^V} = \lambda V. V^{\perp}(x)$, for $V^{\perp}(x) = x$ if $x \in Dom(V)$, $\perp$ otherwise, $\forall x \in \mathcal{X}$
  for each $\lambda x_1 \ldots x_k.\_: A \to op(A_1 \ldots A_k \to A)$,

$$(\lambda x_1 \ldots x_k.t)^{AG^V} = \lambda v_1 \ldots v_k.t(V)$$

  with $Dom(V) = \{x_1, \ldots, x_k\}$ and $V(x_i) = v_i$, $i = 1 \ldots k$.

Each tu parchment presents a typed institution, where the types of each signature Σ are uniformly built as terms over the signature $T\Sigma$, using the type names S of Σ as variables. Analogously, the expressions over Σ are uniformly built as terms over the signature $V\Sigma(S)$, using the names of Σ as variables.

Proposition 5. *Each tu parchment*

$$\mathcal{P} = (\mathbf{Sign}, TN, T\Sigma, \mathcal{TN}, G^T, \nu^T, N, V\Sigma, \mathcal{N}, G^V, \nu^N)$$

defines a typed institution ($\mathbf{Sign}$, *Exp*, *Mod*, $[\![\_]\!]$, $|W_{T\Sigma}(TN(\_) : e)|_e$, τ, $[\_]$) *as follows.*

- *Exp*: $\mathbf{Sign} \to \mathbf{Set}$ *is defined by:*
  On objects: *for each* $\Sigma \in |\mathbf{Sign}|$ *with* $TN(\Sigma) = S$, $N(\Sigma) = N$,

$$Exp(\Sigma) = \{\langle ex, s\rangle \mid ex \in |W_{V\Sigma(S)}(N)|_s \text{ and } s \in |W_{T\Sigma}(S : e)|_e\}$$

  On arrows: *for each arrow* $\sigma: \Sigma \to \Sigma'$ *in* $\mathbf{Sign}$, *with* $TN(\sigma) = f: S \to S'$, $N(\sigma) = h: N \to N'_{|V\Sigma(f)}$,

$$Exp(\sigma)(ex, s) = \langle W_{V\Sigma(f_{T\Sigma})}(\eta^{V\Sigma(S')}_{N'}{}_{|V\Sigma(f_{T\Sigma})} \cdot h)(ex), |W_{T\Sigma}(f : e)|_e(s)\rangle,$$

- *Mod*: $\mathbf{Sign}^{op} \to \mathbf{Cat}$ *is defined by*
  On objects: *for each* $\Sigma \in |\mathbf{Sign}|$ *with* $TN(\Sigma) = S$, $N(\Sigma) = N$, $Mod(\Sigma)$ *is the discrete category whose objects are*

$$\{\langle m^T, m^N\rangle \mid m^T: S \to |W_{T\Sigma}(\mathcal{TN} : e)|_e, m^N: N \to |W_{V\Sigma(\mathcal{TN})}(\mathcal{N})_{|V\Sigma(m^T)}|\}$$

  On arrows: *for each arrow* $\sigma: \Sigma \to \Sigma'$ *in* $\mathbf{Sign}$, *with* $TN(\sigma) = f: S \to S'$, $N(\sigma) = h: N \to N'_{|V\Sigma(f)}$, *and each* $\langle m'^T, m'^N\rangle \in |Mod(\Sigma')|$

$$Mod(\sigma)(\langle m'^T, m'^N\rangle) = \langle m'^T \cdot f, m'^N_{|V\Sigma(f)} \cdot h\rangle.$$

- $[\![\_]\!]\colon |Mod| \times Exp \stackrel{\cdot\cdot}{\longrightarrow} \mathcal{V}$, *with* $\mathcal{V} = \bigcup_{s \in |W_{T\Sigma}(\mathcal{TN}:e)|_e} |G^V|_s$, *is the extranatural transformation whose components are defined as follows:*
  for each $\Sigma \in |\mathbf{Sign}|$, *with* $TN(\Sigma) = S$, $N(\Sigma) = N$, *each model* $\langle m^T, m^N \rangle$ *over* Σ *and each* $s \in |W_{T\Sigma}(S : e)|_e$

$$[\![\langle \_, s \rangle]\!]_\Sigma^{\langle m^T, m^N \rangle} = eval^{G^V, \nu^N}_{V\Sigma(\mathcal{TN}) \mid |W_{T\Sigma}(m^T:e)|_e(s)} \cdot |W_{V\Sigma(m^T)}(m^N)|_s.$$

- $\tau\colon Exp \stackrel{\cdot}{\longrightarrow} |W_{T\Sigma}(TN(\_) : e)|_e$ *is the natural transformation whose components are the projections on the second component;*
- $[\_]\colon |Mod| \times |W_{T\Sigma}(TN(\_) : e)|_e \stackrel{\cdot\cdot}{\longrightarrow} \mathcal{T}$, *with* $\mathcal{T} = |G^T|_e$ *is the extranatural transformation defined, for each signature* Σ *and* Σ*-model* $\langle m^T, m^N \rangle$, *by:*

$$[\_]_\Sigma^{\langle m^T, m^N \rangle} = eval^{G^T, \nu^{T^{T\Sigma}} \circ m^T}_{T\Sigma}.$$

In the following we will use *models of a tu parchment* over a signature Σ for the models over Σ in the typed institution presented by the tu parchment.

Example 3. Let us sketch another example of tu parchment, $\mathcal{FUN}$, presenting a toy functional programming language **FUN**, whose modules have the form

$$\begin{array}{l} tid_1 = te_1, \ldots, tid_n = te_n \\ f_1(x_1^1 : te_1^1; \ldots; x_{n_1}^1 : te_{n_1}^1) : te_1' = exp_1 \\ \ldots \\ f_m(x_1^m : te_1^m; \ldots; x_{n_m}^m : te_{n_m}^m) : te_m' = exp_m \end{array}$$

where *tid* (possibly decorated, as for other metavariables) ranges over type identifiers, f over function identifiers, *te* over type expressions defined by:

$$te ::= \quad tid \mid \mathtt{int} \mid \langle l_1 : te_1, \ldots, l_n : te_n \rangle \mid \mathtt{array}\ [0..N]\ \mathtt{of}\ te,$$

(with N any positive integer constant and l_i field identifiers), x over variables, *exp* over expressions defined by the grammar

$$\begin{array}{rl} exp ::= & x \mid \mathtt{zero} \mid \mathtt{succ}(exp) \mid \mathtt{pred}(exp) \mid \langle l_1 : exp_1, \ldots, l_n : exp_n \rangle \mid [\,] \mid \\ & exp_1[exp_2/exp_3] \mid exp.l \mid exp_1[exp_2] \mid f(exp_1, \ldots, exp_n) \mid \\ & \mathtt{if}\ exp_1 = exp_2\ \mathtt{then}\ exp_3\ \mathtt{else}\ exp_4. \end{array}$$

Then, signatures in **FSign** are interfaces of **FUN** modules, i.e. pairs of the form $\langle t_1 \ldots t_n, f_1(te_1^1; \ldots; te_{n_1}^1) : te_1', \ldots, f_m(te_1^m; \ldots; te_{n_m}^m) : te_m' \rangle$; *FTN* and *FN* gives the sets of elements in the first and second component, respectively.

The type language $FT\Sigma$ has sorts n, e and one operation of sort e for each constructor of type expressions but for the first production, corresponding to the type names provided by the module. Thus,

$$\begin{array}{l} \mathtt{int}\colon \to e \\ \langle l_1 : \_, \ldots, l_k : \_ \rangle\colon e^k \to e \\ \{\mathtt{array}\ [0..N]\ \mathtt{of}\ \_\colon e \to e \mid N \geq 0\} \end{array}$$

moreover, there is a family of operations of sort n (one for each $k \geq 0$)

$$\mathtt{fun}(\_, \ldots, \_, \_)\colon e^k \times e \to n.$$

For each $S \in |\mathbf{Set}|$, the signature $FV\Sigma(S)$ has sorts $|W_{FT\Sigma}(S : e)|_{n \uplus e}$ and a family of overloaded operations (one for each possible choice of types) for each constructor of expressions, for instance, denoting $|W_{FT\Sigma}(S : e)|_e$ by TE:

$\{x\colon \to te \mid te \in TE\}$
$\{\mathtt{zero}\colon \to \mathtt{int}\}$
$\{\langle l_1 : \_, \ldots, l_n : \_\rangle\colon te_1 \times \ldots \times te_n \to \langle l_1 : te_1, \ldots, l_n : te_n\rangle \mid te_1, \ldots, te_n \in TE\}$
$\{\_(\_, \ldots, \_)\colon \mathtt{fun}(te_1 \ldots te_n, te) \times te_1 \times \ldots \times te_n \to te \mid te_1, \ldots, te_n \in TE\}$

and a family of overloaded operations corresponding to function definitions

$$\{\mathtt{fun}(x_1 : te_1; \ldots; x_n : te_n) : te = \_ \colon te \to \mathtt{fun}(te_1 \ldots te_n, te) \mid te, te_1, \ldots, te_n \in TE\}$$

The universes $\mathcal{FTN}$ and $\mathcal{FN}$ are empty, i.e. the universal languages of the two levels consist of the ground terms over $FT\Sigma$ and $FV\Sigma(\emptyset)$ and, accordingly, the valuations $F\nu^T$ and $F\nu^N$ are the empty maps. Indeed, in this case a model is a programming module; hence it associates with the (type or value) names specified in the interface some (type or value) expressions. The algebras FG^T and FG^V correspond to the (straightforward) denotational semantics of the language; we omit them for sake of brevity.

2 Implementation and Derivation in TU Parchments

In this section, we express in the framework of tu parchments two relations between specification or programming modules playing an important role in the practice: *derivation* (a module $\overline{m}$ is defined using the primitives provided by another module m in the same formalism) and *implementation* (a module m is replaced by a "less abstract" module m', possibly changing the formalism). Moreover, we prove that implementation can be propagated "for free" to derived models, i.e. if m' implements m, and $\overline{m}$ is derived from m, then it is possible to construct an implementation $\overline{m}'$ for $\overline{m}$, provided that we are able to translate the linguistic constructs of the two formalisms.

Notation 4: Abbreviations. Let us fix for the rest of the section tu parchments

$$\mathcal{P} = (\mathbf{Sign}, TN, T\Sigma, \mathcal{TN}, G^T, \nu^T, N, V\Sigma, \mathcal{N}, G^V, \nu^N)$$
$$\mathcal{P}' = (\mathbf{Sign'}, TN', T\Sigma', \mathcal{TN}', G'^T, V'^T, N', V\Sigma', \mathcal{N}', G'^V, V'^N);$$

signatures Σ and $\overline{\Sigma}$ in $\mathcal{P}$ and Σ' in $\mathcal{P}'$, with $TN(\Sigma) = S$, $TN(\overline{\Sigma}) = \overline{S}$, $TN'(\Sigma') = S'$, $N(\Sigma) = N$, $N(\overline{\Sigma}) = \overline{N}$ and $N'(\Sigma') = N'$. Moreover, let $m = \langle m^T, m^N\rangle$, $\overline{m} = \langle \overline{m}^T, \overline{m}^N\rangle$, and $m' = \langle m'^T, m'^N\rangle$ be models over Σ, $\overline{\Sigma}$ and Σ', respectively. Finally, the composition of f and g in $\mathbf{Set}_{T\Sigma}$ ($\mathbf{Set}_{T\Sigma'}$) will be denoted by $f \circ g$ ($f \star g$) and the unit $\eta_e^{T\Sigma}$ and counit $\epsilon_e^{T\Sigma}$ ($\eta_e^{T\Sigma'}$, $\epsilon_e^{T\Sigma'}$) by η and ϵ (η', ϵ').

2.1 Derived Models

The derivation relation concerns modules within the same formalism, say m and $\overline{m}$, and holds whenever all the (type and value) names of $\overline{m}$ are defined, via a pair of maps $\chi = \langle \chi^T, \chi^N\rangle$, by expressions built over the (type and value) names of m. In this case, it is clear that the semantics of (the model corresponding to) $\overline{m}$ can be obtained by evaluating such expressions through m.

Our notion of derivation roughly corresponds to what is called implementation by *constructors* in the literature (see e.g. [7]), characterized by the fact that sorts and operations in a specification $\overline{SP}$ are defined in terms of sorts and operations of another specification SP by means of an enrichment $SP+$ of SP together with a signature morphism from $\overline{SP}$ to $SP+$. Anyway, in our approach it is not necessary to have a signature morphism, and the two steps are incorporated at a more concrete level by the mapping χ from names to expressions.

This situation happens in programming languages whenever the module $\overline{m}$ *uses* m; anyway it makes sense also in common institutions. For instance, considering many-sorted algebras, given mappings $\chi^T: \overline{S} \to S$ and $\chi^N: \overline{O} \to O$ which associate with each operation in $\overline{O}$ an open term over Σ, we can derive from an algebra A over $\Sigma = \langle S, O \rangle$ a new algebra $\overline{A}$ over $\overline{\Sigma} = \langle \overline{S}, \overline{O} \rangle$ in such a way that sort renaming χ^T is preserved. Note that in this case (as in most algebraic examples), since the type language is trivial, derivation at the level of types is just sort renaming; on the contrary, in programming languages, type names in $\overline{m}$ can be mapped in type expressions constructed over type names of m.

Definition 6. The model $\overline{m}$ is *derived* from m via

$$\chi = \begin{cases} \chi^T: \overline{S} \to |\underset{T\Sigma}{W}(S:e)|_e \\ \chi^N: \overline{N} \to |\underset{V\Sigma(S)}{W}(N)_{|V\Sigma(\chi^T)|} \end{cases} \quad \text{iff} \quad \begin{array}{l} \overline{m}^T = m^T \circ \chi^T \\ \overline{m}^N = |\underset{V\Sigma(m^T)}{W}(m^N)_{|V\Sigma(\chi^T)|} \cdot \chi^N \end{array}$$

Example 4. Let us consider the parchment $\mathcal{ALG}$ introduced in Example 2 and see a simple example of a derivation. We first define a model for the standard specification of the *stack* data-type, m_S.

Stacks of natural numbers (of maximal length K) are represented by a model $m_S = \langle m_S^T, m_S^N \rangle$ over Σ_S (having sorts *stack* and *elem* and the usual operations *empty*, *push*, *pop*, *top*); m_S^T associates $\mathbb{N} \cup \{errelem\}$ with *elem*, $\mathbb{N}^K \cup \{errstack\}$ with *stack*; m_S^N associates with operations their standard interpretations (*errstack* is used as result of *pop* on an empty stack and *push* on a full stack and *errelem* as result of *top* on an empty stack).

Then, we can derive from it a model for the richer signature $\overline{\Sigma_S}$, that is the enrichment of Σ_S by *swap*: *stack* $\to$ *stack*, swapping the topmost two elements, if any, via the renaming χ that is the identity on all symbols of Σ_S but for *swap* on which it yields $\lambda s.push(top(pop(s)), push(top(s), pop(pop(s))))$.

2.2 Implementation

Having a uniform way of building types and typed expressions allows to distinguish within an implementation step two parts: a global part, translating the type and value languages of the source framework into the corresponding components of the target, and a local part, dealing with the details of the particular models and names. This is very important, because in that way the global part can be reused. In order to be able to define the global part, we need a functor relating the corresponding substitution categories.

Lemma 7. *Each arrow* $\phi^T : T\Sigma \to T\Sigma'$ *in* **AlgSign*** *induces a functor* $F_{\phi^T} : \mathbf{Set}_{T\Sigma} \to \mathbf{Set}_{T\Sigma'}$*, defined by:*

- $F_{\phi^T}(X) = X$ *for all* $X \in |\mathbf{Set}_{T\Sigma}| = |\mathbf{Set}| = |\mathbf{Set}_{T\Sigma'}|$;
- $F_{\phi^T}(f) = G^{T\Sigma}(\epsilon_{F^{T\Sigma'}(Y)_{|\phi^T}} \cdot F^{T\Sigma}(\eta'_Y)) \cdot f$ *for all* $f \in \mathbf{Set}_{T\Sigma}(X,Y)$.

Moreover, if $f = \eta_e^{T\Sigma}{}_Y \cdot f'$ *for some* $f: X \to Y$ *in* **Set**, *then* $F_{\phi^T}(f) = \eta_e^{T\Sigma'}{}_Y \cdot f'$.

Definition 8. A *translator* $\phi = \langle \phi^T, \phi^N \rangle$ of $\mathcal{P}$ into $\mathcal{P}'$ consists of an arrow $\phi^T : T\Sigma \to T\Sigma'$ in **AlgSign***, and a natural transformation $\phi^N : V\Sigma \dot{\longrightarrow} V\Sigma' \cdot F_{\phi^T}$ s.t. $Sorts(\phi^N_S) = |W_{\phi^T}((\eta_e^{T\Sigma'})_S : e)|_{n \uplus e}$.

For each translator $\phi = \langle \phi^T, \phi^N \rangle$ and each arrow $f: S \to S'$ let us denote by ϕ_f the morphism $V\Sigma'(f_{T\Sigma'}) \cdot \phi^N_S$ translating the *higher level* signature $V\Sigma(S)$ into the *lower level* signature $V\Sigma'(S')$.

Example 5. We define a translator $\tilde{\phi}$ of $\mathcal{ALG}$ into $\mathcal{FUN}$ as

$$\langle \tilde{\phi}^T : AT\Sigma \to FT\Sigma, \tilde{\phi}^N : AV\Sigma \dot{\longrightarrow} FV\Sigma \cdot F_{\tilde{\phi}^T} \rangle,$$

where $\tilde{\phi}^T$ gives the translation from the high-level to the low-level type language, and is the identity on sorts, whose translation is fixed, and maps $op(\_ \ldots \_ \to \_): e^k \times e \to n$ to $\mathbf{fun}(\_ \ldots, \_): e^k \times e \to n$

Moreover, for each set S, $\tilde{\phi}^N_S$ gives the translation from the high-level to the low-level language of (value) expressions with types constructed starting from the type names S, and is analogously defined.

Definition 9. The model m' is an *implementation* of m w.r.t. the translator $\phi = \langle \phi^T, \phi^N \rangle$ of $\mathcal{P}$ into $\mathcal{P}'$, via ψ and *abs*, with

- $\psi = \begin{cases} \psi^T : S \to S' \\ \psi^N : N \to N'_{|\phi_\psi} \text{(where } \phi_\psi \text{ is a short notation for } \phi_{\psi^T}) \end{cases}$
- $abs = \{abs_s\}_{s \in |W_{T\Sigma}(S:e)|_e}$ where each $abs_s : D_s \to |W_{V\Sigma(T\mathcal{N})}(\mathcal{N})_{|V\Sigma(m^T)}|_s$, for some $D_s \subseteq |W_{V\Sigma'(T\mathcal{N}')}(\mathcal{N}')_{|V\Sigma'(m'^T) \cdot \phi_\psi}|_s$ with embedding ι_s, satisfying the following *semantic coherency* property:

  for all $t_1, t_2 \in |W_{V\Sigma'(T\mathcal{N}')}(\mathcal{N}')_{|V\Sigma'(m'^T) \cdot \phi_\psi}|_s$ if $eval^{G'^V, V'^N}_{V\Sigma'(T\mathcal{N}')}(t_1) = eval^{G'^V, V'^N}_{V\Sigma'(T\mathcal{N}')}(t_2)$, then both ($t_1 \in D_s$ iff $t_2 \in D_s$) and if $t_1, t_2 \in D_s$, then $abs_s(t_1) = abs_s(t_2)$;

iff the diagram in Fig. 1 commutes, where $\biguplus$ is the coproduct in $SSet(|W_{T\Sigma}(S : e)|_e)$.

Example 6. We show how to express a standard implementation example, i.e. stacks (formalized by the model m_S in Example 4) by pairs array and length (formalized by a model m_R in **FUN**). Note that in our framework the high-level (implemented) and the low-level (implementing) data-types are allowed to stay "in two different worlds"; indeed, the first is expressed by a many-sorted algebra and the second by a functional program.

The model m_R corresponds to the following module

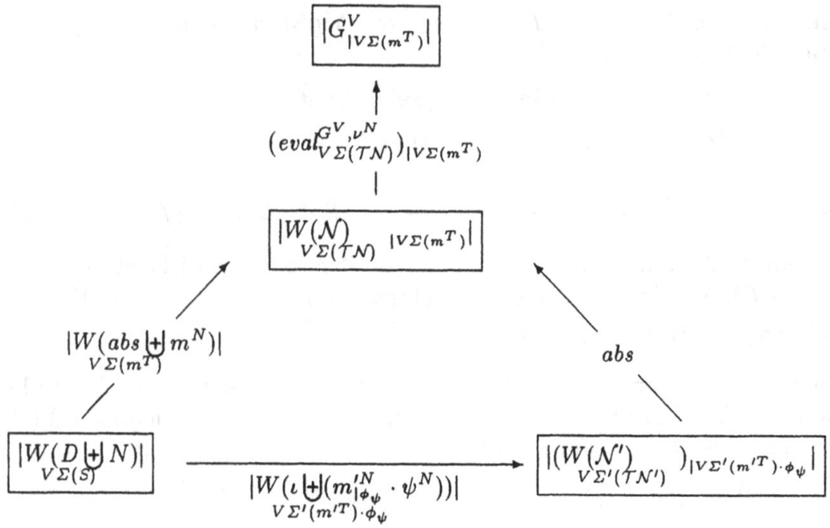

Fig. 1. Implementation Diagram

```
myStack = ⟨elems : array [0..K] of int, length : int⟩. myElem = int
myEmpty : myStack = ⟨elems : [ ], length : zero⟩
myPush(s : myStack, n : myElem) : myStack = if s.length = pred(zero) then s else
    if s.length = K then ⟨elems : [ ], length : pred(zero)⟩ else
        ⟨elems : s.elems[n/s.length], length : succ(s.length)⟩
myPop(s : myStack) : myStack = if s.length = pred(zero) then s else
    if s.length = zero then ⟨elems : [ ], length : pred(zero)⟩ else
        ⟨elems : s.elems, length : pred(s.length)⟩
myTop(s : myStack) : myElem = if s.length = pred(zero) then pred(zero) else
    if s.length = zero then pred(zero) else
        s.elems[pred(s.length)]
```

We define now $\tilde{\psi}$ and $\widetilde{abs}$ s.t. m_R is an implementation of m_S (from Example 4) w.r.t. the translator $\tilde{\phi}$ (from Example 5) via $\tilde{\psi}$ and $\widetilde{abs}$.

The component $\tilde{\psi} = \langle \tilde{\psi}^T, \tilde{\psi}^N \rangle$ gives the correspondence from names in the high-level model to names in the low-level model. In this case $\tilde{\psi}^T$ maps the sorts *stack*, *elem* into the type identifiers **myStack**, **myElem**, respectively, and the operations in Σ_S into the corresponding function identifiers in the module.

The component $\widetilde{abs}$ is the so-called *abstraction map*. It is a family of functions indexed over all the expression types which one can build starting from the type names *stack*, *elem*. In this case $|\underset{T\Sigma}{W}(\{stack, elem\} : e)|_e = \{stack, elem\}$; hence $\widetilde{abs}$ has only the following two components (denoting by s' the type $\langle \texttt{elems} : \texttt{array}\ [0..K]\ \texttt{of int}, \texttt{length} : \texttt{int} \rangle$):

$$\widetilde{abs}_{elem} \colon \widetilde{D}_{elem} \subseteq |W_{FV\Sigma(\emptyset)}|_{\texttt{int}} \to |\underset{AV\Sigma(\mathcal{A}T\mathcal{N})}{W(\mathcal{AN})}|_{\mathbf{N}}$$

$$\widetilde{abs}_{stack} \colon \widetilde{D}_{stack} \subseteq |W_{FV\Sigma(\emptyset)}|_{s'} \to |\underset{AV\Sigma(\mathcal{A}T\mathcal{N})}{W(\mathcal{AN})}|_{(\mathbf{N}^k \cup \{errstack\})}$$

which are defined as follows

- for each $exp \in |W_{FV\Sigma(\emptyset)}|_{\text{int}}$, $exp \in \tilde{D}_{elem}$ iff $eval^{FG^V}_{FV\Sigma(\emptyset)}(exp) = \lambda\rho.z$ for some $z \in \mathbb{N}$; in this case, $\widetilde{abs}_{elem}(exp) = z$
- for each $exp \in |W_{FV\Sigma(\emptyset)}|_{s'}$, $exp \in \tilde{D}_{stack}$ iff $eval^{FG^V}_{FV\Sigma(\emptyset)}(exp) = \lambda\rho.\langle \texttt{elems} : a, \texttt{length} : l\rangle$ and $l \in \{-1 \ldots K\}$; in this case, $\widetilde{abs}_{stack}(exp) = errstack$ if $l = -1$, $a(l-1) \cdot \ldots \cdot a(0)$ otherwise.

Note that the semantic coherency property is guaranteed by the definition, that is based on the low-level semantics.

The intuitive interpretation of the implementation diagram in this particular case is the following. Take (left-bottom corner) a term in the high-level language constructed using as variables the names in the high-level model m_S and the elements of $\tilde{D}$ (i.e. the terms in the low-level universal language which are used for the implementation). An example of such a term is $w = push(\langle \texttt{elems} : [\,], \texttt{length} : \texttt{zero}\rangle, \texttt{zero})$. Then, we obtain terms in the high-level universal language having the same semantics in the following two ways:

- first mapping the names in m_S to the corresponding names in m_R via $\tilde{\psi}$; for instance on w we get $\texttt{myPush}(\langle \texttt{elems} : [\,], \texttt{length} : \texttt{zero}\rangle, \texttt{zero})$. Then, mapping these names to their associated representation in the low-level universal language, corresponding to an expansion of each function call to its body, getting, for instance, on our example

  ```
  myPush(s : myStack, n : myElem) : myStack =
      if s.length = pred(zero) then s else
          if s.length = K then ⟨elems : [ ], length : pred(zero)⟩ else
              ⟨elems : s.elems[n/s.length], length : succ(s.length)⟩
                  (⟨elems : [ ], length : zero⟩, zero)
  ```

  and finally abstracting the result;
- using the abstraction map to translate the low-level terms and m_S to interpret high-level names into the high-level universal language.

2.3 Deriving Implementations

We prove now that, given an implementation m' of m w.r.t. ϕ via ψ and abs, it is possible to construct an implementation $\overline{m}'$ for any model $\overline{m}$ derived from m. Referring to our working example, that means that, having implemented stacks (m_S) by records (m_R) w.r.t. $\tilde{\phi}$ via $\tilde{\psi}$ and $\widetilde{abs}$, we get "for free" an implementation also for any derived operation we can express in terms of the stack primitives, e.g. *swap* mentioned in Example 4.

This is possible under the assumption that, given a signature $\overline{\Sigma}$ in the high-level formalism (e.g. in $\mathcal{ALG}$), a signature $\overline{\Sigma}'$ in the low-level formalism (e.g. in $\mathcal{FUN}$) can be found with the same names of $\overline{\Sigma}$, up to isomorphism: in the example, a module interface in $\mathcal{FUN}$ where type and function identifiers are exactly (module some coding) sorts and operations of $\overline{\Sigma}$. Note that if the translation of the linguistic constructs building the types is not injective, then it is possible that several high-level terms $t_1, \ldots, t_k$ of sort n are translated into one low-level term t of sort n; in that case the names provided by $\overline{\Sigma}'$ of type t have to represent the names provided by $\overline{\Sigma}$ of each type t_i. Thus, the names provided by $\overline{\Sigma}'$ of type t are the disjoint union of the names provided by $\overline{\Sigma}$ of all the t_i's.

Theorem 10. *Let us assume that*

- $\overline{m}$ *is derived from* m *via* $\chi = \langle \chi^T, \chi^N \rangle$
- m' *is an implementation of* m *w.r.t. the translator* $\phi = \langle \phi^T, \phi^N \rangle$ *of* $\mathcal{P}$ *into* $\mathcal{P}'$, *via* $\psi = \langle \psi^T, \psi^N \rangle$ *and* abs.

If there exists a signature $\overline{\Sigma}'$ *in* $\mathcal{P}'$, *with* $TN'(\overline{\Sigma}') = \overline{S}'$ *and* $N'(\overline{\Sigma}') = \overline{N}'$ *s.t.*

1. *there exists an isomorphism* $\gamma: \overline{S} \to \overline{S}'$ *in* **Set**
2. *there exists an isomorphism* $\delta: \uplus\overline{N} \to \overline{N}'$, *where, for each sort* $s' \in |W_{T\Sigma'}(\overline{S}' : \epsilon)|_n$, *the set* $(\uplus\overline{N})_{s'}$ *is the coproduct of* $\{\overline{N}_s \mid \phi_\gamma(s) = s'\}$; *in the following we will denote by* $\uplus f$ *the family of functions with each* s' *component defined as the* pairing *of (that is the unique arrow whose compositions with the injections yield)* $\{f_s \mid \phi_\gamma(s) = s'\}$ *for each family of functions* $f_s: \overline{N}_s \to X_{s'}$

then the following properties hold

1. $\overline{m}' = \langle \overline{m}'^T, \overline{m}'^N \rangle$ *is a model of* $\overline{\Sigma}'$ *in* $\mathcal{P}'$ *where* $\overline{m}'^T = m'^T \star \chi'^T$, $\overline{m}'^N = |W_{V\Sigma'(m'^T)}(m'^N)_{|V\Sigma'(\chi'^T)}| \cdot \chi'^N$ *and*
$$\chi' = \begin{cases} \chi'^T = \psi^T_{T\Sigma'} \star F_{\phi^T}(\chi^T) \star \gamma^{-1}_{T\Sigma'} \\ \chi'^N = \uplus(|W_{\phi_\psi}(\psi^N)_{|V\Sigma(\chi^T)}| \cdot \chi^N) \cdot \delta^{-1} \end{cases}$$
2. $\overline{m}'$ *is derived from* m' *via* χ'
3. $\overline{m}'$ *is an implementation of* $\overline{m}$ *w.r.t.* ϕ^T *and* ϕ^N, *via* $\overline{\psi}$ *and* $\overline{abs} = abs_{|V\Sigma(\chi^T)}$, *with definition domain* $\overline{D} = D_{|V\Sigma(\chi^T)}$ *and embedding* $\overline{\iota} = \iota_{|V\Sigma(\chi^T)}$, *where*
$$\overline{\psi} = \begin{cases} \overline{\psi}^T = \gamma \\ \overline{\psi}^N = \delta_{|\phi_\gamma} \cdot inj \end{cases}$$
and inj *is an indexed set morphism, whose sort component is* $Sorts(\phi_\gamma)$ *and each* $inj_s: \overline{N}_s \to (\uplus\overline{N})_{\phi_\gamma(s)}$ *is the injection (in the coproduct).*

Conclusions and Further Work We have proposed a new metaframework for representing and relating formalisms, in the spirit of the theory of institutions. The overall aim is to provide a framework general enough for including common specification formalisms, but at a more concrete level, in such a way that also non-axiomatic languages are covered. In particular, within our framework it is possible to express two notions closer to programming languages, i.e. *derivation* and *implementation* and to show that implementation can be canonically extended to derived modules.

Moreover, the notion of implementation presented here is a generalization of the concrete data-type implementation introduced by Hoare [4] and allows to relate individual values. Having such relationship, it should be possible to formalize *external calls*, corresponding to models where the implementation of some name is not given, but it is imported from another model in a different language, and *program annotations*, corresponding to using the high-level logic to state properties on the elements of models in the low-level framework, whose semantics is given through the abstraction map.

Other promising applications of tu parchments are the development of fragments of typed equational first-order logic, based on the notion of typed expression evaluation providing an obvious semantics for equality, and the combinations of typed institutions through their syntactic representations, much in the spirit of [6].

Finally we plan to generalize the present approach in a forthcoming extended version allowing languages to be represented by axiomatic specifications instead of plain signatures, in order to capture static semantics constraints. Since the results presented here are based only on the existence of free objects, any choice of specifications preserving such property will carry on the results.

A somehow extended version of this paper, including proofs, is reachable from the web pages of the authors[3].

Acknowledgments. We warmly thank the anonymous referee for his/her careful reading and helpful suggestions.

References

1. M. Bidoit and A. Tarlecki. Behavioural satisfaction and equivalence in concrete model categories. In H. Kirchner, editor, *CAAP '96 - 20th Coll. on Trees in Algebra and Computing*, number 1059 in LNCS, pages 241–256, Berlin, 1996. Springer Verlag.
2. J. A. Goguen and R. M. Burstall. A study in the foundations of programming methodology: Specifications, institutions, charters and parchments. In D. Pitt et al., editor, *Category Theory and Computer Programming*, number 240 in LNCS, pages 313–333, Berlin, 1985. Springer Verlag.
3. J.A. Goguen and R.M. Burstall. Institutions: Abstract model theory for computer science. *Journ. ACM*, 39:95–146, 1992.
4. C.A.R. Hoare. Proofs of correctness of data representations. *Acta Informatica*, 1:271–281, 1972.
5. T. Mossakowski. Using limits of parchments to systematically construct institutions of partial algebras. In M. Haveraaen, O. Owe, and O.-J. Dahl, editors, *11th WADT*, number 1130 in LNCS, pages 379–393, Berlin, 1996. Springer Verlag.
6. T. Mossakowski, A. Tarlecki, and W. Pawłowski. Combining and representing logical systems. In *Category Theory and Computer Science '97*, number 1290 in LNCS, pages 177–196, Berlin, 1997. Springer Verlag.
7. F. Orejas, M. Navarro, and A. Sánchez. Implementation and behavioural equivalence: A survey. In M. Bidoit and C. Choppy, editors, *8th WADT*, number 655 in LNCS, pages 93–125. Springer Verlag, Berlin, 1993.

[3] `http://www.disi.unige.it/person/{CerioliM,ZuccaE}/`

A Method for Fortran Programs Reverse Engineering Using Algebraic Specifications

Sophie Cherki[1] and Christine Choppy[2]

[1] LRI, C.N.R.S. U.R.A. 410 & Université de Paris-Sud, Bât. 490, F-91405 Orsay Cedex, France
e-mail : Sophie.Cherki@lri.fr

[2] IRIN, Université de Nantes & Ecole Centrale, F-44072 Nantes Cedex 03, France
e-mail : Christine.Choppy@irin.univ-nantes.fr

Abstract

When maintenance is neglected within program development, it is usually done in an empirical way and this leads to program deterioration. To cover up this problem in the legacy systems framework, a broader approach can be followed which first involves reverse engineering. Reverse engineering generates abstract descriptions of a program from its source. These descriptions are used to generate the improved program source (inconsistencies removal, optimization of the source code, updating the documentation), and to assist maintenance tasks.

We present here a reverse engineering method which systematically leads to structural and functional descriptions of a Fortran program using algebraic specifications. We emphasize the problems arisen by the Fortran programming language and by some kind of computing styles (few possibilities for type construction, bad use of global variables, errors within the source code, etc). It is important to notice that this process also leads to the detection of code defaults : it is thus possible to remove them before the code generation following the reverse engineering.

Keywords : reverse engineering, algebraic specification, maintenance, typing, legacy systems.

1 Introduction.

In the software engineering literature, the software life cycle picture sometimes ends with a maintenance phase. Maintenance denotes any improvement or bug removal that takes place after the software is delivered. For many years, maintenance was not considered really as part of the software life cycle, it was not studied, and was not taken into account within the software development. Subsequently, the resulting softwares [12] had a poor structure, little and/or obsolete documentation, and it appeared that maintenance had caused deterioration, in particular due to the introduction of new errors and side effects. Moreover, the studies on maintenance costs [15, 10] showed that new approaches were urgently needed. Re-engineering seems to be a promising approach for software maintenance that would reduce the maintenance costs while insuring a longer activity for legacy systems. One of the practical ways to re-engineering starts with a reverse engineering step (cf. Figure 1).

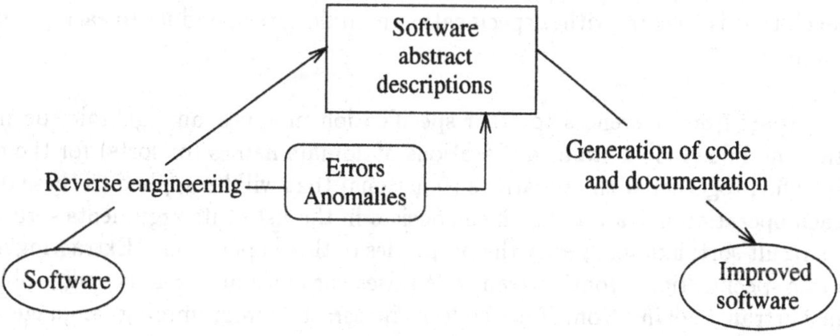

Figure 1. A re-engineering process.

It is well known that the first step to program maintenance is to understand what the program does and how it does it, and that this program comprehension activity is quite time consuming. Reverse engineering is a way to provide an understanding of the system concepts so that it will be possible to modify and restructure it as appropriate [3]. One of the aims of reverse engineering is to analyse a system, and identify its components and their relationships, in order to create more abstract representations for this system. In particular, the source code is used to extract abstract descriptions relative to [8]:

1. the choice of representation for the program data structures
2. the program structure (its components and their relationships)
3. the program functionalities
4. the program application domain.

We present in this paper a process for extracting a formal specification that reflects the program structure and functionalities expressed with algebraic specifications.

It is clear that the work to be done in order to extract the formal specification varies greatly with the programming language used, and also with the style of programming adopted in the software to be processed. We have worked with a software component of an industrial application written in Fortran 77 that was first specified manually [1] in order to study how to mechanize the a posteriori specification process. One of the characteristics of this software is that it is structured into modules (called routines in Fortran) which correspond to a single functionality or a group of functionalities.

One of the advantages of algebraic specifications is that they can be expressed in a modular way: a complex specification is a graph of specification modules assembled together using specification building primitives (e.g. enrichment, parameterization, renaming) [5]. When the software to be processed is well structured, it seems appropriate to reflect this structure at the specification level, that is to start with a specification module corresponding to each program module.

Apart from references to other specification modules, an algebraic specification module [4, 5] contains declarations of domain names (or sorts) for the data specified together with operations (functions) that will be applied to these data; each operation has a profile which consists in the list of its arguments sorts and its result sort; axioms specify the properties of these operations. Extracting algebraic specifications from Fortran code raises various issues, e.g. typing problems in Fortran, moving from some code in imperative programming language to a specification in a functional style, extracting the specification axioms from the code, ... Additional issues are raised by the programming style; for instance, global variables may be used as parameters (or the reverse), and this may blur the structure of the data.

Obviously, some informations are needed when achieving this specification extraction process:

- Since our approach is modular, the specification extraction process is applied to a software module assuming that all the information (e.g. the specification) concerning the components referred to in this module is available.
- Although there may be some ways to extract this as well, we require that the information relative to the data structure representation is available.
- Code schematas are used to represent unit actions (cf. section 3.1) that constitute the program module functionalities.

In the following, we shall describe the specification extraction process that is decomposed into two phases: the signature extraction (the sorts and operations declarations), and the axioms extraction. In the conclusion, we compare our approach with existing reverse engineering approaches. The appendix contains a code example that is used as an illustration throughout the paper.

2 Signature extraction.

The specification signature describes the domains of the data used within the program module and the operations that are performed on these data.

2.1 Sorts extraction.

In a Fortran 77 program module, constants and variables declarations, instructions, and reference to other program units may be found. There are no type declarations, although there may be anonymous array types found in the constants and variables declarations. However, the corresponding specification module should contain a signature (sorts - or domains names -, and operations names), axioms specifying some properties, and possibly references to other specification modules.

The sorts or domain names represent the types of the data used within the program module. In most programming languages (Pascal, Ada, C, object oriented languages, etc.), the domains may be identified through the types declarations and the typed variables declarations (although some ambiguities may occur due to overloading, polymorphism and the use of pointers). Fortran 77 has a rather restricted facility for types since, apart from the predefined types, only anonymous array types can be used. As a result, the Fortran 77 programs may lack information on the type of the data in various ways:

- the predefined types of Fortran 77 (booleans, integers, etc.) are "overloaded", that is they are used for various purposes (for instance, in Example 2.1, an integer is used as an address for a stack - while a type stack cannot be defined);
- pieces of data that would be fields of a record or of a class in other programming languages are merely handled on separate instructions (cf. Example 2.2).

Example 2.1. In the Fortran program module in Figure 2 of the Appendix (taking into account the declarations in the STAGLO file), both variables *I* and *NUMPL* are declared of type integer. However, a closer study of the program functionalities described in the example 2.3 will show that *NUMPL* denotes in fact the address of a stack.

Example 2.2. Let us consider another Fortran program with the following declarations:

integer ISIZE, TABEL(30)

According to these variable declarations, one may assume that *ISIZE* denotes an integer, and *TABEL* an array of integers. However, if both variables are systematically used or defined at the same time, or are both arguments of the same subprograms, there is a possibility that these variables would denote two components of a data (for instance, they might denote the height and the elements of a stack).[1]

In order to be able to retrieve a relevant sort information for the algebraic specification we make use of:

[1] In fact, a global analysis of assignements and parameter passings may help exhibiting similar behaviour for a group of variables. This may be a starting point for identifying the data structure representation used in the code.

- the information on the data structure representation (e.g. stating that stacks are referred to using their integer address, or using both an array and an integer)
- the data flow analysis (in order to propagate the above information through e.g. assignments and subprogram calls).

Using a modular approach, the information available about the underlying modules is propagated through parameter passing.

Example 2.3. In the Fortran program module in Figure 2 of the Appendix, there are the following statements[2]:

call INITPL(SIZPL, NUMPL, ...)
PLTMP = NUMPL

The available information for INITPL is that its parameters are respectively an integer and a stack. Then, we can deduce that SIZPL is an integer, and that NUMPL and PLTMP are stacks.

Note 2.4. When using assignments for deducing the typing information, one should keep in mind that programming languages usually allow implicit coercions (e.g. real-number = integer-number). However, implicit coercions are limited to some predefined basic types, and in that case the variables declarations can be used to find out the correct typing.

In this process of typing each variable or parameter, a distinction should be made with some particular parameters that are used within some dynamic type control (for instance, they are used to provide an address where the control should be directed whenever some arguments do not have the expected type). Within the context of algebraic specifications, it is not relevant to reflect this in the specification.

Example 2.5. In the Fortran program module in Figure 2 of the Appendix, the parameters *ER* and * are used for the type control of the other parameters. Therefore, it is not relevant to try to provide a type for them, or to reflect them at the specification level.

To sum this up, the sorts of the specification module associated to a Fortran module are obtained through the analysis of:

- the variable declarations
- information provided by the data flow analysis
- the information about the data representation that is provided for data that are created in this module (cf. Note 2.9) - and propagated otherwise.

---

[2] In Fortran 77 the assignment is denoted by "=".

2.2 Functionalities extraction.

In order to get the signature of the specification module associated to a given program module, we need to determine the operations and their profiles. We start with the module overall functionality and associate to it a specification operation. This operation profile will be determined using the input and output parameters associated sorts.

The variables that are used within the program module may be global variables, local variables, and parameters. Assuming that their use within the program module takes into account these three categories, the profile of the program module associated operation will be deduced from the parameters associated sorts. Here, a first issue is to determine the input/output status of the parameters. As a matter of fact, some languages like Ada provide exactly what is needed within the subprogram declarations.

Example 2.6. In the following subprogram declaration :
procedure ELPLZ (PLNUM : in out STACK; PLPOS : in NATURAL);
the input parameters *PLNUM* and *PLPOS* respectively denote a stack and an integer, and *PLNUM* is also an output parameter. The specification operation associated to *ELPLZ* takes elements of sort Stack and Natural as arguments, and an element of sort Stack as a result.

With Fortran code, this picture may be blurred in various ways. First, there is no way to mark a distinction between input and output parameters. Then, we already mentioned in the previous section that some parameters are used for type control and should not be reflected at the specification level.

Example 2.7. In the Fortran subroutine declaration in Figure 2 in the Appendix:
subroutine ELPLZ (PLNUM, PLPOS, ER, *)
there is no information about the input/output status of the parameters *PLNUM, PLPOS* (and *ER,* * should not be taken into account - cf. example 2.5 in section 2.1).

Another issue is to determine which are the "true" parameters. There is, in the code examined, a heavy use of global variables. Some of these global variables play in fact the role of parameters. There may also be "false" parameters (for instance, if one forgot to remove them when the code was modified). As a consequence, in order to find out what are the true input and output parameters of the program module, both its formal parameters and the global variables should be examined as regards their definitions and their uses. The use of these variables values outside of the module is also analysed so as to determine whether they should be classed under output parameters:

- When a variable is first defined within the module and its value is used after exiting the module, it is an output parameter.
- When a variable is first defined within the module and its value is not used after exiting the module, it is a local variable.

- When a variable is first used within the module, then modified and its value is used after exiting the module, it is an input output parameter.
- When a variable is first used within the module, and it is not modified or its value is not used after exiting the module, it is an input parameter.

Example 2.8. Let us take as an example the Fortran subroutine in Figure 2 of the appendix. The formal parameter *PLNUM* is first used, then modified, and its value is used after exiting the subroutine: it is therefore an input output parameter of *ELPLZ*. The formal parameter *PLPOS* is only used, thus it is an input parameter of *ELPLZ*. The available information is that *PLNUM* denotes a stack, and *PLPOS* a natural. The specification operation associated to *ELPLZ* is therefore declared as follows[3]:

op-Elplz (_,_) : Stack, Nat → Stack

Note 2.9. Let us note that while determining the "true" input and output parameters of a program module, we also get some information on the kind of functionality. When a variable is first used within the module and its value is not used after exiting the module, the module functionality includes *reading* or *comparing* the value denoted by this variable. When a variable is first used within the module, then modified and its value is used after exiting the module, the module functionality includes the *modification* of the data denoted by this variable. When a variable is first defined within the module and its value is used after exiting the module, the module functionality includes the *creation* of the data denoted by this variable. In the case of data creation, information on the data representation should be provided (cf. section 2.1).

Operations in algebraic specifications are functions, therefore some encoding for procedures is needed. This is straightforward when there is a single output parameter. When there are several output parameters, a new sort is built from the cartesian product of the output parameters associated sorts. The selectors on the components of the cartesian product are also provided.

Example 2.10. In the program module of Figure 2 in the appendix, there is a call to the subroutine *MAKPOP*. *MAKPOP* is declared as follows:

subroutine MAKPOP (PLNUM, NUMEL, PLEND, ER, *)

Its input parameter *PLNUM* denotes a stack, and its output parameters *PLNUM, NUMEL, PLEND* respectively denote a stack, a stack element, and a boolean. The operation associated to *MAKPOP* has thus the following profile: op-Makpop (_) : Stack → MakpopResult where the sort MakpopResult is generated by: < _,_,_ > : Stack, Elem, Bool → MakpopResult and is equipped with three selectors: sel1-Makpop (_) : MakpopResult → Stack etc. Auxiliary functions may also be provided to reflect the functionalities associated with producing each of the three parts of the result.

In summary, the specification functions associated to a program module are obtained in the following way:

[3] The "_" denote the argument's place.

- a function is associated with the program module overall functionality
- its profile is determined after a careful data flow analysis provides information on (i) which are the "true" parameters, (ii) what is the parameters input/output status
- auxiliary functions are used when several results are produced by the program modules.

3 Building the axioms.

The aim here is to build the axioms describing the properties of the specification operation associated to the program module (i.e. its global functionality). This is worked out in two steps:

1. The first step is to obtain an expression of the program module results in terms of its input parameters and of simpler functionalities.
2. The second step is to find out simplifications that will help turning often complicated expressions into simpler legible axioms.

3.1 Expressing outputs in terms of inputs and simpler functionalities.

The first global expression is obtained in terms of the program module global functionality.

Example 3.1. Considering the program module in Figure 2 of the Appendix, one gets the expression:

$$P_{out} = \text{op-Elplz}(P_{in}, pos_{in})$$

where P_{out} denotes the resulting stack, and P_{in} and pos_{in} the input stack and integer.

The global functionality will be then expressed in terms of simpler ones that we call *unit actions*. Unit actions correspond to definitions and uses of data denoted by variables in the program. Unit actions are represented by code schemata stored in a library and described by means of associated equations (cf. a library example in section 6.2 of the Appendix). Each equation expresses the value of a variable in the code schema as a term built from predefined operations and variables of the code schema.

The code is decomposed into code fragments that are identified with code schemata, thus providing the corresponding equations for the unit actions. The expression for the global functionality in terms of simpler ones is obtained by assembling together these equations.

Identifying code fragments with code schemata turns out not to be a simple superposition:

- Obviously, some renaming of both the variables and the subprograms called is needed. Then, the order of a subprogram parameters may be different from the order used for the corresponding specification operation.

- Some statements in the code fragment that do not interfere with the unit action should be forgotten. This may be the case :
  - when some useless code (used for instance when debugging) was not removed;
  - when intricated code fragments represent various unit actions. Using slicing techniques (cf. [17, 6]) the code could be decomposed into fragments corresponding to single unit actions.

Example 3.2. Let us consider the following code fragment from the example Figure 2 of the Appendix.

```
L8    10 CONTINUE
L9       CALL MAKPOP (PLNUM, NUMEL, PLEND, ER, *100)
L10      IF (PLEND .EQ. 0) THEN
L11      I = I + 1
L12    5 FORMAT (10X,'I=',I5,' NUMEL=',I6,2X)
L13      CALL MAKPUS (PLTMP, NUMEL, NUMDE, ER, *100)
L14      GOTO 10
L15      ENDIF
```

This may be identified with the following code schema:

```
LB CONTINUE
   CALL POP (STACKA, ELEMA, EMPTYA)
   IF (EMPTYA .EQ. 0) THEN
     CALL PUSH (ELEMA, STACKB, HEIGHTB)
     GOTO LB
   ENDIF
```

For this, the following renaming where used:

10	PLNUM	NUMEL	PLEND	PLTMP	NUMDE
LB	STACKA	ELEMA	EMPTYA	STACKB	HEIGHTB

variables renaming

MAKPOP (PLNUM, NUMEL, PLEND, ER, *100)	MAKPUS (PLTMP, NUMEL, NUMDE, ER, *100)
POP (STACKA, ELEMA, EMPTYA)	PUSH (ELEMA, STACKB, HEIGHTB)

subprogram calls: renaming and parameter reordering

The ordering of the subprogram MAKPUS first two parameters is changed when expressed by PUSH. The statements L11 and L12 that do not define any of the code schema variables are forgotten.

The corresponding unit action is described by the following equations [4](cf. 6.2):

[4] The following operations will be used: build (which puts its first stack argument on top of its second stack argument), transpose (reverses the stack elements order), mpop (pops n times - where n is its second argument), transfer (transfer(S,n) produces a stack of the reversed top n elements of S), process (performs some computation on its integer argument).

stackb = build (transpose (stacka), stackb)
stacka = emptyst
elema = top (build (transpose (stacka), stackb))
emptya = true
heightb = height (build (transpose (stacka), stackb))

In these equations, variables in the right-hand sides denote data before the unit action is applied, and data in the left-hand side denote data after the unit action took place. Susbtituting the code variables for the code schema variables, one gets equations that represent the unit action performed in the code fragment. The set of code schemata used to extract all ELPLZ unit actions is provided in section 6.2 of the Appendix.

Note 3.3. A unit action may be represented in different ways, for instance because the programming language provides various looping mechanisms. Thus, there may be in the library several code schemata associated with a unit action. For instance, the above unit action is also available under the following code schema:

```
CALL HEIGHT (STACKA, HEIGHTA)
DO CPT = 1, HEIGHTA
   CALL POP (STACKA, ELEMA, EMPTYA)
   CALL PUSH (ELEMA, STACKB, HEIGHTB)
ENDDO
```

Putting together the equations for the various unit actions of the program module, we get an expression of the module output in terms of its inputs and of simpler functionalities.

Example 3.4. The following expression results from the equations for ELPLZ unit actions.

```
(*) op-Elplz (plnum, plpos) =
  build (transpose (pop (mpop (build (transpose (plnum), emptyst), pred (plpos)))),
      push (process (top (mpop (build (transpose (plnum), emptyst), pred (plpos)))),
          build (transfer (build (transpose (plnum), emptyst), pred (plpos)), emptyst)))
```

The processing described in this section leads to an expression of the program functionality that may not be very legible. In the following section, the corresponding axioms are derived after performing some simplifications.

3.2 Simplifications and axioms derivation.

We used mainly two ways for potential expressions simplifications: one is subterms reductions, and the other is to decompose the axiom into several simpler ones which apply on different cases for the variables[5]. The subterms reductions may stem from some programming problems.

[5] The full specifications and proofs are given in [2].

The first problem is related to some programming constraints: since Fortran is an imperative programming language, modifying a data may require the use of local variables that should be initialized.

Example 3.5. In the *ELPLZ*, the initialization of the local variable *PLTMP* leads to the subterm build (transpose (plnum), emptyst) that is found on three occurrences in the above expression (*). Proving the property *build (P, emptyst) = P* leads to the simplified expression :

```
(*1)   op-Elplz (plnum, plpos) =
           build (transpose (pop (mpop (transpose (plnum), pred (plpos)))),
                 push (process (top (mpop (transpose (plnum), pred (plpos)))),
                       transfer (transpose (plnum), pred (plpos))))
```

The second problem comes from the fact that the processed code was not formally specified when developed. In particular, this can be shown by the fact that the unit actions decomposition is not always "ideal"; this may be associated with the fact that a program module may yield several results, or with the fact that the code is not optimized.

Example 3.6. In the expression (*1), the subterm on the last line transfer (transpose (plnum), pred (plpos)) specifies that the stack plnum is transposed (its elements are thus stacked in reverse order) and then partially transferd (part of the elements - up to a given position - are stacked back in the initial order)[6]. The subterm mpop (plnum, height (plnum) - pred (plpos)) provides the same result in one action (and prevents from moving around elements that do not need to). This is expressed by the following properties: *pop (mpop (P, n)) = mpop (P, succ (n))* and *transfer (transpose (P), n) = mpop (P, height (P) - n)*. The expression (*1) is thus simplified into:

```
(**)   op-Elplz (plnum, plpos) =
           build (transpose (mpop (transpose (plnum), succ (pred (plpos)))),
                 push (process (top (mpop (transpose (plnum), pred (plpos)))),
                       mpop (plnum, height (plnum) - pred (plpos))))
```

The following simplifications are obtained through looking for particular values of the input data. These values are found in the axioms specifying the operations (here called "predefined", i.e. "defined before") found in the expression to be simplified.

Example 3.7. Among the available axioms for the operations used in expression (**), some take into account the particular case for the empty stack: *mpop (emptyst, n) = emptyst*, *transpose (emptyst) = emptyst*, and *build (emptyst, P) = P*. Using these axioms, we get:

(Ax1) op-Elplz (emptyst, plpos) = push (process (top (emptyst)), emptyst)

[6] Note that this implies that some of these elements are moved twice, which is not necessary.

Let us note that, when considering particular values as above, one may find some discrepancies between the resulting expression of the current axiom and the predefined operations specification. In that case, either the predefined operation specification should be corrected, or some incoherence in the code is revealed. The comments in the code may be useful to precisely determine what was intended, and what should the current axiom express accordingly. The code defects are not propagated in the specification, thus they will be removed from the new code generated after the reverse engineering phase.

Example 3.8. For the empty stack case, the (Ax1) expression contains the subterm *top (emptyst)* while *top* is not defined in that case. One question is whether the specification of *top* should be changed: the answer is no. If *top (emptyst)* did yield some value, the result when the input stack is empty would be a stack of height 1. The comments of *ELPLZ* do not agree with this result when stating that *ELPLZ* processes the element located at a given height of the input stack. It could then be expected that *ELPLZ* should not modify an empty stack, and that the expression (Ax1) should be op-Elplz (emptyst, plpos) = emptyst. The code should also be corrected along these lines.

Finally, the axiom extraction is completed with conditional axioms the premises of which exclude the particular cases considered before, and that may contain a recursive expression.

Example 3.9. The work done on particular cases lead to single out the empty stack case, and the case where the processing position is outside of the stack (cf. details in [2]). Expressing the non empty stack as *push (e, P)* and a non null position as *succ (pos)* such that *succ (pos) > height (push (e, P))* leads, after simplifications, to:

pos > height (P) ⇒
op-Elplz (push (e, P), succ (pos)) = push (e, op-Elplz (P, succ (pos)))

This leads to the axioms below for *op-Elplz*, which are definitely more legible than expression (∗) given at the end of the preceding section.

op-Elplz (emptyst, pos) = emptyst
op-Elplz (push (e, P), 0) = emptyst
pos > height (P) ⇒ op-Elplz (push (e, P), succ (pos)) = emptyst
pos = height (P) ⇒ op-Elplz (push (e, P), succ (pos)) = push (0, P)
0 < pos < height (P) ⇒ op-Elplz (push (e, P), succ (pos)) = push (e, op-Elplz (P, succ (pos)))

4 Related works.

In order to compare our approach with other works on software reverse engineering, we present some related works. [7] compares methods for abstract data type and abstract state detection techniques that rely on heuristics based on "simple" characteristics of the program (e.g. same module heuristic). [14] makes use of program plans, slicing and tools based on ASF+SDF in order to deal with year 2000 problems in COBOL programs. In [13, 8], the program is first rewritten in a language that is independent from a given programming language, then "events" are extracted in order to describe commonly used programming structures (variable declarations, loops, etc.), and finally "maps" (inference rules producing higher level concepts from events) are used to obtain some informal algorithm for the program. [15, 16] describes an inverse engineering process using some code transformations to obtain a program algorithm in a pseudo-language independent from the programming language. The work done here relates to models, and does not abstract up to algebraic specifications. The approach described in [18, 9] looks for "architectural styles" in the program, and extracts objects and abstract data types from the code. This method could not be applied to Fortran since little facility for type construction is provided. The method for object extraction of [11] is applied to Fortran, but under the hypotheses that data are always represented through local variables, through a COMMON file, or a group of parameters.

Our approach is original in two respects:

- We describe the program functionalities in terms of a modular algebraic specification, thus providing both the signature part and the axioms part.
- The code we start from is in Fortran that is a difficult context for identifying the abstract domains of the data.

5 Conclusion.

In this paper, we described how to extract information from the code so as to generate the corresponding algebraic specification. The first problem to be solved is how to extract the *signature* of the specification module associated to a program module, the main problems being:

- the identification of the domains of the data used within the Fortran program (within the context of Fortran or any programming language that offers little type construction facility)
- the identification of the program module "true" parameters (when some declared parameters may not play a parameter role, and some global variables may play a parameter role).

In order to achieve the signature extraction, it is also necessary to use the data representation information. Since our approach is modular, we assume that the specification and the information on the imported program modules is available.

The second problem to be solved is the extraction of the *axioms* for the specification operation associated to the program module main functionality. Our approach is to decompose the code into fragments; these code fragments are then identified with code schemata available from a library in order to extract the corresponding unit actions. The composition of these unit actions expresses the program module functionality in terms of the input data and of predefined operations. The identification of code fragments with code schemata requires usually to perform some renaming and sometimes to forget instructions that do not interfere with the unit action itself. Finally, we describe how to simplify the resulting expression of the program functionality in order to get more simple and legible axioms. This last part of the process may reveal some code defects (useless instructions, lack of explicit processing for some particular parameter values, etc.) that should not be propagated to the specification level.

The specification extraction process described above was applied to a Fortran industrial software. Our aim now is to explore to which extent this specification extraction work may be automatized, how it can be instrumented, and how and when the necessary information (data representation, "true" parameters, unit actions) may be automatically inferred. We may also consider exploiting systematically other sources of information, namely the program comments and identifiers, in order to show up some inconsistencies in the code. Since the Fortran programming language turns out to be a rather difficult context for the abstract data type specification extraction, we think that adapting our process to other programming languages should not raise much difficulties.

Acknowledgements

The authors thank the anonymous referees for their helpful comments.

6 Appendix.

6.1 Example of a program module.

To illustrate the process described in this paper, the Fortran program module *ELPLZ* is used (cf. Figure 2).

```
L1         SUBROUTINE ELPLZ (PLNUM, PLPOS, ER, *)
L2  %      INCLUDE (STAGLO)
L3         INTEGER*4 I, PLNUM, PLPOS, PLTMP, J, K, ELTMP
L4         SIZPL = 4
L5         CALL INITPL (SIZPL, NUMPL, ER, *100)
L6         PLTMP = NUMPL
L7         I = 0
L8      10 CONTINUE
L9         CALL MAKPOP (PLNUM, NUMEL, PLEND, ER, *100)
L10        IF (PLEND .EQ. 0) THEN
L11        I = I + 1
L12      5 FORMAT (10X,'I=',I5,' NUMEL=',I6,2X)
L13        CALL MAKPUS (PLTMP, NUMEL, NUMDE, ER, *100)
L14        GOTO 10
L15        ENDIF
L16        DO 20, I=1,PLPOS-1
L17        CALL MAKPOP (PLTMP, NUMEL, PLEND, ER, *100)
L18        CALL MAKPUS (PLNUM, NUMEL, NUMDE, ER, *100)
L19     20 CONTINUE
L20        CALL MAKPOP (PLTMP, NUMEL, PLEND, ER, *100)
L21        ELTMP = NUMEL
L22        CALL INTZ (ELTMP, ER, *100)
L23        NUMEL = ELTMP
L24        CALL MAKPUS (PLNUM, NUMEL, NUMDE, ER, *100)
L25     30 CONTINUE
L26        CALL MAKPOP (PLTMP, NUMEL, PLEND, ER, *100)
L27        IF (PLEND .EQ. 0) THEN
L28        I = I + 1
L29        CALL MAKPUS (PLNUM, NUMEL, NUMDE, ER, *100)
L30        GOTO 30
L31        ENDIF
L32        CALL DESTPL (PLTMP, ER, *100)
L33    100 CONTINUE
L34        CALL DIRER (ER,*200)
L35        RETURN
L36    200 CONTINUE
L37        WRITE(6,*) 'ERROR FROM ELPLZ, ER=',ER
L38        RETURN 1
L39        END
```

Figure 2. ELPLZ Fortran program module.

The include file *STAGLO* (line L2) is made up of the following declarations :

```
IMPLICIT NONE
INTEGER*4 NUMPL, SIZPL, NUMDE, PLEND, ER, NUMEL
```

The formal parameters *PLNUM, PLPOS, ER, ** in this program module denote respectively the address of the input-output stack, the input position (from the bottom of the stack), the error code returned upon exiting *ELPLZ* and the address of the message returned by the calling program when a typing error is detected.

The role of *ELPLZ* is to perform some processing on the stack element located at the given position.

An example is given in the Figure 3 below : at the level 4 of the stack, the 6 became a 0.

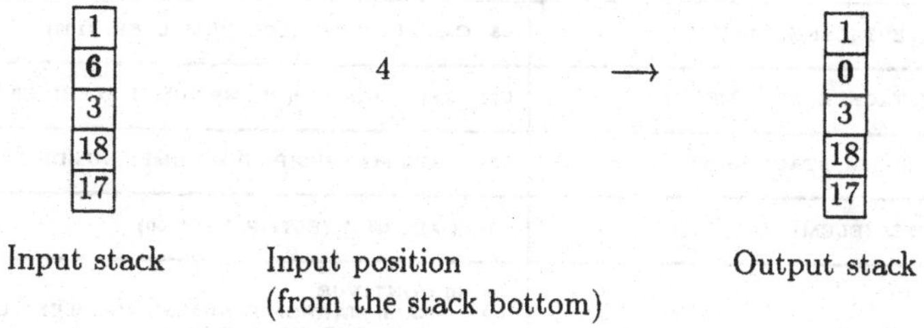

Figure 3. Example of applying *ELPLZ*.

Note 6.1. The program module *ELPLZ* given here is a simplified version of the actual code examined (where the stack elements were more complex data, and so was the data processing).

6.2 Unit actions extraction.

Code schemata and associated code fragments. In order to extract unit actions from the module *ELPLZ* the code schemata in Figure 4 were used to identify code fragments.

Note that the same code schema is used both for code in L8-L15 and for code in L25-L31.

Code schemata and associated equations. In order to describe unit actions from the module *ELPLZ* the equations associated with the code schemata in Figure 4 were used.

Code schemata	code fragments
VARA = VARB	L4 SIZPL = 4 L6 PLTMP = NUMPL L7 I = 0 L21 ELTMP = NUMEL L23 NUMEL = ELTMP
call CREATEST (LENG, STACK)	L5 CALL INITPL (SIZPL, NUMPL, ER, *100)
call POP (STACK, ELEM, EMPTY)	L20 CALL MAKPOP(PLTMP,NUMEL,PLEND,ER,*100)
call PUSH (ELEM, STACK, HEIT)	L24 CALL MAKPUS(PLNUM,NUMEL,NUMDE,ER,*100)
call PROCESS (ELEM)	L22 CALL INTZ (ELTMP, ER, *100)

LB continue
call POP (STACKA, ELEMA, EMPTYA)
if (EMPTYA .eq. 0) then
call PUSH (ELEMA, STACKB, HEIGHTB)
goto LB
endif
	L8 10 CONTINUE
	L9 CALL MAKPOP(PLNUM,NUMEL,PLEND,ER,*100)
	L10 IF (PLEND .EQ. 0) THEN
	L11 I = I + 1
	L12 5 FORMAT (10X,'I=',I5,' NUMEL=',I6,2X)
	L13 CALL MAKPUS(PLTMP,NUMEL,NUMDE,ER,*100)
	L14 GOTO 10
	L15 ENDIF
	
	L25 30 CONTINUE
	L26 CALL MAKPOP(PLTMP,NUMEL,PLEND,ER,*100)
	L27 IF (PLEND .EQ. 0) THEN
	L28 I = I + 1
	L29 CALL MAKPUS(PLNUM,NUMEL,NUMDE,ER,*100)
	L30 GOTO 30
	L31 ENDIF

do LB, CPT=1,HEIT call POP (STACKA, ELEMA, EMPTYA) call PUSH (ELEMA, STACKB, HEIGHTB) LB continue	L16 DO 20, I=1,PLPOS-1 L17 CALL MAKPOP(PLTMP,NUMEL,PLEND,ER,*100) L18 CALL MAKPUS(PLNUM,NUMEL,NUMDE,ER,*100) L19 20 CONTINUE

Figure 4. Code schemata used to extract unit actions from *ELPLZ.*

Code schemata	Equations
VARA = VARB	vara = varb
call CREATEST (LENG, STACK)	stack = emptyst
call POP (STACK, ELEM, EMPTY)	stack = sel1-Makpop (op-Makpop (stack)) = pop (stack) elem = sel2-Makpop (op-Makpop (stack)) = top (stack) empty= sel3-Makpop(op-Makpop(stack)) = Pempty?(stack)
call PUSH (ELEM, STACK, HEIT)	stack = sel1-Makpus (op-Makpus (elem, stack)) = push (elem, stack) heit = sel2-Makpus (op-Makpus (elem, stack)) = succ (height (stack))
call PROCESS (ELEM)	elem = process (elem)

LB continue
call POP (STACKA, ELEMA, EMPTYA)
if (EMPTYA .eq. 0) then
call PUSH (ELEMA, STACKB, HEIGHTB)
goto LB
endif
	stacka = emptyst
	stackb = build (transpose (stacka), stackb)
	emptya = true
	elema = top (build (transpose (stacka), stackb))
	heightb = height (build (transpose (stacka), stackb))

do LB, CPT=1,HEIT call POP (STACKA, ELEMA, EMPTYA) call PUSH (ELEMA, STACKB, HEIGHTB) LB continue	stacka = mpop (stacka, heit) stackb = build (transfer (stacka, heit), stackb) emptya = (height (stacka) ≤ heit) elema = top (build (transfer (stacka, heit), stackb)) heightb = height (build (transfer (stacka, heit), stackb))

References

1. A. Audette. *Etude de l'applicabilité des spécifications algébriques à un logiciel existant dans le domaine de la CFAO.* Master's thesis of Ecole Polytechnique de Montréal, Canada, 1994. (Supervised by C. Choppy).
2. S. Cherki and C. Choppy. *Une méthode de rétroingénierie de programmes Fortran utilisqnt les spécifications algébriques.* LRI and IRIN Research report, 1997.
3. E.J. Chikofsky and J.H. Cross. Reverse Engineering and Design Recovery : A Taxonomy. *IEEE Software : Maintenance, Reverse Engineering and Design Recovery,* 7(1):13–17, 1990.
4. H. Ehrig and B. Mahr. *Fundamentals of Algebraic Specification 1 - Equations and Initial semantics.* EATCS Monographs on Theoretical Computer Science, Springer Verlag, 1985.
5. H. Ehrig and B. Mahr. *Fundamentals of Algebraic Specification 2 - Module specification and constraints.* EATCS Monographs on Theoretical Computer Science, Springer Verlag, 1990.
6. K.B. Gallagher and J.R. Lyle. Using program slicing in software maintenance. *IEEE Transactions on Software Engineering,* 17(8):751–761, 1991.
7. J.-F. Girard, R. Koschke, and G. Schied. Comparison of abstract data type and abstract state encapsulation detection techniques for architectural understanding. In *Fourth Working Conference on Reverse Engineering,* pages 66–75. IEEE Computer Society, 1997.
8. M.T. Harandi and J.Q. Ning. Knowledge-Based Program Analysis. *IEEE Software : Maintenance, Reverse Engineering and Design Recovery,* 7(1):74–81, 1990.
9. D.R. Harris, H.B. Reubenstein, and A.S. Yeh. Reverse Engineering to the Architectural Level. In *17th International Conference on Software Engineering, Seattle, Washington, USA.* ACM, 1995.
10. M. Munro. Maintenance is not a soft option. *Computer Weekly,* pages 28–29, 1989.
11. C.L. Ong and W.T. Tsai. Class and object extraction from imperative code. *Journal of Object-Oriented Programming,* 6(1):58–68, 1993.
12. W.M. Osborne and E.J. Chikofsky. Fitting pieces to the maintenance puzzle. *IEEE Software : Maintenance, Reverse Engineering and Design Recovery,* 7(1):11–12, 1990.
13. C. Rich and L.M. Wills. Recognizing a Program's Design : A Graph-Parsing Approach. *IEEE Software : Maintenance, Reverse Engineering and Design Recovery,* 7(1):82–89, 1990.
14. A. van Deursen, S. Woods, and A. Quilici. Program plan recognition for year 2000 tools. In *Fourth Working Conference on Reverse Engineering,* pages 124–133. IEEE Computer Society, 1997.
15. M. Ward. Abstracting a Specification from Code. *Journal of Software Maintenance : Research and Practice,* 5:101–122, 1993.
16. M. Ward. Reverse Engineering through Formal Transformation : Knuths 'Polynomial Addition' Algorithm. *The Computer Journal,* 37(9):795–813, 1994.
17. M. Weiser. Program slicing. *IEEE Transactions on Software Engineering,* SE-10(4):352–357, 1984.
18. A.S. Yeh, D.R. Harris, and H.B. Reubenstein. Recovering Abstract Data Types and Object Instances from a Conventional Procedural Language. In *2nd Working Conference on Reverse Engineering, Toronto.* IEEE, 1995.

Coalgebra Semantics for Hidden Algebra: Parameterised Objects and Inheritance

Corina Cîrstea*

Oxford University Computing Laboratory
Wolfson Building, Parks Road, Oxford OX1 3QD, UK

Abstract. The theory of hidden algebras combines standard algebraic techniques with coalgebraic techniques to provide a semantic foundation for the object paradigm. This paper focuses on the coalgebraic aspect of hidden algebra, concerned with signatures of destructors at the syntactic level and with finality and cofree constructions at the semantic level. Our main result shows the existence of cofree constructions induced by maps between coalgebraic hidden specifications. Their use in giving a semantics to parameterised objects and inheritance is then illustrated. The cofreeness result for hidden algebra is generalised to abstract coalgebra and a universal construction for building object systems over existing subsystems is obtained. Finally, existence of final/cofree constructions for arbitrary hidden specifications is discussed.

1 Introduction

Algebraic techniques have been intensively studied over the last decades. Their suitability for the specification of data types is due to the availability of effective definition and proof techniques based on induction. Recent work on coalgebras (the formal duals of algebras) [Rei95, Jac95, Jac96, Rut96, Jac97, JR97] suggests their suitability for the specification of dynamical systems. The theory of coalgebras provides a notion of observational indistinguishability as *bisimulation*, a characterisation of abstract behaviours as elements of *final coalgebras* and *coinduction* as a definition/proof principle for system behaviour.

Hidden algebra, introduced in [GD94] and further developed in [MG94, GM97] combines algebraic and coalgebraic techniques in order to provide a semantic foundation for the object paradigm. It is an extension of the theory of many sorted algebras that uses both *constructor* and *destructor* operations and a loose behavioural semantics over a fixed data universe for (the states of) objects. Its coalgebraic nature, emerging from the observational character of the approach, has already been exploited in [MG94] where (coinductive) proof techniques for behavioural satisfaction were developed. The present paper further investigates the relationship between hidden algebra and coalgebra, focusing on the semantic level and in particular on cofree constructions. Their suitability as semantics for the specification techniques used in hidden algebra is emphasised.

* Research supported by an ORS Award and an Oxford Bursary.

The structure of the paper is as follows. Section 2 gives a brief account of the theory of coalgebras as well as an outline of hidden algebra. Section 3 focuses on the coalgebraic aspect of hidden algebra: hidden algebras are mapped to coalgebras (by forgetting the constructors) in such a way that behavioural congruences correspond to bisimulation equivalences on the associated coalgebras. Consequently, coinduction can be used both as a definition principle for object behaviour and as a proof principle for behavioural equivalence. Existence of *final algebras* for coalgebraic hidden specifications is also obtained. The main result of the paper concerns the existence of *cofree hidden algebras* induced by maps between *coalgebraic* (destructor) hidden specifications. Such maps correspond to reusing specifications either horizontally by importation or vertically by refinement. In certain cases, the cofree construction corresponds to a reuse of implementations along the underlying reuse of specifications. A generalisation of a cofreeness result in [Rut96], concerning the existence of cofree object systems over given subsystems is sketched in the last part of Section 3. Section 4 illustrates the use of cofree constructions in giving semantics to the importation of coalgebraic hidden modules, parameterised modules and inheritance. Cofree constructions provide canonical ways to build implementations for more structured/specialised specifications from implementations of the specifications they are built on. Section 5 generalises the final/cofree coextension semantics in Section 3 by considering arbitrary hidden specifications. In this case, the semantics is given by final/cofree *families* of hidden algebras. Section 6 summarises the results presented and briefly outlines future work.

2 Preliminaries

This section gives an account of the basic ideas and concepts in coalgebraic specifications, emphasising their duality to algebraic specifications. A brief introduction to hidden algebra (a combination of algebraic and coalgebraic techniques intended as a specification framework for objects) is also given.

2.1 Algebra and Coalgebra

Algebra and its associated inductive techniques have been successfully used for the specification of data types. The emphasis there is on how the values of a data type are generated, using constructor operations going into the type. Data types are presented as **F-algebras**, i.e. tuples (A, α), with A an object and $\alpha : \mathsf{F}A \to A$ a morphism in some category C, with $\mathsf{F} : \mathsf{C} \to \mathsf{C}$. Among F-algebras, **initial** ones $\iota : \mathsf{F}I \to I$ (least fixed points of F) are most relevant – their elements denote closed programs. Initial algebras come equipped with an **induction principle** stating that no proper subalgebras exist for initial algebras. This principle constitutes the main technique used in algebraic specifications for both definitions and proofs: defining a function on the initial algebra by induction amounts to defining its values on all the constructors; and proving that two functions on the initial algebra coincide amounts to showing that they agree on all the constructors. **Free constructions** are also relevant for data types: they provide least extensions of algebras of a data type to algebras of another and have been used to give semantics to parameterised data types, see e.g. [EM85].

The theory of coalgebras [Rut96, JR97], having its roots in automata theory and transition system theory [Rut95] and concerned with dynamical systems, can be viewed as a dualisation of the theory of algebras. Object systems are coalgebraically defined by specifying how their states can be observed, using destructor operations going out of the object types. Object types appear as G-coalgebras, i.e. tuples (C, β), with C an object and $\beta : C \to \mathsf{G}C$ a morphism in some category C, with $\mathsf{G} : \mathsf{C} \to \mathsf{C}$. **Final** G-coalgebras $\zeta : Z \to \mathsf{G}Z$ (greatest fixed points of G) are in this case relevant – they incorporate all G-behaviours. The unique coalgebra homomorphism from a coalgebra to the final one maps object states to their behaviour. A **bisimulation** between two coalgebras is a relation on their carriers, carrying itself coalgebraic structure. Bisimulations relate states that exhibit the same behaviour. Final coalgebras come equipped with a **coinduction principle** stating that no proper bisimulations exist between a final coalgebra and itself; that is, two elements of a final coalgebra having the same behaviour coincide. Coinduction can be used both in definitions, to define functions into the final coalgebra by giving coalgebraic structure to their domains, and in proofs, to show equality of two elements of the final coalgebra by exhibiting a bisimulation that relates them. Finally, **cofree constructions** are relevant for object types as they provide least restrictive (co)extensions of coalgebras of an object type to coalgebras of another.

2.2 Hidden Algebra

This section provides an outline of hidden algebra. For a detailed presentation of the approach the reader is referred to [GM97].

Hidden algebra extends many sorted algebra to support the specification of objects with hidden states, only accessible through specified interfaces. The fundamental distinction between data values and object states is reflected in the use of visible sorts/operations with standard semantics for data and of hidden sorts/operations with loose behavioural semantics for objects.

A fixed data universe, given by an algebra D (the **data algebra**) of a many sorted signature (V, Ψ) (the **data signature**) is assumed, with the additional constraint that each element of D is named by a constant in Ψ. For convenience, we take $D_v \subseteq \Psi_{[],v}$ for each $v \in V$.

Definition 1. A **(hidden) signature over** (V, Ψ, D) is a pair (H, Σ) with H a set of **hidden sorts** and Σ a $V \cup H$-sorted signature satisfying: (i) $\Sigma_{w,v} = \Psi_{w,v}$ for $w \in V^*$, $v \in V$ and (ii) **monadicity:** for $\sigma \in \Sigma_{w,s}$, at most one sort appearing in w (by convention, the first one) is hidden.

$\Sigma \setminus \Psi$-operations having exactly one hidden-sorted argument are called **destructors**, while those having only visible-sorted arguments are called **constructors**.

Definition 2. A **(hidden) signature map** $\phi : (H, \Sigma) \to (H', \Sigma')$ is a many sorted signature morphism $\phi : (V \cup H, \Sigma) \to (V \cup H', \Sigma')$ such that $\phi\restriction_{(V,\Psi)} = id_{(V,\Psi)}$ and $\phi(H) \subseteq H'$. A **(hidden) signature morphism** is a hidden signature map such that if $\sigma' \in \Sigma'_{h'w',s'}$ with $h' \in \phi(H)$, then $\sigma' = \phi(\sigma)$ for some $\sigma \in \Sigma$.

Signature maps specify arbitrary (vertical) structure, while signature morphisms specify horizontal structure (importation of hidden modules). Imported hidden sorts are protected by signature morphisms, in that no new destructor operations are added for them by the target signature.

Definition 3. A **(hidden) Σ-algebra** is a many sorted $(V \cup H, \Sigma)$-algebra A such that $A\restriction_\Psi = D$. A **(hidden) Σ-homomorphism** between Σ-algebras A and B is a many sorted Σ-homomorphism $f : A \to B$ such that $f_v = id_{D_v}$ for $v \in V$. Σ-algebras and Σ-homomorphisms form a category $\mathsf{HAlg}(\Sigma)$. Hidden signature maps $\phi : \Sigma \to \Sigma'$ induce reduct functors $\mathsf{U}_\phi : \mathsf{HAlg}(\Sigma') \to \mathsf{HAlg}(\Sigma)$.

Hidden algebra takes a behavioural approach to objects: their states can only be observed through experiments; indistinguishability of states by experiments is captured by behavioural equivalence.

Definition 4. Given a signature (H, Σ), a **Σ-context** for sort $s \in V \cup H$ is an element of $T_\Sigma[z]_v$ with z an s-sorted variable and $v \in V$. Given a Σ-algebra A, **behavioural equivalence on** A (denoted $\sim_A$) is defined by: $a \sim_{A,s} a'$ iff $c_A[a] = c_A[a']$ for all contexts c for s, with $s \in V \cup H$ and $a, a' \in A_s$.

Satisfaction of equations is also behavioural – one only requires the two sides of an equation to look the same under any observation rather than coincide.

Definition 5. A **(hidden) specification** is a triple (H, Σ, E) with (H, Σ) a hidden signature and E a set of Σ-equations. A Σ-algebra A **behaviourally satisfies** a (conditional) Σ-equation e of form $(\forall X)\, l = r \textbf{ if } l_1 = r_1, \ldots, l_n = r_n$ (written $A \models\!\!\!\equiv_\Sigma e$) if and only if for any assignment $\theta : X \to A$, $\bar{\theta}(l) \sim_A \bar{\theta}(r)$ whenever $\bar{\theta}(l_i) \sim_A \bar{\theta}(r_i)$, $i = 1, \ldots, n$. Given sets E and E' of Σ-equations, we write $E \models\!\!\!\equiv_\Sigma E'$ if $A \models\!\!\!\equiv_\Sigma E$ implies $A \models\!\!\!\equiv_\Sigma E'$ for any Σ-algebra A.

[MG94] gives a characterisation of behavioural equivalence as greatest *behavioural congruence* (congruence which coincides with equality on visible sorts) and uses it to obtain a coinductive-like proof technique for behavioural equivalence.

We restrict our attention to specifications whose equations have visible-sorted conditions only. To each such specification (Σ, E) one can associate another specification $(\Sigma, \overline{E})$ (by letting $\overline{E} = \{c[e] \mid e \in E,\ c \in T_\Sigma[z] \text{ appropriate for } e\}$), such that $A \models\!\!\!\equiv_\Sigma E$ iff $A \models\!\!\!\equiv_\Sigma \overline{E}$ iff $A \models_\Sigma \overline{E}$.

Definition 6. Let (Σ, E) and (Σ', E') be hidden specifications. A hidden signature map $\phi : \Sigma \to \Sigma'$ defines a **specification map** $\phi : (\Sigma, E) \to (\Sigma', E')$ if and only if $E' \models\!\!\!\equiv_{\Sigma'} \phi(\overline{E})$. A specification map whose underlying signature map is a signature morphism is called a **specification morphism**.

Given a specification map $\phi : (\Sigma, E) \to (\Sigma', E')$, the functor U_ϕ induced by $\phi : \Sigma \to \Sigma'$ maps hidden (Σ', E')-algebras to hidden (Σ, E)-algebras.

Theorem 7. *The category* Spec *of hidden specifications and specification maps is finitely cocomplete. Pushouts in* Spec *preserve specification morphisms.*

We note in passing that the constraint on hidden signature morphisms is used in [GD94] to obtain an institution of hidden algebras. Moreover, specification morphisms $\phi : (\Sigma, E) \to (\Sigma', E')$ satisfy $E' \models_{\Sigma'} \phi(E)$, i.e. they are the theory morphisms of this institution. A different institution may be obtained by considering hidden signature maps and a slightly different notion of sentence, given by a Σ-equation together with a subsignature of Σ for the contexts under which the equation is expected to hold. This is the institution that underlies our treatment of parameterisation in Section 4.1.

3 Coalgebra and Hidden Algebra

This section focuses on the coalgebraic nature of hidden algebra. First we illustrate how viewing hidden algebras as coalgebras provides both a characterisation of abstract behaviours by means of final coalgebras and a coalgebraic definition of behavioural equivalence as greatest bisimulation. Next, we prove the existence of cofree constructions induced by maps between coalgebraic hidden specifications. Such constructions provide canonical ways to (co)extend algebras along specification maps by restricting the behaviour as little as possible. Finally, we present a generalisation of a result in [Rut96] concerned with cofree object systems over given subsystems.

3.1 Basic Results

A closer look at the definition of behavioural equivalence reveals that only destructor operations are relevant. Hence, in investigating the coalgebraic aspect of hidden algebra we can restrict our attention to signatures of destructors.

Definition 8. A hidden signature Σ is a **coalgebraic/destructor signature** if all $\Sigma \setminus \Psi$-operations are destructors.

Proposition 9. *Let Δ be the destructor subsignature of Σ. Then Σ-behavioural equivalence is the greatest behavioural Δ-congruence.*

Proof. By monadicity together with the data algebra being fixed.

Proposition 10. *For a coalgebraic signature Δ, $\mathsf{HAlg}(\Delta) \simeq \mathsf{G}_\Delta$-$\mathsf{Coalg}$, where*

$$\mathsf{G}_\Delta : \mathsf{Set}^H \to \mathsf{Set}^H, \ \ \mathsf{G}_\Delta(X)_h = \prod_{\delta \in \Delta_{hw,s}} X_s^{D_w}, \ \ h \in H \ \ (\text{with } X_v = D_v \text{ if } v \in V)$$

Proof. Δ-algebras A correspond to G_Δ-coalgebras $\alpha : C \to \mathsf{G}_\Delta C$ with $C_h = A_h$ for $h \in H$ and α_h mapping $a \in A_h$ and $\delta \in \Delta_{hw,s}$ to $\delta_A(a, \_) : D_w \to A_s$. Also, Δ-homomorphisms $f : A \to A'$ correspond to G_Δ-homomorphisms $g : C \to C'$ with $g_h = f_h$ for $h \in H$. Moreover, the above is a one-to-one correspondence.

Corollary 11. *There exists a final Δ-algebra F_Δ, having hidden carriers:*

$$F_{\Delta,h} = \prod_{v \in V} [L_\Delta[z_h]_v \to D_v], \quad h \in H$$

(with $L_\Delta[z_h]$ consisting of "local" Δ-contexts for sort h, i.e. contexts containing only one occurrence of the hidden variable) and Δ-operations:

- $\delta_{F_\Delta}((s_v)_{v \in V}, \bar{d}) = s_{v'}(\delta(z_h, \bar{d}))$ *for* $\delta \in \Delta_{hw,v'}$
- $\delta_{F_\Delta}((s_v)_{v \in V}, \bar{d}) = (s'_v)_{v \in V}$ *with* $s'_v(c) = s_v(c[\delta(z_h, \bar{d})])$ *for* $\delta \in \Delta_{hw,h'}$

Moreover, behavioural equivalence on a Δ-algebra coincides with bisimilarity on its associated coalgebra.

The elements of F_Δ correspond to *abstract behaviours* (functions mapping experiments to data values); the unique homomorphism from an arbitrary Δ-algebra to F_Δ maps hidden states to their behaviour; two hidden states are behaviourally equivalent if and only if they are mapped to the same element of F_Δ.

We note that signature maps $\phi : (H, \Delta) \to (H', \Delta')$ induce natural transformations $\eta : \mathsf{U} \circ \mathsf{G}_{\Delta'} \Rightarrow \mathsf{G}_\Delta \circ \mathsf{U}$ (where $\mathsf{U} : \mathsf{Set}^{H'} \to \mathsf{Set}^H$ is the reindexing functor induced by $\phi : H \to H'$) given by: $(\eta_X)_h((f_{\delta'})_{\delta' \in \Delta'_{\phi(h)w',s'}}) = (f_{\phi(\delta)})_{\delta \in \Delta_{hw,s}}$ for $f_{\delta'} \in X_s^{D_w}$, $h \in H$. This observation will be used in Section 3.3.

In algebraic specifications, equations induce relations on the carriers of algebras and quotients of algebras by *least congruences* containing such relations are of interest. Dually, in coalgebra one is interested in *greatest invariants* (subcoalgebras) contained in given predicates on the carriers of coalgebras [Jac97]. Such predicates can be specified in hidden algebra using *state equations*, i.e. equations in one hidden variable – the induced predicates consist of those states for which the equations are behaviourally satisfied.

Definition 12. A hidden specification (H, Δ, E) is **coalgebraic** if (H, Δ) is coalgebraic and all the equations in E are state equations.

3.2 Cofree Coextensions

In algebraic specifications, free constructions provide least extensions of algebras along morphisms between data type specifications. Dually, in coalgebraic specifications *cofree constructions* are of interest – they provide least restrictions of coalgebras along maps between coalgebraic hidden specifications.

Given categories C and D and a functor $\mathsf{U} : \mathsf{D} \to \mathsf{C}$, a **cofree construction** w.r.t. U on a C-object C consists of a D-object C^* and a C-morphism $\epsilon_C : \mathsf{U}C^* \to C$ which is *couniversal*: given any D-object D and C-morphism $f : \mathsf{U}D \to C$, there exists a unique D-morphism $\bar{f} : D \to C^*$ such that $\epsilon_C \circ \mathsf{U}\bar{f} = f$. If C^* and ϵ_C exist for each C-object C, the mapping $C \mapsto C^*$ extends to a functor $\mathsf{F} : \mathsf{C} \to \mathsf{D}$ in such a way that the C-morphisms ϵ_C define a natural transformation $\epsilon : \mathsf{U} \circ \mathsf{F} \to \mathsf{Id}_{\mathsf{C}}$. Moreover, F is a right adjoint to U with counit ϵ.

This section proves the existence of cofree hidden algebras w.r.t. forgetful functors induced by coalgebraic specification maps. [Rut96] formulates a similar result in an abstract setting where C and D are categories of coalgebras of endofunctors on Set and U is induced by a natural transformation between such endofunctors. Here we extend this result to the case when the underlying categories of C and D are distinct. This extension appears as a generalisation of the cofreeness result for hidden algebra and provides a canonical way of building structured systems over existing subsystems.

The cofree construction for hidden algebra dualises, to a certain extent, the free construction [TWW82] for many sorted algebra. When cofreely coextending a (Δ, E)-algebra A along a specification map $\phi : (\Delta, E) \to (\Delta', E')$, instead of using the elements of A to *generate* the elements of a Δ'-algebra, one views them as information that can be *extracted* from elements of a Δ'-algebra (finality replaces initiality). Also, quotienting by least congruences is replaced by taking greatest invariants. The construction amounts to:

1. first, building the final algebra F'_A of an enriched signature Δ'_A containing destructor operations that give A-states as result
2. next, taking the greatest Δ'_A-invariant of F'_A for which the above destructors agree with the Δ-structure of A
3. finally, taking the greatest Δ'_A-invariant induced by the equations $\overline{E'}$.

2 ensures that the Δ-reduct of the cofree coextension has a Δ-homomorphism into A, 1 ensures that the cofree coextension is final among all Δ'-algebras having this property, while 3 ensures behavioural satisfaction of E'.

Theorem 13 Cofreeness. *Let $\phi : (\Delta, E) \to (\Delta', E')$ be a coalgebraic specification map. The reduct functor $\mathsf{U}_\phi : \mathsf{HAlg}(\Delta', E') \to \mathsf{HAlg}(\Delta, E)$ has a right adjoint $\mathsf{C}_\phi : \mathsf{HAlg}(\Delta, E) \to \mathsf{HAlg}(\Delta', E')$.*

Proof. We first define the action of C_ϕ on objects. Let $A \models_\Delta E$. In order to temporarily view the A-states as data, a visible sort $\bar{h}$ is added to Ψ for each $h \in H$, resulting in a data signature $\Psi^\oplus$; also, operations $s_h : h \to \bar{h}$ and $s_h : \phi(h) \to \bar{h}$ are added to Δ and Δ' respectively, resulting in signatures $\Delta^\oplus$ and $\Delta'^\oplus$ with inclusions $\iota_A : (\Psi^\oplus, \Delta, D_A) \hookrightarrow (\Psi^\oplus, \Delta^\oplus, D_A)$ and $\iota'_A : (\Psi^\oplus, \Delta', D_A) \hookrightarrow (\Psi^\oplus, \Delta'^\oplus, D_A)$ (where D_A denotes the extension of D to a $\Psi^\oplus$-algebra interpreting each $\bar{h}$ as A_h). Then $\phi : (\Psi, \Delta, D) \to (\Psi, \Delta', D)$ extends to $\phi_A : (\Psi^\oplus, \Delta^\oplus, D_A) \to (\Psi^\oplus, \Delta'^\oplus, D_A)$ by letting $\phi_A\restriction_\Delta = \phi$, $\phi_A(s_h) = s_h$ for each $h \in H$.

Now let F_A and F'_A be the final $(\Psi^\oplus, \Delta^\oplus, D_A)$- and $(\Psi^\oplus, \Delta'^\oplus, D_A)$-algebras. A can also be made into a $(\Psi^\oplus, \Delta^\oplus, D_A)$-algebra by defining $(s_h)_A$ as id_{A_h}. By finality, there exist unique $(\Psi^\oplus, \Delta^\oplus, D_A)$-homomorphisms $g : \mathsf{U}_{\phi_A} F'_A \to F_A$ and $l : A \to F_A$. Moreover, l faithfully embeds A into F_A: $l_h(a_1) = l_h(a_2) \Rightarrow$

$(s_h)_{F_A}(l_h(a_1)) = (s_h)_{F_A}(l_h(a_2)) \Rightarrow (s_h)_A(a_1) = (s_h)_A(a_2) \Rightarrow a_1 = a_2$. Define $\mathsf{C}_\phi A$ to be the greatest $(\Psi^\oplus, \Delta'^\oplus, D_A)$-invariant of F'_A such that $g\restriction_{\mathsf{U}_{\phi_A}\mathsf{C}_\phi A}$ factors through l and such that $\mathsf{U}_{\iota'_A}\mathsf{C}_\phi A \models_{\Delta'} \overline{E'}$.

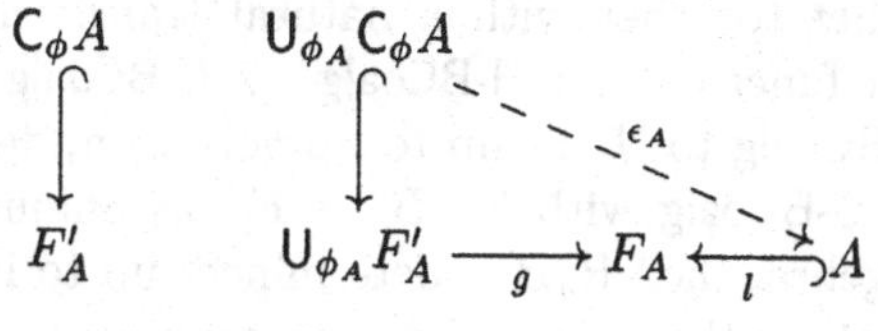

The action of C_ϕ on a (Δ, E)-homomorphism $f : A \to B$ is defined as follows. First, f is used to make A and F_A into $(\Psi^\oplus, \Delta^\oplus, D_B)$-algebras and F'_A into a $(\Psi^\oplus, \Delta'^\oplus, D_B)$-algebra. Finality of F_B and F'_B gives unique $(\Psi^\oplus, \Delta^\oplus, D_B)$- and $(\Psi^\oplus, \Delta'^\oplus, D_B)$-homomorphisms $! : F_A \to F_B$ and $!' : F'_A \to F'_B$. It then follows by maximality of $\mathsf{C}_\phi B$ that $!'\restriction_{\mathsf{C}_\phi A}$ factors through the inclusion of $\mathsf{C}_\phi B$ into F'_B (since $g'\restriction_{\mathsf{U}_{\phi_B}!'(\mathsf{C}_\phi A)}$ factors through l' and $\mathsf{U}_{\iota'_B}!'(\mathsf{C}_\phi A) \models_{\Delta'} \overline{E'}$). Hence, $\mathsf{C}_\phi f$ can be defined as $!'\restriction_{\mathsf{C}_\phi A}$.

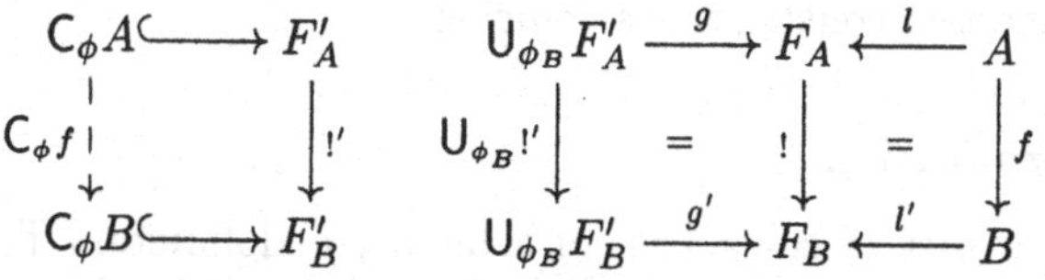

It is straightforward to check that C_ϕ is a functor.

Lemma 14 Adjunction. *C_ϕ is right adjoint to U_ϕ.*

Proof. For $A \models_\Delta E$, the A-component ϵ_A of the counit $\epsilon : \mathsf{U}_\phi \circ \mathsf{C}_\phi \Rightarrow \mathsf{Id}$ is the unique factorisation of $g\restriction_{\mathsf{U}_{\phi_A}\mathsf{C}_\phi A}$ through l (recall that l is faithful). Hence, ϵ_A is a Δ-homomorphism.

It remains to prove couniversality of ϵ_A. Given $B \models_{\Delta'} E'$, the unique extension of a Δ-homomorphism $f : \mathsf{U}_\phi B \to A$ to a Δ'-homomorphism $\bar{f} : B \to \mathsf{C}_\phi A$ is obtained by first using f to make B into a $(\Psi^\oplus, \Delta'^\oplus, D_A)$-algebra (with unique $(\Psi^\oplus, \Delta'^\oplus, D_A)$-homomorphism $f' : B \to F'_A$) and then observing that uniqueness of $(\Psi^\oplus, \Delta^\oplus, D_A)$-homomorphisms into F_A gives $(\mathsf{U}_{\phi_A} f'); g = f; l$, which implies that $g\restriction_{\mathsf{U}_{\phi_A} Im(f')}$ factors through l; also, $\mathsf{U}_{\iota'_A} Im(f') \models_{\Delta'} \overline{E'}$, since $B \models_{\Delta'} \overline{E'}$. Hence, by maximality of $\mathsf{C}_\phi A$, $Im(f')$ is a $(\Psi^\oplus, \Delta'^\oplus, D_A)$-invariant of $\mathsf{C}_\phi A$ and $\bar{f} : B \to \mathsf{C}_\phi A$ can be defined as f'. Then $f = (\mathsf{U}_\phi \bar{f}); \epsilon_A$ follows by uniqueness of $(\Psi^\oplus, \Delta^\oplus, D_A)$-homomorphisms into F_A. Also, uniqueness of $\bar{f}$ follows from uniqueness of $(\Psi^\oplus, \Delta'^\oplus, D_A)$-homomorphisms into a subalgebra of the final $(\Psi^\oplus, \Delta'^\oplus, D_A)$-algebra.

Theorem 13 now follows from Lemma 14.

Remark. [Jac96] presents a cofreeness result for categories of *behaviour coalgebras*. Objects of such a category G-BCoalg are coalgebras of an endofunctor G : Set → Set, while morphisms between them are given by functions that only commute with the coalgebra structure up to bisimulation. Because of this weaker notion of morphism, an isomorphism class in G-BCoalg is given by an isomorphism class in Set together with a function into the carrier of the final G-coalgebra. The cofree construction is also set-theoretic: given functors $\mathsf{G}, \mathsf{H} : \mathsf{Set} \to \mathsf{Set}$ together with a natural transformation $\eta : \mathsf{H} \Rightarrow \mathsf{G}$ (inducing a forgetful functor U_η : H-BCoalg $\to$ G-BCoalg), the right adjoint R_η : G-BCoalg $\to$ H-BCoalg to U_η is, up to isomorphism, determined by a pullback in Set: if $B \in$ G-BCoalg with $b : B \to F_\mathsf{G}$ as unique G-homomorphism into the final G-coalgebra, then $\mathsf{R}_\eta B$ is determined, up to isomorphism, by the pullback in Set of b along the unique G-homomorphism $! : \mathsf{U}_\eta F_\mathsf{H} \to F_\mathsf{G}$, while the counit is obtained by pulling back ! along b. The inclusion of categories G-Coalg $\hookrightarrow$ G-BCoalg preserves final objects, hence the two cofree constructions are isomorphic in G-BCoalg. The advantage of the construction in [Jac96] over the standard construction stands in reducing the number of bisimilar states (while still implementing the same behaviour). Moreover, the construction in [Jac96] supports the reuse of implementations (the G-structure of B is used in defining the G-structure of its cofree coextension). With our construction, this only happens for Δ-algebras that are extensional (behavioural equivalence is equality), case in which the two constructions coincide.

3.3 A Generalisation

In [Rut96], categories of coalgebras of arbitrary endofunctors $\mathsf{T}, \mathsf{S} : \mathsf{Set} \to \mathsf{Set}$ and forgetful functors U_η : S-Coalg $\to$ T-Coalg induced by natural transformations $\eta : \mathsf{S} \Rightarrow \mathsf{T}$ are considered (U_η maps an S-coalgebra $\gamma : C \to \mathsf{S}C$ to the T-coalgebra $\eta_C \circ \gamma : C \to \mathsf{T}C$) and existence of cofree coalgebras w.r.t. U_η is proved, under the assumption that for any set C, the endofunctor $\mathsf{S} \times C$ on Set (mapping a set X to the set $\mathsf{S}X \times C$) has a final coalgebra.

In the case of one-sorted specifications with no equations, our result can be viewed as an instance of the result in [Rut96] – according to a remark in Section 3.1, the signature map underlying ϕ induces a natural transformation between the endofunctors associated to Δ' and Δ. But our result also applies to specification maps whose underlying signature maps are not surjective on hidden sorts, suggesting a generalisation of the result in [Rut96] to the case when the categories underlying S and T are distinct. This generalisation involves a functor U between these categories and a natural transformation $\eta : \mathsf{U} \circ \mathsf{S} \Rightarrow \mathsf{T} \circ \mathsf{U}$. Existence of a cofree functor w.r.t. U_η is proved under similar assumptions.

Theorem 15. *Let* C *and* D *be categories with binary products and* $\mathsf{U} : \mathsf{D} \to \mathsf{C}$ *be a functor that preserves binary products and has a right adjoint right inverse* R. *Let* $\mathsf{T} : \mathsf{C} \to \mathsf{C}$, $\mathsf{S} : \mathsf{D} \to \mathsf{D}$ *be endofunctors and* $\eta : \mathsf{U} \circ \mathsf{S} \Rightarrow \mathsf{T} \circ \mathsf{U}$ *be a natural transformation (inducing a forgetful functor* U_η : S-Coalg $\to$ T-Coalg*). If the functors* $\mathsf{S} \times \mathsf{R}C$ *and* $\mathsf{T} \times C$ *have final coalgebras for any* C*-object* C*, then* U_η *has a right adjoint* C_η.

Proof. U_η maps an S-coalgebra $\gamma : D \to \mathsf{S}D$ to the T-coalgebra $\mathsf{U}\gamma ; \eta_D : \mathsf{U}D \to \mathsf{TU}D$ (a T-subsystem $\mathsf{U}\gamma ; \eta_D$ is extracted from the S-system γ). A canonical way to build S-systems over T-subsystems is given by the functor C_η, defined on a T-coalgebra $\gamma : C \to \mathsf{T}C$ as follows.

1. Let $\delta : F \to \mathsf{T}F \times C$ be the final $\mathsf{T} \times C$-coalgebra.
2. Let $! : \langle \gamma, id \rangle \to \delta$ be the unique $\mathsf{T} \times C$-homomorphism of $\langle \gamma, id \rangle$ into δ.
3. Let $\delta' : F' \to \mathsf{S}F' \times \mathsf{R}C$ be the final $\mathsf{S} \times \mathsf{R}C$-coalgebra. Then $\langle \eta_{F'}, id \rangle \circ \mathsf{U}\delta'$ is a $\mathsf{T} \times C$-coalgebra with $!' : \langle \eta_{F'}, id \rangle \circ \mathsf{U}\delta' \to \delta$ as unique $\mathsf{T} \times C$-homomorphism into δ.
4. Let $\gamma' : C' \to \mathsf{S}C' \times \mathsf{R}C$ be the greatest $\mathsf{S} \times \mathsf{R}C$-invariant of δ' such that $!' \lceil_{\mathsf{U}C'}$ factors through $!$ in $(\mathsf{T} \times C)$-Coalg and let $\epsilon_C : \mathsf{U}C' \to C$ be the unique factorisation (as $!$ is monic). Define $\mathsf{C}_\eta \gamma$ as $\pi_1 \circ \gamma'$.

The construction is illustrated in the diagram below.

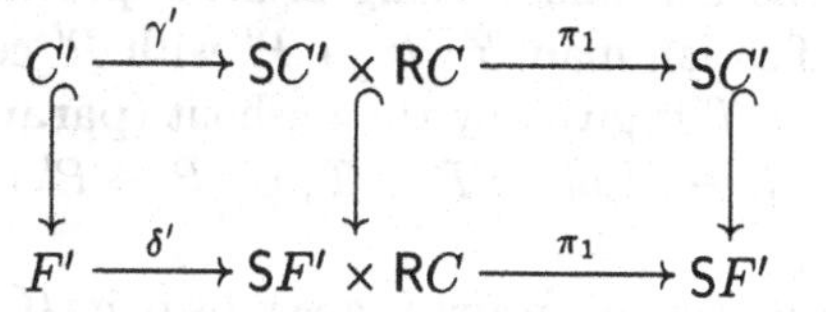

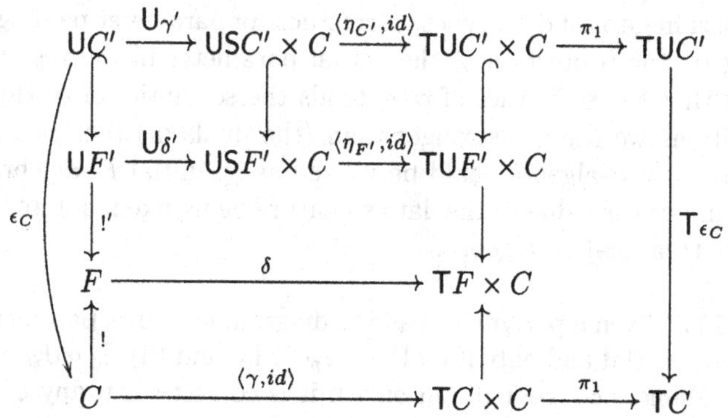

Then, C_η is right adjoint to U_η with counit ϵ: any T-homomorphism $f : \mathsf{U}_\eta \tau \to \gamma$ with $\tau : D \to \mathsf{S}D$ an S-coalgebra induces a $\mathsf{S} \times \mathsf{R}C$-structure on D such that f becomes a $\mathsf{T} \times C$-homomorphism. Uniqueness of $\mathsf{T} \times C$-homomorphisms into F together with maximality of γ' are then used to define an S-homomorphism $\bar{f} : \tau \to \mathsf{C}_\eta \gamma$ such that $\mathsf{U}_\eta \bar{f}; \epsilon_C = f$, in the same way as this was done in Theorem 13.

Remark. By letting $\mathsf{C} = \mathsf{Set}^H$, $\mathsf{D} = \mathsf{Set}^{H'}$, $\mathsf{R} : \mathsf{Set}^H \to \mathsf{Set}^{H'}$ with $(\mathsf{R}A)_{h'} = \prod_{h'=\phi(h)} A_h$, $\mathsf{T} = \mathsf{G}_\Delta$, $\mathsf{S} = \mathsf{G}_{\Delta'}$ and $\eta : \mathsf{U} \circ \mathsf{S} \Rightarrow \mathsf{T} \circ \mathsf{U}$ as in Section 3.1, we obtain Theorem 13 for the case when $E = E' = \emptyset$.

4 Semantics by Cofree Constructions

In this section, cofree functors are used to give semantics to parameterisation and inheritance in coalgebraic hidden algebra.

4.1 Parameterisation

Cofree functors C_ϕ induced by specification morphisms $\phi : P \to T$ provide an appropriate semantics for the importation of coalgebraic hidden modules: supplied with a P-algebra A, the cofree construction provides the most general T-algebra that exhibits the P-behaviour of A. A theory of parameterised modules with cofree constructions as semantics can be developed for coalgebraic hidden algebra in the same style as this was done for data types [EM85] using free constructions. Moreover, a semantic characterisation of correctness of parameter passing in terms of persistence of the cofree functors can be given.

Definition 16. A **coalgebraic parameterised specification** is a specification morphism $\phi : P \hookrightarrow T$ with both P and T coalgebraic. A **parameter passing morphism** for ϕ is a specification map $\psi : P \to P'$ with P' coalgebraic. The **instantiation** of P with ψ in T is given by the pushout (**parameter passing diagram**) $\phi' : P' \to T'$, $\psi' : T \to T'$ of $\phi : P \to T$, $\psi : P \to P'$ in Spec.

The semantics of parameter passing diagrams is given by pairs $(\mathsf{C}_\phi, \mathsf{C}_{\phi'})$ of cofree functors induced by the specification morphisms ϕ and ϕ' (see Theorem 7). As in the case of parameterised data types, correctness of parameter passing is defined by requiring (i) the protection of the actual parameter in the result specification and (ii) that the semantics of ϕ' extends the semantics of ϕ. However, the actual conditions we use are stronger than (the duals of) the ones in [EM85], because there, any P-algebra could be viewed as an initial P'-algebra for some P', whereas in our case, due to the data signature being fixed, not any P-algebra is isomorphic to a final P'-algebra.

Definition 17. Given a parameter passing diagram as above, parameter passing is **correct w.r.t.** ψ if and only if (i) $\mathsf{U}_{\phi'} \circ \mathsf{C}_{\phi'} \simeq \mathsf{Id}$, and (ii) $\mathsf{C}_\phi \circ \mathsf{U}_\psi \simeq \mathsf{U}_{\psi'} \circ \mathsf{C}_{\phi'}$. Parameter passing is **correct** if and only if it is correct w.r.t. any ψ.

Standard compositionality results use **amalgamations** to define the semantics of combined specifications purely on the semantic level [EM85]. Existence of amalgamations in hidden algebra amounts to pushouts in Spec being transformed by the functor $\mathsf{HAlg} : \mathsf{Spec} \to \mathsf{Cat}^{\mathsf{op}}$ into pullbacks in $\mathsf{Cat}^{\mathsf{op}}$.

Lemma 18. *Hidden algebra has amalgamations.*

Proof. By pushouts in Spec being pushouts of the underlying many sorted specifications, together with many sorted amalgamations preserving hidden algebras.

Definition 19. A parameterised specification ϕ is **persistent** if and only if C_ϕ is persistent ($\mathsf{U}_\phi \circ \mathsf{C}_\phi \simeq \mathsf{Id}$).

Lemma 20. *Given a parameter passing diagram as above, if ϕ is persistent then ϕ' is persistent.*

Proof. A consequence of amalgamations being pullbacks is that the functor $\mathsf{Id} \oplus_{\mathsf{U}_\psi} (\mathsf{C}_\phi \circ \mathsf{U}_\psi) : \mathsf{HAlg}(P') \to \mathsf{HAlg}(T')$ (with $\oplus$ denoting amalgamation) is right adjoint to $\mathsf{U}_{\phi'}$ with identity as counit. The conclusion then follows by any two right adjoints being naturally isomorphic.

Theorem 21. *Parameter passing is correct for ϕ if and only if ϕ is persistent.*

Proof. If ϕ is persistent then, by Lemma 20, ϕ' is persistent, hence (i) of Definition 17 holds. (ii) follows from $\mathsf{C}_{\phi'}$ being isomorphic to $\mathsf{Id} \oplus_{\mathsf{U}_\psi} (\mathsf{C}_\phi \circ \mathsf{U}_\psi)$, which gives $\mathsf{U}_{\psi'} \circ \mathsf{C}_{\phi'} \simeq \mathsf{C}_\phi \circ \mathsf{U}_\psi$. The converse follows by taking ψ the identity.

Example 1 Channels. Channels consisting of a sender and a receiver can be specified by parameterising the receiver by the sender. A sender is simply a stream that uses its `send` method to send values `vals`. An alternating sender is a sender that alternates the values it sends. A receiver receives values from a sender `sen` using its `rec` method and stores them in `valr`. The pushout semantics of instantiating `REC` with `ASEN` is a specification denoted `REC[ASEN]` which consists of `REC` together with the equation for alternating streams.

```
obj SEN is pr NAT .
  sort Sen .
  op vals : Sen -> Nat .
  op send : Sen -> Sen .
endo
obj ASEN is using SEN .
  var S : Sen .
  eq vals(send(send(S)) = vals(S) .
endo
```

```
th REC[X :: SEN] is
  sort Rec .
  op valr : Rec -> Nat .
  op sen : Rec -> Sen .
  op rec : Rec -> Rec .
  var R : Rec .
  eq sen(rec(R)) = send(sen(R)) .
  eq valr(rec(R))=vals(sen(R)) .
endth
```

Now consider a `SEN`-algebra `A` implementing alternating streams: $\mathtt{Sen_A} = \mathbb{N} \times \mathbb{N}$, $\mathtt{vals_A(n_1,n_2)} = \mathtt{n_1}$, $\mathtt{send_A(n_1,n_2)} = \mathtt{(n_2,n_1)}$. In constructing its cofree coextension $\mathtt{A}^*$ along $\mathtt{SEN} \hookrightarrow \mathtt{REC}$ we follow the three steps outlined in Section 3.2. First, we build the final $\mathtt{REC} \cup \{\mathtt{s} : \mathtt{Sen} \to \mathtt{Sen_A}\}$-algebra `A1`, having carriers $\mathtt{Sen_{A1}} = \{\mathtt{f} \mid \mathtt{f} : \{\mathtt{send}\}^* \to \mathbb{N} \times \mathtt{Sen_A}\}$, $\mathtt{Rec_{A1}} = \{(\mathtt{g},\mathtt{h}) \mid \mathtt{g} : \{\mathtt{rec}\}^* \to \mathbb{N}, \mathtt{h} : \{\mathtt{rec}\}^*\mathtt{sen}\{\mathtt{send}\}^* \to \mathbb{N} \times \mathtt{Sen_A}\}$. A sender state `f` assigns a sender value and a $\mathtt{Sen_A}$-state to each experiment consisting of a finite number of `send`s. Similarly, a receiver state `(g,h)` assigns a receiver value to each experiment consisting of a finite number of `rec`s, as well as a sender value and a $\mathtt{Sen_A}$-state to each experiment consisting of a finite number of `rec`s followed by `sen` and then by a finite number of `send`s. Second, the greatest subalgebra of `A1` for which examining the $\mathtt{Sen_A}$-state commutes with the `SEN`-operations is taken, resulting in a `REC`-algebra `A2` having carriers $\mathtt{Sen_{A2}} = \mathtt{Sen_{A1}}$ (the second component of `f` on the empty sequence of `send`s uniquely determines `f`) and $\mathtt{Rec_{A2}} = \{(\mathtt{g},\mathtt{h}) \mid \mathtt{g} : \{\mathtt{rec}\}^* \to \mathbb{N}, \mathtt{h} : \{\mathtt{rec}\}^* \to \mathtt{Sen_A}\}$. Finally, imposing the `REC`-equations results in a `REC`-algebra $\mathtt{A}^*$ having carriers: $\mathtt{Sen_{A^*}} = \mathtt{Sen_A}$, $\mathtt{Rec_{A^*}} = \mathbb{N} \times \mathtt{Sen_A}$ (the values of `g` and `h` on the empty sequence of `rec`s uniquely determine `g` and `h`) and operations: $\mathtt{vals_{A^*}} = \mathtt{vals_A}$, $\mathtt{send_{A^*}} = \mathtt{send_A}$, $\mathtt{valr_{A^*}(n,n_1,n_2)} = \mathtt{n}$, $\mathtt{sen_{A^*}(n,n_1,n_2)} = \mathtt{(n_1,n_2)}$, $\mathtt{rec_{A^*}(n,n_1,n_2)} = \mathtt{(n_1,n_2,n_1)}$. `A*` uses the implementation provided by `A` for its sender part.

4.2 Inheritance

Class inheritance (with non-monotonic overriding) can be specified in hidden algebra using (partial) specification maps. Here we use a specification of bank accounts to emphasise the suitability of cofree constructions as a semantics for inheritance.

Example 2 Bank Accounts. Bank accounts `ACC` are specified using a `bal`(ance) attribute and methods for `dep`(ositing)/`with`(drawing) a given amount. More specialised accounts – a history account that maintains a `his`(tory) of the transactions made into the account and a savings account from which withdrawals are only allowed if the account is not in saving state – are then derived from `ACC`. The former specialisation corresponds to inheritance with monotonic overriding, while the latter non-monotonically overrides the `with` method[2].

```
obj ACCSIG is pr INT .
  sort Acc .
  op bal : Acc -> Int .
  ops dep, with : Acc Nat -> Acc .
endo
obj ACC is pr ACCSIG .
  var N : Nat .
  var A : Acc .
  eq bal(dep(A,N)) = bal(A) + N .
  eq bal(with(A,N)) = bal(A) - N .
endo
obj HACC is pr LIST[INT] .
  ex ACC * (sort Acc to HAcc) .
  op his : HAcc -> List .
  var N : Nat .
  var H : HAcc .
  *** monotonic overriding
  eq his(dep(H,N)) = N;his(H) .
  eq his(with(H,N)) = (-N);his(H)
endo
```

```
obj SACC is
  ex ACCSIG * (sort Acc to SAcc) .
  op start, end : SAcc -> SAcc .
  op sav? : SAcc -> Bool .
  var N : Nat .
  var S : SAcc .
  *** monotonic overriding
  eq bal(dep(S,N)) = bal(S) + N .
  eq sav?(dep(S,N)) = sav?(S) .
  *** non-monotonic overriding
  ceq bal(with(S,N)) = bal(S) - N
   if sav?(S) == false .
  ceq bal(with(S,N)) = bal(S)
   if sav?(S) == true .
  eq sav?(with(S,N)) = sav?(S) .
  eq bal(start(S)) = bal(S) .
  eq sav?(start(S)) = true .
  eq bal(end(S)) = bal(S) .
  eq sav?(end(S)) = false .
endo
```

The semantics of the inheritance relation between `HACC` and `ACC` is given by the cofree functor induced by the specialisation of `ACC` to `HACC`. For the inheritance relation between `SACC` and `ACC`, the semantics is given by the composition of the forgetful functor induced by hiding the non-monotonically overridden operation `with` with the cofree functor induced by the specialisation of `ACC` without the `with` method to `SACC`.

[2] In general, only *defined* operations should be non-monotonically overridden. Given a coalgebraic specification (Δ, E), the operations in $\Delta' \subseteq \Delta$ are **defined** if in any (Δ, E)-algebra, behavioural $\Delta \setminus \Delta'$-equivalence is a Δ-congruence. A similar approach is taken in [Jac96], where in addition to a "core" part, a class specification may contain "definable" functions which do not contribute to the meaning of the specification and can therefore be arbitrarily overridden.

Now consider an ACC-algebra A given by: $\mathtt{Acc_A} = \mathtt{Int}$, $\mathtt{bal_A(I)} = \mathtt{I}$, $\mathtt{dep_A(I,J)} = \mathtt{I+J}$, $\mathtt{with_A(I,J)} = \mathtt{I-J}$. Its cofree coextensions to a HACC-algebra HA and a SACC-algebra SA are given below.

$\mathtt{HAcc_{HA}} = \mathtt{Acc_A} \times \mathtt{IntList}$
$\mathtt{bal_{HA}(I,L)} = \mathtt{I}$
$\mathtt{his_{HA}(I,L)} = \mathtt{L}$
$\mathtt{dep_{HA}((I,L),J)} = \mathtt{(I+J, J;L)}$
$\mathtt{with_{HA}((I,L),J)} = \mathtt{(I-J,(-J);L)}$

$\mathtt{SAcc_{SA}} = \mathtt{Acc_A} \times \{\mathtt{true}, \mathtt{false}\}$
$\mathtt{bal_{SA}(I,B)} = \mathtt{I}$
$\mathtt{sav?_{SA}(I,B)} = \mathtt{B}$
$\mathtt{dep_{SA}((I,B),J)} = \mathtt{(I+J,B)}$
$\mathtt{with_{SA}((I,false),J)} = \mathtt{(I-J,false)}$
$\mathtt{with_{SA}((I,true),J)} = \mathtt{(I,true)}$
$\mathtt{start_{SA}(I,B)} = \mathtt{(I,true)}$
$\mathtt{end_{SA}(I,B)} = \mathtt{(I,false)}$

The counit of the adjunction provides coercion operations that map states in the subclass to states in the superclass. In both of the above cases, the coercions are projections extracting the superclass attributes. Also in both cases, the superclass implementation is reused by the subclass.

5 Combining Algebra with Coalgebra

We have illustrated the relevance of final/cofree constructions to coalgebraic hidden specifications and maps between them. Not surprisingly, the existence of final/cofree hidden algebras does not generalise to arbitrary hidden specifications – there is no universal way of interpreting the constructors in either a final or a cofree algebra. However, final/cofree *families* of hidden algebras exist.

The notion of *final family of objects* generalises the notion of final object: given a category C, a family $(F_j)_{j\in J}$ of C-objects is **final** if and only if, for any C-object C, there exist unique $j \in J$ and C-morphism $f : C \to F_j$. Similarly, the notion of *couniversal family of morphisms* [Die79] generalises the notion of couniversal morphism: given a functor $\mathsf{U} : \mathsf{D} \to \mathsf{C}$ and a C-object C, a family of C-morphisms $\epsilon_{C,j} : \mathsf{U}C_j^* \to C$ with C_j^* an object of D for each $j \in J$ is a **couniversal family of morphisms from** U **to** C if and only for any D-object D and C-morphism $f : \mathsf{U}D \to C$, there exist unique $j \in J$ and D-morphism $\bar{f} : D \to C_j^*$ such that $U\bar{f}; \epsilon_{C,j} = f$. If for every C there exists a couniversal family of morphisms from U into C, then U_ϕ is said to have a **right multiadjoint**.

Now let Σ denote a hidden signature with $\Sigma = \Gamma \cup \Delta$ as splitting into hidden subsignatures of constructors and destructors respectively and observe that signature maps preserve such splittings. Also, let F_Δ denote the final hidden Δ-algebra and I_Γ denote the initial hidden Γ-algebra (given by the free many sorted Γ-algebra over D). Finally, let $\mathsf{Set}_D^{V\cup H}$ denote the category of $V \cup H$-sorted sets with $(D_v)_{v\in V}$ as V-components and $V \cup H$-sorted functions with $(id_v)_{v\in V}$ as V-components.

Theorem 22. *For any hidden signature Σ there exists a final family of hidden Σ-algebras.*

Proof. Let $I, F \in \mathsf{Set}_D^{V \cup H}$ be the carriers of I_Γ and F_Δ respectively and let $J = \{j \mid j : I \to F \text{ in } \mathsf{Set}_D^{V \cup H}\}$. Each $j \in J$ uniquely induces a Σ-structure F_j on F such that $F_j \restriction_\Delta = F_\Delta$ and such that j defines a Γ-homomorphism from I_Γ to $F_j \restriction_\Gamma$. Then $(F_j)_{j \in J}$ is a final family of hidden Σ-algebras.

Therefore, the category of hidden Σ-algebras can be sliced into subcategories C_j, $j \in J$, with each C_j having a final object F_j. This justifies using the family $(F_j)_{j \in J}$ as final-like semantics for Σ.

Theorem 23. *Let $\phi : \Sigma \to \Sigma'$ be a hidden signature map. The functor U_ϕ : $\mathsf{HAlg}(\Sigma') \to \mathsf{HAlg}(\Sigma)$ has a right multiadjoint.*

Proof. Let $\phi_\Delta : \Delta \to \Delta'$ denote the restriction of ϕ to destructor subsignatures. For a hidden Σ-algebra A, let $(A \restriction_\Delta)^*$ denote the cofree coextension of $A \restriction_\Delta$ along ϕ_Δ and let J_A denote the family of Σ'-algebras A_j^* such that $A_j^* \restriction_{\Delta'} = (A \restriction_\Delta)^*$ and such that the function underlying $\epsilon_{A \restriction_\Delta} : \mathsf{U}_{\phi_\Delta}(A \restriction_\Delta)^* \to A \restriction_\Delta$ defines a Σ-homomorphism $\epsilon_{A,j} : \mathsf{U}_\phi A_j^* \to A$. Then, the family $(\epsilon_{A,j})_{j \in J_A}$ is a couniversal family of morphisms from U_ϕ to A.

Theorems 22 and 23 can be extended from hidden signatures to *split* hidden specifications. A hidden specification (Σ, E) is called **split** if and only if $E = E_\Delta \cup E_\Sigma$ with E_Δ consisting of state Δ-equations and E_Σ consisting of Σ-equations with visible-sorted variables only. Final families of (Σ, E)-algebras exist for any split specification (Σ, E) – the sub-family $J' \subseteq J$ consisting only of those F_js which behaviourally satisfy E is considered. Also, if (Σ, E) and (Σ', E') are split hidden specifications and $\phi : (\Sigma, E) \to (\Sigma', E')$ is a specification map such that $\phi \restriction_{(\Delta, E_\Delta)} : (\Delta, E_\Delta) \to (\Delta', E'_{\Delta'})$ is also a specification map, then U_ϕ has a right multiadjoint – for each Σ-algebra A, the sub-family $J'_A \subseteq J_A$ consisting only of those Σ'-algebras A_j^* which behaviourally satisfy E' is considered.

6 Conclusions and Future Work

We have investigated the coalgebraic nature of hidden algebra, concentrating on semantical aspects such as finality and cofree constructions. We have proved the existence of cofree hidden algebras along maps between coalgebraic hidden specifications and emphasised their relevance in giving semantics to parameterisation and inheritance. Also, we have sketched a possible generalisation of a cofreeness result from [Rut96]. Finally, the final/cofree semantics has been lifted from coalgebraic to arbitrary hidden algebra.

With the current definition of hidden signatures, hidden constants (operations from visible sorts to hidden sorts) are the only constructor operations allowed. In practice however, new objects can be created by putting together existing objects (e.g. by tupling), suggesting a generalisation of the theory of hidden algebras that allows arbitrary constructors. One expects to still be able to reason coalgebraically about behavioural equivalence, hence Proposition 9 must hold for generalised hidden signatures (preservation of Δ-behavioural equivalence by constructors can be achieved either by imposing it as a constraint on algebras or by fully specifying the Δ-behaviour of the constructors). The extension of the results in this paper to generalised hidden algebra remains to be studied.

The integration of the algebraic and coalgebraic aspects of hidden algebra also deserves further study, perhaps along the lines of [Mal96] where objects are viewed as *algebra-coalgebra pairs*, or [TP97] where a similar notion called *bi-algebra* is considered.

Acknowledgements I would like to thank my supervisor, Dr Grant Malcolm, for his guidance and his comments on several drafts of this paper.

References

[Die79] Y. Diers. Familles universelles de morphismes. *Annales de la Société Scientifique de Bruxelles*, 93(3):175–195, 1979.

[EM85] H. Ehrig and B. Mahr. Fundamentals of algebraic specification 1: Equations and initial semantics. In *EATCS Monographs on TCS*. Springer, 1985.

[GD94] J. Goguen and R. Diaconescu. Towards an algebraic semantics for the object paradigm. In H. Ehrig and F. Orejas, editors, *Recent Trends in Data Type Specification*, number 785 in LNCS. Springer, 1994.

[GM97] J. Goguen and G. Malcolm. A hidden agenda. to appear, 1997.

[Jac95] B. Jacobs. Mongruences and cofree coalgebras. In V.S. Alagar and M. Nivat, editors, *Algebraic Methods and Software Technology*, number 936 in LNCS. Springer, 1995.

[Jac96] B. Jacobs. Inheritance and cofree constructions. In P. Cointe, editor, *European Conference on Object-Oriented Programming*, number 1098 in LNCS. Springer, 1996.

[Jac97] B. Jacobs. Invariants, bisimulations and the correctness of coalgebraic refinements. Technical Report CSI-R9704, University of Nijmegen, 1997.

[JR97] B. Jacobs and J. Rutten. A tutorial on (co)algebras and (co)induction. *Bulletin of the EATCS*, 62:222–259, 1997.

[Mal96] G. Malcolm. Behavioural equivalence, bisimilarity and minimal realisation. In M. Haveraaen, O. Owe, and O.-J. Dahl, editors, *Recent Trends in Data Type Specifications*, number 1130 in LNCS. Springer, 1996.

[MG94] G. Malcolm and J. Goguen. Proving correctness of refinement and implementation. Technical Monograph PRG-114, Oxford University, 1994.

[Rei95] H. Reichel. An approach to object semantics based on terminal coalgebras. *Mathematical Structures in Computer Science*, 5, 1995.

[Rut95] J. Rutten. A calculus of transition systems (towards universal coalgebra). Technical Report CS-R9503, CWI, 1995.

[Rut96] J. Rutten. Universal coalgebra: a theory of systems. Technical Report CS-R9652, CWI, 1996.

[TP97] D. Turi and G. Plotkin. Towards a mathematical operational semantics. In *Proceedings LICS*, 1997.

[TWW82] J. Thatcher, E. Wagner, and J. Wright. Data type specification: Parameterization and the power of specification techniques. *ACM Transactions on Programming Languages and Systems*, 4(4), 1982.

A Completeness Result for Equational Deduction in Coalgebraic Specification*

Andrea Corradini

Dipartimento di Informatica, Corso Italia 40, 56125, Pisa
email: `andrea@di.unipi.it`

Abstract. The use of coalgebras for the specification of dynamical systems with a hidden state space is receiving more and more attention in the years, as a valid alternative to algebraic methods based on observational equivalences. However, to our knowledge, the coalgebraic framework is still lacking a complete equational deduction calculus which enjoys properties similar to those stated in Birkhoff's completeness theorem for the algebraic case.
In this paper we present a sound and complete equational calculus for a restricted class of coalgebras. We compare our notion of coalgebraic equation to others in the literature, and we hint at possible extensions of our framework.

1 Introduction

In recent years there has been a growing interest in the theory of coalgebras, motivated by the fact that they are particularly suitable to specify, in an implementation independent way, a wide class of systems, typically discrete dynamical systems with a hidden state space [Rut96, Jac96a, Rei95, HS95]. Examples include various kinds of transition systems, deterministic and nondeterministic automata, (concurrent) objects, hybrid systems, and (possibly) infinite data structures like streams and trees.

In the theory of *algebraic specification*, an abstract data type is specified by a set of operations (constructors) which determine how values of the carrier are built up, and a set of formulas (in the simplest case equations) stating which values should be identified. In the standard initial semantics the defining equations impose a congruence on the initial algebra.

Dually, the coalgebraic specification of a class of systems is characterized by a set of operations, sometimes called destructors, which tell us how a *state* (i.e., an element of the carrier) can be observed and transformed to successor states. Also in this case it is often convenient to impose additional defining conditions which restrict the range of possible observations and transitions. In this paper we will stick to the case where such conditions are expressed by mean of defining equations only. Using the standard final semantics, defining equations determine

* Research carried on while the author was visiting the CWI, Amsterdam, supported by the EC Fixed Contribution Contract n. EBRFMBICT960840.

a sub-coalgebra of the final coalgebra, containing only the behaviours of interest. In particular, as explained in [HR95], properties expressible by equations in this framework are *safety* or *invariant* properties, i.e., properties that must hold in any possible state of a coalgebra.

A natural question is whether for equations in a coalgebraic framework there is a complete calculus of deduction, enjoying properties similar to those of the classical rules of equational deduction in an algebraic framework, as stated by Birkhoff's completeness theorem. This is the main topic of this paper, where we present some preliminary results.

Before summarizing the contents of the paper, it is worth recalling that many authors agree on the fact that it is useful to employ coalgebraic techniques for the specification of systems with a hidden state space, but to stick to *algebraic* techniques for the specification of the involved data. This is consistent with the purely algebraic approaches which use initial semantics for the specification of data structures, and final semantics (based on behavioural equivalences) for state spaces (see for example [ONS93, BHW95] and the references therein). The results in this paper are based instead on a purely coalgebraic approach: we leave as a topic of future research to investigate how far they can be generalized to a hybrid coalgebraic/algebraic framework.

We start introducing in Section 2 the class of coalgebras we are concerned with. They are presented in an algebraic style, by providing a *co-signature*, i.e., a signature satisfying strong constraints on sorts and operations. In particular, sorts include one single "hidden sort", corresponding to the carrier of the coalgebra, and other "visible" sorts for inputs and outputs, which are given a fixed interpretation. In transition system terminology, such coalgebras can model deterministic, non-terminating transition systems with inputs and outputs, but cannot model in a direct way possibly terminating or non-deterministic systems for which more general coalgebras would be needed.

Clearly, to speak about "equations" in a coalgebraic setting one first needs to understand what "terms" are. In the related literature [HR95, Jac96a], the terms used in coalgebraic equations are standard "algebraic" terms built from constructors and destructors, with the restriction that only one variable of the hidden sort can appear in an equation. We are more strict in this respect, because of our commitment to a pure coalgebraic framework. Firstly, we only allow destructors and not constructors; thus visible sorts will be interpreted as sets without any algebraic structure defined on them. Coalgebraic terms, built only over destructors, have for us a precise interpretation as the basic *experiments* or *observations* that one can make on the states of a coalgebra. As such, they will be further constrained to include only constants of input sorts, and to be of visible sort, because the result of an observation cannot be a hidden state.

In the general case, using an object-based terminology, an observation on a state of a coalgebra consists of performing a sequence of (possibly parametrized) methods (or transitions), followed by a (possibly parametrized) attribute which delivers the observed result. As an equation is just a pair of terms denoting observations, such equation is valid in a coalgebra if the two observations return the same result for all the states of the carrier. This notion of validity is used in a standard way in Section 4 to define the class of models of a *coalgebraic specification* (i.e., a co-signature and a set of equations). Next the main result of the paper is presented, namely a set of equational deduction rules which is shown to be sound and complete, in the sense that an equation can be deduced from a set of equations E if and only if it is valid in all models for E. For the proof of the theorem, it turned out to be useful to have at hand an explicit description of the final coalgebra for a given co-signature, which is introduced in Section 3. The leading idea there is that an element of the final coalgebra is characterized by all the experiments or observations one can perform on it, thus it is just a function from the set of possible observations to corresponding output values.

In Section 5 we compare the expressive power of our notion of coalgebraic equations with those proposed in the related literature. It is easily shown that our equations are less expressive, because of the strong restrictions we impose on terms. Nevertheless, we hint in the concluding section at possible extensions of our main result that should allow us to recover greater expressive power. Some of these extensions are worked out in the full paper [Cor97], where the reader can also find the proofs which are missing here.

2 Coalgebras

In this section we introduce the class of coalgebras which will be considered in the rest of the paper. Coalgebras will be introduced in an "algebraic style", by providing a set of sorts and operator names satisfying suitable restrictions. This style of presentation is essentially borrowed from works by Bart Jacobs [Jac96b, Jac96a] (see also [HR95, Rei95]), and it is related to the more usual functorial definition of coalgebras in [Cor97]. The restrictions are so strong, that we prefer to introduce a new terminology instead of calling sorts and operations a "signature", as one would expect.

Definition 1 (coalgebraic signatures and coalgebras). A *(one sorted) coalgebraic signature* or *co-signature* is a triple $\Pi = \langle S, OP, [\![\_]\!] \rangle$, where S, the *sorts*, OP, the *operators*, and $[\![\_]\!]$ the *interpretation of visible sorts* are as follows:

- S is a triple $S = \langle X, \{I_1, \ldots, I_k\}, \{O_1, \ldots, O_h\} \rangle$ where X is the *hidden sort*, I_j is an *input sort* for $j \in \underline{k}$,[2] and O_j is an *output sort* for $j \in \underline{h}$. The sets of input and output sorts do not need to be disjoint, and their elements are also called *visible sorts*.
- OP is a pair of sets $OP = \langle \{m_1, \ldots m_l\}, \{a_1, \ldots, a_n\} \rangle$, where $m_j : X \times I_{k_j} \to X$ is a *method* for $j \in \underline{l}, k_j \in \underline{k}$, and $a_j : X \times I_{k_j} \to O_{h_j}$ is an *attribute* for $j \in \underline{n}, k_j \in \underline{k}, h_j \in \underline{h}$.
- $[\![\_]\!]$ is a function mapping each visible sort to a non-empty set. For each visible sort V and each element $v \in [\![V]\!]$, $\underline{v} : V$ will be a constant denoting element v.

[2] For a natural number n, by $\underline{n}$ we denote the set $\{1, \ldots, n\}$; thus $\underline{0} = \emptyset$.

A *Π-coalgebra* A consists of a set X_A, the *carrier*, a function $m_{j_A} : X_A \times [\![I_{k_j}]\!] \rightarrow X_A$ for each method, and a function $a_{j_A} : X_A \times [\![I_{k_j}]\!] \rightarrow [\![O_{h_j}]\!]$ for each attribute. The class of all Π-coalgebras is denoted $Coalg(\Pi)$.

Thus all the sorts appearing in a co-signature have a fixed interpretation, but for the hidden sort. The separation between sorts and their interpretation is just for conceptual clarity, and we will simply ignore it in the following by denoting both the sort and the corresponding set by the same symbol, and dropping the third component of a co-signature. Such co-signatures are "one sorted" because only one hidden sort is allowed; in our opinion the notions and results presented in this paper should lift smoothly to the many sorted case, as in the algebraic case, but we did not check the details yet.

Allowing exactly one input argument (or parameter) for methods and attributes is not a restriction, because the input set can be a cartesian product. For example, a method with n parameters $m : X \times I_1 \times \ldots \times I_n \rightarrow X$ will be denoted by $m : X \times \prod_{j \in \underline{n}} I_j \rightarrow X$, and set $\prod_{j \in \underline{n}} I_j$ will be listed as an input sort. If $n = 0$, since the empty product yields a one-element set $\mathbf{1}$, the parameterless method $m : X \times \mathbf{1} \rightarrow X$ will be denoted by $m : X \rightarrow X$.

In object-based terminology, the coalgebras just introduced are expressive enough to specify parametric methods and attributes. In transition system terminology, such coalgebras can model deterministic, non-terminating transition systems with inputs and outputs, but cannot model in a direct way possibly terminating or non-deterministic systems: this could be obtained for example by allowing methods to be relations instead of total functions.

Definition 2 (coalgebra homomorphisms, sub-coalgebras). Given two Π-coalgebras A and B, a *homomorphism* $f : A \rightarrow B$ is a function between their carriers $f : X_A \rightarrow X_B$ such that for each method $m : X \times I \rightarrow X$ it holds $m_B(f(x), v) = f(m_A(x, v))$ for all $x \in X_A$ and $v \in I$, and for each attribute $a : X \times I \rightarrow O$ it holds $a_B(f(x), v) = a_A(x, v)$ for all $x \in X_A$ and $v \in I$.

A Π-coalgebra Z is *final* if for each Π-coalgebra A there is only one homomorphism to Z, which we will denote $!_A : A \rightarrow Z$.

A is a *sub-coalgebra* of B if $X_A \subseteq X_B$, and the inclusion $A \hookrightarrow B$ is a homomorphism. A subset $S \subseteq X_B$ of the carrier of a Π-coalgebra B is the carrier of a sub-coalgebra of B iff for all methods $m : X \times I \rightarrow X$ and for all $x \in S, v \in I$, we have $m_B(x, v) \in S$. In this case the coalgebraic structure on S is obtained by restricting all functions of B to the subcarrier S.

Example 1 (deterministic transition systems). Let TS be the co-signature $TS = \langle\langle X, \mathbf{1}, O\rangle, \langle next : X \rightarrow X, val : X \rightarrow O\rangle\rangle$,[3] where $O = \{o_0, o_1, \ldots, o_n\}$ (we assume that $n > 1$). A TS-coalgebra can be interpreted as a non-terminating, deterministic transition system: the method *next* returns for each state in the carrier the successor state; the attribute *val* returns an observation in O for each state. As we have no way to fix an initial state, one should think of a

[3] To improve readability, a singleton set is denoted by its only element.

TS-coalgebra as specifying the collection of all the transition sequences starting from all possible states. Here are some examples of TS-coalgebras:

1. $SW = \langle\{\mathtt{ON},\mathtt{OFF}\}, next_{SW} : \{\mathtt{ON} \mapsto \mathtt{OFF}, \mathtt{OFF} \mapsto \mathtt{ON}\}, val_{SW} : \{\mathtt{ON} \mapsto o_1, \mathtt{OFF} \mapsto o_0\}\rangle$. This coalgebra represents a system which loops forever between the two states, producing an infinite sequence (a *stream*) of alternating o_0 and o_1. Clearly, the first element of the stream depends on the state we choose to start with.
2. For every $0 < m \leq n$, let TS_m be the coalgebra $\langle \mathbb{N}, next_m : i \mapsto i+1, val_m : i \mapsto o_{(i \bmod m)}\rangle$. As a system, TS_m never passes twice through the same state, and outputs a cyclic stream where the elements $o_0, o_1, \ldots, o_m$ are repeated forever in that order. Note that the possible output streams of TS_2 are the same as those of SW; actually it is easy to check that there is a homomorphism from TS_2 to SW mapping even numbers to $\mathtt{OFF}$ and odd numbers to $\mathtt{ON}$.
3. Let $O^{\mathbb{N}}$ be the set of all streams of elements of O, w range over $O^{\mathbb{N}}$ and o range over O. Then the coalgebra Z_{TS} is defined as $\langle O^{\mathbb{N}}, next_Z : o \cdot w \mapsto w, val_Z : o \cdot w \mapsto o\rangle$. It is easy to see that Z_{TS} is a final TS coalgebra: if A is a TS-coalgebra, and $x \in X_A$, define $!_A(x) = w$, where $w \in O^{\mathbb{N}}$ is the stream of values returned in A starting from state x. In fact, this mapping satisfies the conditions of Definition 2 and it can be shown that it is the only one.

Example 2 (bank accounts). Let BA be the co-signature $\langle\langle X, \{\mathbb{Z}, \mathbf{1}\}, \mathbb{Z}\rangle, \langle ch : X \times \mathbb{Z} \to X, bal : X \to \mathbb{Z}\rangle\rangle$, where $\mathbb{Z}$ is the set of integers. A BA-coalgebra can be interpreted as a collection of (very rudimentary) ***bank account*** states. At a state x of the carrier, two operations are possible: to see the *bal*ance attribute of the state, $bal(x)$, which is an integer, or to *ch*ange the account state to a new state $ch(x,z)$ using an input integer z. Here are some examples of BA-coalgebras:

1. $BA_1 = \langle \mathbb{Z}, ch_1 : \langle z, z'\rangle \mapsto z + z', bal_1 : z \mapsto z\rangle$. This coalgebra models a correct bank account, which records in the state the total amount of the deposited (or withdrawn) money, and returns it when bal is applied.
2. $BA_0 = \langle \mathbf{1} = \{*\}, ch_0 : \langle *, z'\rangle \mapsto *, bal_0 : * \mapsto 0\rangle$. This models an account consisting of a single state, which always returns 0.
3. Let w range over $\mathbb{Z}^*$, the set of finite strings over $\mathbb{Z}$, and let $\Sigma(w)$ denote the sum of all integers in w. Then $BA_h = \langle \mathbb{Z}^*, ch_h : \langle w, z\rangle \mapsto w \cdot z, bal_h : w \mapsto \Sigma(w)\rangle$ is again a correct account, which also records in the state the history of all deposited sums.
4. Let $Z_{BA} = \langle\{\psi : \mathbb{Z}^* \to \mathbb{Z}\}, ch_Z : \langle \psi, z\rangle \mapsto \lambda w \,.\, \psi(z \cdot w), bal_Z : \psi \mapsto \psi(\epsilon)\rangle$. We will see in Section 3 that this is a final BA-coalgebra.

3 On the Structure of Final Coalgebras

In Universal Algebra, even if the initial algebra of a signature is only determined up to isomorphism, we usually have a concrete representation of its elements in mind which are the ground terms built over the signature. Furthermore in an *algebraic specification*, i.e., a pair $\langle \Sigma, E \rangle$ where E is a set of equations over Σ, terms (possibly non-ground) play a fundamental rôle, as they appear in equations. Similarly, an explicit description of the elements of the final coalgebra of a given co-signature will provide the ingredients for defining equations and the background of the proofs of the main theorem in the next section. The existence of final coalgebras is ensured by the fact that a co-signature determines the class of coalgebras for a *restricted polynomial functor*, as shown in [Cor97], and by the fact that every polynomial functor is bounded [Bar93, Rut96]. The leading idea in what follows is that the elements of the hidden sort of the final coalgebra are functions from sets of contexts of visible sort to elements of the corresponding sorts. In other words, one determines the identity of a state of the final coalgebra by looking at all possible observations over that state.

Definition 3 (contexts, transitions, observations). Let Π be a co-signature. A *context* c *of sort* Y *over* Π, denoted $c : Y$, is a well-sorted term of sort Y built from the operators of Π and constants of input sorts, *containing exactly one occurrence of variable, which is denoted* x, *and which must be of hidden sort.*[4]

A context of hidden sort is also called a *transition sequence*; the *empty context* is $x : X$; and a *transition* is a transition sequence having only the empty context as proper sub-context. The set of transitions for Π will be denoted $\mathcal{T}rans(\Pi)$, and the set of transition sequences $\mathcal{T}rans^*(\Pi)$. An *observation* $c : O$ is a context of output sort; the set of observations for Π will be denoted $\mathcal{O}bs(\Pi)$. If $c : Y$ is a context and $t : X$ is a transition sequence, by $c[t/x]$ we denote the context (of sort Y) obtained by replacing the only occurrence of x in c by t.[5]

Given a Π-coalgebra $A = \langle X_A, \{m_{j_A}\}_{j \in \underline{l}}, \{a_{j_A}\}_{j \in \underline{n}} \rangle$, every context $c : Y$ over Π determines a function $[c]_A$, the *interpretation of* c *in* A, having X_A as domain and defined in the following way:

1. $[x]_A : X_A \to X_A$ is the identity function;
2. If $c : X$ is a non-empty transition sequence, then it must be of the form $c = m(c', \underline{v})$, for a transition sequence $c' : X$, a method m, and a constant of input sort $\underline{v} : I$. In this case function $[c]_A : X_A \to X_A$ is inductively defined as $[c]_A(y) = m_A([c']_A(y), v)$ for all $y \in X_A$.
3. Similarly, if $c : O$ is an observation, then it must be of the form $c = a(c', \underline{v})$, for a transition sequence $c' : X$, an attribute a and a constant of input sort $\underline{v} : I$. In this case function $[c]_A : X_A \to O$ is defined as $[c]_A(y) = a_A([c']_A(y), v)$ for all $y \in X_A$.

[4] In an algebraic framework contexts are usually defined as terms "with a hole", i.e., with a single occurrence of a placeholder, denoted $[]$. Thus *ground* contexts where $[]$ is of hidden sort correspond exactly to those in our definition, if we regard variable x as playing the rôle of $[]$. Our notation is justified by the desire of being as far as possible consistent with the related literature (see the discussion in Section 5).

[5] There is an obvious bijection between sequences of elements of $\mathcal{T}rans(\Pi)$ and elements of $\mathcal{T}rans^*(\Pi)$, mapping the empty sequence to the empty context, and sequence $c \cdot c'$ to $c'[c/x]$. Hence the name "transition sequences" for the elements of $\mathcal{T}rans^*(\Pi)$.

Proposition 4 (properties of context interpretation). *The following useful properties hold:*

1. *If $c \in Trans^*(\Pi)$ and $c' \in Obs(\Pi)$, then $[c']_A([c]_A(y)) = [c'[c/x]]_A(y)$ for all $y \in X_A$.*
2. *If $f : A \to B$ is a homomorphism and $c \in Trans^*(\Pi)$, then $f([c]_A(y)) = [c]_B(f(y))$ for all $y \in X_A$.*
3. *If $f : A \to B$ is a homomorphism and $c \in Obs(\Pi)$, then $[c]_A(y) = [c]_B(f(y))$ for all $y \in X_A$.* □

Example 3. Let TS be the co-signature of Example 1. Then the following are legal contexts, with the associated sort: $x : X$ (the empty context), $next(x) : X$ (a transition), $val(next(next(x))) : O$ (an observation). Making reference to the TS-coalgebra SW introduced in that example, we have for example that $[next(x)]_{SW} = next_{SW}$, and $[val(next(next(x)))]_{SW} = val_{SW} \circ next_{SW} \circ next_{SW} = \{\texttt{ON} \mapsto o_1, \texttt{OFF} \mapsto o_0\}$.

Let now BA be the co-signature of Example 2. We have the following:

- $bal(ch(x,z)) : \mathbb{Z}$ is a well-sorted term, but it is not a context because it contains two variables.
- $bal(x) + \underline{5} : \mathbb{Z}$ is not a context, because operator '+' does not belong to the co-signature.
- $bal(ch(x,\underline{5})) : \mathbb{Z}$ is a legal observation. Making reference to the BA-coalgebras of Example 2, we have $[bal(ch(x,\underline{5}))]_{BA_1} = z \mapsto z+5$, $[bal(ch(x,\underline{5}))]_{BA_0} = * \mapsto 0$, $[bal(ch(x,\underline{5}))]_{BA_h} = w \mapsto \Sigma(w)+5$, $[bal(ch(x,\underline{5}))]_{Z_{BA}} = \psi \mapsto \psi(5)$.
- $ch(x, bal(x)) : X$ is not a context because there are two occurrences of the variable x (see Section 5).

Theorem 5 (structure of the final coalgebra). *Let Π be a co-signature as in Definition 1, and let Z_Π be defined as*[6]

$$Z_\Pi = \prod_{(c:O)\in Obs(\Pi)} O$$

Spelling out the definition of dependent product, we obtain the following equivalent definition of Z_Π, that we shall use along the paper:

$$Z_\Pi = \{\psi : Obs(\Pi) \to \coprod_{i\in\underline{h}} O_i \mid \forall (c:O) \in Obs(\Pi) \,.\, \psi(c) \in O\}$$

That is, Z_Π is the set of all functions having the set of observations as domain, and mapping each observation of sort O to a value in O.

Next, for each method $m : X \times I \to X$ in Π define $m_Z : Z_\Pi \times I \to Z_\Pi$ as $m_Z(\psi, v) = \lambda c \,.\, \psi(c[m(x,\underline{v})/x])$, and for each attribute $a : X \times I \to O$ in Π define $a_Z : Z_\Pi \times I \to O$ as $a_Z(\psi, v) = \psi(a(x,\underline{v}))$. These operations, that are clearly well-defined, turn Z_Π into the carrier of a Π-coalgebra (that will be denoted in the same way). Then

[6] This compact notation using dependent products was suggested by Bart Jacobs.

1. *Z_Π is a final Π-coalgebra;*
2. *For each observation $c : O \in \mathcal{O}bs(\Pi)$, the interpretation of c in Z_Π, $[c]_Z : Z_\Pi \to O$ is given by $[c]_Z : \psi \mapsto \psi(c)$.* □

Example 4 (final coalgebras). For the co-signature TS of Example 1 the set of observations is clearly $\mathcal{O}bs(TS) = \{val(x) : O, val(next(x)) : O, \ldots, val(next^n(x)) : O, \ldots\}$ which is isomorphic to $\mathbb{N}$. Thus by the last result, the carrier of the final coalgebra is given by $\{\psi : \mathcal{O}bs(TS) \to O\} \cong \{\psi : \mathbb{N} \to O\} \cong O^{\mathbb{N}}$, thus as expected it is isomorphic to the carrier of Z_{TS}.

For the co-signature BA, the possible observations are $\mathcal{O}bs(BA) = \{bal(x)\} \cup \{bal(ch(x, z_1)) \mid z_1 \in \mathbb{Z}\} \cup \{bal(ch(ch(x, z_1), z_2)) \mid z_1, z_2 \in \mathbb{Z}\} \cup \ldots \cong \mathbb{Z}^*$, showing that coalgebra Z_{BA} of Example 2 is final.

4 Equational deduction in coalgebras

In this section we present the main result of the paper, by introducing an equational calculus for coalgebras and proving its soundness and completeness. We use an algebraic terminology, avoiding to prefix all nouns to be introduced with "co-", even if this would be probably more correct.

Definition 6 (terms). For a co-signature Π, the set of *Π-terms* contains all observations for Π and all constants of output sort, formally: $Terms(\Pi) = \{t : O \mid (t : O) \in \mathcal{O}bs(\Pi)\} \cup \{\underline{o} : O \mid o \in O \text{ and } O \text{ is an output sort of } \Pi\}$.

Given a Π-coalgebra A, every term $t : O$ induces a function $[t]_A : X_A \to O$, defined as in Definition 3 if $t : O$ is an observation, while if it is a constant $\underline{o} : O$, then $[\underline{o}]_A : y \mapsto o$ for all $y \in X_A$.

Definition 7 (equations and validity). An *equation over Π* is a pair $\langle t_1 : O, t_2 : O\rangle$ of terms of the same sort, usually written $t_1 =_O t_2$.

Given a Π-coalgebra A and an element of its carrier $y \in X_A$, we say that equation $t_1 =_O t_2$ *holds for y in A* if $[t_1]_A(y) = [t_2]_A(y)$, denoted $y, A \models t_1 =_O t_2$, or simply $y \models t_1 =_O t_2$ if A is clear from the context. The same equation is *valid* in A, denoted $A \models t_1 =_O t_2$, if it holds in A for all elements of the carrier, i.e., if $[t_1]_A$ and $[t_2]_A$ are the same function from X_A to O. These notions extend in the expected way to a set of equations E: we write $A \models E$ if $A \models e$ for all $e \in E$. Also, if $\mathcal{A}$ is a class of algebras, we write $\mathcal{A} \models E$ if $A \models E$ for all $A \in \mathcal{A}$.

A *coalgebraic specification* is a pair $\langle \Pi, E\rangle$, where E is a set of equations over the co-signature Π. Given a coalgebraic specification $\langle \Pi, E\rangle$, the class of its *models* $Coalg(\Pi, E)$ is defined as $Coalg(\Pi, E) = \{A \in Coalg(\Pi) \mid A \models E\}$

Definition 8 (rules of coalgebraic equational deduction). The *deduction rules of coalgebraic equational logic* are the following, where t, t', t'' range over $Terms(\Pi)$, O, O' over output sorts, and all such symbols are universally quantified if they don't appear in the premises; furthermore, all equations are assumed to be well-sorted:

[reflexivity, symmetry, transitivity]

$$\frac{}{\emptyset \vdash t =_O t} \qquad \frac{E \vdash t =_O t'}{E \vdash t' =_O t} \qquad \frac{E_1 \vdash t =_O t', E_2 \vdash t' =_O t''}{E_1 \cup E_2 \vdash t =_O t''}$$

[unity] For all output sorts O such that $card(O) = 1$,

$$\frac{}{\emptyset \vdash t =_O t'}$$

[contradiction] For all output sorts O and $o_1, o_2 \in O$ with $o_1 \neq o_2$,

$$\frac{E \vdash \underline{o_1} =_O \underline{o_2}}{E \vdash t =_{O'} t'}$$

[forward closure] For all transitions $c \in \mathcal{T}rans(\Pi)$,

$$\frac{E \vdash t =_O t'}{E \vdash t[c/x] =_O t'[c/x]}$$

where term $t[c/x] : O$ is defined as $\underline{o} : O$ if $t = \underline{o}$, and as the term obtained by substituting c for the only occurrence of x in t, if t is an observation.

Given a set of equations E and an equation e, we write $E \vdash e$ if there is a proof of $E' \vdash e$, with $E' \subseteq E$, using only the above rules and the following one:

[axiom]

$$\frac{e \in E}{\{e\} \vdash e}$$

In the following we shall sometimes denote by $\hat{E}$ the *theory* of E, i.e., the set $\hat{E} = \{e \mid E \vdash e\}$.

Let us briefly comment the rules just introduced. The first five rules express structural properties of equality: *reflexivity*, *symmetry* and *transitivity* are standard and do not need any comment; and obviously *unity* and *contradiction* are sound rules, due to the fixed interpretation of output sorts. In particular, *contradiction* allows one to generate any possible equation in the case of an "inconsistent" specification which equates two distinct constants.[7]

The last rule, *forward closure*, is the only one closely related to the coalgebraic structure: it states that if two observations deliver the same result when applied to each state, then they deliver the same result also when applied to a state reachable through a transition c, essentially because the carrier of the coalgebra is closed with respect to the application of methods. It is worth stressing that this rule can be considered as the dual of the *congruence* rule used in algebraic specification, stating that for every operator f of arity n, if $E \vdash t_i = t'_i$ for $i \in \underline{n}$, then $E \vdash f(t_1, \ldots, t_n) = f(t'_1, \ldots, t'_n)$.

[7] The *soundness* of a rule equivalent to *contradiction* is remarked also by Lawrence Moss in [Mos97] in a much richer logical framework. Theorem 11 below shows that, as far as an equational calculus is concerned, such a rule contributes in an essential way to completeness. A similar rule was suggested in a private discussion by Bart Jacobs.

Example 5. Let TS be the co-signature of Example 1, and let $E_1 = \{val(x) =_O val(next(next(x)))\}$ and $E_2 = \{val(next(x)) =_O \underline{o_0}\}$ be two sets of equations.

By repeated use of rules *forward closure*, *transitivity* and *reflexivity* it is easy to see that $E_1 \vdash val(next^n(x)) =_O val(next^m(x))$ if and only if $|n-m|$ is even. Furthermore, $E_2 \vdash val(next^n(x)) =_O \underline{o_0}$ for all $n \geq 1$. Note that rule *unity* cannot be used because we assumed that $card(O) > 1$.

Theorem 9 (soundness of coalgebraic equational deduction). *Let Π be a co-signature and let E be a set of equations over Π. Then for all equations $t =_O t'$ over Π,*

$$E \vdash t =_O t' \quad \Longrightarrow \quad Coalg(\Pi, E) \models t =_O t'$$

Proof. We proceed by induction on the depth of the proof that $E \vdash t =_O t'$. If the last rule used is *axiom*, the statement holds by definition of $Coalg(\Pi, E)$. If it is *reflexivity*, *symmetry* or *transitivity*, then the statement follows by the standard properties of equality. If the last rule applied is *unity*, then set O is a singleton, and for all $A \in Coalg(\Pi, E)$ the fact that $[t]_A = [t']_A$ as functions from X_A to O follows by finality of O.

If the last rule applied is *contradiction*, then we first show that $Coalg(\Pi, E)$ contains only the empty coalgebra. In fact, suppose that $A \in Coalg(\Pi, E)$ is such that $A \models \underline{o_1} =_O \underline{o_2}$ and $y \in X_A$; this implies that $[\underline{o_1}]_A(y) =_O [\underline{o_2}]_A(y)$, i.e., $o_1 = o_2$, contradicting the premise of the rule. Now, from the definition of validity it follows immediately that every equation is valid in the empty coalgebra; thus it is the only coalgebra in $Coalg(\Pi, E)$, and it follows that $Coalg(\Pi, E) \models e$ for every equation e, showing the soundness of the rule.

If the last rule is *forward closure*, suppose by induction hypothesis that $A \models t =_O t'$, and that $c \in Trans(\Pi)$ is a transition. By definition, for all $y \in X_A$ we have $y \models t =_O t'$, which implies that $[c]_A(y) \models t =_O t$, i.e., $[t]_A([c]_A(y)) = [t']_A([c]_A(y))$ for all $y \in X_A$. Then the statement follows by Proposition 4 (2), because for each term t, $[t]_A([c]_A(y)) = [t[c/x]]_A(y)$. □

Next we show that the class of models of a coalgebraic specification $Coalg(\Pi, E)$ has a final object, which is a sub-coalgebra of the final Π-coalgebra, and will play a central role in the proof of the completeness result below.

Proposition 10 (final model of a coalgebraic specification). *Let Π be a co-signature and let E be a set of equations over Π. Let Z_Π be the final coalgebra for Π (as for Theorem 5) and let*

$$Z_{\Pi\downarrow E} = \{\psi \in Z_\Pi \mid \psi \models \hat{E}\}$$

where $\hat{E}$ is the theory of E. Then $Z_{\Pi\downarrow E}$ is the carrier of a final coalgebra in $Coalg(\Pi, E)$.

Proof. We first show that $Z_{\Pi\downarrow E}$ is the carrier of a sub-coalgebra of Z_Π, then that it belongs to $Coalg(\Pi, E)$, and lastly that it is a final coalgebra there.

For the first point, let $m : X \times I \to X$ be a method in Π, m_Z be its interpretation in Z_Π, $\psi \in Z_{\Pi\downarrow E}$ and $v \in I$. By Definition 2 we have to show that $m_Z(\psi, v) \in Z_{\Pi\downarrow E}$, i.e., that $m_Z(\psi, v) \models e$ for all e such that $E \vdash e$. In fact, suppose that $E \vdash t =_O t'$; then $[t]_Z(m_Z(\psi, v)) =$ [using Definition 3] $= [t[m(\psi, \underline{v})/x]]_Z(\psi) =$ [by *forward closure*] $= [t'[m(\psi, \underline{v})/x]]_Z(\psi) = [t']_Z(m_Z(\psi, v))$.

Next, for all $e \in E$, $E \vdash e$ holds by rule *axiom*, thus by the very definition of $Z_{\Pi\downarrow E}$ it belongs to $Coalg(\Pi, E)$. Finally, let A be a coalgebra in $Coalg(\Pi, E)$, let $!_A : X_A \to Z_\Pi$ be the unique homomorphism to the final Π-coalgebra, and let $t =_O t'$ be an equation in $\hat{E}$. By the soundness result above we have that for all $y \in X_A$ it holds that $[t]_A(y) = [t']_A(y)$, thus $[t]_Z(!_A(y)) = [t']_Z(!_A(y))$ by point 3 of Proposition 4, and since this holds for any equation in $\hat{E}$, we have $!_A(y) \models \hat{E}$, and thus $!_A(y) \in Z_{\Pi\downarrow E}$. Therefore $!_A : X_A \to Z_{\Pi\downarrow E}$ is a well-defined homomorphism. Its uniqueness follows from the finality of Z_Π. □

Alternative characterizations of the carrier of the final coalgebra satisfying a set of equations are proposed in [HR95] and [Jac96b]. In [HR95] the set $Z_{\Pi\downarrow E}$ is determined as the collection of all elements $!_A(y) \in Z_\Pi$ for $A \in Coalg(\Pi, E)$ and $y \in X_A$; in [Jac96b], instead, the same set is determined as the carrier of the largest subcoalgebra contained in the set $\{\psi \in Z_\Pi \mid \psi \models E\}$, which in general is not the carrier of a coalgebra.

Theorem 11 (completeness of coalgebraic equational deduction). *Let Π be a co-signature and let E be a set of equations over Π. Then for all equations $t =_O t'$ over Π,*

$$Coalg(\Pi, E) \models t =_O t' \quad \Longrightarrow \quad E \vdash t =_O t'$$

Proof. Let $Z_{\Pi\downarrow E}$ be the final coalgebra in $Coalg(\Pi, E)$ (as for Proposition 10). We prove by contradiction that for each equation $t =_O t'$, $Z_{\Pi\downarrow E} \models t =_O t'$ implies that $E \vdash t =_O t'$; clearly this implies completeness.

Suppose, by absurd, that $Z_{\Pi\downarrow E} \models t =_O t'$ and $E \not\vdash t =_O t'$. We can immediately deduce that (1) O is not a singleton (otherwise $E \vdash t =_O t'$ would hold by *unity*), and that (2) for each output sort O', it does not hold that $E \vdash \underline{o} =_{O'} \underline{o'}$ for two distinct elements $o, o' \in O'$ (otherwise $E \vdash t =_O t'$ would hold by *contradiction*).

Next we show that $Z_{\Pi\downarrow E}$ is not empty. For this, we use the explicit structure of Z_Π provided by Theorem 5 to determine one of its elements for which all equations derivable from E hold. For each output sort O, let $\hat{o} \in O$ be an arbitrarily chosen but fixed element (which exists by non-emptyness). Now, let function $\hat{\psi} : \mathcal{O}bs(\Pi) \to \coprod_{j \in \underline{h}} O_j$ be defined as follows:

$$\hat{\psi}(c) = \begin{cases} o & \text{if } c : O \text{ and } \underline{o} : O \in [c]_{\hat{E}} \\ \hat{o} & \text{if } c : O \text{ and there is no constant in } [c]_{\hat{E}} \end{cases}$$

Function $\hat{\psi}$ is a well-defined element of Z_Π: in fact, all terms in $[c]_{\hat{E}}$ are of the same sort, because the equations in E are well-sorted and the rules of deduction preserve this property; furthermore, it is not possible that two distinct constants are in $[c]_{\hat{E}}$, because this would contradict (2) above. Next, it is immediate to see that function $\hat{\psi}$ actually belongs to $Z_{\Pi\downarrow E}$, because its definition is consistent over all $\hat{E}$-equivalence classes. Thus $Z_{\Pi\downarrow E}$ is not empty.

Let us now proceed by case analysis on the structure of t and t', i.e., the left- and right-hand side of the equation. If both are constants of sort O, then they must be the same because $Z_{\Pi\downarrow E}$ is not empty, but then $E \vdash t =_O t'$ by reflexivity, yielding a contradiction.

Suppose now that $t : O$ is an observation, while $t' = \underline{o} : O$ is a constant, and consider the equivalence class $[t]_{\hat{E}}$. Obviously, such class cannot contain $\underline{o}$ (otherwise $E \vdash t =_O t'$), neither can it contain a constant $\underline{o'} \neq \underline{o}$. In this last case, in fact, we would have $E \vdash t =_O \underline{o'}$, which implies (by soundness) $Z_{\Pi\downarrow E} \models t =_O \underline{o'}$, and thus $Z_{\Pi\downarrow E} \models \underline{o} =_O \underline{o'}$, which is absurd by non-emptyness. Now let ψ be an arbitrary element of $Z_{\Pi\downarrow E}$ (which exists by non-emptyness), let o' be an element of O different from o (which exists by (1) above), and define function ψ' as follows:

$$\psi'(c) = \begin{cases} \psi(c) & \text{if } c \notin [t]_{\hat{E}} \\ o' & \text{if } c \in [t]_{\hat{E}}. \end{cases}$$

By construction we have that $\psi' \in Z_{\Pi\downarrow E}$, and clearly $\psi' \not\models t =_O \underline{o}$, contradicting the hypothesis that $Z_{\Pi\downarrow E} \models t =_O t'$.

Finally, suppose that both t and t' are observations, and consider the equivalence classes $[t]_{\hat{E}}$ and $[t']_{\hat{E}}$. By arguments similar as above, we can deduce that both classes contain at most one constant, which is necessarily of sort O. Furthermore, it is not possible that both contain a constant. For, if the two constants were equal, we could infer that $E \vdash t =_O t'$ (by reflexivity and transitivity); and if they were distinct, we could obtain that $Z_{\Pi\downarrow E} \models \underline{o} =_O \underline{o'}$ for $o \neq o'$, which is impossible by non-emptyness. Now using the same technique as in the last case, we can easily find an element $\psi' \in Z_{\Pi\downarrow E}$ such that $\psi' \models t =_O o$ and $\psi' \models t' =_O o'$ for $o \neq o'$, which contradicts the hypothesis $Z_{\Pi\downarrow E} \models t =_O t'$. □

Example 6. Let E_1 and E_2 be the equations of Example 5. Making reference to the TS-coalgebras of Example 1 example, it is easy to check, for example, that

- $SW \models E_1$
- $TS_1 \models E_1$ and $TS_2 \models E_1$
- $TS_m \not\models E_1$ for $m > 2$
- $TS_1 \models E_2$ and $TS_m \not\models E_2$ for $m > 1$

Furthermore, if Z_{TS} is the final TS-coalgebra having carrier $O^{\mathbf{N}}$, then its subcoalgebra $Z_{TS\downarrow E_1}$, as for the proof of Theorem 11, has as carrier the set $\{(o_i \cdot o_j)^\omega \mid i, j \in \{0, \ldots, n\}\}$,[8] while $Z_{TS\downarrow E_2}$ has as carrier $\{o \cdot o_0^\omega \mid o \in O\}$.

[8] If w is a sequence, w^ω denotes the stream obtained by concatenating infinitely many times w.

5 Coalgebraic specifications in the related literature

We discuss here how our approach is related to some simple examples of coalgebraic specifications taken from the literature. In general, we will see that less restrictive kinds of equations are often used, for which, however, the topic of finding a complete calculus of deduction has not been addressed. So we shall hint at possible extensions of our approach which should allow us to capture such more general formats of equations.

Equations as introduced in Definition 7, i.e., pairs of observations, can be found for example in [Rei95, Jac96b]. But in the same papers one finds immediately other equations which are not legal according to our definition. This can happen for various reasons, that we shall discuss in turn.

Equations with variables of input sorts. As an example, in [Jac96a] the co-signature[9] BA of Example 2 is extended with attribute $name : X \rightarrow String$ and with method $ch\text{-}name$: $X \times String \rightarrow X$ with the obvious meaning. Then, among others, equation $bal(ch\text{-}name(s, x)) = bal(x)$ is considered, stating that the change of the owner's name should not affect the balance of the account. In this equation s is a variable of type $String$, which is not allowed according to our definition. This fact is not really problematic because such an equation can safely be considered as an *equation scheme*, representing the set of equations obtained by replacing s with all possible constants of type $String$.

Equations containing algebraic operators. In [Jac96b] the following equation over BA is considered: $bal(ch(x, z)) = bal(x) + z$. Even replacing the variable $z : \mathbb{Z}$ with a constant, such an equation would not be legal because "+" is not an operator of the co-signature, but it belongs to the algebraic structure of $\mathbb{Z}$. This is a quite a common situation in literature, as mentioned in the Introduction, where algebraic and coalgebraic aspects are mixed in a specification. Sticking to our pure coalgebraic setting, the equation above could be replaced by an equivalent, infinite set of *conditional* equations like

$$bal(x) = \underline{z'} \quad \Longrightarrow \quad bal(ch(x, \underline{z})) = \underline{z' + z}$$

where $\underline{z'}$, $\underline{z}$, and $\underline{z' + z}$ are three constant of sort $\mathbb{Z}$ related in the expected way.

Terms with multiple occurrences of the state variable. As stressed in Example 3, a term like $bal(ch(x, bal(x))) : X$ is not a context, according to our definition, because it contains two occurrences of variable x. Our intuitive reason for this restriction is that contexts should be basic experiments, and $bal(ch(x, bal(x)))$ is not basic, as it consists of two experiments: first observe $bal(x)$, then change state according to the result and observe again the balance. The study of a generalization of the calculus to admit such terms is left as future work. As for the previous point, one possibility would be to use conditional

[9] The examples from other papers we refer to in this section are often presented with a different terminology. We take the freedom of recasting them according to our syntax.

equations. For example, equation $bal(ch(x, bal(x))) = 2 * bal(x)$ would become the conditional equation scheme

$$bal(x) = \underline{z} \implies bal(ch(x, \underline{z})) = \underline{2 * z}.$$

Conditional equations. As soon as the examples become a bit larger, conditional equations are needed: see for example, besides of the considerations in the last two points, the specification of a simple database system in [Rei95], or that of a memory in [Jac97]. Our hope is that the results presented in this paper generalize smoothly to the conditional case, as it happens in the algebraic case (see [Sel72]): this is a topic for future work.

Equations of hidden sort. In [HR95] the authors consider an equation of hidden sort which we can recast as equation $e \equiv (next(next(x)) =_X x)$ over co-signature TS. Such an equation is said to be *valid* in a TS-coalgebra A if $[next(next(x))]_A = [x]_A$, and *behaviourally valid* if for all $y \in A$ we have $!_A([next(next(x))]_A(y)) = !_A([x]_A(y))$, i.e., the interpretations in A of the two transition sequences $next(next(x))$ and x map each state to *bisimilar* states.[10] Making reference to Example 1, we have for example that $SW \models e$, $TS_2 \models_{beh} e$, but $TS_2 \not\models e$. In [Jac97] a different syntax is used for equations that are required to be behaviourally valid, namely $t \underline{\leftrightarrow} t'$.

As far as behavioural validity is concerned, the above equation e is equivalent to the infinite set of equations $E = \{t[next(next(x))/x] =_O t \mid (t : O) \in \mathcal{O}bs(TS)\}$, which are legal according to Definition 7. Therefore under this interpretation an equation of hidden sort can be regarded as a syntactic representation of an infinite set of equations.

If instead the above equation is interpreted as true equality of elements of the carrier of a coalgebra, then it does not fit in the formal framework introduced in the previous section. Nevertheless, as shown in the full version of this paper, [Cor97], the calculus of deduction can be extended to a calculus for *generalized* equations, i.e., equations as in Definition 7 but possibly of hidden sort as well. Just to to give a hint about the result presented there, such a complete calculus is obtained by allowing for equations of hidden sort in all rules of Definition 8, and by adding the following rule:

[substitution] For all context $c : Y$ over Π (i.e., observation or transition sequence),

$$\frac{E \vdash c_1 =_X c_2}{E \vdash c[c_1/x] =_Y c[c_2/x]}$$

[10] For the class of coalgebras considered in this paper, we can safely define as "bisimilar" two states which are mapped, by the unique homomorphism, to the same element of the final coalgebra. As a consequence, since homomorphisms preserve observations, bisimilar states cannot be distinguished through observations. For the more general definition of bisimulation of coalgebras and the relationship with the notion of bisimulation in process algebra we refer the reader to [Rut96].

6 Conclusions

In this paper we presented a sound and complete calculus for equational deduction for a class of coalgebras, and we discussed the relationship between our calculus and the use of equations in coalgebraic specification in the related literature. Taking into account the considerations in the last section, we see various directions for possible generalizations of the results presented here. A first extension is presented in [Cor97]: it allows one to consider equations of hidden sorts as well, as mentioned in the previous section. Quite surprisingly, such a generalization of the completeness result will be introduced through motivations and techniques which are apparently not very much related to the notion of "true equality in hidden sorts", but are instead related to the problem of finding concepts that dualize the notion of variables and substitutions in algebraic specification.

Secondly, we would like to consider *conditional* equations as well, looking for a complete calculus in the line of [Sel72]: conditional equations seem sufficiently expressive to capture most kinds of coalgebraic equations we have found in literature (always sticking to a purely coalgebraic setting).

Thirdly, one could consider a wider class of coalgebras, for example those for functors closed under coproducts and powerset, which cannot be defined via cosignatures (see the discussion in [Cor97], Section 2.1). Some preliminary efforts to consider the whole class of polynomial functors (thus including coproducts) showed that the main difficulties should not be in finding the rules of the calculus, but instead in designing a satisfactory syntax for introducing such coalgebras and for presenting the structure of the final coalgebra along the line of Section 3.

Fourthly, we would like to allow for equations between terms containing algebraic components as well, like constructors or variables. However, it is not clear at all whether the result presented here could be extended in a meaningful way to a hybrid algebraic and coalgebraic framework.

Last but not least, we would like to investigate more the expressive power of equations in a coalgebraic framework. A first fact in this direction is proved in [Cor97], showing that every class of coalgebras determined by a set of equations is a "covariety", according to a definition by Jan Rutten [Rut96], but that not every covariety is equational. This fact, together with the observation that coalgebraic equations only allow to express safety or invariant properties, suggest that the expressive power of equations for coalgebras is weaker than that for algebras (and the same can be said about conditional equations as well). Thus on the one hand a natural question arises: which class of formulas has for coalgebras the same defining power that equations have for algebras, that is, is able to characterize *all* covarieties? An answer to this can be found, most probably, by adapting results recently presented in [Mos97], using some fragment of infinitary modal logic. On the other hand, if the emphasis is placed on the possibility of mechanizing an effective coalgebraic calculus of deduction, then one might prefer to avoid infinitary formulas at all, trying to find out which interesting properties of systems can be described by other kinds of finitary formulas, and whether some sound and complete calculus can be found for them.

Acknowledgements Jan Rutten introduced me to the Theory of Coalgebras at the beginning of my stage at CWI. Bart Jacobs pointed to me the open problem of looking for a dual version of Birkhoff's Completeness Theorem. I want to thank both of them together with Ulrich Hensel for the many stimulating discussions on this and on other topics related to coalgebras.

References

[Bar93] M. Barr. Terminal coalgebras in well-founded set theory. *Theoret. Comput. Sci.*, 114:299–315, 1993.

[BHW95] M. Bidoit, R. Hennicker, and M. Wirsing. Behavioural and Abstractor Specifications. *Science of Computer Programming*, 25:146–186, 1995.

[Cor97] A. Corradini. A Complete Calculus for Equational Deduction in Coalgebraic Specification. Technical Report SEN-R9723, CWI, Amsterdam, 1997.

[HR95] U. Hensel and H. Reichel. Defining equations in terminal coalgebras. In E. Astesiano, G. Reggio, and A. Tarlecki, editors, *Recent Trends in Data Type Specification*, volume 906 of *LNCS*, pages 307–318. Springer Verlag, 1995.

[HS95] U. Hensel and D. Spooner. A view on implementing processes: Categories of circuits. In M. Haveraaen, O. Owe, and O. Dahl, editors, *Recent Trends in Data Types Specification*, volume 1130 of *LNCS*, pages 237–255. Springer Verlag, 1995.

[Jac96a] B. Jacobs. Inheritance and cofree constructions. In P. Cointe, editor, *European Conference on object-oriented programming*, volume 1098 of *LNCS*, pages 210–231. Springer Verlag, 1996.

[Jac96b] B. Jacobs. Objects and classes, co-algebraically. In B. Freitag, C.B. Jones, C. Lengauer, and H.-J. Schek, editors, *Object-Orientation with Parallelism and Persistence*, pages 83–103. Kluwer Acad. Publ., 1996.

[Jac97] B. Jacobs. Invariants, Bisimulations and the Correctness of Coalgebraic Refinement. In *Proceedings AMAST'97*, LNCS. Springer Verlag, 1997.

[Mos97] L.S. Moss. Coalgebraic Logic. *Annals of Pure and Applied Logic*, 1997. To appear.

[ONS93] F. Orejas, M. Navarro, and A. Sánchez. Implementation and behavioural equivalence: a survey. In M. Bidoit and C. Choppy, editors, *Recent Trends in Data Type Specification*, volume 655 of *LNCS*, pages 93–125. Springer Verlag, 1993.

[Rei95] H. Reichel. An Approach to Object Semantics based on Terminal Coalgebras. *Mathematical Structures in Computer Science*, 5:129–152, 1995.

[Rut96] J.J.M.M. Rutten. Universal coalgebra: a theory of systems. Technical Report CS-R9652, CWI, 1996.

[Sel72] A. Selman. Completeness of calculii for axiomatically defined classes of algebras. *Algebra Universalis*, 2:20–32, 1972.

Specifying with Defaults: Compositional Semantics

F. Miguel Dionísio[1] and Udo W. Lipeck[2]

[1] Dept. of Mathematics, I.S.T., Technical Uni. of Lisbon, Portugal
fmd@math.ist.utl.pt
[2] Inst. für Informatik, Uni. Hannover, Germany
ul@informatik.uni-hannover.de

Abstract. We present an abstract specification theory that formalizes *non-monotonic* composition constructs from specification languages providing explicit non-monotonic mechanisms as a specification facility.
This theory generalizes the institutional framework from Goguen and Burstall by adding defeasibility mechanisms to a given institution. The denotation of a specification module consists of *defaults* (formulas organized by priority) that are assumed to be true in the absence of explicit information to the contrary. In other words defaults are assumed to be true unless they are *overridden* by other defaults of higher priority. Formulas that cannot be overriden are called axioms. Such structures of axioms and prioritized defaults are called *hierarchic specifications*.
The abstract specification theory of hierarchic specifications consists in formalizing, independently of the underlying logic, the structuring operations of hierarchic specifications. These operations are defined both on the syntactical and semantical levels by canonical constructions on corresponding syntactical and semantical *categories*, and account for the modular construction of hierarchic specifications by combining, reusing and modifying (with overriding) previously specified modules.

Introduction

Classical Specification Theory. The theory of institutions from Goguen and Burstall ([14]) constitutes an "abstract specification theory"[3] in that it provides a formal description of structuring operations among specifications. Such specification operations provide the means to control the complexity of a system, by describing it as a combination of simpler parts: "Complexity is a fundamental problem in programming methodology: large programs, and their large specifications, are very difficult to produce, to understand, to get right, and to modify. A basic strategy for defeating complexity is to break large systems into smaller pieces that can be understood separately, and that when put back together give the original system"[13].

The basic idea motivating the institutional framework is that specifications, i.e. rigorous descriptions of parts of a system, denote logical theories. Structuring

[3] The term "abstract specification theory" is taken from [11].

operations denote canonical operations among those theories. The formalization of a specification language consists in the choice of the appropriate logic (institution) and the corresponding characterization of the structuring operations.

The chosen paradigm dictates the choice of the underlying logic. For instance temporal logic(s) are used to give semantics to object oriented specification languages ([22, 10, 11, 21, 23]). The main contribution of the theory of institutions is the fact that the compositional constructs are *independent* of the underlying logic. The specification language Clear ([7]) can be used to build large specifications from theories from *any* logical system.

Non-monotonic Specification Theory. We present an abstract specification theory that formalizes *non-monotonic* composition constructs, and extends the institutional framework.

There are several reasons for using non-monotonic formalisms in (the semantics of) specification languages. The first is that actual systems, reasoning in the presence of incomplete information, use such mechanisms: planning systems, diagnose systems and truth maintenance systems, for instance. The second is that non-monotonic logics provide the formalization of the way actual systems store and process their information: the several database and knowledge base completions modeling the fact that in such systems only positive information is kept; the theory of belief revision setting the general rules for the addition of new information (inconsistent with the previous knowledge state); the frame rule modeling the minimal change of properties after the occurrence of an action. And, finally, the third is that the specification process improves in modularity and reusability if such mechanisms are available. Non-monotonic formalisms give formal grounds to "the requirement to re-use specification modules as far as possible, i.e. not only to include the same components in different contexts, but also to prefer modification of given parts over new definitions. To reduce development costs, software should be designed in a "differential" way - select a module from the library, refine it by adding new functions, and modify it by *overriding* some old ones" [5].

The form of reasoning known as default reasoning is fruitful in formalizing the non-monotonic aspects referred. Default reasoning is reasoning in the presence of incomplete information: in the absence of evidence to the contrary, assume the "default". For instance we can assume (and specify) that, by default, a book (in a library) can be borrowed. If, however, this book is a reserved book this conclusion can be defeated by explicit information stating that *reserved books* cannot be borrowed.

Defaults are formalized here by formulas organized by priority. A default with more important priority overrides a conflicting default of less important priority: from the point of view of the later the more important default is "evidence to the contrary". Axioms are formulas that cannot be overridden. Prioritized defaults have been introduced in [6] and further studied in [3, 20]. Their impact in specification is stated in [4, 5].

Axioms and prioritized defaults are the modularization units used in the theory of composition developed here. This means that we want to formalize specification languages that use default mechanisms and we take prioritized defaults and axioms as the denotation of such a specification modules. Constructions involving specification modules are interpreted as operations involving the corresponding denotations.

For instance the specification of *reserved books* is obtained from the module *books* (reuses it). A new priority level is added, more important than those of *books*, with the formula stating that *reserved books* cannot be lent. All other properties of *book* will hold for *reserved books* since they are not contradicted by the more important formula. Only the difference between *reserved books* and *books* must be stated. This construction is given by a canonical operation involving the specification *books* and the "difference" between *reserved books* and *books*.

Organization. The paper is organized as follows. Firstly, in section 1 we define the notion of theory and establish the Galois connection properties of hierarchic specifications. These are fundamental for the definition of composition of hierarchic specifications, presented in section 2. Composition of hierarchic specifications comprehends the definition of the syntactic and semantic categories that formalize it, the relation between these categories and the establishment of sufficient conditions for the existence of composition constructions. In section 3 we conclude the paper.

1 Theories and Galois Connections

The first step towards an *abstract* theory of composition is the definition of the notion of *theory* of the syntactical entities to be composed, in our case hierarchic specifications. Theories are taken here as the syntactical counterpart of the semantics. In the classical case the theory of a presentation A (set of formulas) is the set of formulas entailed by A. The theory of A has the same semantics as A (the same class of models) and has the property that it is the biggest (most formulas) presentation having the same semantics as A. In this way the theory of A can alternatively be characterized as a special representative of the equivalence class of the presentations having the same semantics as A.

This is the property that we choose for defining the notion of theory for hierarchic specifications: we look for a special representative of the equivalence class of the hierarchic specifications having the same *semantics* as a given one. Such theories are in a one to one relation to the semantics of hierarchic specifications and provide the needed abstraction. When defining composition of hierarchic specifications these will be compared via their theories, or, which is the same, via their semantics.

It is clear from these considerations that the notion of *semantics* (of hierarchic specifications) is the key to abstraction. In the classical case the semantics of a presentation is the class of its models. An important property of this semantics is the fact that it is *denotational* w.r.t. union of presentations. This means that the semantics of the union of presentations depends only on the semantics of each argument presentation (the semantics of the union is the intersection of the classes of models). This is a necessary property of any theory of composition: the syntactical constructions (union in this case) must have a corresponding semantic construction.

In the case of hierarchic specifications the "standard" semantics, the lexicographic preference (presented below in section 1.2), does not have enough information to account for the syntactic operations we are concerned with, namely union of axioms and defaults and "addition" of priority levels. For this reason a new semantics (with more structure) of hierarchic specifications is introduced.

Moreover, a Galois connection relating the syntactical and semantical constructions of hierarchic specifications is displayed. This generalizes the corresponding property of classical presentations and means that relations between hierarchic specifications ("inclusion" see below) are mirrored by relations between the corresponding semantics (again "inclusion", but in the opposite direction). Moreover operations among hierarchic specifications ("unions" and "intersections") are also mirrored by operations among the corresponding semantics (unions are mapped to intersections and intersections to unions). The reverse is also true: relations and operations among the semantics of hierarchic specifications are also mirrored by relations and operations among the corresponding theories. The Galois connection properties are fundamental for formalizing composition of hierarchic specifications.

We begin by recalling needed institutional notions. Then we study the special case of specifications (i.e. hierarchic specifications with one only priority level) and define the corresponding notion of theory. The Galois connection is displayed for this case. Finally, we generalize these notions for hierarchic specifications. The properties of specifications and the corresponding theory of composition have been presented in [8]. The properties of hierarchic specifications and the corresponding theory of composition have been defined in [9].

Institutions. We recall the definitions of institution ([14]), presentation model and theory (of a presentation) in the following.

Note that Set is the category of sets and functions, Cat the category[4] of categories and functors between them (and Cat^{op} its opposite quasicategory). Given a category (or quasicategory) $\mathcal{C}$, the class of its objects is denoted by $|\mathcal{C}|$.

Definition 1. An *institution* consists of (a) a category Sign whose objects are called *signatures*, (b) a functor $\mathsf{Sen} : \mathsf{Sign} \to \mathsf{Set}$ that assigns to each signature the set of its *formulas*, (c) a functor $\mathsf{Mod} : \mathsf{Sign} \to \mathsf{Cat}^{op}$ giving for each signature Σ a category whose objects are called *Σ-interpretation structures* and whose morphisms are the Σ-interpretation structure morphisms, and (d) a relation $\models_\Sigma \subseteq |\mathsf{Mod}(\Sigma)| \times \mathsf{Sen}(\Sigma)$, called *$\Sigma$-satisfaction* such that for every morphism $\phi : \Sigma_1 \to \Sigma_2$ the condition $m_2 \models_{\Sigma_2} \mathsf{Sen}(\phi)(f)$ iff $\mathsf{Mod}(\phi)(m_2) \models_{\Sigma_1} f$, called the

[4] In fact it is a quasicategory (see [1]). But this distinction is not relevant to the constructions presented here and we will not refer it further.

Satisfaction Condition, holds for each model m_2 of $|\mathsf{Mod}(\Sigma_2)|$ and each formula f of $\mathsf{Sen}(\Sigma_1)$.

Given an institution $\mathcal{I}$ a *presentation* is a pair $P = (\Sigma, A)$ where $\Sigma \in |\mathsf{Sign}^{\mathcal{I}}|$ is a signature from $\mathcal{I}$ and $A \subseteq \mathsf{Sen}^{\mathcal{I}}(\Sigma)$ is a set of formulas from Σ. A Σ-presentation A is a presentation (Σ, A). The interpretation structures that satisfy all formulas in a given presentation are said to *satisfy* the presentation and called *models* of the presentation. Presentations and their models are related by the function $^\bullet$ that assigns to a Σ-presentation A the class $A^\bullet$ of all Σ-interpretation structures that are models of A. In the reverse direction the function $^\bullet$ assigns to a class $\mathcal{M} \subseteq |\mathsf{Mod}^{\mathcal{I}}(\Sigma)|$ of Σ-interpretation structures the set $\mathcal{M}^\bullet$ of all Σ-formulas that are satisfied in each interpretation structure from $\mathcal{M}$. $\mathcal{M}^\bullet$ is called the *theory* of $\mathcal{M}$. The *theory* of a Σ-presentation A is the set $A^{\bullet\bullet}$ of the Σ-formulas satisfied in the models of A.

These two functions form a Galois connection (see [14]).

1.1 Specifications

Specifications are a special case of hierarchic specifications with one only priority level (of defaults). The corresponding concepts are simplified versions of those needed for hierarchic specifications. They are a good introduction to the concepts needed later.

Specifications consist of two sets of formulas (of a given signature), the set of axioms and the set of defaults. The semantics of a specification is a preference relation (i.e. a pre-order) that organizes the models of the axioms by how well they satisfy the defaults. In this way the "best" models are the maximal ones in this preference relation. The maximal models are those models of the axioms that satisfy most defaults. The relation between syntax (specifications) and semantics (their preference relations) takes the form of a Galois connection. This connection generalizes the Galois connection for the classic case: more axioms imply less models. But the new structure of defaults obeys a formally similar rule: addition of defaults implies less relations of preference between those models.

Syntax and Semantics. The definition of specification and the corresponding semantics is stated formally in the following.

Definition 2. A *specification* (from an institution $\mathcal{I}$) is a triple $S = (\Sigma, A, D)$ where $\Sigma \in |\mathsf{Sign}^{\mathcal{I}}|$ is a signature from $\mathcal{I}$, $A \subseteq \mathsf{Sen}^{\mathcal{I}}(\Sigma)$ is a set of formulas from Σ, the set of *axioms* from S and $D \subseteq \mathsf{Sen}^{\mathcal{I}}(\Sigma)$ is a set of formulas from Σ, the set of *defaults* from S. A *Σ-specification* (A, D) is a specification (Σ, A, D).

The semantics of a specification is a relation on the models of its axioms, representing that some of these models are better than other since they satisfy more of the defaults ([16, 24, 4]). This relation is a pre-order.

Definition 3. A *pre-order* (from $\mathcal{I}$) is a triple $\mathcal{R} = (\Sigma, \mathcal{M}, \sqsubseteq)$ where $\Sigma \in |\mathsf{Sign}^{\mathcal{I}}|$ is a signature from $\mathcal{I}$, $\mathcal{M} \subseteq |\mathsf{Mod}^{\mathcal{I}}(\Sigma)|$ is a class of interpretation structures of the signature Σ, $\sqsubseteq \subseteq \mathcal{M} \times \mathcal{M}$ is a reflexive and transitive relation among those interpretation structures. A *Σ-pre-order* $(\mathcal{M}, \sqsubseteq)$ is a pre-order $(\Sigma, \mathcal{M}, \sqsubseteq)$.

The pre-order induced by a specification relates the models of the axioms by how well they satisfy the defaults.

Definition 4. The pre-order induced by a specification $S = (\Sigma, A, D)$ is the pre-order $S^\star = (\Sigma, A^\bullet, \sqsubseteq)$ with the same signature as S, the models of the axioms from S as class of interpretation structures, and the relation $\sqsubseteq$ among those models defined by $m \sqsubseteq n$ iff for all $d \in D$, if $m \models d$ then $n \models d$.

Consequences. The equivalence classes of the preference $S^\star$ of a specification S are those classes of models of the axioms that satisfy precisely the same defaults. The maximal equivalence classes of $S^\star$ are the classes of models of the axioms that satisfy most defaults. The set of formulas $\mathcal{E} = [m]^\bullet$ of the formulas holding in all interpretation structures from a *maximal* equivalence class $[m]$ of $S^\star$ is called an *extension* of S. A formula belonging to all extensions of S is called a *skeptical* consequence of S. A formula belonging to some extension of S is called a *credulous* consequence of S. The usual notion of consequence is the notion of skeptical consequence ([16, 24, 15]). The notion of credulous consequence, introduced in [18], is important in the context of abduction ([17]). We do not commit to one or the other since both can be derived from the chosen semantics.

Theories and Galois Connection. The theory of a specification S is a special representative of the equivalence class of those specifications having the same preference ($S^\star$) as S. This theory is the "biggest" of such specifications, i.e. the one with most axioms and defaults.

A more interesting characterization can be motivated through an analogy with the classical case. The theory of a presentation A is the set of such formulas that can be added to A without changing the class of models. So we are now looking for formulas (both axioms and defaults) that can be added to a specification S without changing its semantics, namely its preference relation. On the axiom side the situation is like the classical case: these are the classical consequences of the axioms in S. The new defaults are called "implicit" in S and are the formulas that can be added as defaults to S without changing its preference. Such formulas d must satisfy the following property: if $m \sqsubseteq n$ and $m \models d$ then $n \models d$ (where $\sqsubseteq$ is the preference of S). Otherwise (see the definition 4 of preference relation) when adding the default d to S the new preference would be different from $S^\star$ since m and n would no longer be related.

In general a sentence d is an implicit default of a pre-order $\mathcal{R}$ if, whenever it is satisfied by an interpretation structure m from $\mathcal{R}$ it is also satisfied by all interpretation structures better (according to $\mathcal{R}$) than m.

Definition 5. The set of *defaults implicit* in a Σ-pre-order $\mathcal{R} = (R, \sqsubseteq)$, denoted by $\mathcal{R}^\circ$, is the set of Σ-formulas

$$\mathcal{R}^\circ = \{d \in \mathsf{Sen}^{\mathcal{I}}(\Sigma) \ : \ \text{for all } m_1, m_2 \in R \text{ if } m_1 \sqsubseteq m_2 \text{ and } m_1 \models d \text{ then } m_2 \models d\}.$$

Definition 6. The function * assigns to a Σ-specification S its induced *preference relation*, the Σ-pre-order $S^\star$, and the function * assigns to a Σ-pre-order $\mathcal{R}$ the specification $\mathcal{R}^\star = (|\mathcal{R}|^\bullet, \mathcal{R}^\circ)$. The specification $\mathcal{R}^\star$ is called the *theory* of $\mathcal{R}$. The specification $S^{\star\star}$ is called the *theory* of the specification S.

As in the classical case the functions * relating specifications with their semantics form a Galois connection. This is stated formally in the following theorem. We note the formal similarity between the axiom and default parts: when adding an axiom a to S the new class of models is obtained from the class of models of S by deleting those that do not satisfy a. When adding a default d the new preference relation is obtained from that of S by deleting the relation pairs (m, n) that "do not satisfy" the default d: i.e. such that $m \models d$ but $n \not\models d$.

The binary relation $\Subset$ denotes inclusion of the sets of axioms and defaults when relating specifications. When relating pre-orders it denotes inclusion of classes of models and relations (seen as classes of relation pairs). The operations $\Cup$ and $\Cap$ are pairwise union and intersection. The pairwise union of pre-orders is the union of the classes of models and the *transitive* closure of the union of the pre-orders (in order to ensure that the result is a pre-order).

Theorem 7. *Let S, S' and $S_n, n \in N$, be Σ-specifications, $\mathcal{R}, \mathcal{R}'$ and $\mathcal{R}_n, n \in N$, be Σ-pre-orders (N is some set of indexes). Then (1) $S \Subset S'$ implies $S^\star \Supset S'^\star$, (2) $\mathcal{R} \Subset \mathcal{R}'$ implies $\mathcal{R}^\star \Supset \mathcal{R}'^\star$, (3) $S \Subset S^{\star\star}$ and (4) $\mathcal{R} \Subset \mathcal{R}^{\star\star}$.*

Proof. Proofs can be found in [9].

The Galois connection properties imply a one to one relation between theories and the semantics of specifications and also that the theory of S is the biggest (w.r.t. $\Subset$) specification having the same preference as S.

1.2 Hierarchic Specifications

In this section we generalize the concepts and properties of specifications to hierarchic specifications. We note that the "standard" semantics of hierarchic specifications, its lexicographic preference, does not have enough information to explain composition. For this purpose we introduce new, more structured semantics of hierarchic specifications. With these notions we define the theory of a hierarchic specification and display a corresponding Galois connection.

Syntax. Hierarchic specifications consist of axioms and sets of defaults organized by priorities. The priority structure is a partial order.

Definition 8. A *hierarchic specification* is a tuple $S = (\Sigma, A, (H, \preceq), \Delta)$ consisting of: a signature $\Sigma \in |\mathsf{Sign}^{\mathcal{I}}|$, a set of axioms $A \subseteq \mathsf{Sen}^{\mathcal{I}}(\Sigma)$, a well-founded partial order $(H, \preceq)$ of priority, with *non-empty* set of priority levels H and a function Δ assigning to each priority level $h \in \mathcal{H}$ a set of defaults $\Delta(h) \subseteq \mathsf{Sen}^{\mathcal{I}}(\Sigma)$. To refer to the set of defaults from S at level h we use the notation $\mathsf{df}(S, h) = \Delta(h)$.

Following [20] we adopt the convention that the lower priority levels are the most important. This is justified by the structure of classes and subclasses in object oriented specifications: the lower priority levels correspond to subclasses. Its defaults have usually more priority than those of the superclasses since they express more specific properties.

Lexicographic Semantics. The overall meaning of a hierarchic specification has to give preference to the defaults of lower (better) priority levels. This is achieved formally by assigning to a hierarchic specification a pre-order among the models of its axioms that compares them according to how well they satisfy the defaults. This is now meant in the sense that models are preferred because either they satisfy more defaults or they strictly satisfy better defaults. The intended preference organizes the models in a way reminiscent of the lexicographic ordering of words in dictionaries (see Ryan [19, 20]). We define the *lexicographic preference* induced by a hierarchic specification.

Definition 9. The *lexicographic preference* induced by a hierarchic specification S is the pre-order $\mathsf{lex}^\circ(S)$ with: the same signature as S, the models of the axioms from S as class of interpretation structures, and the relation $\sqsubseteq^\circ$ among those models defined by $m \sqsubseteq^\circ n$ iff for every priority level $h \in H$ if $\mathsf{df}(S,h)(m) \not\subseteq \mathsf{df}(S,h)(n)$ then there is $h' \prec h$ with $\mathsf{df}(S,h')(m) \subset \mathsf{df}(S,h')(n)$ (note that this last inclusion is strict). The set $\mathsf{df}(S,h)(m)$ is the set of defaults from S at level h that are satisfied by the interpretation structure m.

The lexicographic preference of a hierarchic specification S has been shown (in [2]) to be canonical in the sense that it is the only relation among those models satisfying certain conditions. The consequences of a hierarchic specification are defined in terms of the maximal equivalence classes of $\mathsf{lex}^\circ(S)$ in the same way as for specifications.

1.3 Semantics for Composition

The lexicographic preference just presented is the main semantics for defining logical properties of a hierarchic specification. However, it has not enough information to account for composition of hierarchic specifications. This means that hierarchic specifications having the same lexicographic preference do not, in general, behave in the same way with respect to composition. This is illustrated in the following example.

Example 1. Consider two hierarchic specifications S_1 and S_2 with the same priority structure consisting of three priority levels related as follows: $h_1 \prec h_2 \prec h_3$. Assume that S_1 has the set $\{d\}$ of defaults at level h_1 and the empty set of defaults in the other levels. Let S_2 have the same set $\{d\}$ of defaults at level h_3 and the empty set of defaults in the other levels. The hierarchic specifications S_1 and S_2 have the same lexicographic preference (since there is only one default d). Consider now that we add, in both specifications, the default $\neg d$ at level h_2.

In S_1 this has no effect: the previous default d overrides $\neg d$. But in S_2 the opposite happens: d is overridden by $\neg d$. The resulting specifications have different lexicographic preferences.

In the following we present two equivalent semantics, the *hierarchies of lexicographic and differential preferences* that have enough structure to explain composition of hierarchic specifications.

From any of these semantics the lexicographic preference of the original specification can be derived. This means that the new semantics still contains the logical information. The new structure is needed to explain the compositional operations of hierarchic specifications. We have shown in [9] that the new semantics are as abstract as possible, i.e. they have the structure needed to explain composition and not more.

We define firstly hierarchies of pre-orders, that consist of a class of interpretation structures and pre-orders organized by priority.

Definition 10. A *hierarchy of pre-orders* is a tuple $\mathcal{H} = (\Sigma, \mathcal{M}, (H, \preceq), \Theta)$ consisting of: a signature $\Sigma \in |\mathsf{Sign}^{\mathcal{I}}|$, a class of interpretation structures $\mathcal{M} \subseteq \mathsf{Mod}^{\mathcal{I}}(\Sigma)$, a well-founded partial order $(H, \preceq)$ of priority, with *non-empty* set of priority levels H and a function Θ assigning to each priority level $h \in H$ a reflexive and transitive binary relation $\Theta(h) \subseteq \mathcal{M} \times \mathcal{M}$ over the class of interpretation structures $\mathcal{M}$. A *Σ-hierarchy of pre-orders* $(\mathcal{M}, (H, \preceq), \Theta)$ is the hierarchy $(\Sigma, \mathcal{M}, (H, \preceq), \Theta)$.

The first structured semantics of hierarchic specifications is its hierarchy of lexicographic preferences. In short the hierarchy of lexicographic preferences of a hierarchic specification S assigns to each priority level h the lexicographic preference of the substructure of S obtained by restricting S to the levels h and under h. In this way each priority level h has the information of the interaction of the defaults from h with those at more important levels.

Definition 11. The *hierarchy of lexicographic preferences* induced by a hierarchic specification S, denoted by $S^{\oplus}$ is the hierarchy of pre-orders with: (a) the same signature as S, (b) the models of the axioms from S as class of interpretation structures, (c) the same well-founded partial order $(H \preceq)$ as S, (d) the function $\sqsubseteq^{\oplus}$ that to each priority level $h \in H$ assigns the relation $\sqsubseteq_h^{\oplus}$ defined by $m \sqsubseteq_h^{\oplus} n$ iff for every priority level $h' \preceq h$ if $\mathsf{df}(S, h')(m) \not\subseteq \mathsf{df}(S, h')(n)$ then there is $h'' \prec h'$ with $\mathsf{df}(S, h'')(m) \subset \mathsf{df}(S, h'')(n)$ (this last inclusion is strict). Recall that the set $\mathsf{df}(S, h)(m)$ is the set of defaults from S at level h that are satisfied by the interpretation structure m.

Clearly the lexicographic preference of S can be derived from the hierarchy of lexicographic preferences of S. The lexicographic preference of S is the preference assigned by $S^{\oplus}$ to the existing or imagined top priority level (since it computes the interaction of all levels). This preference is the intersection of all partial preferences occurring in $S^{\oplus}$.

The hierarchy of differential preferences has the same information content as the hierarchy of lexicographic preferences (see [9]) in the sense that each of them can be derived from the other. This information, however is displayed in a way technically convenient for the purpose of composition[5]. We motivate this semantics as follows. The composition operations of hierarchic specifications comprehend addition of axioms and defaults at each priority level (and also combinations of priority structure). The addition of axioms is treated classically. The addition of defaults at a particular priority level is the interesting operation in this context. Let h be a priority level from a hierarchic specification S and d a formula that we want to add at level h. The question is how the semantics of $S + d$ relates to the semantics of S. Consider firstly that the semantics of S is the hierarchy of lexicographic preferences just defined. At level h we have the lexicographic preference corresponding to the substructure of S obtained by restricting the priority levels to h and the levels under h (more important than h). Formal similarity with the concepts presented for specifications would suggest that the new preference would be obtained from this by deleting the pairs (m, n) that do not agree with the new default d, i.e. such that $m \models d$ but $n \not\models d$. This is, however, not true. In fact the defaults at more important (lower levels) have to be taken into account and it may be the case that n is preferred to m because it strictly satisfies better defaults than m. In this case the new default d at level h has no effect. The pairs that can be deleted are those pairs (m, n) that are equivalent in more important levels (i.e. satisfy precisely the same defaults from more important levels) and do not agree with d.

The hierarchy of differential preferences assigns to each priority level h the following preference. Firstly only interpretation structures that are equivalent at levels more important than h can be related. Two such levels m and n are related iff if $m \models d'$ then $n \models d'$ for every default d' from the *local set of defaults at level* h. In this way the addition of a default d at level h has the effect of deleting all those pairs (of the differential preference) that do not agree with d. This implies a formal similarity with the previous constructions for specifications that is of convenience.

Definition 12. The *hierarchy of differential preferences* of a hierarchic specification S, denoted by $S^{\ominus}$ is the hierarchy of pre-orders with: (a) the same signature as S, (b) the models of the axioms from S as class of interpretation structures, (c) the same well-founded partial order $(H \preceq)$ as S, (d) the function $\sqsubseteq^{\ominus}$ that to each priority level $h \in H$ assigns the relation $\sqsubseteq^{\ominus}_h$ defined by $m \sqsubseteq^{\ominus}_h n$ iff $\mathsf{df}(S, h)(m) \subseteq \mathsf{df}(S, h)(n)$ and $\mathsf{df}(S, h')(m) = \mathsf{df}(S, h')(n)$ for every priority level $h' \prec h$.

1.4 Theories and Galois Connection

Having defined the semantics of hierarchic specifications as its hierarchy of differential preferences we need now the reverse direction, namely to assign to such a semantics a hierarchic specification. The motivation for this other direction

[5] Note that the theory that we develop here on top of the hierarchy of differential preferences could be rewritten in terms of the hierarchy of lexicographic preferences.

is that the hierarchic specification assigned to $S^{\ominus}$ should be the *theory* of S, a canonical representative of the equivalence class of the hierarchic specifications having the same semantics as S. This theory should be the biggest hierarchic specification among such equivalent specifications. This means the hierarchic specification having the most axioms and, at each priority level, the most defaults. These are the defaults that can be added at each level without changing the semantics of a specification, i.e. the *defaults implicit* at each priority level h, and correspond to the defaults that do not change the differential preference *locally* at that level (i.e. the defaults implicit in the differential preference at level h in the sense of definition 5). The reason why the interaction of lower levels need not be taken into account is that such interaction is already coded in the differential preference: interpretation structures can only be related by preference if they were equivalent at levels below.

Definition 13. Let $\mathcal{H}$ be a Σ-hierarchy of pre-orders. The corresponding *theory*, denoted by $\mathcal{H}^{\ominus}$ is the Σ-hierarchic specification with: (a) the same well-founded partial order $(H, \preceq)$ as $\mathcal{H}$, (B) the Σ-formulas satisfied by each interpretation structure from $\mathcal{H}$ as set of axioms, (c) the function Δ defined by $\Delta(h) = R_h^{\circ}$ that to each priority level $h \in H$ assigns the set of the defaults implicit in the Σ-pre-order R_h assigned by $\mathcal{H}$ to the level h.

The *theory* of a hierarchic specification S is the hierarchic specification $S^{\ominus\ominus}$.

The functions assigning to a hierarchic specification its hierarchy of differential preferences and to a hierarchy of pre-orders form again a Galois connection, if the hierarchies of pre-orders are restricted to *hierarchies of differential pre-orders* (any hierarchy of differential preferences is an hierarchy of differential pre-orders). Hierarchies of differential pre-orders satisfy the property that the pre-orders at each level are contained in the equivalence classes of pre-orders at lower levels.

Definition 14. Let $\mathcal{H}$ be a hierarchy of pre-orders and $(H, \preceq)$ its priority structure. $\mathcal{H}$ is said to be a *hierarchy of differential pre-orders* iff $\equiv_{h'} \supseteq \sqsubseteq_h$ for every $h' \prec h$. The relation $\equiv_{h'}$ is the equivalence of the pre-order at level h' and $\sqsubseteq_h$ is the pre-order at level h.

The Galois connection is stated formally in the next theorem. The relation $\Subset$ among hierarchic specifications of the same priority structure denotes inclusion of the sets of axioms and pointwise (at each priority level) inclusion of sets of defaults. When relating hierarchies of pre-orders of the same priority structure it denotes inclusion of classes of models and pre-orders at each level (seen as classes of relation pairs). The operations $\Cup$ and $\Cap$ are pointwise union and intersection, i.e. union of the axioms and, at each priority level, of the defaults from the argument hierarchic specifications at that level. The situation with union and intersection of hierarchies of pre-orders is similar. Care again has to be taken with the union of pre-orders (at each level): the transitive closure of the union is taken in order to ensure that the result is a pre-order.

Theorem 15. *Let S, S' be Σ-hierarchic specifications and $\mathcal{H}, \mathcal{H}'$ be Σ-hierarchies of* differential *pre-orders. Then (1) $S \Subset S'$ implies $S^{\ominus} \Supset S'^{\ominus}$, (2) $\mathcal{H} \Supset \mathcal{H}'$ implies $\mathcal{H}^{\ominus} \Subset \mathcal{H}'^{\ominus}$, (3) $S \Subset S^{\ominus\ominus}$ and (4) $\mathcal{H} \Subset \mathcal{H}^{\ominus\ominus}$.*

The Galois connection properties imply a one to one relation between theories and the semantics of hierarchic specifications: $S_1^{\ominus} = S_2^{\ominus}$ iff $S_1^{\ominus\ominus} = S_2^{\ominus\ominus}$. Since the hierarchy of differential preferences is equivalent to the hierarchy of lexicographic preferences (see [9]) we also have $S_1^{\oplus} = S_2^{\oplus}$ iff $S_1^{\ominus\ominus} = S_2^{\ominus\ominus}$. This means that the notion of theory is the same independently of which of these two semantics is chosen. Moreover the theory $S^{\ominus\ominus}$ of S is the biggest (w.r.t. $\Subset$) specification having the same hierarchy of differential (or lexicographic) preferences as S.

2 Composition of Hierarchic Specifications

In the previous section we have provided hierarchic specifications with a notion of theory. This notion corresponds to an abstraction of the particular way such specifications are written since it declares equivalent the specifications having the same theory (or equivalently the same semantics). Moreover a notion of inclusion (of meaning) between specifications (defined by inclusion of theories) has been defined.

We extend these notions to account for composition of hierarchic specifications. Special care is taken in order that the syntactical concepts and operations defined have a corresponding semantics. The formalizations generalize the classical theory of composition of presentations ([14]) to hierarchic specifications and are inspired by the formalization of composition of hierarchic specifications presented in [5] .

To the notion of equivalence of specifications is added a notion of *independence of representation*. This corresponds to identify as equivalent specifications that only differ because they have been written with different symbols. Independence of representation is formalized by the notion of isomorphism in the category hieSpec of hierarchic specifications. Renaming of priority level names is also taken into account.

Composition of hierarchic specifications is understood as the addition of syntactical entities such as axioms, defaults, priority levels and relations between them. This composition is formalized by canonical constructions in the category hieSpec of hierarchic specifications. The constructions depend on the existence of signature symbols that are used in the axioms and defaults of the specifications result (i.e. they depend on the constructions of the underlying category Sign of signatures). They depend furthermore on the existence of a priority structure that expresses the combination of the priority structures of the hierarchic specifications involved.

The category of hierarchic specifications is mirrored on the semantic side by the semantic category hiePref. The semantics of the composition of specifications is obtained from the semantics of the parameter specifications by canonical constructions in these categories.

We begin by defining the category of partial orders, that accounts for the composition of priority structure. After we define the category of hierarchic specifications and the corresponding mirror semantic category. We then investigate the relation between the syntactic and semantic constructions and display sufficient conditions for the existence of such constructions.

Category of Partial Orders. Composition of hierarchic specifications is, first of all, composition of their structure of priority. Composition of partial orders is formalized by colimits in the category StPart. In this category we restrict the partial order morphisms to those that *strictly* respect the orderings. This means that levels that are *strictly* related by priority will remain strictly related by priority. This formalization models the construction of a specification by adding to it more relations between (possibly more) priority levels (and axioms and defaults) and rejects the possibility of identifying levels strictly related by priority. However, unrelated levels may be identified.

Definition 16. The category StPart of partial orders consists of: **objects:** partial orders, and **morphisms:** a (strict) partial order morphism $\phi : (H, \preceq) \to (H', \preceq')$ is a function $\phi : H \to H'$ that respects $\prec$ (the strict relation corresponding to $\preceq$): if $h_1 \prec h_2$ then $\phi(h_1) \prec' \phi(h_1)$.

The category StPart is *not* cocomplete as opposed to the category Part of partial orders and partial order morphisms. In particular it does not have all coequalizers. It is however easy to check that the colimits in StPart, when they exist, coincide with the corresponding colimits in Part.

Category of Hierarchic Specifications. The category of hierarchic specifications relates hierarchic specifications via their theories by axiom and default preserving signature and partial order morphisms.

Recall that, given a signature $\sigma : \Sigma_1 \to \Sigma_2$, the function $\mathsf{Sen}(\sigma) : \mathsf{Sen}(\Sigma_1) \to \mathsf{Sen}(\Sigma_2)$ sends each formula from the language $\mathsf{Sen}(\Sigma_1)$ to a formula in the language $\mathsf{Sen}(\Sigma_2)$. We will denote the function $\mathsf{Sen}(\sigma)$ by $\hat{\sigma}$. With this convention and letting $\mathsf{ax}(S)$ denote the set of axioms from S and $\mathsf{df}(S, h)$ the set of defaults from S at level h the definition of the category hieSpec follows.

Definition 17. The category hieSpec of hierarchic specifications consists of:

- **objects:** all hierarchic specifications, and
- **morphisms:** a hierarchic specification morphism $(\sigma, \phi) : S_1 \to S_2$ from the Σ_1-hierarchic specification S_1 to the Σ_2-hierarchic specification S_2 is: a signature morphism $\sigma : \Sigma_1 \to \Sigma_2$ from the signature of S_1 to that of S_2 and a *strict* partial order morphism $\phi : \mathsf{po}(S_1) \to \mathsf{po}(S_2)$ from the partial order of priority of S_1 to that of S_2, such that
  - $\hat{\sigma}(\mathsf{ax}(S_1^{\ominus\ominus})) \subseteq \mathsf{ax}(S_2^{\ominus\ominus})$, i.e. the image by $\hat{\sigma}$ of the axioms from the theory of S_1 (the consequences of the axioms from S_1) are axioms from the theory of S_2 and

- $\hat{\sigma}(\mathsf{df}(S_1^{\ominus\ominus}, h)) \subseteq \mathsf{df}(S_2^{\ominus\ominus}, \phi(h))$ for each level $h \in \mathsf{po}(S_1)$, i.e. the image by $\hat{\sigma}$ of the defaults from the theory of S_1 at level h (the defaults implicit in S_1 at level h) have to be defaults from the theory of S_2 at level $\phi(h)$.

Categories of Hierarchies of Pre-orders. We define the category of hierarchies of pre-orders and the category of those hierarchies of pre-orders that are the semantics (hierarchies of differential preferences) of some hierarchic specification. Firstly we have to define the notion of the reduct of a pre-order w.r.t. a signature morphism.

Definition 18. Recall that, given a signature morphism $\sigma : \Sigma_1 \to \Sigma_2$ and an interpretation structure $m_2 \in \mathsf{Mod}(\Sigma_2)$ the *reduct* of m_2 w.r.t. $\sigma : \Sigma_1 \to \Sigma_2$ is the Σ_1-interpretation structure $\mathsf{Mod}(\sigma)(m_2)$. The reduct $\mathsf{Mod}(\sigma)(m_2)$ will be denoted by $\check{\sigma}(m_2)$. Furthermore, given a Σ_2-pre-order $R = (|R|, \sqsubseteq)$ the *reduct relation* denoted by $\check{\sigma}(R)$ is the pair $(\check{\sigma}(|R|), \check{\sigma}(\sqsubseteq))$ where $\check{\sigma}(|R|)$ is the class $\{\check{\sigma}(m) : m \in |R|\}$ of the reducts of the interpretation structures participating in R and $\check{\sqsubseteq} = \check{\sigma}(\sqsubseteq)$ is the smallest pre-order[6] among those interpretation structures that satisfies $\check{\sigma}(m_2) \mathrel{\check{\sqsubseteq}} \check{\sigma}(n_2)$ if $m_2 \sqsubseteq n_2$, that relates the reducts of the Σ_2-interpretation structures from $|R|$ if they were related by R.

Let $|\mathcal{H}|$ denote the class of interpretation structures participating in the hierarchy of pre-orders $\mathcal{H}$ and $\mathsf{rl}(\mathcal{H}, h)$ denote the pre-order assigned by $\mathcal{H}$ to the level h. The definition of the category **hiePre** of hierarchies of pre-orders follows.

Definition 19. The category **hiePre** of Σ-hierarchies of pre-orders consists of:

- **objects:** all hierarchies of pre-orders, and
- **morphisms:** A morphism $(\overleftarrow{\sigma, \phi}) : \mathcal{H}_2 \to \mathcal{H}_1$ from the Σ_2-hierarchy of pre-orders $\mathcal{H}_2$ to the Σ_1-hierarchy of pre-orders $\mathcal{H}_1$ is a signature morphism $\sigma : \Sigma_1 \to \Sigma_2$ from the signature of $\mathcal{H}_1$ to that of $\mathcal{H}_2$ and a partial order morphism $\phi : \mathsf{po}(\mathcal{H}_1) \to \mathsf{po}(\mathcal{H}_2)$ from the partial order of $\mathcal{H}_1$ to that of $\mathcal{H}_2$ such that
  - $\check{\sigma}(|\mathcal{H}_2|) \subseteq |\mathcal{H}_1|$, i.e. the reducts of the interpretation structures participating in $\mathcal{H}_2$ are interpretation structures participating in $\mathcal{H}_1$ and
  - $\check{\sigma}(\mathsf{rl}(\mathcal{H}_2, \phi(h_1))) \subseteq \mathsf{rl}(\mathcal{H}_1, h_1)$ for every $h_1 \in |\mathsf{po}(\mathcal{H}_1)|$, i.e. whenever two interpretation structures are related by the pre-order from $\mathcal{H}_2$ at level $\phi(h_1)$ their reducts are related by the pre-order from $\mathcal{H}_1$ at the level h_1. Note that this implies the reducts to be related in all pre-orders from levels h_1' with $\phi(h_1') = \phi(h_1)$.

The category of those hierarchies of pre-orders that are the differential semantics of some specification is defined as expected.

Definition 20. The category **hiePref** is the full subcategory of **hiePre** with objects the Σ-hierarchies of pre-orders $\mathcal{H}$ such that $\mathcal{H} = \mathcal{H}^{\ominus\ominus}$.

---

[6] The smallest relation that satisfies $\check{\sigma}(m_2) \mathrel{\check{\sqsubseteq}} \check{\sigma}(n_2)$ if $m_2 \sqsubseteq n_2$, is not, in general, a Σ_1-pre-order, since it may fail to be transitive. The reduct pre-order $\check{\sigma}(R)$ is the transitive closure of this relation.

Syntax and Semantics. Hierarchic specifications are related in the category hieSpec via their theories, or equivalently, via their meaning. This fact is made explicit by the following lemma: hierarchic specifications are related in hieSpec precisely in the same way that their semantics are related in hiePref .

Lemma 21. *There is a* hieSpec *morphism* $(\sigma, \phi) : S_1 \to S_2$ *iff there is a* hiePre *(or equivalently a* hiePref*) morphism* $\overleftarrow{(\sigma, \phi)} : S_2^{\ominus} \to S_1^{\ominus}$.

The syntactic and semantic categories (and also their constructions) are related by the functors hieSem : hieSpec $\to$ hiePref^{op} and hieSyn : hiePref^{op} $\to$ hieSpec. Their definition follows.

Definition 22. The functor hieSem : hieSpec $\to$ hiePref^{op} associates to each hierarchic specification S its hierarchy of differential preferences $S^{\ominus}$ and to each hieSpec morphism $(\sigma, \phi) : S_1 \to S_2$ the hiePref morphism $\overleftarrow{(\sigma, \phi)} : S_2^{\ominus} \to S_1^{\ominus}$.

The functor hieSyn : hiePref^{op} $\to$ hieSpec associates to each hierarchy of differential preferences $S^{\ominus}$ its theory $S^{\ominus\ominus}$ and to each hiePref morphism $\overleftarrow{(\sigma, \phi)}$: $S_2^{\ominus} \to S_1^{\ominus}$ the hieSpec morphism $(\sigma, \phi) : S_1^{\ominus\ominus} \to S'^{\ominus\ominus}_2$.

The semantics of the composition of hierarchic specifications is obtained by combining the semantics of the argument specifications. That is to each colimit of a diagram $\mathcal{D}$ in hieSpec there corresponds a limit in hiePref involving the semantics of the hierarchic specifications in $\mathcal{D}$: the limit of the image of $\mathcal{D}$ via hieSem. The reverse is also true. In this way it is possible to define on the semantics side a composition of hierarchic specifications and then check what is its syntactical expression (in particular the resulting hierarchic specification).

Theorem 23. *The image by* hieSyn *of a limit in* hiePref *is a colimit in* hieSpec *and the image by* hieSem *of a colimit in* hieSpec *is a limit in* hiePref.

Existence of Constructions. The category StPart of partial orders is not cocomplete. Therefore, hieSpec does not have all colimits and hiePref does not have all limits. However, when the underlying category Sign of signatures is cocomplete, all colimits of (small) diagrams exist in hieSpec provided that the corresponding combination of partial orders exists. The same is true for the limits in hiePref.

Existence of colimits in hieSpec is presented in the following theorem.

Theorem 24. *Let* $\mathcal{D}$ *be a diagram in* hieSpec. *Then* $\mathcal{D}$ *has a colimit in* hieSpec *if the category* Sign *of signatures is cocomplete and the diagram* $\mathcal{P}(\mathcal{D})$ *has a colimit in* StPart, *where* $\mathcal{P}$: hieSpec $\to$ StPart *is the forgetful functor sending each hierarchic specification to its priority structure (i.e. the composition of the corresponding partial orders of priority is defined).*

Existence of limits in hiePre and hiePref is ensured by theorems 23 and 24.

3 Conclusions

We have presented an institution independent theory of composition of hierarchic specifications that generalizes the classical theory from Goguen and Burstall. Composition is formalized on the syntactic and semantic levels by canonical operations on appropriate categories. For this purpose a new semantics of hierarchic specifications, the hierarchy of differential preferences, has been introduced. A corresponding notion of theory has been defined and a Galois connection between hierarchic specifications and their semantics has been established.

The category of hierarchic specifications is obtained by comparing hierarchic specifications via their theories by axiom and default preserving signature and priority structure morphisms. Canonical (co-)constructions in this category represent compositions of hierarchic specifications and have been shown to exist provided that both a result signature and a result priority structure exist.

The category of hierarchic specifications is mirrored on the semantic side by the category of their semantics. This implies, in particular, that the composition of hierarchic specifications can be interpreted on the semantic side by corresponding semantic constructions. In other words we have provided a denotational semantics for the composition of hierarchic specifications.

The theory just presented formalizes structuring constructs from specification languages providing explicit non-monotonic mechanisms. The identification of the most important structuring constructs (such as parameterization in this setting of hierarchic specifications) is the main subject of future work.

References

1. J. Adámek, H. Herrlich, and G. Strecker. *Abstract and Concrete Categories.* John Wiley & Sons, New York, 1990.
2. H. Andréka, M. Ryan, and P.-Y. Schobbens. Operators and laws for combining preference relations. In R. J. Wieringa and R. Feenstra, editors, *IS-CORE'94 - Selected papers*, pages 191–206. World Scientific Publishers, 1995.
3. S. Braß. Deduction with supernormal defaults. In P. Schmitt G. Brewka, K. Jantke, editor, *Nonmonotonic and Inductive Logic - Second International Workshop, 1991*, pages 153–174, Berlin, 1992. Springer[1].
4. S. Braß and U. W. Lipeck. Semantics of inheritance in logical object specifications. In Claude Delobel, Michael Kifer, and Yoshifumi Masunaga, editors, *Deductive and Object-Oriented Databases, 2nd Int. Conf. (DOOD'91)*, number 566 in LNCS, pages 411–430. Springer, 1991[1].
5. S. Braß, M. Ryan, and U. W. Lipeck. Hierarchical defaults in specifications. In G. Saake and A. Sernadas, editors, *Information Systems — Correctness and Reusability, Workshop IS-CORE '91*, number 91-03 in Informatik-Bericht, pages 179–201. TU Braunschweig, 1991[1].
6. G. Brewka. *Nonmonotonic Reasoning: Logical Foundations of Commonsense.* Cambridge Tracts in Theoretical Computer Science. Cambridge University Press, Cambridge, 1991.

[1] Available by ftp at ftp://ftp.informatik.uni-hannover.de/papers

7. R. M. Burstall and J. A. Goguen. Putting theories together to make specifications. In Raj Reddy, editor, *Proceedings of the 5th International Joint Conference on Artificial Intelligence (IJCAI)*, pages 1045–1058, Cincinnati, Ohio, 1977. Department of Computer Science, Carnegie-Mellon University.
8. F. M. Dionísio, U. W. Lipeck, and S. Braß. Composition of default specifications. In R. J. Wieringa and R. Feenstra, editors, *IS-CORE'94 - Selected papers*, pages 207–221. World Scientific Publishers, 1995[2].
9. F. Dionísio. *Composition of Hierarchic Default Specifications.* PhD thesis, University of Hannover, 1997[2].
10. J. Fiadeiro and T. Maibaum. Describing, structuring and implementing objects. In J. W. de Bakker, W. P. de Roever, , and G. Rozenberg, editors, *Foundations of Object-Oriented Languages*, number 489 in LNCS, pages 275–310. Springer, 1991.
11. J. Fiadeiro and T. Maibaum. Temporal theories as modularization units for concurrent system specification. *Formal Aspects of Computing*, 4:239–272, 1992.
12. P. Gärdenfors, editor. *Belief Revision.* Cambridge Press, Cambridge, England, 1992.
13. J. A. Goguen. A categorial manifesto. Technical report prg-72, Programming Research Group, University of Oxford, 1989.
14. J. A. Goguen and R. M. Burstall. Institutions: Abstract model theory for specification and programming. *Journal of the ACM*, 39(1):95–146, 1992.
15. D. Makinson. Five faces of minimality. *Studia Logica*, 52:339–379, 1993.
16. J. McCarthy. Circumscription - a form of non-monotonic reasoning. *Artificial Intelligence*, 13:27–39, 1980.
17. D. Poole. A logical framework for default reasoning. *Artificial Intelligence*, 36:27–47, 1988.
18. R. Reiter. On closed world data bases. In H. Gallaire and J. Minker, editors, *Logic and Data Bases*, pages 55–76. Plenum Press, New York, 1978.
19. M. Ryan. Defaults and revision in structured theories. In *Proc. Sixth Annual IEEE Symposium on Logic in Computer Science (LICS)*, pages 362–373, Los Alamitos, CA, 1991. IEEE Computer Society Press.
20. M. Ryan. Representing defaults as sentences with reduced priority. In B. Nebel and W. Swartout, editors, *Proc. Second International Conference on Principles of Knowlage Representation and Reasoning (KR'92).* Morgan Kaufmann, 1992.
21. A. Sernadas, J. F. Costa, and C. Sernadas. An institution of object behaviour. In H. Ehrig and F. Orejas, editors, *Recent Trends in Data Type Specification*, pages 337-350. Springer, 1994. LNCS 785[3].
22. A. Sernadas, J. Fiadeiro, C. Sernadas, and H.-D. Ehrich. Abstract object types: A temporal perspective. In B. Banieqbal, H. Barringer, and A. Pnueli, editors, *Temporal Logic in Specification*, pages 324-350. Springer, 1989[3].
23. A. Sernadas, C. Sernadas, and J. F. Costa. Object specification logic. *Journal of Logic and Computation*, 5(5):603–630, 1995[3].
24. Y. Shoham. Nonmonotonic logics: meaning and utility. In *Proceedings of IJCAI-87*, pages 388–392, Milan, 1987.

[2] Available by ftp at ftp://ftp.cs.math.ist.utl.pt/pub/DionisioFM and also at the address ftp://ftp.informatik.uni-hannover.de/papers

[3] Available by ftp at ftp://ftp.cs.math.ist.utl.pt/pub/SernadasA

An Inductive View of Graph Transformation*

F. Gadducci[1] and R. Heckel[2]

[1] TUB, Fachbereich 13 Informatik, Franklinstraße 28/29, 10587 Berlin, Germany, (gfabio@cs.tu-berlin.de).

[2] Università di Pisa, Dipartimento di Informatica, Corso Italia 40, 56124 Pisa, Italy, (reiko@di.unipi.it).

Abstract. The dynamic behavior of rule-based systems (like *term rewriting systems* [24], *process algebras* [27], and so on) can be traditionally determined in two orthogonal ways. Either *operationally*, in the sense that a way of embedding a rule into a state is devised, stating explicitly how the result is built: This is the role played by (the application of) a substitution in term rewriting. Or *inductively*, showing how to build the class of all possible reductions from a set of basic ones: For term rewriting, this is the usual definition of the rewrite relation as the minimal closure of the rewrite rules. As far as *graph transformation* is concerned, the operational view is by far more popular: In this paper we lay the basis for the orthogonal view. We first provide an inductive description for graphs as arrows of a freely generated *dgs-monoidal* category. We then apply 2-categorical techniques, already known for term and term graph rewriting [29, 7], recasting in this framework the usual description of graph transformation via double-pushout [13].

1 Introduction

The *theory of graph transformation* [30] basically studies a variety of formalisms which extend the theories of formal languages and term rewriting, respectively, in order to deal with structures more general than strings and terms. In both of these "classical" rewrite formalisms, there are two different ways of defining the rewrite relation $\rightarrow_{\mathcal{R}}$ for a given rewrite system $\mathcal{R}$: The *operational definition* states e.g. that a term rewrite rule $l \rightarrow r \in \mathcal{R}$ is applicable to a term t if an instance of l occurs as a subterm in t. Then, this subterm may be removed and replaced by a corresponding instance of r, leading to a derived term s. Equivalently, we may give an *inductive definition* where the rewrite relation is the smallest relation which contains $\mathcal{R}$ and is closed under substitution and context. While the operational definition is clearly more intuitive, and well-suited for implementation purposes, the inductive one plays an important role in the theory of term rewriting, since it allows for the development of analysis techniques based on structural induction.

* Research partly supported by the EC TMR Network GETGRATS (General Theory of Graph Transformation Systems) through the Technical University of Berlin and the University of Pisa.

In the *double-pushout* (DPO) approach to graph transformation [15, 13] (and in most of the other approaches) the operational definition is by far more popular. Inductive definitions of DPO graph transformation have been given in [1, 12], but they do not have the same role as in the theory of term rewriting. One reason may be that, unlike for strings and terms, there is no straightforward inductive definition of graphs. Rather, each possible interpretation suggests a different choice of the basic operations. In [1], for example, a hyper-graph is considered as *a set of edges glued by means of vertices*. The operations for building graphs are disjoint union, and renaming and fusion of nodes. In [12], a graph is described in a logical style by *a set of edge predicates over node variables*, using (partly non-disjoint) union, and renaming and binding of variables.

In this paper, graphs are seen as *distributed states* consisting of several local components connected through interfaces. The distributed structure is made explicit by regarding a graph as an *arrow* of a category whose objects are sets of interface vertices: The source and target object of an arrow representing a local state are the interfaces through which it is connected to other local states. Then, arrow composition represents the composition of two local states over a common interface. Our inductive definition of graphs is thus based on the most basic operation in the DPO approach: The *gluing of graphs*, categorically described as a pushout in the category of graphs. This category of *ranked graphs* (i.e., graphs with interfaces) is defined in Section 2.

The algebraic structure of this category is axiomatized in Section 3 by the notion of *dgs-monoidality*: It is shown that the category of ranked graphs is isomorphic to the free dgs-monoidal category generated from a single arrow.

Lifting this structure of graphs to transformations in Section 4, they immediately get a distributed flavor, that is, transformations of complex graphs are build from transformations of simpler components. In fact, this notion of rewriting is closely related to distributed graph transformation in the sense of [14]. Technically speaking, the transformations are given by the cells of a 2-category freely generated from basic cells which represent the rules of the system. Such 2-categorical models are well-known for term rewriting: See e.g. [28, 31]. More recently, they have been applied to term graph rewriting [7].

2 Graphs

This section introduces (ranked) graphs as isomorphism classes of (ranked) concrete graphs. This presentation departs slightly from the standard definition (see for example [13]), because our main concern is the algebraic structure of graphs.

Definition 1 (directed concrete graphs). A *(directed) concrete graph* d is a four-tuple $d = \langle E, N, s, t \rangle$, where E is a set of *edges*, N is a set of *nodes*, $s, t : E \to N$ are functions called respectively the *source* and *target* function (and we shall often denote these components by $N(d)$, $E(d)$, s_d and t_d, respectively). A node n is *isolated* if it has neither ingoing nor outgoing arcs. A graph is *discrete* if E is empty (or, equivalently, all nodes are isolated).

Let d and d' be two concrete graphs. A *concrete graph morphism* $f : d \to d'$ is a pair of functions $\langle f_n, f_e \rangle$ such that $f_n : N(d) \to N(d')$ and $f_e : E(d) \to E(d')$. These functions must preserve source and target, i.e., for each edge $e \in E(d)$, $s_{d'}(f_e(e)) = f_n(s_d(e))$, and $t_{d'}(f_e(e)) = f_n(t_d(e))$.

Concrete graphs and their morphisms form a category we denote **DCG**. □

In the following, for each $i \in \mathbb{N}$ we denote by $\underline{i}$ the set $\{1, \ldots, i\}$ (thus $\underline{0} = \emptyset$).

Definition 2 (ranked concrete graphs and graphs). An (i, j)*-ranked concrete graph* (or also, a *concrete graph of rank* (i, j)) is a triple $g = \langle r, d, v \rangle$, where d is a concrete graph, $r : \underline{i} \to N(d)$ is a function called the *root mapping*, and $v : \underline{j} \to N(d)$ is a function called the *variable mapping*. Node $r(l)$ is called the l*-th root* of d, and $v(k)$ is called the k*-th variable* of d, for each admissible j, k.

Two (i, j)-ranked concrete graphs $g = \langle r, d, v \rangle$ and $g' = \langle r', d', v' \rangle$ are *isomorphic* if there exists a *ranked concrete graph isomorphism* $\phi : g \to g'$, i.e., a concrete graph isomorphism $\phi : d \to d'$ such that $\phi \circ r = r'$ and $\phi \circ v = v'$. A (i, j)*-ranked graph* G (or *with rank* (i, j)) is an isomorphism class of (i, j)-ranked concrete graphs. We shall often write G^i_j to recall that G has rank (i, j). □

The idea of equipping graphs with lists of distinguished nodes in order to define composition operations on them is not new (see for example [1, 16]): Roughly, for a graph G^i_j, the components $\underline{i}, \underline{j}$ represent *discrete interfaces*, through which graphs can be equipped with a compositional structure. And, dealing with isomorphism classes of concrete graphs, we disregard the concrete identity of nodes and edges when manipulating graphs.

We introduce now two operations on ranked graphs. The *composition* of two ranked graphs is obtained by gluing the variables of the first one with the roots of the second one, and it is defined only if their number is equal. This operation allows us to define a category having ranked graphs as arrows. Next the *union* of graphs is introduced: It is always defined, and it is a sort of disjoint union where roots and variables are suitably renumbered. This second operation provides the category of graphs with a monoidal structure, made explicit in Section 3.3.

Definition 3 (the category of graphs). Let $G'^j_k = [\langle r', d', v' \rangle]$ and $G^i_j = [\langle r, d, v \rangle]$ be two ranked graphs. Their *composition* is the ranked graph $H^i_k = G'^j_k; G^i_j$ defined as $H^i_k = [\langle in_d \circ r, d'', in_{d'} \circ v' \rangle]$, where $\langle d'', in_d, in_{d'} \rangle$ is a pushout of $\langle v : \underline{j} \to d, r' : \underline{j} \to d' \rangle$ in **DCG** (set $\underline{j}$ regarded as a discrete concrete graph).

DG denotes the category having as objects underlined natural numbers, and as arrows from $\underline{j}$ to $\underline{i}$ all (i, j)-ranked graphs. Arrow composition is defined as in Definition 3, and the identity on $\underline{i}$ is the graph G^i_{id} of rank (i, i) having i nodes, and where the k-th root is also the k-th variable, for all $k \in \underline{i}$. □

The well-definedness of graph composition easily follows from the uniqueness of pushouts, up-to isomorphism. Then it is easy to check that **DG** is a well-defined category, because composition is associative, and the identity laws hold.

Definition 4 (union of ranked graphs). Let $G^i_j = [\langle r, d, v\rangle]$ and $G'^k_l = [\langle r', d', v'\rangle]$ be two ranked graphs. Their *union* or *parallel composition* is the graph of rank $(i+k, j+l)$ $G^i_j \oplus G'^k_l = [\langle r'', d \uplus d', v''\rangle]$, where $\langle d \uplus d', p_0, p_1\rangle$ is the coproduct of d and d' in **DCG**, and $r'' : \underline{i+k} \to d \uplus d'$ and $v'' : \underline{j+l} \to d \uplus d'$ are the morphisms induced by the universal property. □

Example 1 (graphs, composition and union). Figure 1 shows four graphs. Nodes are graphically represented by a dot, "•", from (to) where the edges leave (arrive); the list of numbers on the left (right) represent pointers to the variables (roots): A dashed arrow from j to a node indicates that it is the j-th variable (root). For example, graph G_2 has rank $(1, 4)$, with five nodes, two of them isolated: There is only one "root node" (pointed by 1) with two outgoing edges to "variable nodes" (pointed by 1 and 4).

These graphical conventions make easy the operation of composition, that can be performed by matching the roots of the first graph with the variables of the second one, and then by eliminating them. For example, graph $G_1 ; G_2$ is the composition of G_1 and G_2, of rank $(1, 2)$. The last graph is $G_1 \oplus G_2$, the union of G_1 and G_2, of rank $(5, 6)$. □

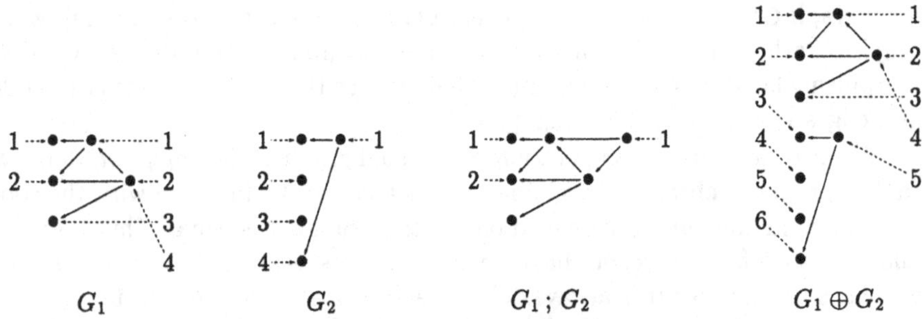

Fig. 1. Two graphs, their composition and their union.

We are now ready to show the main result of this section, namely, that every graph can be built from a small set of atomic ones.

Definition 5 (atomic graphs). An *atomic graph* is one of those graphs depicted in Figure 2. □

Theorem 6 (decomposition of graphs). *Every graph can be obtained as the value of an expression containing only atomic graphs as constants and composition and union as operators.* □

Space limitations force us to omit the proof. Nevertheless, a similar one for (acyclic) term graph can be actually found in [9], carried out by induction on

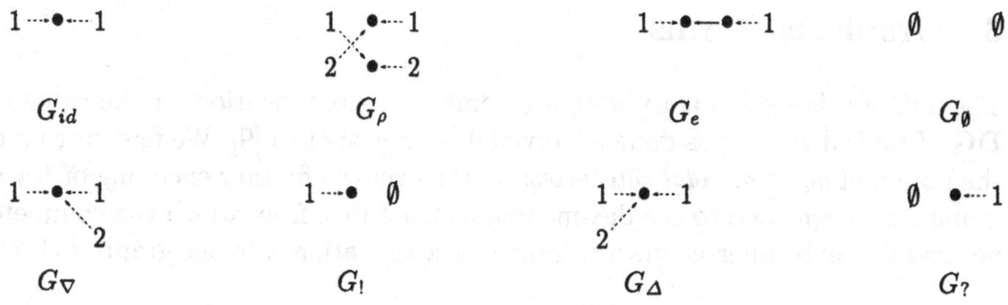

Fig. 2. Atomic graphs.

the number of nodes of a given graph. In this version, the main technical tool is the use of a suitable composition of G_∇ and G_Δ to induce cycles, as shown in Example 2: Such a remark is the basis of our finitary description of *traced monoidal categories*, to be found in next section.

Example 2 (a few graphs). In this example we want to use the operators for building a few simple graphs. The picture below shows the construction of the (2, 2)-ranked graph $G_\uparrow$ as the value of the expression $G_\nabla \oplus G_{id}; G_{id} \oplus G_e \oplus G_{id}; G_{id} \oplus G_\Delta$ (with "$\oplus$" binding stronger than ";"). Here, G_{id} is used to create isolated nodes which are "visible" (so to say) from both interfaces, G_e provides the edge, and G_∇ and G_Δ are used to glue the source and the target of this edge with the two additional vertices:

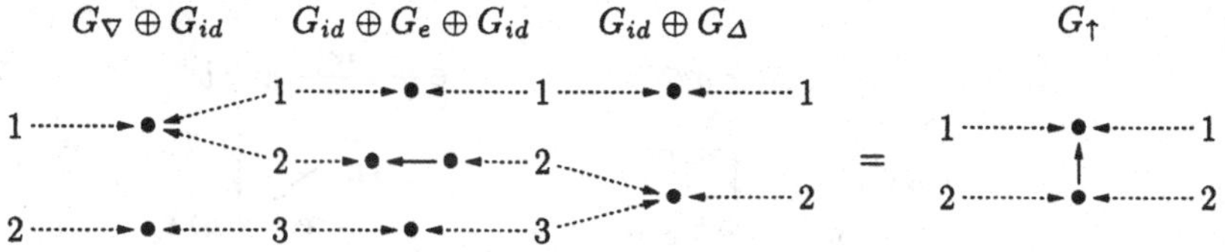

The "downward" edge between two visible nodes is obtained by applying permutations, like in $G_\rho; G_\uparrow; G_\rho$.

Loops are created by gluing together vertices, that is, $G_c = G_\nabla; G_\uparrow; G_\Delta$ is the graph of rank (1, 1):

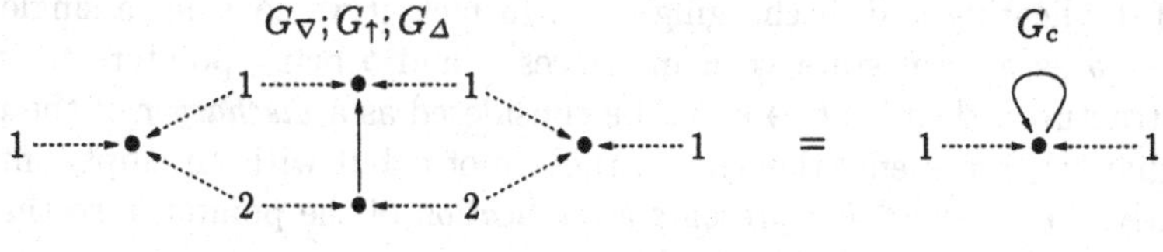

In order to restrict e.g. the right interface, we may compose with $G_!$: Then $G_c; G_!$ yields the (0, 1)-ranked loop. □

3 Graphs as Terms

The aim of this section is to present a complete axiomatization for the category **DG** of ranked graphs, as done for (cyclic) term graphs in [9]. We first introduce the notion of *dgs-monoidal categories*; we then show a finitary encoding of *traced monoidal categories* into the dgs-monoidal structures, from which the completeness result can be inferred via a folklore characterization of hyper-graphs [33, 21].

3.1 On (d)gs-monoidal categories

In this section we introduce *dgs-monoidal* categories, an extension of *gs-monoidal* ones [9, 7], which are used for our equational presentation of graphs.

Definition 7 (gs-monoidal categories). A *gs-monoidal category* **C** is a six-tuple $\langle \mathbf{C}_0, \otimes, e, \rho, \nabla, ! \rangle$, where $\langle \mathbf{C}_0, \otimes, e, \rho \rangle$ is a symmetric strict monoidal category and $! : Id \Rightarrow e : \mathbf{C}_0 \to \mathbf{C}_0$, $\nabla : Id \Rightarrow \otimes \circ D : \mathbf{C}_0 \to \mathbf{C}_0$ are two transformations (D is the diagonal functor), such that $!_e = \nabla_e = id_e$ and satisfying the *coherence* axioms

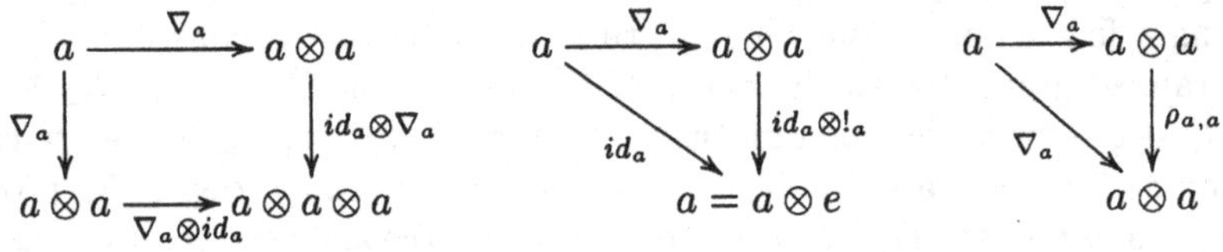

and the *monoidality* axioms

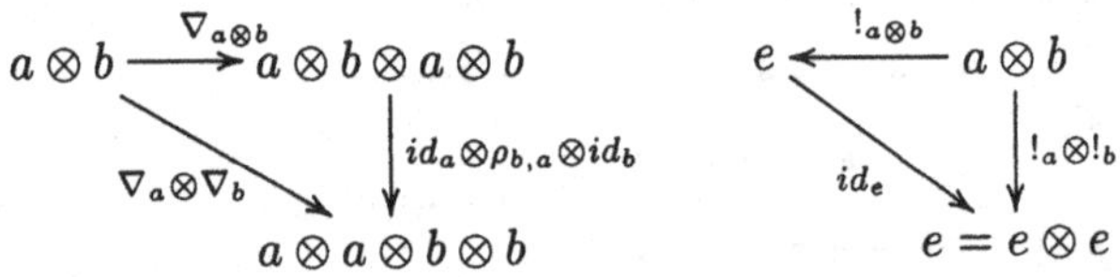

A *gs-monoidal functor* $F : \mathbf{C} \to \mathbf{C}'$ is a symmetric monoidal functor such that $F(!_a) = !'_{F(a)}$ and $F(\nabla_a) = \nabla'_{F(a)}$. The category of small gs-monoidal categories and their functors is denoted by **GS-Cat**. □

Introduced in [9], *gs-monoidal categories* can be roughly considered as monoidal categories, enriched by two transformations that allows for a controlled way of duplicating and discharging data. In fact, if we consider a single arrow $t : a \to b$ as a local state with interfaces a and b being pointers to internal data structures, then $!_b : b \to e$ can be considered as a *discharger* of the pointer b, so that $t; !_b$ represents the same structure of t, but with an empty interface. Similarly, $\nabla_b : b \to b \otimes b$ represents a *duplication* of the pointer b, so that $t; \nabla_b$ can be seen as a *shared* instance of t.

These structures fill the gap between monoidal and cartesian categories: It can be considered categorical folklore [18, 25, 19, 7] that, equipping a monoidal category with suitable *natural* transformations, we obtain a cartesian category: See e.g. [9] for a recollection. In our case, an instance of the theorem is obtained simply requiring the naturality of ∇ and $!$: Such naturality forces a correspondence between pointers and underlying structures (so that e.g. $t; !_b = !_a$: deleting a pointer is the same as deleting the structure), and allows to recast the usual notion of *algebraic theory* [26].

Definition 8 (dgs-monoidal categories). A *dgs-monoidal category* $\mathbf{C}$ is a eight-tuple $\langle \mathbf{C_0}, \otimes, e, \rho, \nabla, !, \Delta, ? \rangle$, such that both the six-tuples $\langle \mathbf{C_0}, \otimes, e, \rho, \nabla, ! \rangle$ and $\langle (\mathbf{C_0})^{op}, \otimes, e, \rho, \Delta, ? \rangle$ are gs-monoidal categories (where $(\mathbf{C_0})^{op}$ is the dual category of $\mathbf{C_0}$), and satisfying

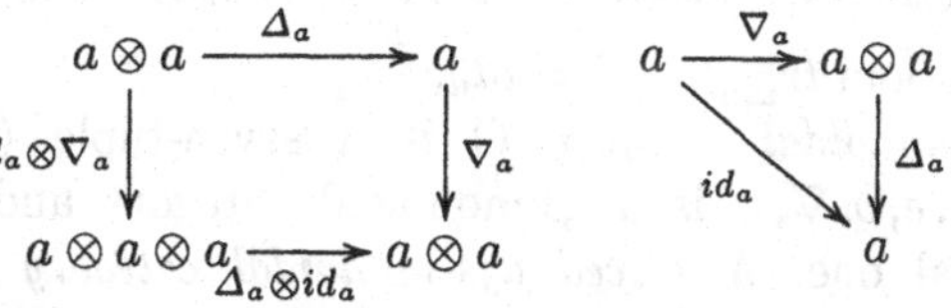

$a \otimes a$
Δ_a
a
$id_a \otimes \nabla_a$
∇_a
$a \otimes a \otimes a$
$\Delta_a \otimes id_a$
$a \otimes a$
a
∇_a
$a \otimes a$
id_a
Δ_a
a

A *dgs-monoidal functor* $F : \mathbf{C} \to \mathbf{C'}$ is a gs-monoidal functor such that also F^{op} is gs-monoidal. The category of small dgs-monoidal categories and their functors is denoted by **DGS-Cat**. □

Equivalent notions of dgs-monoidal categories have surfaced quite frequently in recent years. A (bicategorical) presentation is used as a description of the (bi)category of relations already in [5], which forms the basis for some recent work on the categorical description of *circuits* [21, 17]: Arrows are processes, and the bicategorical structure allows to relate specifications which are equivalent but show a different internal structure (a different *implementation*, so to say).

Intuitively, the arrow $?_a$ can be interpreted as the NEW operator, creating a pointer to a new name of sort a; while Δ_a corresponds to a *matching*, forcing the equality of two pointers. Some of the additional axioms can then be explained as *housekeeping* operations on the set of pointers of a structure. For example, the axiom $\nabla_a; \Delta_a$ simply states that duplicating the pointers to a structure, and then equating them, results in an unchanged set of pointers. Instead, since $?_a$ creates a new name, and $!_a$ deletes only the associated pointer, *without* destroying the new name, then $?_a; !_a$ is in general different from id_e.

3.2 On the notion of feedback

Traced monoidal categories have been studied *per se* as a categorical tool [20]. However, they already surfaced in the literature related to algebraic theories. In fact, there is a strong connection between traced categories and *iteration theories* (that is, algebraic theories with an equational characterization of (least) fix-point [2, 3]) as it is pointed out in the works on *flownomial calculus* [32, 33].

Definition 9 (traced monoidal categories). A *traced monoidal category* $\mathbf{C}$ is a five-tuple $\langle \mathbf{C_0}, \otimes, e, \rho, tr \rangle$, where $\langle \mathbf{C_0}, \otimes, e, \rho \rangle$ is a symmetric strict monoidal category,[3] which is equipped with a family of functions $tr^u_{a,b} : \mathbf{C}[a \otimes u, b \otimes u] \to \mathbf{C}[a, b]$, satisfying the *naturality* axioms

$$tr^u_{a,b}(f; (id_b \otimes g)) = tr^u_{a,b}((id_a \otimes g); f) \qquad tr^u_{c,d}((h \otimes id_u); f; (l \otimes id_u)) = h; tr^u_{a,b}(f); l$$

the *vanishing* axioms

$$tr^e_{a,b}(f) = f \qquad tr^v_{a,b}(tr^u_{a\otimes v, b\otimes v}(f)) = tr^{u\otimes v}_{a,b}(f)$$

the *superposing* axiom

$$tr^u_{a\otimes c, b\otimes d}((id_a \otimes \rho_{c,u}); (f \otimes g); (id_b \otimes \rho_{u,d})) = tr^u_{a,b}(f) \otimes g$$

and the *yanking* axiom $tr^u_{u,u}(\rho_{u,u}) = id_u$.

A *traced gs-monoidal category* $\mathbf{C}$ is a seven-tuple $\langle \mathbf{C}_0, \otimes, e, \rho, \nabla, !, tr\rangle$, such that $\langle \mathbf{C}_0, \otimes, e, \rho, \nabla, !\rangle$ is a gs-monoidal category and $\langle \mathbf{C}_0, \otimes, e, \rho, tr\rangle$ is a traced monoidal one. A *traced dgs-monoidal category* $\mathbf{C}$ is a nine-tuple $\langle \mathbf{C}_0, \otimes, e, \rho, \nabla, !, \Delta, ?, tr\rangle$ such that $\langle \mathbf{C}_0, \otimes, e, \rho, \nabla, !, \Delta, ?\rangle$ is a dgs-monoidal category, $\langle (\mathbf{C}_0)^{op}, \otimes, e, \rho, tr\rangle$ is a traced monoidal one, and satisfying

$$tr^u_{e,u}(\nabla_u) = ?_u \qquad\qquad tr^u_{u,e}(\Delta_u) = !_u$$

A *traced monoidal* (*gs-monoidal, dgs monoidal*) functor is a symmetric monoidal (gs-monoidal, dgs-monoidal) functor such that $F(tr^u_{a,b}(t)) = tr^{F(u)}_{F(a),F(b)}(F(t))$. The category of small traced monoidal (gs-monoidal, dgs-monoidal) categories and their functors is denoted by **Tr-Cat** (**TrGS-Cat** and **TrDGS-Cat**). □

A correspondence theorem is proved in [9], namely the existence of an isomorphism between the category of term graphs over a signature Σ and the free traced gs-monoidal category over Σ. As suggested in [21], the dgs-monoidal is the proper one to deal instead with *term hyper-graphs*: In fact, they prove a correspondence theorem between (a logical presentation of) term hyper-graphs (called circuits) over a given *hyper-signature* and (an equivalent version of) traced dgs-monoidal categories. The encoding result proved in the next section (see Theorem 11) will allow us to state our completeness result as a corollary of this property.

3.3 A finite description for traces

We open the section showing that **DG** can be equipped with a dgs-monoidal structure. We then prove the encoding of a traced category into the dgs-monoidal one, deriving this way our completeness result.

Proposition 10 (DG is dgs-monoidal). *The category* **DG** *of ranked graphs has a dgs-monoidal structure, given by the eight-tuple*

$$\langle \mathbf{DG}, \oplus, G_{id}, G_\rho, G_\nabla, G_!, G_\Delta, G_? \rangle$$

where the auxiliary arrows are (defined as) the atomic graphs of Definition 5. □

[3] When introduced in [20], the authors dealt with the more general *balanced monoidal categories* where, basically, the family of symmetries is not unique.

The next theorem shows that dgs-monoidal categories are also traced, via a simple encoding for traces which seems not to be known in the literature.[4]

Theorem 11 (encoding feedback). *Let* $\mathbf{C}$ *be a dgs-monoidal category* $\langle \mathbf{C}_0, \otimes, e, \rho, \nabla, !, \Delta, ? \rangle$. *Then it admits also a traced dgs-monoidal structure: The family "tr" is defined as*

$$tr^u_{a,b}(f) = (id_a \otimes \nabla^h_u); (f \otimes id_u); (id_a \otimes \Delta^h_u)$$

where ∇^h_u, Δ^h_u *denote* $?_u; \nabla_u$ *and* $\Delta_u; !_u$, *respectively.*

Proof. All the additional axioms for traced dgs-monoidal categories are easy to check (by directly substituting the derived operators, via the axioms for dgs-monoidality) except for the first naturality axioms. As an example, the superposing axiom can be simply checked as

$$\begin{array}{lr}
tr^u_{a\otimes c, b\otimes d}((id_a \otimes \rho_{c,u}); (f \otimes g); (id_b \otimes \rho_{u,d})) & = \\
(id_a \otimes id_c \otimes \nabla^h_u); [(id_a \otimes \rho_{c,u}); (f \otimes g); (id_b \otimes \rho_{u,d}) \otimes id_u]; (id_b \otimes id_d \otimes \Delta^h_u) & = \\
(id_a \otimes \nabla^h_u); f; (id_b \otimes \Delta^h_u) \otimes g & = \\
tr^u_{a,b}(f) \otimes g &
\end{array}$$

via the naturality of ρ, and the axioms $\rho_{e,c} = \rho_{e,d} = id_e$.

Let us move now to the first naturality axiom. We need to use a decomposition property, suggested in [21]: For each $f : a \to c$ and $g : c \to b$, the equality $f; g = (id_a \otimes \nabla^h_c); (f \otimes id_c \otimes g); (\Delta^h_c \otimes id_b) = (\nabla^h_c \otimes id_a); (g \otimes id_c \otimes f); (id_b \otimes \Delta^h_c)$ holds. Then, the proof goes in the following way

$$\begin{array}{lr}
tr^u_{a,b}(f; (id_b \otimes g)) & = \\
(id_a \otimes \nabla^h_u); [f; (id_b \otimes g) \otimes id_u]; (id_b \otimes \Delta^h_u) & = \\
\nabla^*_{a\otimes u}; [f; (id_b \otimes g) \otimes !_a \otimes ?_b \otimes id_u]; \Delta^*_{b\otimes u} & = \\
\nabla^*_{a\otimes u}; [(id_a \otimes id_u \otimes \nabla^h_{b\otimes u}); (f \otimes id_b \otimes id_u \otimes g \otimes id_u); & \\
(\Delta^h_{b\otimes u} \otimes id_b \otimes id_u) \otimes !_a \otimes ?_b \otimes id_u]; \Delta^*_{b\otimes u} & = \\
\nabla^*_{a\otimes u}; [!_a \otimes ?_b \otimes u \otimes (\nabla^h_{b\otimes u} \otimes id_a \otimes id_u); (g \otimes id_u \otimes id_b \otimes id_u \otimes f); & \\
(id_b \otimes id_u \otimes \Delta^h_{b\otimes u})]; \Delta^*_{b\otimes u} & = \\
\nabla^*_{a\otimes u}; [!_a \otimes ?_b \otimes u \otimes (g \otimes id_u); f]; \Delta^*_{b\otimes u} & = \\
\nabla^*_{a\otimes u}; [(id_a \otimes g); f \otimes !_a \otimes ?_b \otimes id_u]; \Delta^*_{b\otimes u} & = \\
(id_a \otimes \nabla^h_u); [(id_b \otimes g); f \otimes id_u]; (id_b \otimes \Delta^h_u) & = \\
tr^u_{a,b}((id_a \otimes g); f) &
\end{array}$$

where $\nabla^*_{a\otimes u}$, $\Delta^*_{b\otimes u}$ denote $(id_a \otimes ?_u); \nabla_{a\otimes u}$ and $\Delta_{b\otimes u}; (id_b \otimes !_u)$, respectively. □

We are finally able to provide our main theorem for the section: It relies on the correspondence results given for traced dgs-monoidal theories in the literature, coupled with the properties we proved with Proposition 10 and Theorem 11.

Theorem 12 (DG is a free structure). *Let* d *be a concrete graph with just one node and one arrow, and let* $\mathbf{DGS}(d)$ *be the associated free dgs-monoidal category. Then* $\mathbf{DGS}(d)$ *is isomorphic to* $\mathbf{DG}$ *via a dgs-monoidal functor.* □

[4] We just discovered an equivalent description, provided without proof, in [22].

4 A 2-category for Graph Rewriting

We open this section recalling a few definitions about the *double-pushout* approach to graph transformation. We will then present the basic notions regarding *2-categories*: They will be used to provide an alternative presentation of graph transformation, which we will prove equivalent to the traditional one.

4.1 Basic notions of double-pushout

Historically, the first of the algebraic approaches to graph transformation is the so-called *double-pushout (DPO) approach* introduced in [15], which owes its name to the basic construction used to define a single derivation step, modeled indeed by two gluing diagrams (i.e., pushouts) in the category **DCG** of concrete graphs.

Definition 13 (graph productions). A *graph production* $p : s$ is composed of a *production name* p and of a span of injective graph morphisms $s = \langle L \leftarrow_l K \rightarrow_r R \rangle$, called *production span*. A *graph transformation system* $\mathcal{G}$ is a set of graph productions (with different names). □

Derivation steps in the DPO approach are defined operationally, as double pushout constructions: The left-hand side L contains the items that must be present for an application of the production, the right-hand side R those that are present afterwards, and the context graph K specifies the "gluing items", i.e., the objects which are read during application but are not consumed.

Definition 14 (DPO derivation). A *double-pushout* is a diagram like in Figure 3, where top and bottom are production spans and (1) and (2) are pushouts. If $p : \langle L \leftarrow_l K \rightarrow_r R \rangle$ is a production, a *derivation step* from G to H via production p and "context embedding" d is denoted by $G \Rightarrow_{p/d} H$.

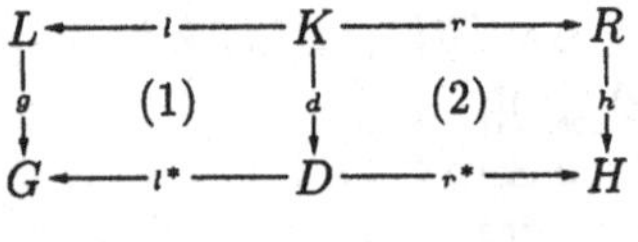

Fig. 3. A DPO derivation step

More abstractly, we denote by $\Rightarrow_{\mathcal{G}} \subseteq |\mathbf{DCG}| \times |\mathbf{DCG}|$ the corresponding *rewrite relation* on graphs, and by $\Rightarrow^*_{\mathcal{G}}$ its reflexive and transitive closure. □

The existence of a derivation is characterized by the gluing conditions, which characterizes the existence (and uniqueness up-to isomorphism) of the *pushout complement*, i.e., the context graph D and morphisms l^* and d such that subdiagram (1) in the left of Figure 3 is a pushout. Operationally speaking, the application of a production $p : \langle L \leftarrow_l K \rightarrow_r R \rangle$ to a graph G consists of three steps. First, the match $g : L \rightarrow G$ has to be chosen, providing an occurrence of L in G, such that the gluing conditions are satisfied. Then, all objects of G matched by $L - l(K)$ are removed. This leads to the context graph D. Finally, the objects of $R - r(K)$ are added to D, to obtain the derived graph H.

4.2 A few notions on 2-categories

In this section we briefly present some notions about 2-categories needed in the rest of the paper. We first recall the basic definitions: For an introduction, we refer the reader to the classical work [23].

Definition 15 (2-categories). A *2-category* $\underline{\mathbf{C}}$ is a four-tuple $\langle Ob_c, \mathbf{C}, *, id\rangle$ such that Ob_c is a set of *2-objects* and, indexed by elements in Ob_c, $\mathbf{C}$ is a family of categories $\mathbf{C}[a,b]$ (the *hom-categories* of $\underline{\mathbf{C}}$), $*$ is a family of functors $*_{a,c}^{b} : \mathbf{C}[a,b] \times \mathbf{C}[b,c] \to \mathbf{C}[a,c]$ and id is a family of objects $id_a \in |\mathbf{C}[a,a]|$, satisfying for each $\alpha \in \mathbf{C}[a,b]$

[category] $\qquad id_a * \alpha = \alpha = \alpha * id_b \qquad (\alpha * \beta) * \gamma = \alpha * (\beta * \gamma)$

where for the sake of readability the indexes of $*$ are dropped, and the arrow id_{id_a} (the identity on the object id_a) is denoted by the object itself. □

The *category* axioms impose a categorical structure to the pair $\mathbf{C}_u = \langle Ob_c, |\mathbf{C}[a,b]|\rangle$ (denoted as the *underlying category* of $\underline{\mathbf{C}}$), where the objects are the elements of the set of 2-objects, and the elements of the hom-set $\mathbf{C}_u[a,b]$ are the objects of the hom-categories $\mathbf{C}[a,b]$. Then, roughly, a 2-category can be simply described as a category $\mathbf{C}$ such that, given any two objects a, b, the hom-set $\mathbf{C}[a,b]$ is actually a category. In fact, we denote as *arrows* and *cells* of the 2-category $\underline{\mathbf{C}}$ the objects and arrows of the hom-categories, respectively: By $\alpha : f \Rightarrow g : a \to b$ we mean that α is a cell in $\mathbf{C}[a,b]$ from f to g, depicted as

$$a \underset{g}{\overset{f}{\Downarrow \alpha}} b.$$

Definition 16 (2-functors). Let $\underline{\mathbf{C}}, \underline{\mathbf{D}}$ be 2-categories. A *2-functor* $F : \underline{\mathbf{C}} \to \underline{\mathbf{D}}$ is a pair $\langle F_o, F_m\rangle$, where $F_o : Ob_c \to Ob_d$ is a function and F_m is a family of functors $F_{a,b} : \mathbf{C}[a,b] \to \mathbf{C}[F_o(a), F_o(b)]$.

We denote **2Cat** the category of (small) 2-categories and 2-functors. □

The paradigmatic example of 2-category is $\underline{\mathbf{Cat}}$, having small categories as objects, functors as arrows, and natural transformations as cells. As far as rewriting is concerned, it is well-known that a suitable class of 2-categories, *algebraic 2-theories*, can be used for describing term rewriting (see e.g. [31, 28], and the more recent [34, 11]), while in [7, 8] similar categorical models have been proposed for term graph rewriting, *(traced) gs-monoidal 2-theories*.

A *computad* [35] is a category equipped with a graph structure over hom-sets (informally, a set of cells not closed under composition), which intuitively represent a rule based system, the states being the arrows and the rules the cells. The main fact for our analysis is that from a computad a (structured) 2-category is generated in a free way, by closing the cells under all relevant operations.

Definition 17 (computads). A *computad* $\mathbf{C}_S$ is a pair $\langle \mathbf{C}, S\rangle$, where $\mathbf{C}$ is a category and S is a set of "cells" over the hom-sets of $\mathbf{C}$. A *computad morphism* $\langle F, h\rangle : \mathbf{C}_S \to \mathbf{D}_T$ is a pair such that $F : \mathbf{C} \to \mathbf{D}$ is a functor and $h : S \to T$ is a function, preserving source and target of the cells in the expected way.

Computads and their morphisms form a category, denoted **Comp**. □

Proposition 18 (free 2-categories). *Let V_2 :* **2Cat** $\to$ **Comp** *be the forgetful functor mapping a 2-category to the underlying computad (simply forgetting cell composition): It admits a left adjoint F_2 :* **Comp** $\to$ **2Cat**. □

Intuitively, the free functor F_2 composes the cells of a computad in all the possible ways, both horizontally and vertically, imposing further equalities in order to satisfy the axioms of a 2-category.

4.3 From DPO derivations to cells

In this section we show how from a graph transformation system we can obtain a suitable computad, such that the cells of the freely generated 2-category faithfully describe the derivations of the system.

Definition 19 (discrete graph production). A graph production $p : \langle L \leftarrow_l K \to_r R\rangle$ is *discrete* if K is discrete. A graph transformation system is discrete if all its productions are so. □

Given a production $p : \langle L \leftarrow_l K \to_r R\rangle$, the associated *discrete production* is obtained taking the discrete graph K_d underlying the context graph K. It is well known that, from the point of view of the rewrite relation, there is no difference between a graph transformation system and the associated discrete instance.

Definition 20 (induced discrete productions). Given a production p : $\langle L \leftarrow_l K \to_r R\rangle$, we denote as γ_p the associated *discrete production*, obtained by taking the discrete graph K_d underlying the context graph K and restricting l and r. Given a graph transformation system $\mathcal{G}$, we denote by $\mathcal{G}_d$ the corresponding discrete one obtained by "making discrete" the productions of $\mathcal{G}$. □

Proposition 21 (discrete rewrites). *Let $\mathcal{G}$ be a graph transformation system: Then, $\Rightarrow_{\mathcal{G}} = \Rightarrow_{\mathcal{G}_d}$.* □

From now on we shall assume, without loss of generality, that for each discrete graph K, its set of nodes is formed by natural numbers, so that $\underline{j} = N(K)$ for $j \in \mathbb{N}$. We can now provide the translation from a discrete graph transformation system to a computad.

Definition 22 (discrete computad). Let $\mathcal{G}$ be a discrete graph transformation system. The associated computad $\mathbf{C}_{\mathcal{G}}$ has **DG** as underlying category, and a cell γ_p for each production $p : \langle L \leftarrow_l K \rightarrow_r R\rangle \in \mathcal{G}$, such that

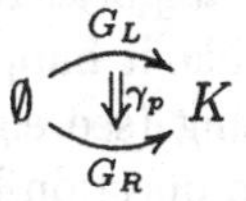

where K is regarded as the set $N(K)$, and G_L and G_R denote the ranked graphs $[\langle l, L, \emptyset\rangle]$ and $[\langle r, R, \emptyset\rangle]$. □

Then, we can finally prove our main theorem.

Theorem 23 (derivation as cells). *Let $\mathcal{G}$ be a graph transformation system. Then there exists a derivation $G \Rightarrow^*_{\mathcal{G}} H$ iff there exists a 2-cell from $[\langle\emptyset, G, \emptyset\rangle]$ to $[\langle\emptyset, H, \emptyset\rangle]$ in $F_2(\mathbf{C}_{\mathcal{G}})[\emptyset, \emptyset]$.*

Proof. Only if. The proof goes via induction on the length of a derivation. It is enough to note that the DPO step of Figure 3 is modeled via the cell $\gamma_p * [\langle\emptyset, D, k\rangle]$, depicted as

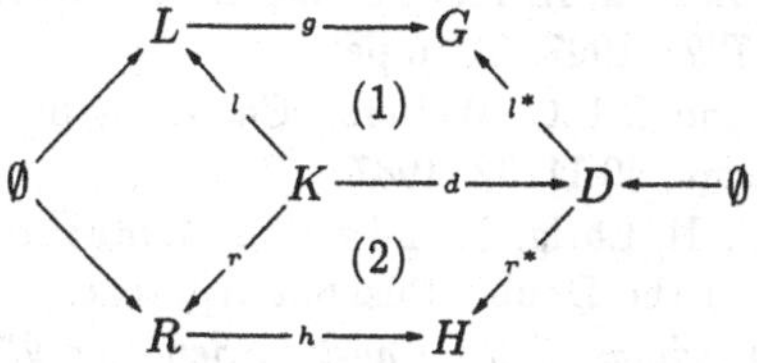

If (outline). By structural induction, with induction base the set of cells γ_p associated to the productions, we prove that the rewrite relation is preserved by composition, both inside each hom-category and with respect to each component of the family "$*$" of composition functors. □

5 Further Works

In our opinion, future work lies in the field of *concurrent semantics* of graph transformation. In fact, despite the correspondence we got is faithful at the level of the rewrite relation, we cannot recast the notions of *parallel derivations* and *shift equivalence* by means of the 2-categorical structure. Any solution should take into account three different aspects of the problem. First, the dgs-monoidal structure should be lifted to the 2-categorical level, as for similar results with *iteration 2-theories* [10, 4], also obtaining a generalized version of Theorem 11. This would provide us with the syntax of parallel productions and derivations.

Second, the structure of the interfaces should be enriched, at least by "isolated" edges, in order to overcome the restriction to productions with discrete interface graphs. In fact, while "making discrete" a production does not mean any harm for the rewrite relation (cf. Proposition 21), it reduces the set of possible parallel derivations.

Finally, the most delicate point is to ensure that the equivalence induced on cells by the coherence axioms of 2-categories resembles the usual *shift equivalence* on abstract graph derivations [6], like it happens for *permutation equivalence* in categorical models of term rewriting (see e.g. [11]). Here, particular attention should be paid to the isomorphism question between graphs: A problem that is at the basis of the *standard isomorphisms* solution in [6].

References

1. M. Bauderon and B. Courcelle. Graph expressions and graph rewritings. *Mathematical Systems Theory*, 20:83–127, 1987.
2. S. Bloom and Z. Ésik. *Iteration Theories.* EATCS Monographs on Theoretical Computer Science. Springer Verlag, 1993.
3. S. Bloom and Z. Ésik. Solving polinomials fixed point equations. In *Mathematical Foundations of Computer Science*, volume 841 of *LNCS*, pages 52–67. Springer Verlag, 1994.
4. S.L. Bloom, Z. Ésik, A. Labella, and E.G. Manes. Iteration 2-theories. In *Proceedings AMAST'97*, 1997. To appear.
5. A. Carboni and R.F.C. Walters. Cartesian bicategories I. *Journal of Pure and Applied Algebra*, 49:11–32, 1987.
6. A. Corradini, H. Ehrig, M. Löwe, U. Montanari, and F. Rossi. Abstract Graph Derivations in the Double-Pushout Approach. In H.-J. Schneider and H. Ehrig, editors, *Proceedings of the Dagstuhl Seminar 9301 on Graph Transformations in Computer Science*, volume 776 of *LNCS*, pages 86–103. Springer Verlag, 1994.
7. A. Corradini and F. Gadducci. A 2-categorical presentation of term graph rewriting. In *Proceedings CTCS'97*, volume 1290 of *LNCS*. Springer Verlag, 1997.
8. A. Corradini and F. Gadducci. Rewriting cyclic structures. draft, 1997.
9. A. Corradini and F. Gadducci. An algebraic presentation of term graphs, via gs-monoidal categories. *Applied Categorical Structures*, 1998. To appear. Available at `http://www.di.unipi.it/~gadducci/papers/aptg.ps`.
10. A. Corradini and F. Gadducci. Rational term rewriting. In M. Nivat, editor, *Proceedings FoSSaCS'98*, LNCS. Springer Verlag, 1998. To appear.
11. A. Corradini, F. Gadducci, and U. Montanari. Relating two categorical models of term rewriting. In *Rewriting Tecniques and Applications*, volume 914 of *LNCS*, pages 225–240. Springer Verlag, 1995.
12. A. Corradini and U. Montanari. An Algebra of Graphs and Graph Rewriting. In *Proceedings of the 4th Summer Conference on Category Theory and Computer Science (CTCS '91)*, volume 530 of *LNCS*, pages 236–260. Springer Verlag, 1991.
13. A. Corradini, U. Montanari, F. Rossi, H. Ehrig, R. Heckel, and M. Löwe. Algebraic Approaches to Graph Transformation I: Basic Concepts and Double Pushout Approach. In G. Rozenberg, editor, *Handbook of Graph Grammars and Computing by Graph Transformation.* World Scientific, 1997.
14. H. Ehrig, P. Boehm, U. Hummert, and M. Löwe. Distributed parallelism of graph transformation. In *13th Int. Workshop on Graph Theoretic Concepts in Computer Science*, volume 314 of *LNCS*, pages 1–19. Springer Verlag, 1988.

15. H. Ehrig, M. Pfender, and H.J. Schneider. Graph-grammars: an algebraic approach. In *Proceedings IEEE Conf. on Automata and Switching Theory*, pages 167–180, 1973.
16. A. Habel. *Hyperedge replacement: Grammars and languages*, volume 643 of *LNCS*. Springer Verlag, 1992.
17. U. Hensel and D. Spooner. A view on implementing processes: Categories of circuits. In M. Haveraaen, O. Owe, and O. Dahl, editors, *Recent Trends in Data Types Specification*, volume 1130 of *LNCS*, pages 237–255. Springer Verlag, 1995.
18. H.-J. Hoenke. On partial recursive definitions and programs. In *International Conference on Fundamentals of Computations Theory*, volume 56 of *LNCS*, pages 260–274. Springer Verlag, 1977.
19. B. Jacobs. Semantics of weakening and contraction. *Annals of Pure and Applied Logic*, 69:73–106, 1994.
20. A. Joyal, R. Street, and D. Verity. Traced Monoidal Category. *Mathematical Proceedings of the Cambridge Philosophical Society*, 119:425–446, 1996.
21. P. Katis, N. Sabadini, and R.F.C. Walters. Bicategories of processes. *Journal of Pure and Applied Algebra*, 115:141–178, 1997.
22. P. Katis, N. Sabadini, and R.F.C. Walters. Span(graph): A categorical algebra of transition systems. to appear Proceedings AMAST'97, 1997.
23. G.M. Kelly and R.H. Street. Review of the elements of 2-categories. In G.M. Kelly, editor, *Proceedings of the Sydney Category Seminar*, volume 420 of *Lecture Notes in Mathematics*, pages 75–103. Springer Verlag, 1974.
24. J.W. Klop. Term rewriting systems. In S. Abramsky, D. Gabbay, and T. Maibaum, editors, *Handbook of Logic in Computer Science*, volume 1, pages 1–116. Oxford University Press, 1992.
25. Y. Lafont. Equational reasoning with 2–dimensional diagrams. In *Term Rewriting, French Spring School of Theoretical Computer Science*, volume 909 of *LNCS*, pages 170–195. Springer Verlag, 1995.
26. F.W. Lawvere. Functorial semantics of algebraic theories. *Proc. National Academy of Science*, 50:869–872, 1963.
27. R. Milner. *Communication and Concurrency*. Prentice Hall, 1989.
28. A.J. Power. An abstract formulation for rewrite systems. In *Proceedings Category Theory in Computer Science*, volume 389 of *LNCS*, pages 300–312. Springer Verlag, 1989.
29. J. Power. A 2-categorical parsing theorem. *Journal of Algebra*, 129:439–445, 1990.
30. G. Rozenberg, editor. *Handbook of Graph Grammars and Computing by Graph Transformation*. World Scientific, 1997.
31. D.E. Rydehard and E.G. Stell. Foundations of equational deductions: A categorical treatment of equational proofs and unification algorithms. In *Proceedings Category Theory in Computer Science*, volume 283 of *LNCS*, pages 114–139. Springer Verlag, 1987.
32. G. Stefanescu. On flowchart theories: Part II. The nondeterministic case. *Theoret. Comput. Sci.*, 52:307–340, 1987.
33. G. Stefanescu. Algebra of flownomials. Technical Report SFB-Bericht 342/16/94 A, Technical University of München, Institut für Informatik, 1994.
34. J.G. Stell. *Categorical Aspects of Unification and Rewriting*. PhD thesis, University of Manchester, 1992.
35. R. Street. Categorical structures. In M. Hazewinkel, editor, *Handbook Of Algebra*, pages 529–577. Elsevier, 1996.

On Combining Semi-Formal and Formal Object Specification Techniques

Martin Gogolla, Mark Richters
University of Bremen, Informatics Department
Postfach 330440, D-28334 Bremen, Germany

Abstract. In the early phases of software development it seems profitable to freely mix semi-formal and formal design techniques. Formal techniques have their strength in their ability to rigorously define desired software qualities like functionality, whereas semi-formal methods are usually said to be easier to understand and to be more human-nature oriented. We propose a new approach in order to combine these two areas by exploiting how constructs of the formal specification language TROLL light are related to the graphical elements of the UML approach.

1 Introduction

As a first step and in order to explain the roles of and the relationships between semi-formal and formal specifications, let us start with a very simple software process model. We assume software development starts somehow with stating the requirements of the software system. These requirements are then formalized in the specification phase and are made more precise with respect to implementation in the design phase. In the implementation phase the design is translated to concrete code, and in the maintenance phase the software is adjusted to evolving user needs. In all phases backtracking to earlier phases is possible.

Now consider the different description techniques used in these phases in reverse order. Maintenance and implementation mainly have to handle programming language code. Design and specification deal with specification language texts. Requirements are usually formulated in natural language thus utilizing informal description techniques. The main link between informal and formal descriptions are often so-called semi-formal techniques. Frequently, diagrams are employed to visualize certain system aspects. The transition from informal to formal descriptions is one of the most crucial points in software development because all later phases rely on the first formal specification. We concentrate on this topic in translating a formal specification language into the Unified Modeling Language UML [BJR97]. Thus we take advantage of the visual power (and the informal semantics) of UML in order to represent aspects of TROLL light specifications having equivalent formal semantics [GH95] as the corresponding UML constructs.

Our approach is very similar to [LR97]. The intention there however seems to be to have a new alternative, complete, and graphical specification formalism whereas we only want to visualize certain aspects with a well-established notation. Earlier work related to ours is [JWH+94]. That approach however considers one of the UML predecessors OMT [RBP+91]. It does not treat for instance activity diagrams, and handles object interaction aspects in a different way. The object specification language GNOME [RS95] also had a graphical representation but it did not draw strong connections to established techniques. Close to our intention is also work [BHH+97] in the direction of giving a formal semantics for UML.

The structure of the rest of this paper is as follows. In Sect. 2 we explain our view on the general relationship between semi-formal and formal specifications. Section 3 introduces a special case of this, namely the way we combine TROLL light specifications with the UML software development approach. The paper ends with some concluding remarks.

2 Relating semi-formal and formal specifications

A closer look to the early software development phases where semi-formal and formal descriptions are mixed is displayed in Fig. 1. We do not assume that the development steps mentioned before are carried out in the sketched order but that requirement and specification steps are freely interchanged. In our approach semi-formal descriptions of software systems are built with the central aim of achieving a formal specification. They concentrate on certain particular aspects of the system to be described. An aspect is obtained by considering only very special properties of the system to be developed (an aspect in this sense is not be understood as a component or module of the system). For example, one aspect in this sense may be the structural system aspect (what objects are in the system and how are they related) and another one may be the dynamic system aspect (how do these objects evolve in time). These aspects correspond to what is known in the database field as a view on the underlying basic unit. There may be many such aspects or views, e.g. an alpha, beta, ..., omega view. These views in turn can be subject to focusing on certain properties only (views can be built on views and so on). For example, in Fig. 1 we have views alpha-1, ..., alpha-n which rely on the alpha view.

In the software development process, of course, one does a lot of transformation respectively modification of semi-formal and formal descriptions, and it is challenging to ask to what extent the previously achieved results can be used further. In the first line one has to relate formal descriptions and semi-formal ones. Translations in both directions can make sense. In going from formal to semi-formal descriptions you give for instance later system users the chance to understand the planned system functionality or you allow for designers to communicate on a short cut basis. In going from semi-formal to formal specifications you make precise what has been achieved for instance in discussions with later system users. Most interesting is a situation where you already have a certain development state with consistent formal and semi-formal descriptions and you

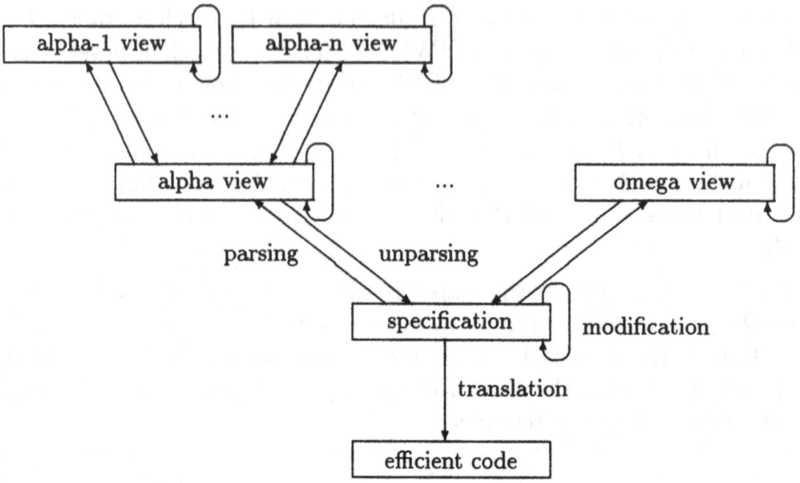

Fig. 1. Semi-formal views on a specification

make modifications respectively further developments on the semi-formal level. The challenging question then of course is to consistently incorporate these newly added design decisions into the already well-developed formal specification.

So, the overall aim of our approach is to give support for the development of formal specifications by means of graphical description techniques. But on the other hand we do not insist on pure graphical terms in the sense that we expect that everything must be expressed in a graphical notion, i.e. we do not expect that the views cover the complete specification. There will be certain parts that are easier to express in textual form, for example complex formulas as they appear in constraints. But again, certain aspects of the formula could be emphasized in a graphical way.

3 Relationship between TROLL light and UML

3.1 General overview

The above considerations are now made more precise. Our approach as depicted in Fig. 2 starts from the specification language TROLL light [GCH93, CGH95, HCG94], a language with which we have worked now for quite a while and for which (among other things) an algebraic semantics [GH95] is known. Related object specification languages are TROLL [JSHS91], TELOS [KMSJ88], Mondel [BBE+90], CMSL [Wie91], ALBERT [DDP93], TQL [MT94] and RML [GMB94], among others. Two aspects (in the above sense of the word) of a TROLL light specification are the structural and dynamic system aspect. Other interesting aspects (e.g. an identifier aspect concerning identifier visibility) could be thought of. The structural system aspect (in OMT terminology the object model[1]) mainly concerns single system states whereas the dynamic system aspect (in OMT terms the dynamic model) explains how the system states evolve in time. Two possible views on the structure aspect could be a detailed single type view where one considers only one object type with all its structural attributes, and, in contrast to this, a clustered view on a complex structure model may abstract from the properties of single object types and may build clusters of object types belonging together by some given criteria. Only the relationships between such clusters may be represented in this clustered view. For the dynamic aspect different views may be established as well. In the detailed view we may look at a single object's evolution in all its aspects, for instance with completely given preconditions of allowed events. In contrast to this, in a substate view we may leave out preconditions and look at possible substates of a given section from the detailed view.

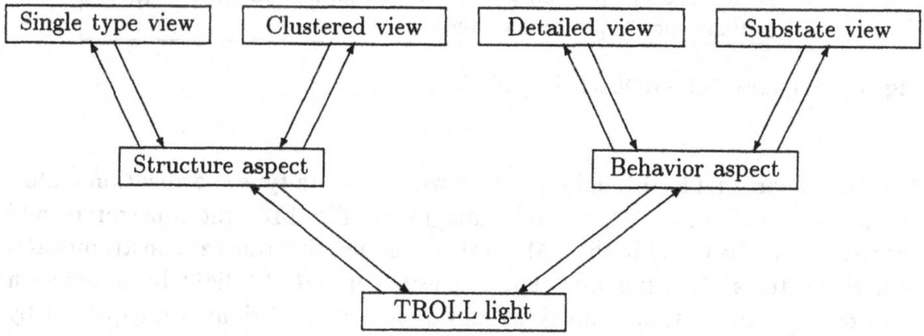

Fig. 2. Two UML views on a TROLL light specification

As indicated in Fig. 3, for representing aspects semi-formally we choose the well-developing UML [BJR97] approach. In TROLL light specifications we have a signature part giving names to data types and object types (templates). Instances of data types (specified with a data type specification language like ACT ONE [EFH83], Extended ML [ST86], SPECTRUM [BFG+93], or CASL [CoF97]) are data values whereas instances of object types are objects whose structural properties (its attributes) are given by data values and which can change in the course of time. Subobject relationships constitute part-of associations between objects, and events are the things happening to objects (its operations). UML refers to this signature part of a TROLL light specification as static structure diagrams (OMT calls this the object model) where data types and classes (corresponding to templates) are given. TROLL light object-valued

[1] The use of the word model in this place is different from the usual use in the formal specification community. Here it refers to an aspect of a specification, not to a model in the sense of a formal interpretation (for instance an algebra) belonging to a signature and satisfying certain axioms.

attributes realize UML associations, and TROLL light subobject relationships correspond to UML aggregations. For the dynamic part (in UML the state diagrams, in OMT the dynamic model) the TROLL light signature mentions the events corresponding to UML operations and UML events.

	TROLL light	UML
Signature	Data types	Data types in structure diagrams
	Templates	Classes in structure diagrams
	Attributes	Attributes, Associations in structure diagrams
	Subobjects	Aggregations in structure diagrams
	Events	Operations, Events in structure diagrams
Axioms	Valuation	Data types in structure and activity diagrams
	Constraints	Constraints in structure diagrams
	Derivation	Derived items in structure diagrams
	Interaction	Action transitions, Conditions in state diagrams
	Behavior	State diagrams

Fig. 3. Comparison of TROLL light and UML

The axiom part of TROLL light together with the data type specification realize in the valuation rules UML activity diagrams. TROLL light constraints and derivations can be found in the UML static structure diagrams as constraints and as derived items. The dynamic aspect reflected in TROLL light by interaction rules representing action transitions and action preconditions are captured by details in the UML state diagrams. These are formally given in a TROLL light text as behavior patterns.

3.2 Drafting TROLL light concepts

Before presenting the details of our translation we explain central TROLL light concepts by means of an example describing a car rentals application scenario. One of the central object types is the booking template in Fig. 4. Booking objects use the data types String, Int, and Bool and the object types Customer and Car. They have object-valued attributes TheCustomer and TheCar of object sort customer and car, respectively. The convention is that (data and object) types start with upper-case letters whereas corresponding sorts begin with lower-case ones. Furthermore, there are data-valued attributes IsClosed, BookingCat, StartDay and EndDay giving a switch for whether there is still something to do with the booking or not, a booking category describing the car's desired equipment, and the required start and end day of the car usage. There are derived attributes (whose values are not stored but are calculated from other items) IsOpen and IsCurrent. The things to possibly change the state of a booking are the events (or operations), in this particular case a birth event createBooking, an event makeCurrent for actually assigning a car to a booking and an event to close the current booking.

```
TEMPLATE Booking
DATA TYPES String, Int, Bool;
TEMPLATES Customer, Car;
ATTRIBUTES
  TheCustomer : customer; TheCar : car; BookingCat : int;
  IsClosed : bool; StartDay, EndDay : int; -- YYYYMMDD
  DERIVED IsOpen : bool; DERIVED IsCurrent : bool;
EVENTS
  BIRTH createBooking(aCustomer : customer, aCat : int,
                      aStartDay : int, anEndDay : int);
  makeCurrent(aCar : car);
  close;
CONSTRAINTS
  1 <= BookingCat AND BookingCat <= 5;
  19900101 <= StartDay AND StartDay <= EndDay AND EndDay <= 20001231;
VALUATION
  [ createBooking(aCustomer,aCat,aStartDay,anEndDay) ]
      TheCustomer = aCustomer, BookingCat = aCat,
      StartDay = aStartDay, EndDay = anEndDay, IsClosed = FALSE;
  [ makeCurrent(aCar) ] TheCar = aCar;
  [ close ] TheCar = NIL, IsClosed = TRUE;
DERIVATION
  IsOpen = UNDEF(TheCar) AND NOT(IsClosed);
  IsCurrent = DEF(TheCar) AND NOT(IsClosed);
BEHAVIOR
  PROCESS Booking = ( createBooking -> BookingOpen );
  PROCESS BookingOpen = ( { BookingCat <= CarCat(aCar) }
                          makeCurrent(aCar) -> BookingCurrent );
  PROCESS BookingCurrent = ( close -> BookingClosed );
  PROCESS BookingClosed = ( );
END TEMPLATE
```

Fig. 4. TROLL light Booking template

The possible object states are restricted in the constraints section. For demonstration purposes, we have modeled the booking category and the dates as integer values. A more sophisticated choice would be to introduce appropriate data types for these items. But nevertheless in both cases we need to require somewhere that the start day lies before (or is equal to) the end day of the booking. One place to do this is the constraints section described here, another choice would be a precondition in the behavior pattern to be described below. The valuations determine how the events (if they occur) effect the attribute values. The derivation section gives rules for computing the derived attribute values from other given things. In contrast to the simple derivations here, quite complex rules are also allowed in this place. For example, the Company template as given in Fig. 5 computes in the DeliverCarsMenu derived attribute of sort bag(tuple(int,int,bag(tuple(string,int)))) all open booking numbers and booking categories together with the collection of all free car numbers and car categories which fit to the booking. Last, as again shown in Fig. 4, the possible occurrence of events is described in a CSP-like notation in the behavior patterns. There is always one process with the name of the template to be defined which is the starting process, so here a booking's life begins with a createBooking event; after that the makeCurrent event is allowed (provided that the category of the given car fits the category of the booking); at last the event close may appear leading to a state where no further events are possible, but observations on the attributes can still be made. This Booking template example, however, does not mention the subobjects and the interaction section of a template.

```
TEMPLATE Company
DATA TYPES String, Int;
TEMPLATES  Customer, Car, Booking;
SUBOBJECTS Customers(CustomerNo : int) : customer;
           Cars(CarNo : string) : car;
           Bookings(BookingNo : int) : booking;
ATTRIBUTES DeliverCarsMenu:bag(tuple(int,int,bag(tuple(string,int))));
...

EVENTS deliverCar(aBookingNo : int, aCarNo : string);
...

DERIVATION DeliverCarsMenu =
  SELECT BookingNo(B), BookingCat(Booking(B)),
    ( SELECT CarNo(Car(C)), CarCat(Car(C))
      FROM   C IN Cars
      WHERE  Available(Car(C)) AND
             CarCat(Car(C)) >= BookingCat(Booking(B)) )
  FROM   B IN Bookings
  WHERE  StartDay(Booking(B)) = Today AND IsOpen(Booking(B));
...

INTERACTION deliverCar(aBookingNo,aCarNo) >>
  Cars(aCarNo).pickUp,
  Bookings(aBookingNo).makeCurrent(Cars(aCarNo));
...

END TEMPLATE
```

Fig. 5. Section of TROLL light Company template

In the subobjects section it is possible to introduce non-sharable components of the current object (sharing is possible via object-valued attributes). For example, the Company template as given in Fig. 5 describes the structure of the overall system object containing all other objects. Therefore, this template mentions the customers, cars and bookings as its subobjects. Regarding object identification, from the Company point of view a booking is identified by a booking number (given as an integer number parameter for the subobject symbol Bookings). From a conceptual point of view, the subobject symbol Bookings gives a means for accessing booking objects from company objects. From a technical point of view, the symbol Bookings can be looked at and used in three different ways: It can be used as (1) an infinite array (notation: `Bookings(BookingNo : int): booking`) with an integer index giving for each integer a booking object (in each state there is only a finite number of booking objects different from the undefined booking), (2) a map (notation: `Bookings : map(BookingNo : int, Booking : booking)`) from integers to bookings, or (3) a set of tuples (notation: `Bookings : set(tuple(BookingNo : int, Booking : booking))`) with two components of sort integer and booking and with the additional constraint that the tuple set also establishes a function (`FORALL (B1, B2 IN Bookings) BookingNo(B1) = BookingNo(B2) IMPLIES Booking(B1) = Booking(B2)`).

In the interaction section, communication patterns between objects are specified. For instance, it is possible to require the simultaneous occurrence of an event in a component subobject and an event in the enclosing aggregated object. In the Company template, the deliverCar event is, for instance, synchronized by interaction rules with the pickUp event of a corresponding car object and a makeCurrent event of a corresponding Booking object.

3.3 Representing TROLL light descriptions with UML

In Fig. 6 and Fig. 7 we have displayed the UML representation of a structure model corresponding to the attributes, subobjects, and events sections of appropriate TROLL light templates for our scenario. Templates together with their data-valued attributes and events are pictured as rectangles. Associations (object-valued attributes and subobject relationships) between object types are denoted as lines. As displayed by the diamond, all Customer, Booking, and Car objects are aggregates (subobjects) of an object[2] belonging to template Company. Multiplicity is denoted as adornments at association edges. For example, the stars * close to Customer, Booking and Car denote that a company can consist of many such objects whereas the adornments 1 close to the diamonds near company expresses that each customer, booking, and car belongs to exactly one company. Also the references of a booking to its customer and car are

[2] This object may be viewed as "the" system object in the sense that all other objects appear as components respectively subobjects of this distinguished object. In the TROLL light specification, however, the company template is just a normal template. But its role as the system object template becomes evident when the TROLL light animation system is invoked by telling it that the session should start with exactly one object of sort company.

given. In general, multiplicity is denoted by sets of integer intervals of the form low..high. The adornments * and 1 are abbreviations for 0..* and 1..1 where the star * expresses an unbounded cardinality.

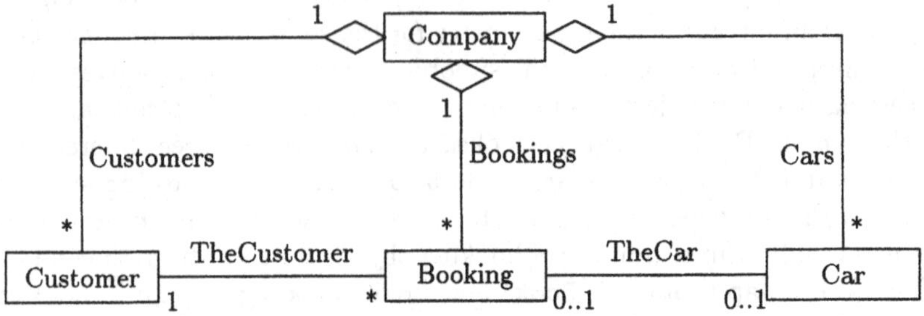

Fig. 6. Translation of templates, subobjects and attributes

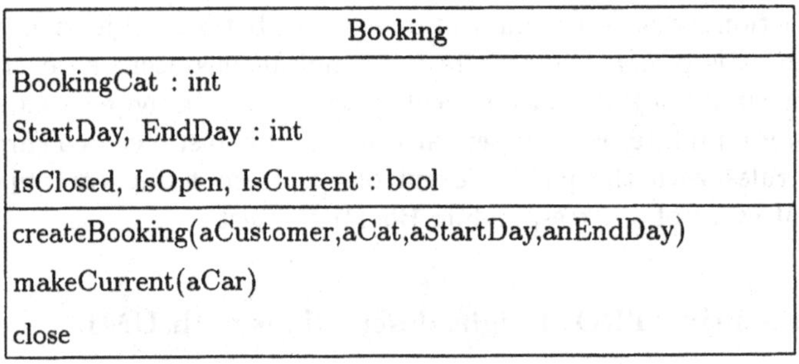
Booking
BookingCat : int
StartDay, EndDay : int
IsClosed, IsOpen, IsCurrent : bool
createBooking(aCustomer,aCat,aStartDay,anEndDay)
makeCurrent(aCar)
close

Fig. 7. Translation of TROLL light attributes and events

The dynamic aspect is visualized by the statechart-like diagram in Fig. 8. Here, the behavior patterns are presented. For example, the process description of Company is given by the upper left side of the diagram. States correspond to process names and transitions to event occurrences. For reasons of simplicity, the rather involved event preconditions are not displayed. They could be displayed as a separate view concentrating on single templates as this is shown in Fig. 9 for the Booking template. The interaction rules could be shown as additional adornments of edges like deliverCar / { pickUp, makeCurrent } because the event deliverCar in the DeliverCars state calls for event pickUp in Car and for event makeCurrent in Booking. The general possibility of concurrent event occurrences is represented as an orthogonal composition of and-states [HG96].

Please note, that not all parts in all aspects are visualized. We think for instance complex formulas as they appear in constraints or derivation rules should still be given as texts in linear form only. For example, the derived attribute De-

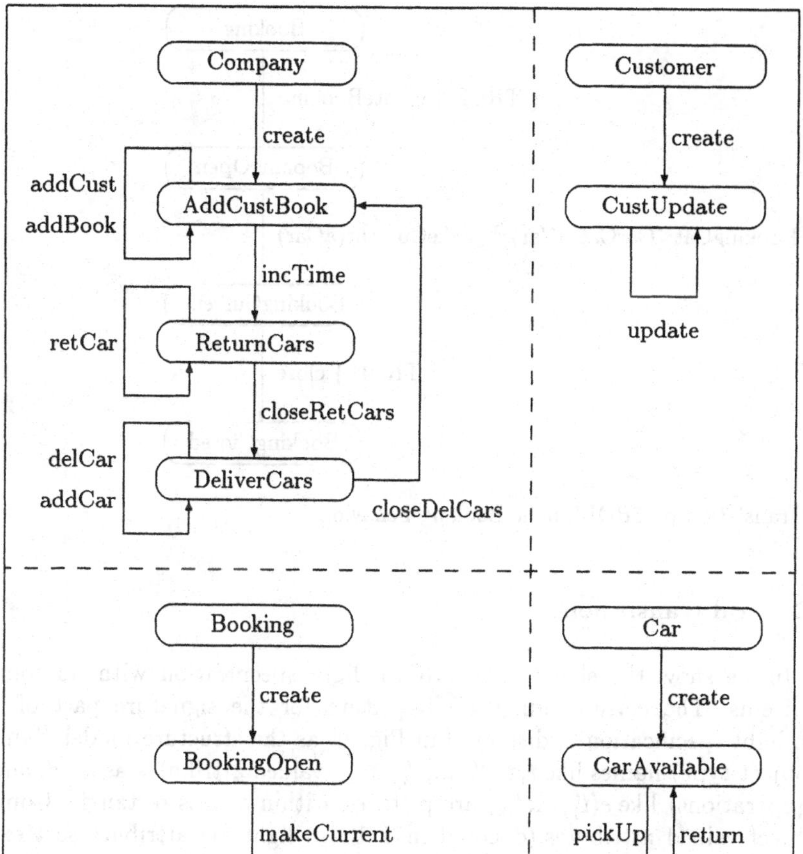

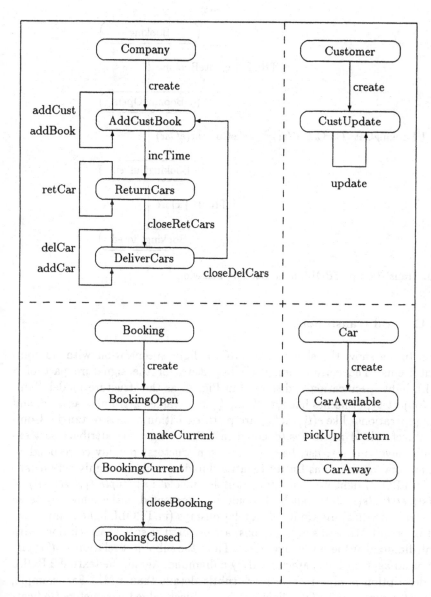

Fig. 8. Translation of TROLL light behavior patterns

liverCarsMenu with its rather complex definition is easier to understand when presented as a text.

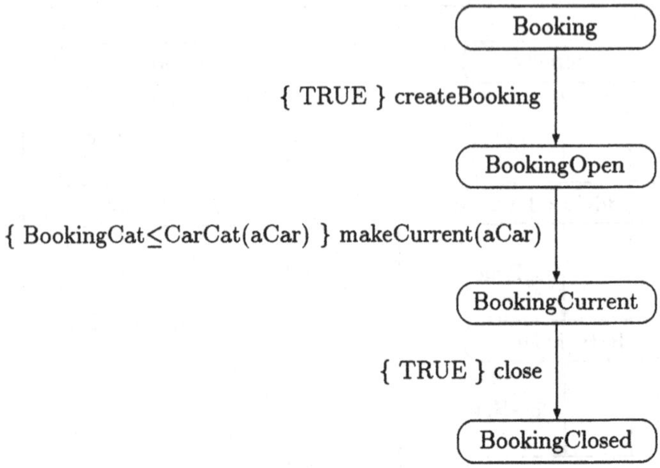

Fig. 9. Translation of TROLL light Booking behavior

3.4 General translation

In Fig. 10 we show the shape of a TROLL light specification with its component items. The corresponding UML picture for the signature part of a TROLL light specification is displayed in Fig. 11 as the structure model. Template (object type) names like $t, t', t'_1, ..., t'_n$, data-valued attributes as $a : d$, and events (operations) like $e(t'_1, ..., t'_m)$ are pictured within a class rectangle. Complex, object-valued attributes (denoted in TROLL light via attribute sort expressions involving set, list, bag, or tuple constructors) roughly correspond to UML associations given as Entity-Relationship diamonds.[3] Roughly here means that, in TROLL light we could, for example, have $a' : set(tuple(t_1, list(t_2)))$ or $a' : list(set(tuple(t_1, t_2)))$ but both would be represented in the same way as an UML structure diagram. Component relationships (in TROLL light terminology subobjects, in UML terms aggregations) are pictured as special associations with a small diamond at the side of aggregate. Thus, the subobject relation $s : t'$ is pictured as an aggregation characterized by a diamond. Again, the textual TROLL light description is able to give more subtle details than UML. For example, subobject symbols in TROLL light can have object-valued parameters (indicating existential dependence) that cannot be represented in UML. The translation does not touch the TROLL light constraints (where states are restricted by a formula φ) and derivation sections (where an attribute a'' is specified to have always the value of a term τ) which have textual representations in the UML diagrams.

The behavioral model representing the interaction and behavior part is given

[3] Diamonds are chosen in the case of three or more participating classes whereas binary links would be shown as direct lines.

TEMPLATE t
DATA TYPES d
TEMPLATES $t_1, ..., t_n, t', t'_1, ..., t'_m$
ATTRIBUTES $a : d$
$a' : expr'(t_1, ..., t_n)$
SUBOBJECTS $s : t'$
EVENTS $e(p_1 : t'_1, ..., p_m : t'_m)$
CONSTRAINTS φ
DERIVATION $a'' = \tau$
VALUATION $[e(p_1, ..., p_m)]\ a = expr(p_1, ..., p_m)$
INTERACTION $e >> e'$
BEHAVIOR PROCESS x = ($\{c\}$ $e \rightarrow y$)
END TEMPLATE

Fig. 10. Important sections of TROLL light specifications

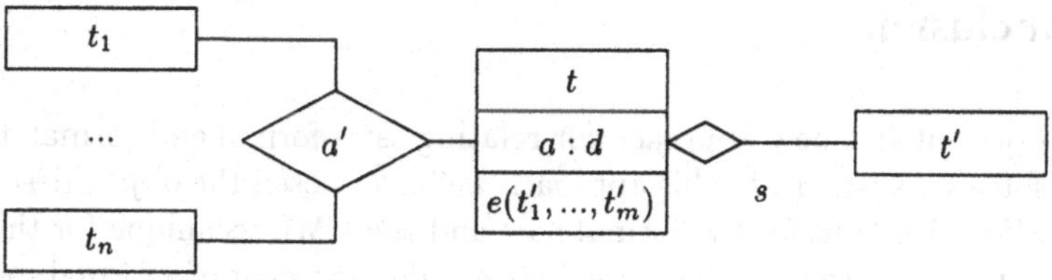

Fig. 11. Translation of TROLL light signatures

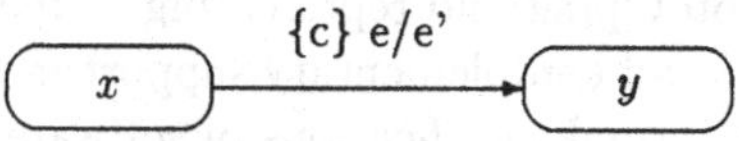

Fig. 12. Translation of TROLL light behavior

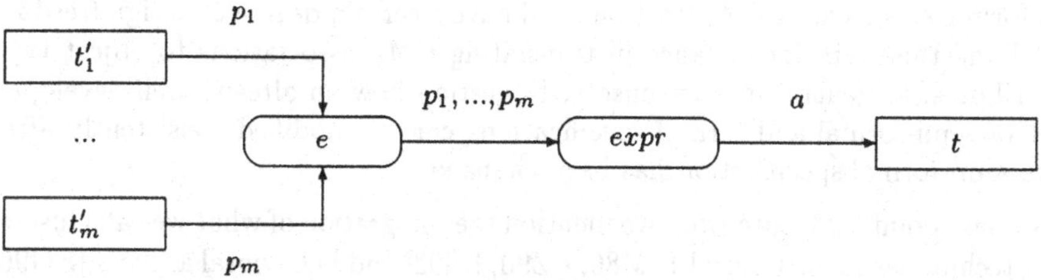

Fig. 13. Translation of TROLL light valuation

in Fig. 12. Process names correspond to states and process transitions to event occurrences (eventually restricted by a precondition). Thus, in the UML picture we have nodes x and y and an edge labeled with the event e causing a transition for the given TROLL light process. If possible, we display in the same diagram the interaction calling mechanism $e >> e'$ by stating the called event after the calling one with a slash as the separator (e/e' in the diagram). This is simple in the case that there are no preconditions. If there are more involved preconditions, then the behavior diagram must distinguish between the different cases by introducing corresponding alternative edges. Without going into details (due to space limitations) we mention that alternatively sequence diagrams can be used to represent event calling. The valuation axioms are reflected in UML by the activity diagrams indicated in Fig. 13. This diagram expresses that (parameter) objects p_i of sort t'_i flow into the event e (classified as an activity). Event e collects their values and passes them to the expression $expr$ of the valuation rule (also classified as an activity) which in turn hands the result of $expr(p_1, ..., p_m)$ to an object of sort t by updating the attribute a. Additionally, the corresponding data type specification (especially with its corresponding function definitions) and the interaction rules will give rise to more details for the activity diagrams.

4 Conclusion

We have presented a new approach for relating semi-formal and formal descriptions of software systems. In this approach we have chosen the object description language TROLL light for the formal part and the UML technique for the semi-formal part. There is a strong relation between development of a formal specification by means of semi-formal specifications and prototyping formal specifications for instance with an animation tool, e.g. the one presented in [RG97]. We think both approaches, i.e. prototyping and representing formal specifications by diagrams, are indispensable and complementary support mechanisms for achieving consistent and complete formal specifications of software systems.

We think it is relatively easy to derive corresponding UML diagrams from TROLL light specifications, whereas it is not that obvious to go from diagrams to formal specifications. At least one will have a certain degree of design freedom in doing this step, for instance in translating UML associations to object type attributes. In general it is an unsolved question how an already well-developed set of semi-formal and formal specifications can be modified consistently after the semi-formal specification has been changed.

As a last point for future work we mention the integration of what we call clustering techniques for structural [FM86, HZ90, RS92] and behavioral aspects [HG96] as a further abstraction mechanism. In larger designs one needs mechanisms for building clusters of design parts in order to achieve a degree of detail that can be understood. A cluster representing a structural or dynamic system aspect is considered to be, for example, a higher level node representing a collection of nodes on the lower level. The benefit of abstracting several nodes to one abstract node lies of course in the improved clearness of the underlying diagram. This problem however is also unsolved on the semi-formal UML level.

References

[BBE+90] G. v. Bochmann, M. Barbeau, M. Erradi, L. Lecomte, P. Mondain-Monval, and N. Williams. Mondel: An Object-Oriented Specification Language. Département d'Informatique et de Recherche Opérationelle, Publication 748, Université de Montréal, 1990.

[BFG+93] M. Broy, C. Facchi, R. Grosu, R. Hettler, H. Hussmann, D. Nazareth, F. Regensburger, O. Slotosch, and K. Stølen. The Requirement and Design Specification Language SPECTRUM — An Informal Introduction (Version 1.0). Technical Report TUM I9311-12, TU München, 1993.

[BHH+97] Ruth Breu, Ursula Hinkel, Christoph Hofmann, Cornel Klein, Barbara Paech, Bernhard Rumpe, and Veronika Thurner. Towards a Formalization of the Unified Modeling Language. In Mehmet Aksit and Satoshi Matsuoka, editors, *Proc. 11th European Conf. Object-Oriented Programming (ECOOP'97)*, pages 344–366. Springer, Berlin, LNCS 1241, 1997.

[BJR97] G. Booch, I. Jacobson, and J. Rumbaugh. *UML Summary (Version 1.0)*. Rational Corporation, Santa Clara, 1997. http://www.rational.com.

[CGH95] S. Conrad, M. Gogolla, and R. Herzig. Safe Derivations in Object Hierarchies. In D. Patel, Y. Sun, and S. Patel, editors, *Int. Conf. Object-Oriented Information Systems (OOIS'94)*, pages 306–319. Springer, London, 1995.

[CoF97] CoFI Task Group on Language Design. CASL - The CoFI Algebraic Specification Language. Technical Report, DAIMI, Aarhus, Danmark, 1997.

[DDP93] E. Dubois, P. Du Bois, and M. Petit. O-O Requirements Analysis: An Agent Perspective. In O.M. Nierstrasz, editor, *Proc. European Conf. on Object-Oriented Programming (ECOOP'93)*, pages 458–481. Springer, Berlin, LNCS 707, 1993.

[EFH83] H. Ehrig, W. Fey, and H. Hansen. ACT ONE: An Algebraic Specification Language with Two Levels of Semantics. Technical Report 83-03, Technische Universität Berlin, 1983.

[FM86] P. Feldman and D. Miller. Entity Model Clustering: Structuring a Data Model by Abstraction. *Computer Journal*, 29(4):348–360, August 1986.

[GCH93] M. Gogolla, S. Conrad, and R. Herzig. Sketching Concepts and Computational Model of TROLL light. In A. Miola, editor, *Proc. 3rd Int. Conf. Design and Implementation of Symbolic Computation Systems (DISCO)*, pages 17–32. Springer, Berlin, LNCS 722, 1993.

[GH95] M. Gogolla and R. Herzig. An Algebraic Semantics for the Object Specification Language TROLL light. In E. Astesiano, G. Reggio, and A. Tarlecki, editors, *Recent Trends in Data Type Specification (WADT'94)*, pages 288–304. Springer, Berlin, LNCS 906, 1995.

[GMB94] S. Greenspan, J. Mylopoulos, and A. Borgida. On Formal Requirements Modeling Languages: RML Revisited. In B. Fadini, editor, *Proc. 16th Int. Conf. on Software Engineering (ICSE'94)*, pages 135–148. IEEE Computer Society Press, 1994.

[HCG94] R. Herzig, S. Conrad, and M. Gogolla. Compositional Description of Object Communities with TROLL light. In C. Chrisment, editor, *Proc. Basque Int. Workshop on Information Technology (BIWIT'94): Information Systems Design and Hypermedia*, pages 183–194. Cépaduès-Éditions, Toulouse, 1994.

[HG96] D. Harel and E. Gery. Executable Object Modeling with Statecharts. In *18th Int. Conf. Software Engineering*, pages 246–257. Springer Verlag, Berlin, 1996.

[HZ90] S. Huffman and R. V. Zoeller. A Rule-Based System Tool for Automated ER Model Clustering. In F. H. Lochovsky, editor, *Proc. 8th Int. Conf. Entity-Relationship Approach*, pages 221–236. Elsevier Science Publishers, 1990.

[JSHS91] R. Jungclaus, G. Saake, T. Hartmann, and C. Sernadas. Object-Oriented Specification of Information Systems: The TROLL Language. Informatik-Bericht 91–04, Technische Universität Braunschweig, 1991.

[JWH+94] R. Jungclaus, R. J. Wieringa, P. Hartel, G. Saake, and T. Hartmann. Combining TROLL with the Object Modeling Technique. In B. Wolfinger, editor, *Innovationen bei Rechen- und Kommunikationssystemen. GI-Fachgespräch FG 1: Integration von semi-formalen und formalen Methoden für die Spezifikation von Software*, pages 35–42. Informatik aktuell, Springer, Berlin, 1994.

[KMSJ88] M. Koubarakis, J. Mylopoulos, M. Stanley, and M. Jarke. TELOS: A Knowledge Representation Language for Requirements Modelling. Technical Report CSRI-222, University of Toronto, 1988.

[LR97] M. Larosa and G. Reggio. A Graphic Notation for Formal Specifications of Dynamic Systems. In P. Lucas, editor, *Proc. 4th Int. Symposium Formal Methods Europe (FME'97)*. Springer, LNCS, 1997.

[MT94] M. Missikoff and M. Toiati. MOSAICO - A System for Conceptual Modeling and Rapid Prototyping of Object-Oriented Database Application. *SIGMOD Record*, 23(2):508–519, 1994.

[RBP+91] J. Rumbaugh, M. Blaha, W. Premerlani, F. Eddy, and W. Lorensen. *Object-Oriented Modeling and Design.* Prentice-Hall, Englewood Cliffs (NJ), 1991.

[RG97] M. Richters and M. Gogolla. A Web-based Animator for Object Specifications in a Persistent Environment. In M. Bidoit and M. Dauchet, editors, *Proc. 7th Int. Conf. Theory and Practice of Software Development (TAPSOFT'97)*, pages 867–870. Springer, Berlin, LNCS 1214, 1997.

[RS92] O. Rauh and E. Stickel. Entity Tree Clustering - A Method for Simplifying ER Designs. In *Proc. Inf. Conf Entity-Relationship Approach (ERA'92)*, 1992.

[RS95] J. Ramos and A. Sernadas. A Brief Introduction to GNOME. Research Report, Section of Computer Science, Department of Mathematics, Instituto Superior Técnico, 1096 Lisboa, Portugal, 1995.

[ST86] D.T. Sannella and A. Tarlecki. Extended ML: An Institution-Independent Framework for Formal Program Development. In *Proc. Workshop on Category Theory and Computer Programming*, pages 364–389. Springer, LNCS 240, 1986.

[Wie91] R.J. Wieringa. Equational Specification of Dynamic Objects. In R.A. Meersman, W. Kent, and S. Khosla, editors, *Object-Oriented Databases: Analysis, Design & Construction (DS-4), Proc. IFIP WG2.6 Working Conf.*, pages 415–438. North-Holland, Amsterdam, 1991.

Modular Aspects of Rewrite-Based Specifications

Bernhard Gramlich*

INRIA Lorraine & CRIN, BP 101, 54602 Villers-lès-Nancy, France
e-mail: gramlich@loria.fr, url: http://www.loria.fr/~gramlich

Abstract. We investigate modular properties of term rewriting systems, the basic operational formalism for equational specifications. First we study sufficient conditions for the preservation of the termination property under disjoint (and more general) combinations of term rewriting systems. By means of a refined analysis of existing approaches we show how to prove several new asymmetric preservation results. For this purpose we introduce two interesting new properties of term rewriting systems related to collapsing reductions: *uniquely collapsing* and *collapsing confluent*. We discuss these properties w.r.t. well-known confluence, consistency and normal form properties, and show that they are modular for left-linear systems, but not in general.

1 Introduction

The study of the modularity behaviour of rewrite systems (w.r.t. to the preservation of important properties) under various types of combinations has become a very active and fruitful area of research (cf. [15], [18], [9] for surveys). This field is of utmost importance for the modular construction of (the operational version of) equational specifications with desirable properties as well as for their structured analysis by a divide-and-conquer approach.

Here we shall be concerned with the question under which conditions disjoint unions of rewrite systems inherit termination from their constituent systems. In particular, we are interested in asymmetric preservation conditions which have only partially been explored up to now. This analysis naturally leads to a thorough investigation of collapsing reductions and the role of (non-)left-linearity partially refining the existing analyses of [25] and [13, 14, 22]. Besides the implications for the preservation of termination under (disjoint) combinations this analysis of collapsing reduction and (non-)left-linearity may have other potential applications, too.

The rest of the paper is organized as follows. In Section 2 we recall some basic terminology about term rewriting and modularity. In Section 3 we give a brief survey of the basic approaches and (some) modularity results for termination and prepare our analysis. The main results of the paper are presented in Sections 4, 5 and 6. Finally we briefly discuss possible extensions (of the results on preservation of termination) for non-disjoint combinations of systems.

* This work was supported by a *Marie Curie Research Fellowship* of the European Community under contract No ERBFMBICT961235.

2 Preliminaries

We assume familiarity with the basic no(ta)tions, terminology and theory of term rewriting (cf. e.g. [4], [10]) but recall some no(ta)tions for the sake of readability. The set of terms over some given signature $\mathcal{F}$ and some (disjoint) countably infinite set $\mathcal{V}$ of variables is denoted by $\mathcal{T}(\mathcal{F},\mathcal{V})$. Positions (in terms) are ordered by the prefix ordering $\leq$ as usual. Concatenation of positions is denoted by juxtaposition. The 'empty' root position is denoted by λ. The set of positions of a term s is denoted by $Pos(s)$. The sets of variable positions and of non-variable, i.e., function symbol, positions of s are denoted by $\mathcal{V}Pos(s)$ and $\mathcal{F}Pos(s)$, respectively. The subterm of s at some position $p \in Pos(s)$ is denoted by s/p. A term rewriting system (TRS) is a pair $(\mathcal{F},\mathcal{R})$ consisting of a signature $\mathcal{F}$ and a set $\mathcal{R}$ of rewrite rules over $\mathcal{F}$, i.e., pairs (l,r) — also denoted by $l \to r$ — with $l,r \in \mathcal{T}(\mathcal{F},\mathcal{V})$. Here we require that l is not a variable, and that all variables of r occur in l. Instead of $(\mathcal{F},\mathcal{R})$ we also write $\mathcal{R}^{\mathcal{F}}$ or simply $\mathcal{R}$ if $\mathcal{F}$ is clear from the context or irrelevant. For reduction steps with the rewrite relation $\to_{\mathcal{R}} = \to$ induced by $\mathcal{R}$ we sometimes add additional information as in $s \to_{p,\sigma,l\to r} t$ with the obvious meaning. Furthermore, we make free use of context notations like $s = C[\sigma(l)] \to_{l\to r} C[\sigma(r)] = t$ or $s = C[s_1,\ldots,s_n]_{p_1,\ldots,p_n}$ where the p_i's indicate the respective positions of the s_i's. If in a reduction sequence (or derivation) $D : s_0 \to^* s_n$ every step is uniquely specified, e.g., by indicating the position p of the redex contracted, the applied rule $l \to r$ (and redundantly the matching substitution σ), then we speak of a labelled derivation. The innermost reduction relation $\underset{i}{\to}$ (induced by $\mathcal{R}$) is given by: $s \underset{i}{\to} t$ if $s = C[\sigma(l)] \to_{l\to r} C[\sigma(r)] = t$ for some $l \to r \in \mathcal{R}$, some context $C[.]$ and some substitution σ such that no proper subterm of $\sigma(l)$ is reducible. A TRS is *non-overlapping* if it has no critical pairs. It is an *overlay* or *overlaying* system if critical overlaps between rules of $\mathcal{R}$ occur only at the root position. An *inside critical pair* is obtained by overlapping some rule into another one properly below the root. A TRS $\mathcal{R}$ is *terminating* or *strongly normalizing* (SN) if $\to_{\mathcal{R}}$ is *terminating*, i.e., if there is no infinite derivation $s_0 \to_{\mathcal{R}} s_1 \to_{\mathcal{R}} s_2 \ldots$. $\mathcal{R}$ is *innermost terminating* if $\underset{i}{\to}$ is terminating. It is *confluent* or *Church-Rosser* (CR) if ${}^*\!\leftarrow \circ \to^* \subseteq \to^* \circ {}^*\!\leftarrow$ or equivalently $\leftrightarrow^* \subseteq \to^* \circ {}^*\!\leftarrow$. Confluence plus termination is also called *completeness* or *convergence*. $\mathcal{R}$ is *locally confluent* or *weakly Church-Rosser* (WCR) if $\leftarrow \circ \to \subseteq \to^* \circ {}^*\!\leftarrow$. The set of *irreducible* terms (or terms *in normal form*) is denoted by NF($\mathcal{R}$). $\mathcal{R}$ has *unique normal forms* (UN) if for all terms s,t, $s \leftrightarrow^* t$ and $s,t \in$ NF($\mathcal{R}$) imply $s = t$. $\mathcal{R}$ has *unique normal forms w.r.t. reduction* (UN$^{\to}$) if for all s,t,u, $s \;{}^*\!\leftarrow u \to^* t$ and $s,t \in$ NF($\mathcal{R}$) imply $s = t$. $\mathcal{R}$ has the *normal form property* (NF) if for all s,t, $s \leftrightarrow^* t$ and $t \in$ NF($\mathcal{R}$) imply $s \to^* t$. $\mathcal{R}$ is *consistent* (CON) if $x \leftrightarrow^* y$ implies $x = y$, and *consistent w.r.t. reduction* (CON$^{\to}$) if $x \;{}^*\!\leftarrow s \to^* y$ implies $x = y$. Local versions of certain properties like termination and confluence (for instance: t is terminating) also make sense, with the obvious interpretation. A rewrite rule $l \to r$ is *collapsing* if r is a variable. A TRS is *non-collapsing* if it contains no collapsing rule. It is *left-linear* if any variable occurs at most once in any left hand side.

Next we recall some basic terminology for analyzing rewriting in disjoint unions (cf. [24], [15]). A property P of rewrite systems is said to be *modular* (for disjoint systems) if, for all disjoint systems $\mathcal{R}_b^{\mathcal{F}_b}$, $\mathcal{R}_w^{\mathcal{F}_w}$ and $\mathcal{R}^{\mathcal{F}}$ with $\mathcal{R}^{\mathcal{F}} = (\mathcal{R}_b \uplus \mathcal{R}_w)^{\mathcal{F}_b \uplus \mathcal{F}_w}$: $P(\mathcal{R}_b^{\mathcal{F}_b}) \wedge P(\mathcal{R}_w^{\mathcal{F}_w}) \iff P(\mathcal{R}^{\mathcal{F}})$. We say that P is modular for (disjoint unions of) TRSs satisfying Q if $(P \wedge Q)$ is modular. Let us assume subsequently that $\mathcal{R}_b^{\mathcal{F}_b}$ and $\mathcal{R}_w^{\mathcal{F}_w}$ are disjoint TRSs with $\mathcal{R}^{\mathcal{F}}$ (or $\mathcal{R} = \mathcal{R}_b \oplus \mathcal{R}_w$) denoting their disjoint union. Furthermore we shall use the abbreviating notations $\mathcal{T} = \mathcal{T}(\mathcal{F}, \mathcal{V})$ and $\mathcal{T}_i = \mathcal{T}(\mathcal{F}_i, \mathcal{V})$ for $i = b, w$. First of all, in order to achieve better readability we introduce the mostly used chromatic terminology. Many definitions, notations and case distinctions are symmetric w.r.t. the two systems. The non-explicit case is therefore often indicated in parentheses (or omitted). Function symbols from $\mathcal{F}_b$ ($\mathcal{F}_w$) are called *black* (*white*). Variables are *transparent*, i.e., have no colour. A term $s \in \mathcal{T}$ is called *black* (*white*) if $s \in \mathcal{T}_b$ ($s \in \mathcal{T}_w$). We say that s is *top black* (*top white, top transparent*) if $root(s) \in \mathcal{F}_b$ ($root(s) \in \mathcal{F}_w$, $root(s) \in \mathcal{V}$). If $s = C[s_1, \ldots, s_n]$ is top black (top white), and the s_i's are the maximal top white (top black) subterms in s, we also write $s = C^b[\![s_1, \ldots, s_n]\!]$ ($s = C^w[\![s_1, \ldots, s_n]\!]$). In this case the s_i's are the white (black) *principal subterms* or *aliens* of s. We use $s = C[\![s_1, \ldots, s_n]\!]$ to denote either of the cases and say that the s_i's are the principal subterms (or aliens) of s. The *rank* of $s \in \mathcal{T}$ is defined by: $rank(s) = 0$ if $x \in \mathcal{V}$, $rank(s) = 1$ if $s \in (\mathcal{T}_b \cup \mathcal{T}_w) \backslash \mathcal{V}$, and $rank(s) = 1 + \max\{rank(s_i) \mid 1 \le i \le n\}$ if $s = C[\![s_1, \ldots, s_n]\!]$ $(n \ge 1)$. The *rank* of a derivation $D : s_1 \to s_2 \to \ldots$ is $\min\{rank(s_i) \mid s_i \text{ occurs in } D\}$. For $s \to t$, if s reduces to t by applying some rule in one of the principal subterms of s, we write $s \xrightarrow{i} t$, otherwise $s \xrightarrow{o} t$. The relations $\xrightarrow{i}$ and $\xrightarrow{o}$ are called *inner* and *outer* reduction, respectively. A rewrite step $s \to t$ is *destructive at level 1* if the root symbols of s and t have different colours, i.e., if either s is top black and t top white or top transparent (i.e., a variable), or s is top white and t top black or top transparent. The rewrite step $s \to t$ is *destructive at level* $n + 1$ if $s = C[\![s_1, \ldots, s_j, \ldots, s_n]\!] \xrightarrow{i} C[s_1, \ldots, t_j, \ldots, s_n]$ with $s_j \to t_j$ destructive at level n. A step $s \to t$ is *destructive* if it is destructive at some level $n \ge 1$. Note that if a rewrite step is destructive then the applied rule must be collapsing. A term $s \in \mathcal{T}$ is called *preserved* if no derivation issuing from s contains a destructive step. We say that s is *inner preserved* if all its principal subterms are preserved.

Due to lack of space we omit numerous basic facts about rewriting in disjoint unions which we shall tacitly use in the sequel (cf. e.g. [24], [15]).

3 Some Known Results

It is well-known that confluence is modular but termination is not ([24]):

Example 1 [24]. The disjoint TRSs

$$\mathcal{R}_b = \{\, f(a, b, x) \to f(x, x, x) \,\} \qquad \mathcal{R}_w = \left\{ \begin{matrix} G(x, y) \to x \\ G(x, y) \to y \end{matrix} \right\}$$

are terminating, but $\mathcal{R} = \mathcal{R}_b \oplus \mathcal{R}_w$ is not, due to the cyclic derivation $f(a, b, G(a, b)) \to_{\mathcal{R}_1} f(G(a, b), G(a, b), G(a, b)) \to^+_{\mathcal{R}_2} f(a, b, G(a, b)) \to \ldots$.

The non-confluence of the second system above is not essential for the existence of such counterexamples, since even completeness is not modular ([23]). A simple counterexample is the following.

Example 2 [5]. The disjoint TRSs

$$\mathcal{R}_b = \left\{ \begin{array}{c} f(a,b,x) \to f(x,x,x) \\ a \to c \\ b \to c \\ f(x,y,z) \to c \end{array} \right\} \qquad \mathcal{R}_w = \left\{ \begin{array}{c} K(x,y,y) \to x \\ K(y,y,x) \to x \end{array} \right\}$$

are terminating and confluent hence complete, but again their disjoint union $\mathcal{R} = \mathcal{R}_b \oplus \mathcal{R}_w$ allows a cycle:

$$f(a,b,K(a,c,b)) \to_{\mathcal{R}_b} f(K(a,c,b)^3) \to^+_{\mathcal{R}} f(a,b,K(a,c,b)).$$

The known positive results for modularity of termination are, roughly speaking, based on three different approaches concerning the essential ideas and proof structures (cf. [9]):

(1) a **general approach** via an abstract structure theorem where the basic idea is to reduce non-termination in the union to non-termination of a slightly extended generic version of one of the systems ([7], [19]),
(2) a **modular approach** via modularity of innermost termination where sufficient criteria for the equivalence of innermost termination and general termination are combined with the modularity of innermost termination ([8]), and
(3) a **syntactic approach** via left-linearity which in essence is based on commutation and uniqueness properties of (collapsing) reduction in left-linear systems ([25], [14, 22]).

Due to lack of space we cannot give a more detailed account of the (numerous) papers on the subject and the above classification. We only mention some results to which we refer later on. Virtually all proofs for showing preservation results for termination under disjoint unions (and more general combinations) rely on properties of minimal counterexamples of the following form: *If the union of two disjoint terminating systems $\mathcal{R}_b$ and $\mathcal{R}_w$ (having some properties) is non-terminating, then a minimal counterexample in the union must enjoy certain properties and, consequently, $\mathcal{R}_b$ and $\mathcal{R}_w$ must satisfy certain (additional) properties.* Since in general the role of $\mathcal{R}_b$ and $\mathcal{R}_w$ in minimal counterexamples need not be symmetric this may naturally entail corresponding (positive) symmetric and asymmetric preservation results. In the literature this observation has been systematically exploited for approach (1) above, and partially also for (3). Here we shall show how to do this for (2) and how to considerably refine the existing analysis for (3). In all obtained new results (on the preservation of termination) one of the systems must be non-collapsing but not necessarily the other one.

The main (symmetric) preservation result corresponding to (2) above is the following (here and subsequently we focus on the non-trivial implication of the corresponding modularity result).

Theorem 1 [8]. *If $\mathcal{R}_b$, $\mathcal{R}_w$ are terminating, confluent and overlaying then $\mathcal{R}_b \oplus \mathcal{R}_w$ is terminating (as well as confluent and overlaying).*

Two symmetric results and an asymmetric one corresponding to approach (3) above are the following.

Theorem 2 [25]. *If $\mathcal{R}_b$, $\mathcal{R}_w$ are terminating, confluent and left-linear, then $\mathcal{R}_b \oplus \mathcal{R}_w$ is terminating (as well as confluent and left-linear).*

Theorem 3 [22, 14]. *If $\mathcal{R}_b$, $\mathcal{R}_w$ are terminating, consistent w.r.t. reduction and left-linear, then $\mathcal{R}_b \oplus \mathcal{R}_w$ is terminating (as well as consistent w.r.t. reduction and left-linear).*

Theorem 4 [25]. *Let $\mathcal{R}_b$, $\mathcal{R}_w$ be terminating TRSs. If $\mathcal{R}_b$ is non-collapsing and $\mathcal{R}_w$ is left-linear and confluent then $\mathcal{R}_b \oplus \mathcal{R}_w$ is terminating.*

4 An Asymmetric Version of the Modular Approach

We shall show that in Theorem 1, if one of the two terminating systems is non-collapsing, confluent and overlaying, then the other system need only be confluent but not overlaying for ensuring termination of $\mathcal{R}_b \oplus \mathcal{R}_w$.

First we introduce some definitions for locally confluent TRSs. In such systems a term is terminating if and only if it is complete (by Newman's Lemma). Hence we can define $\Phi(t) = C[t_1\downarrow, \ldots, t_n\downarrow]$ for $t = C[t_1, \ldots, t_n]$ such that $t_1, \ldots, t_n$ are the (uniquely defined) maximal complete subterms of t. Here $t_i\downarrow$ denotes the unique normal form of t_i. Clearly we have $t \to^* \Phi(t)$.

If $s \to t$ by contracting a terminating redex in s (i.e., $s \to_p t$ for some $p \in Pos(s)$, with s/p terminating), we write $s \to_{\mathrm{sn}} t$.[2] If $s \to t$ by contracting a non-terminating redex in s (i.e., $s \to_p t$ for some $p \in Pos(s)$, with s/p non-terminating), we write $s \to_{\neg\mathrm{sn}} t$. Clearly, every reduction step can be written as $s \to_{\mathrm{sn}} t$ or $s \to_{\neg\mathrm{sn}} t$. We observe that whenever $s \to_{\mathrm{sn}} t$ by a root reduction step then this implies that s is terminating. The relation $\to_{\mathrm{sn}}$ is terminating for any TRS (which is easily proved by structural induction). Furthermore, every infinite derivation contains infinitely many $\to_{\neg\mathrm{sn}}$-steps (which are not $\to_{\mathrm{sn}}$-steps). Contracting a terminating redex is compatible with the transformation Φ, in the following sense.

Lemma 5 [8]. *Let $\mathcal{R}$ be locally confluent. If $s \to_{\mathrm{sn}} t$ then $\Phi(s) \to^* \Phi(t)$.*

Lemma 6. *Suppose $\mathcal{R}$ is a locally confluent TRS. Let $l \to r$ be a rule of $\mathcal{R}$ such that there exists no inside critical peak by overlapping some other (or the same) rule of $\mathcal{R}$ into $l \to r$ properly below the root. Let σ be a substitution such that σl is not complete. Then $\Phi(\sigma l) = (\Phi \circ \sigma)l$ (where $\Phi \circ \sigma$ is the substitution defined by $(\Phi \circ \sigma)x = \Phi(\sigma x)$). In particular, if additionally σx is complete for all $x \in Var(l)$, then $\Phi(\sigma l) = (\Phi \circ \sigma)l = (\sigma\downarrow)l$ and all proper subterms of $\Phi(\sigma l)$ are irreducible.*[3]

---

[2] Here, the acronym 'sn' in the index stands for *strongly normalizing*.

[3] Note that $\sigma\downarrow$ is the normalized substitution defined by $(\sigma\downarrow)(x) = (\sigma x)\downarrow$.

Proof. We have to show that normalization of all maximal complete subterms in σl can be achieved by normalizing all maximal complete subterms in the "substitution part σ of σl". If no subterm of σl is complete we clearly obtain $\Phi(\sigma l) = (\Phi \circ \sigma)l = \sigma l$ by definition of Φ. Hence, we may assume that some subterm of σl is complete. Let $\sigma l = C[t_1, \ldots, t_n]$, $n \geq 1$, where the t_i's are the maximal complete subterms of σl, let's say with $\sigma l/p_i = t_i$. Note that, due to the assumption that σl is not complete, we have $\lambda < p_i$ for all p_i. Now, if p_i is below the position p of some variable x in l then we get $t_i\!\downarrow = (\Phi \circ \sigma)(l)/p_i$ since t_i is also a maximal complete subterm of σx. If p_i is a non-variable position of l then we have $t_i = \sigma(l/p_i)$. Since t_i is complete, for every variable x which occurs in l (strictly) below p_i, σx is also complete. Let $\sigma' x = (\Phi \circ \sigma)(x) = (\sigma x)\!\downarrow$ for these variables. By definition of Φ we get $t_i\!\downarrow = \sigma(l/p_i)\!\downarrow = \sigma'(l/p_i)\!\downarrow$. We still have to show $\sigma'(l/p_i)\!\downarrow = \sigma'(l/p_i)$. From irreducibility of σ' and $\lambda < p_i$ we conclude that $\sigma'(l/p_i)$ must be irreducible, because otherwise there would exist an inside critical pair in $\mathcal{R}$ by overlapping some rule into $l \to r$ properly below the root. Hence we are done. □

Lemma 7. *Suppose $\mathcal{R}$ is a locally confluent TRS. Let $l \to r$ be a rule of $\mathcal{R}$ such that there exists no inside critical pair by overlapping some other (or the same) rule of $\mathcal{R}$ into $l \to r$ properly below the root. Let σ be a substitution such that σl is not complete. If $s = C[\sigma l] \to_{p,\sigma,l\to r} C[\sigma r] = t$ then $\Phi(s) \to^+ \Phi(t)$.*

Proof. Straightforward using Lemma 6. □

Theorem 8. *Let $\mathcal{R}_b$, $\mathcal{R}_w$ be terminating and confluent TRSs such that $\mathcal{R}_b$ is additionally a non-collapsing overlay system. Then $\mathcal{R} = \mathcal{R}_b \oplus \mathcal{R}_w$ is terminating.*

Proof. Suppose for a proof by contradiction that the disjoint union $\mathcal{R} = \mathcal{R}_b \oplus \mathcal{R}_w$ is non-terminating. Consider an infinite ($\mathcal{R}$-) derivation

$$D : s_1 \to s_2 \to s_3 \to \ldots$$

with the additional minimality property that all proper subterms of s_1 are terminating and hence complete. Then D must have the form

$$D : s_1 \to \ldots \to s_n \to_\lambda s_{n+1} \to \ldots,$$

i.e., eventually some step $s_n \to_\lambda s_{n+1}$ is a (first) root reduction step. Clearly, this step is a $\to_{\neg sn}$-step, and all proper subterms of s_n are complete.

Now consider the case that s_n is top white. All black principal subterms of s_n are complete and any derivation issuing from them consists of (complete) top black reducts, since $\mathcal{R}_b$ is non-collapsing. Together with the infinity of D this implies that all s_k, $k \geq n$, are top white and that D contains infinitely many outer $\mathcal{R}_w$-steps. But then, by identifying abstraction of all black principal subterms (i.e., by their replacement by some same fresh variable), we obtain an infinite (pure) $\mathcal{R}_w$-derivation which contradicts termination of $\mathcal{R}_w$.

The other case is that s_n is top black. Since all white principal subterms of s_n are complete, all $\to_{\neg sn}$-steps in D after s_n (including $s_n \overset{o}{\to} s_{n+1}$) must be outer $\mathcal{R}_b$-steps, and there must be infinitely many of these. By definition of Φ

we know that all principal subterms in $\Phi(s_n)$ are irreducible. Hence, applying Φ to D and using Lemma 7 (which is applicable, because $\mathcal{R}_b$ is a locally confluent overlay system) and Lemma 5 we obtain the infinite $\mathcal{R}$-derivation

$$\Phi(s_n) \to^* \Phi(s_{n+1}) \to^* \Phi(s_{n+2}) \to^* \dots$$

where all (proper) reduction steps are outer $\mathcal{R}_b$-steps. As above, identifying abstraction yields an infinite pure $\mathcal{R}_b$-derivation contradicting termination of $\mathcal{R}_b$. Thus we may conclude that $\mathcal{R} = \mathcal{R}_b \oplus \mathcal{R}_w$ must be terminating. □

Let us give an example for illustrating the applicability of Theorem 8.

Example 3. Consider the modified version of Example 1 where

$$\mathcal{R}_b = \{\, f(a,b,x,x) \to f(x,x,x,x) \,\} \qquad \mathcal{R}_w = \left\{ \begin{array}{r} G(x,x) \to x \\ G(A,B) \to A \\ A \to B \end{array} \right\}$$

Both systems are terminating and confluent, and moreover $\mathcal{R}_b$ is a non-collapsing overlay system. Hence, Theorem 8 yields completeness of the disjoint union $\mathcal{R}_b \oplus \mathcal{R}_w$. We note that none of the previous modularity results is applicable here (including the recent ones of [3]). In particular, $\mathcal{R}_w$ is neither overlaying nor left-linear, and $\mathcal{R}_b$ is also not left-linear.

5 Asymmetric Versions of the Syntactic Approach

Example 4. Consider the disjoint TRSs

$$\mathcal{R}_b = \{\, f(x,g(x),y) \to f(y,y,y) \,\} \qquad \mathcal{R}_w = \left\{ \begin{array}{r} H(x,x) \to K(B) \\ H(x,x) \to C \\ K(x) \to x \end{array} \right\}$$

Both systems are terminating as well as their disjoint union. However, we observe that none of the known modularity results applies here. In particular, both systems are not left-linear, and the second one is non-confluent (due to $B \;{}^{+}\!\!\leftarrow H(x,x) \to C$ with B and C distinct normal forms).

Yet, we observe that collapsing reduction is deterministic in $\mathcal{R}_w$ above in the sense that whenever a term s collapses to a variable x then x has a unique *ancestor* occurrence in s. This property is violated in Example 2 where we have for instance $K(z,z,z) \to z$, either by applying the rule $K(x,y,y) \to x$ (in which case the ancestor of z in $K(z,z,z)$ is the first z) or by applying the rule $K(y,y,x) \to x$ (in which case the ancestor of z in $K(z,z,z)$ is the last z). In fact, it turns out that also the property of not loosing the possibility of collapsing reductions is a crucial one. For instance, in the TRS $\{G(x) \to x, G(x) \to A\}$ we can collapse $G(x)$ to x but after reduction of $G(x)$ to A this possibility is lost.

In order to precisely define the first property informally described above we need some auxiliary definitions.

Definition 9. For any labelled derivation $D : s \to^* t$ and any $q \in VPos(t)$ the set $anc(q, D)$ of *ancestors* of q in D is defined recursively as follows. Id D is empty, then $anc(q, D) = \{q\}$. For a labelled one-step derivation $D : s \to_{p,\sigma,l\to r} t$: $anc(q, D) := \{q\}$ if p and q are disjoint, and $anc(q, D) := \{pp_3p_2 \mid l/p_3 = x\}$ if $q = pp_1p_2$, $r/p_1 = x \in \mathcal{V}$ and $\sigma(x)/p_2 \in \mathcal{V}$. For a non-empty labelled derivation $D = D_1; D_2$ from s to t with $D_1 : s \to^* s'$, $D_2 : s' \to_{p,\sigma,l\to r} t$ we define: $anc(q, D) := \bigcup_{q' \in anc(q,D_2)} anc(q', D_1)$. Slightly abusing notation, we denote $\bigcup_{D:s\to^* t} anc(q, D)$ (where the union ranges over all labelled derivations from s to t) by $anc(q, s \to^* t)$. Furthermore, if $t = x$ (and consequently $q = \lambda$), we also write — again slightly abusing notation — $anc(x, s)$ instead of $anc(\lambda, s \to^* x)$. If, for $t/q = x \in \mathcal{V}$, $anc(q, s \to^* t) = \{p\}$, then we say that x in t at (position) q *has a unique ancestor* in s at (position) p. If in this case $t = x$ (and consequently $q = \lambda$), we also write $anc(x, s) = p$ instead of $anc(x, s) = \{p\}$ and say that x *has a unique ancestor* in s at p.

In Example 2 e.g., for the derivation $D : K(x, x, x) \to_{K(x,y,y)\to x} x$ we have $anc(\lambda, D) = \{1\}$, and $anc(x, K(x, x, x)) = \{1, 3\}$.

We remark that rewriting in left-linear TRSs enjoys some nice abstraction properties. In particular, we have the following.

Lemma 10. *Let $\mathcal{R}$ be a left-linear TRS. If $s = C[x, \ldots, x]_{p_1,\ldots,p_n} \to^* x$ such that $p_i \in anc(x, s)$, i.e., $p_i \in anc(\lambda, D)$ for some labelled derivation $D : s \to^* x$, then $C[x_1, \ldots, x_n] \to^* x_i$ (using the same rules at the same positions as in D) where $x_1, \ldots, x_n$ are mutually distinct fresh variables.*

Proof. This is a consequence of the left-linearity of $\mathcal{R}$. □

Definition 11. A TRS $\mathcal{R}^{\mathcal{F}}$ is said to be *uniquely collapsing* (UC) if, for every $s \in \mathcal{T}(\mathcal{F}, \{x\})$, $s = C[x, \ldots, x]_{p_1,\ldots,p_n} \to^* x$ (where all occurrences of x in s are displayed) implies that x has a unique ancestor in s at p_i (for some unique i, $1 \le i \le n$). We say that $\mathcal{R}^{\mathcal{F}}$ is *collapsing confluent* (CCR) if $x \; {}^*\!\!\leftarrow s \to^* t$ implies $t \to^* x$ for every $s, t \in \mathcal{T}(\mathcal{F}, \mathcal{V})$, $x \in \mathcal{V}$.

We observe that collapsing confluence is a restricted version of the normal form property NF (where the only normal forms considered are variables), and can also be expressed more locally as follows.

Lemma 12. *For any TRS $\mathcal{R}^{\mathcal{F}}$ the following assertions are equivalent:*

(1) $\forall s, t \in \mathcal{T}(\mathcal{F}, \mathcal{V}), x \in \mathcal{V} : x \; {}^*\!\!\leftarrow s \to t \Longrightarrow t \to^* x$.
(2) $\forall s, t \in \mathcal{T}(\mathcal{F}, \mathcal{V}), x \in \mathcal{V} : x \; {}^*\!\!\leftarrow s \to^* t \Longrightarrow t \to^* x$.
(3) $\forall t \in \mathcal{T}(\mathcal{F}, \mathcal{V}), x \in \mathcal{V} : x \leftrightarrow^* t \Longrightarrow t \to^* x$.

Proof. Routine. □

Next we define a transformation which enables to abstract from white 'layers' but keeps the information concerning potential (white) 'layer collapses'.

Definition 13. Let $\mathcal{R}_b$, $\mathcal{R}_w$ be disjoint TRSs such that $\mathcal{R}_w$ is uniquely collapsing. The (white) abstraction mapping $\Theta : \mathcal{T}(\mathcal{F}_b \uplus \mathcal{F}_w) \to \mathcal{T}(\mathcal{F}_b \uplus \{G^1, A^0\})$ is recursively defined as follows:

$$\Theta(t) = \begin{cases} t & \text{if } \ t \in \mathcal{T}(\mathcal{F}_b) \\ C^b[\![\Theta(t_1), \ldots, \Theta(t_n)]\!] & \text{if } \ t = C^b[\![t_1, \ldots, t_n]\!] \\ A & \text{if } \ t \in \mathcal{T}(\mathcal{F}_w) \\ A & \text{if } \ t = C^w[\![t_1, \ldots, t_n]\!], C^w[x, \ldots, x] \not\to^*_{\mathcal{R}_w} x \\ G(\Theta(t_i)) & \text{if } \ t = C^w[\![t_1, \ldots, t_n]\!]_{p_1, \ldots, p_n}, \\ & \quad \hat{t} := C^w[x, \ldots, x] \to^*_{\mathcal{R}_w} x, anc(x, \hat{t}) = p_i \end{cases}$$

We observe that Θ above is well-defined, since $\mathcal{R}_w$ is assumed to be uniquely collapsing. For illustration, consider the derivation $f(K(H(a,a)), g(H(a,a)), K(b)) \to_{\mathcal{R}_w} f(H(a,a), g(H(a,a)), K(b)) \to_{\mathcal{R}_b} f(K(b), K(b), K(b))$ in Example 4 (where a, b are assumed to be black constants). Here, abstraction with Θ yields:

$$f(A, g(A), G(b)) = f(A, g(A), G(b)) \to_{\mathcal{R}_b} f(G(b), G(b), G(b)) \, .$$

Lemma 14. *Let $\mathcal{R}_b$, $\mathcal{R}_w$ be disjoint TRSs such that $\mathcal{R}_b$ is non-collapsing and $\mathcal{R}_w$ is uniquely collapsing and collapsing confluent. Let s be a top black term. Then the following properties hold:*

(a) $s \xrightarrow{o}_{\mathcal{R}_b} t \implies \Theta(s) \to_{\mathcal{R}_b} \Theta(t)$.

(b) $s \xrightarrow{i}_{\mathcal{R}_b} t \implies \Theta(s) \to^{=}_{\mathcal{R}_b} \Theta(t)$.

(c) $s \xrightarrow{i}_{\mathcal{R}_w} t \implies \Theta(s) \to^{=}_{\{G(x) \to x\}} \Theta(t)$.

Proof. (a) is straightforward by definition of Θ. (b) is proved by induction on $n = rank(s)$ and case analysis using the non-collapsing property of $\mathcal{R}_b$ and (a). The proof of (c) is also by induction on $n = rank(s)$ and case analysis exploiting that $\mathcal{R}_w$ is uniquely collapsing and collapsing confluent. Note that without collapsing confluence of $\mathcal{R}_w$ an inner $\mathcal{R}_w$-step could be translated into a step using the rule $G(x) \to A$. This is due to the fact that in a top white, white term $C^w[x, \ldots, x]$ the possibility of a collapse to x might be eliminated by a next (possibly non-collapsing) $\mathcal{R}_w$-step. □

Lemma 15. *A TRS $\mathcal{R}$ is terminating if and only if $\mathcal{R} \oplus \{G(x) \to x\}$ is terminating.*

Proof. The 'if-part' of the equivalence is trivial. The 'only-if' part is a consequence of [16, Theorem 6]. Note that an alternate (direct and easy) proof (of the 'only-if' part) is also possible by defining an appropriate semantic interpretation. □

Theorem 16. *Let $\mathcal{R}_b$, $\mathcal{R}_w$ be two disjoint terminating TRSs. If $\mathcal{R}_b$ is non-collapsing and $\mathcal{R}_w$ is uniquely collapsing and collapsing confluent, then $\mathcal{R}_b \oplus \mathcal{R}_w$ is terminating.*

Proof. For a proof by contradiction assume $\mathcal{R}_b$ and $\mathcal{R}_w$ are given as above such that $\mathcal{R}^{\mathcal{F}} = \mathcal{R}_b \oplus \mathcal{R}_w$ is non-terminating. We consider a counterexample of minimal rank

$$D : s_0 \to_{\mathcal{R}} s_1 \to_{\mathcal{R}} s_2 \to_{\mathcal{R}} \cdots$$

where w.l.o.g. we may assume $s_i \in \mathcal{T}(\mathcal{F}) = \mathcal{T}(\mathcal{F}_b \uplus \mathcal{F}_w)$ for all i, $0 \leq i$. Since D is minimal and $\mathcal{R}_b$ is non-collapsing we conclude that all s_i are top black. Now, applying Lemma 14 and exploiting the fact that D contains infinitely many outer $\mathcal{R}_b$-steps we obtain an infinite $(\mathcal{R}_b \oplus \{G(x) \to x\})$-derivation

$$\Theta(D) : \Theta(s_0) \to^* \Theta(s_1) \to^* \Theta(s_2) \to^* \ldots .$$

But by Lemma 15 this implies non-termination of $\mathcal{R}_b$, hence a contradiction. □

In Example 4 the system $\mathcal{R}_b = \{f(x, g(x), y) \to f(y, y, y)\}$ is non-collapsing and $\mathcal{R}_w = \{H(x, x) \to K(B), H(x, x) \to C, K(x) \to x\}$ is easily shown to be both uniquely collapsing and collapsing confluent. Hence, by Theorem 16 we conclude termination of $\mathcal{R}_b \oplus \mathcal{R}_w$.

Example 5. Consider the disjoint TRSs

$$\mathcal{R}_b = \{\, f(a, g(x), y) \to f(y, y, y) \,\} \qquad \mathcal{R}_w = \left\{ \begin{array}{c} H(x, x) \to K(B) \\ K(x) \to C \\ K(x) \to x \end{array} \right\}$$

Both systems are terminating as well as their disjoint union. However, we observe that none of the known modularity results (including Theorem 16 above) applies here. In particular, the second system is not left-linear and not (collapsing) confluent (due to $x \leftarrow K(x) \to C$ with C irreducible). Observe, however, that in contrast to Example 4 the non-collapsing system $\mathcal{R}_b$ is additionally left-linear.

We shall show now that dropping the collapsing confluence condition for $\mathcal{R}_w$ in Theorem 16 is possible provided we additionally require that $\mathcal{R}_b$ is left-linear.

Lemma 17. *Let $\mathcal{R}_b$, $\mathcal{R}_w$ be disjoint TRSs such that $\mathcal{R}_b$ is non-collapsing and $\mathcal{R}_w$ is uniquely collapsing. Let s be a top black term. Then the following properties hold:*

(a) $s \stackrel{o}{\to}_{\mathcal{R}_b} t \implies \Theta(s) \to_{\mathcal{R}_b} \Theta(t)$.
(b) $s \stackrel{i}{\to}_{\mathcal{R}_b} t \implies \Theta(s) \to^{=}_{\mathcal{R}_b} \Theta(t)$.
(c) $s \stackrel{i}{\to}_{\mathcal{R}_w} t \implies \Theta(s) \to^{=}_{\{G(x) \to x, G(x) \to A\}} \Theta(t)$.

Proof. (a) and (b) hold by Lemma 14(a) and (b). (c) follows by induction on $rank(s)$ (cf. the proof of Lemma 14(c)). In particular, collapsing confluence of $\mathcal{R}_w$ is not needed here. □

Lemma 18. *Let $\mathcal{R}$ be a left-linear TRS. Then, $\mathcal{R}$ is terminating if and only if $\mathcal{R} \oplus \{G(x) \to x, G(x) \to A\}$ is terminating.*

Proof. The 'if-part' of the equivalence is trivial. For the 'only-if' part assume that $\mathcal{R}$ is terminating. Let $\mathcal{R}_{G_x} := \{G(x) \to x\}$, $\mathcal{R}_{G_A} := \{G(x) \to A\}$. Obviously, $\mathcal{R}_{G_A}$ is terminating. By assumption and Lemma 15 we know that $(\mathcal{R} \cup \mathcal{R}_{G_x})$ is terminating. Hence, using the *quasi-commutation* approach of Bachmair & Dershowitz [1, Lemmas 1,2], for termination of $\mathcal{R} \cup (\mathcal{R}_{G_x} \cup \mathcal{R}_{G_A}) = (\mathcal{R} \cup \mathcal{R}_{G_x}) \cup \mathcal{R}_{G_A}$ it suffices to show that $(\mathcal{R} \cup \mathcal{R}_{G_x})$ *quasi-commutes over* $\mathcal{R}_{G_A}$, i.e.: $\to_{\mathcal{R}_{G_A}} \circ \to_{\mathcal{R} \cup \mathcal{R}_{G_x}} \subseteq \to_{\mathcal{R} \cup \mathcal{R}_{G_x}} \circ \to^*_{(\mathcal{R} \cup \mathcal{R}_{G_x}) \cup \mathcal{R}_{G_A}}$. The latter property is a consequence of

(1) $\to_{\mathcal{R}_{G_A}} \circ \to_{\mathcal{R}_{G_x}} \subseteq \to_{\mathcal{R}_{G_x}} \circ \to_{\mathcal{R}_{G_A}}$, and
(2) $\to_{\mathcal{R}_{G_A}} \circ \to_{\mathcal{R}} \subseteq \to_{\mathcal{R}} \circ \to^*_{\mathcal{R}_{G_A}}$.

Now, (1) is straightforward by an easy case analysis exploiting the special shape of the rules $G(x) \to x$ and $G(x) \to A$. (2) is also easy by a standard case analysis, but essentially relies on left-linearity of $\mathcal{R}$. Hence, we are done. □

Theorem 19. *Let $\mathcal{R}_b$, $\mathcal{R}_w$ be two disjoint terminating TRSs. If $\mathcal{R}_b$ is non-collapsing and left-linear, and $\mathcal{R}_w$ is uniquely collapsing, then $\mathcal{R}_b \oplus \mathcal{R}_w$ is terminating.*

Proof. For a proof by contradiction assume $\mathcal{R}_b$ and $\mathcal{R}_w$ are given as above such that $\mathcal{R} = \mathcal{R}_b \oplus \mathcal{R}_w$ is non-terminating. We consider a counterexample of minimal rank

$$D : s_0 \to_{\mathcal{R}} s_1 \to_{\mathcal{R}} s_2 \to_{\mathcal{R}} \dots$$

where w.l.o.g. we may assume $s_i \in \mathcal{T}(\mathcal{F}) = \mathcal{T}(\mathcal{F}_b \uplus \mathcal{F}_w)$ for all i, $0 \leq i$. Since D is minimal and $\mathcal{R}_b$ is non-collapsing we conclude that all s_i are top black. Now, applying Lemma 17 and exploiting the fact that D contains infinitely many outer $\mathcal{R}_b$-steps we obtain an infinite $(\mathcal{R}_b \oplus \{G(x) \to x, G(x) \to A\})$-derivation

$$\Theta(D) : \Theta(s_0) \to^* \Theta(s_1) \to^* \Theta(s_2) \to^* \dots .$$

But by Lemma 18 this implies non-termination of $\mathcal{R}_b$, hence a contradiction. □

Using Theorem 19 it is straightforward to prove termination of $(\mathcal{R}_b \oplus \mathcal{R}_w)$ in Example 5: $\mathcal{R}_b = \{f(a, g(x), y) \to f(y, y, y)\}$ is non-collapsing and left-linear, and $\mathcal{R}_w = \{H(x, x) \to K(B), K(x) \to C, K(x) \to x\}$ is easily shown to be uniquely collapsing. Hence Theorem 19 is applicable.

However, let us remark that in general applying the preservation criteria for termination that rely on the properties UC and CCR, i.e., Theorems 16 and 19, is not obvious. In fact, the properties UC and CCR are undecidable in general, even for left-linear terminating TRSs (this can be shown by using the undecidability of PCP, Post's correspondence problem). Consequently, it might be worthwhile looking for interesting decidable (syntactic) conditions ensuring them (this seems to be non-trivial). Yet, we think that the analysis performed is of independent interest because it provides a deeper structural insight into the crucial phenomena causing (non-)termination in disjoint unions of terminating systems.

6 Modularity Criteria for UC and CCR

Next we shall investigate how the newly introduced properties of being uniquely collapsing (UC) and collapsing confluent (CCR) relate to other well-known confluence, normal form and consistency properties. Moreover, we shall study their modularity behaviour.

Lemma 20. *For any TRS the following implications hold and are proper:*

(a) NF $\Longrightarrow$ CCR.
(b) CR $\Longrightarrow$ CCR $\Longrightarrow$ CON.
(c) UC $\Longrightarrow$ CON$^{\rightarrow}$.

Proof. The implication in (a) follows from the respective definitions (and Lemma 12). Furthermore, the TRS $\{a \to b, a \to c\}$ (over the signature $\mathcal{F} = \{a, b, c\}$) is a counterexample to the reverse implication. In (b), the implications also follow from the respective definitions (and making use of Lemma 12). The implications are proper, because for instance $\{g(x) \to x, g(a) \to b\}$ is CCR but not CR, and $\{f(x) \to x, f(x) \to g(x)\}$ is CON but not CCR. Finally, the implication UC $\Longrightarrow$ CON$^{\rightarrow}$ in (c) holds, since any counterexample $x \ {}^{+}\!\!\leftarrow C[x,y]_{p_1,p_2} \to^{+} y$ $(x \neq y)$ to CON$^{\rightarrow}$ immediately yields a counterexample to UC: $x \ {}^{+}\!\!\leftarrow C[x,x]_{p_1,p_2} \to^{+} x$ where p_1, p_2 are distinct ancestors of x in $C[x,x]$. Moreover, $\{f(x,x) \to x\}$ is obviously CON$^{\rightarrow}$ but not UC. □

Neither UC nor CCR is a modular property of TRSs as shown by the following examples.

Example 6. The disjoint TRSs

$$\mathcal{R}_b = \left\{ \begin{array}{c} f(x,x,y,z) \to y \\ f(x,g(x),y,z) \to z \end{array} \right\} \qquad \mathcal{R}_w = \left\{ \begin{array}{c} G(x) \to x \\ G(x) \to A \end{array} \right\}$$

are both uniquely collapsing (as is not difficult to verify). However, their disjoint union $\mathcal{R} = \mathcal{R}_b \oplus \mathcal{R}_w$ is not even consistent w.r.t. reduction since in $\mathcal{R}$ we have for instance:

$$z \leftarrow f(A, g(A), y, z) \ {}^{+}\!\!\leftarrow f(G(g(A)), G(g(A)), y, z) \to y \, .$$

Example 7. The disjoint TRSs

$$\mathcal{R}_b = \{\, f(x,x,y) \to y \,\} \qquad \mathcal{R}_w = \left\{ \begin{array}{c} A \to B \\ A \to C \end{array} \right\}$$

are obviously collapsing confluent, but in $\mathcal{R} = (\mathcal{R}_b \oplus \mathcal{R}_w)$ we have

$$f(B, C, y) \ {}^{+}\!\!\leftarrow f(A, A, y) \to y$$

where $f(B,C,y)$ and y are distinct normal forms. Hence, $\mathcal{R}$ is not collapsing confluent.

Figure 1 summarizes the relationships between the various confluence and normal form properties considered as well as their modularity behaviour. Missing implications do not hold, the boxed properties are modular and the others are not modular (cf. [24], [21], [15], [13], [22, 14] for the relations and (non-)modularity properties not treated above).

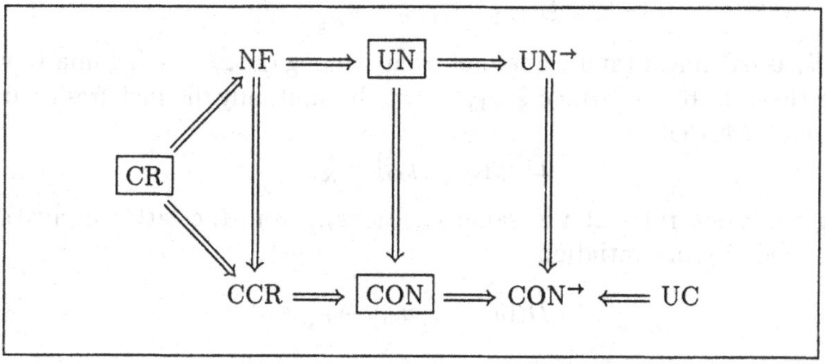

Fig. 1. confluence / normal form properties and their modularity behaviour

The properties NF and $\mathrm{UN}^{\rightarrow}$ are not modular as shown by Middeldorp ([15]). However, modularity can be recovered by imposing left-linearity (LL): $\mathrm{NF} \wedge \mathrm{LL}$ is modular ([15]), and $\mathrm{UN}^{\rightarrow} \wedge \mathrm{LL}$ as well as $\mathrm{CON}^{\rightarrow} \wedge \mathrm{LL}$ are modular ([13]). Interestingly, it turns out that for collapsing confluence (CCR) and the property of being uniquely collapsing (UC) modularity can also be recovered by imposing left-linearity.

Theorem 21. *Collapsing confluence is modular for left-linear TRSs (i.e., $CCR \wedge LL$ is modular).*

Proof. For the non-trivial part of the modularity statement assume that $\mathcal{R}_b$ and $\mathcal{R}_w$ are disjoint left-linear TRSs that are collapsing confluent. First we observe that, by [21, Lemma 3.6] $\leftrightarrow^*_{\mathcal{R}^{\mathcal{F}}}$ is a conservative extension of both $\leftrightarrow^*_{\mathcal{R}_b}$ and $\leftrightarrow^*_{\mathcal{R}_w}$, i.e., for any $i \in \{b, w\}$:

$$(*) \quad \forall s, t \in \mathcal{T}(\mathcal{F}_i, \mathcal{V}) : s \leftrightarrow^*_{\mathcal{R}_i} t \iff s \leftrightarrow^*_{\mathcal{R}} t\,.$$

Now assume that in $\mathcal{R} = \mathcal{R}_b \oplus \mathcal{R}_w$ we have $x \;{}^*_{\mathcal{R}}\!\leftarrow s \rightarrow^*_{\mathcal{R}} t$. Then we must show: $t \rightarrow^*_{\mathcal{R}} x$. To this end we first reduce t to an inner preserved reduct as follows: If $t \in \mathcal{T}(\mathcal{F}_i, \mathcal{V})$ (for some $i \in \{b, w\}$) then t is already inner preserved. In that case we get $x \leftrightarrow^*_{\mathcal{R}_i} t$ by $(*)$, and collapsing confluence of $\mathcal{R}_i$ implies $t \rightarrow^*_{\mathcal{R}_i} x$ as desired. Otherwise, assume w.l.o.g. $t = C^b[\![t_1, \ldots, t_n]\!]$. Since every term has a preserved reduct (cf. [11]), every principal subterm t_i of t can be reduced to a preserved reduct t'_i. Hence we get $t = C^b[\![t_1, \ldots, t_n]\!] \rightarrow^*_{\mathcal{R}} C^b[t'_1, \ldots, t'_n] =: t'$ and $x \leftrightarrow^* t'$. Obviously, t' is of the form $t' = D^b[\![u_1, \ldots, u_m]\!]$ and is inner preserved. Now, according to [21, Lemma 3.13], an abstraction (modulo $\leftrightarrow^*_{\mathcal{R}}$) of the maximal top white special subterms in x and t' yields the following:

$$x \leftrightarrow^*_{\mathcal{R}_b} D^b[x_1, \ldots, x_m]$$

where $x_1, \ldots, x_m$ denote (not necessarily distinct) fresh variables with $x_j = x_k \iff u_j \leftrightarrow^*_{\mathcal{R}} u_k$, for $1 \leq j, k \leq m$.[4] Thus, by collapsing confluence of $\mathcal{R}_b$ we get

$$D^b[x_1, \ldots, x_m] \rightarrow^*_{\mathcal{R}_b} x \,.$$

Since $\mathcal{R}_b$ is left-linear (and x does not occur among the x_i), by Lemma 10 we can replace the variable occurrences $x_1, \ldots, x_m$ by mutually distinct fresh variables $y_1, \ldots, y_m$ such that

$$D^b[y_1, \ldots, y_m] \rightarrow^*_{\mathcal{R}_b} x$$

(using the same rules at the same positions). From the latter derivation we finally obtain by instantiation

$$D^b[u_1, \ldots, u_m] \rightarrow^*_{\mathcal{R}_b} x$$

which together with $t \rightarrow^*_{\mathcal{R}} D^b[u_1, \ldots, u_m]$ yields

$$t \rightarrow^*_{\mathcal{R}} x$$

as desired. Hence we are done. □

In order to illustrate what may go wrong in the above construction for non-left-linear systems consider again Example 7. There we had the derivations

$$f(B, C, y) \; {}_{\mathcal{R}}^{*}\!\!\leftarrow f(A, A, y) \rightarrow_{\mathcal{R}} y$$

in the disjoint union. Now, $f(B, C, y)$ is inner preserved (it is even irreducible). Hence, abstraction (modulo $\leftrightarrow^*_{\mathcal{R}}$) yields (due to $B \leftrightarrow^*_{\mathcal{R}} C$)

$$f(z, z, y) \leftrightarrow^*_{\mathcal{R}_b} y$$

with z a fresh variable. Collapsing confluence of $\mathcal{R}_b = \{f(x, x, y) \rightarrow y\}$ entails $f(z, z, y) \rightarrow^*_{\mathcal{R}_b} y$, in fact even

$$f(z, z, y) \rightarrow_{\mathcal{R}_b} y \,.$$

But now linearization w.r.t. z is impossible, i.e., $f(z_1, z_2, y) \rightarrow_{\mathcal{R}_b} y$ (from which we would get $f(B, C, y) \rightarrow_{\mathcal{R}_b} y$ as desired) does not hold.

Theorem 22. *A left-linear TRSs is uniquely collapsing if and only it is consistent w.r.t. reduction* ($LL \Longrightarrow [UC \iff CON^{\rightarrow}]$).

[4] Note that in order to ensure applicability of [21, Lemma 3.13] as above one has to verify that — in the terminology of [21] — t' is '$\leftrightarrow^*_{\mathcal{R}_b}$-normalized' and $u_1, \ldots, u_m$ are '$\leftrightarrow^*_{\mathcal{R}}$-normalized'. The latter property (which implies the former one) follows from the preservation of the u_j in combination with collapsing confluence of $\mathcal{R}_b$ and of $\mathcal{R}_w$.

Proof. Assume $\mathcal{R}$ is left-linear. The implication UC $\Longrightarrow$ CON$^{\rightarrow}$ holds (even without left-linearity) by Lemma 20(c). Conversely, assuming $\neg$UC we have to show $\neg$CON$^{\rightarrow}$. Hence, consider a counterexample (to UC), i.e., a derivation $s = C[x \ldots, x]_{p_1,\ldots,p_n} \rightarrow^* x$ (with $n \geq 2$) where x has two distinct ancestor occurrences p_i and p_j in s, let's say p_1 and p_n. By left-linearity of $\mathcal{R}$ and Lemma 10 this implies $C[y, x, \ldots, x, z] \rightarrow^* y$ and $C[y, x, \ldots, x, z] \rightarrow^* z$ (for some fresh distinct variables y, z) which yields $\neg$CON$^{\rightarrow}$ as desired. □

In view of Theorem 22, and since confluence implies in particular consistency w.r.t. reduction and collapsing confluence, Theorem 16 above constitutes a generalization of Theorem 4 (due to Toyama, Klop & Barendregt [25]). Furthermore we remark that together with the (non-trivial) modularity of CON$^{\rightarrow} \wedge$ LL which was proved in [13] Theorem 22 above entails the following consequence.

Theorem 23. *Being uniquely collapsing is modular for left-linear TRSs (i.e., UC $\wedge$ LL is modular).*

7 Extensions to Non-Disjoint Unions

Extensions of our asymmetric preservation results for termination are possible for instance for *composable* TRS ([17, 18, 20]) where constructors may be shared as well as the defining rules for all shared defined symbols. In such combinations the problematic 'layer-coalescing' reductions are not only possible by application of collapsing rules, but also by application of 'shared function symbol lifting' rules. Taking this effect into account, Theorem 8 extends in a natural way to composable TRSs by requiring 'layer-preservation' ([20]) instead of the non-collapsing property of one of the involved systems (cf. [9, Theorem 5.4.12]). Similar extensions (to composable systems) of the presented symmetric and asymmetric preservation results for the syntactic approach seem also possible along the same line of reasoning by replacing the non-collapsing requirement by layer-preservation and by forbidding shared function symbol lifting rules. Whether extensions to certain hierarchical combinations are possible remains to be clarified (cf. e.g. [12], [2], [6]).

Acknowledgements: I would like to thank the anonymous referees for some useful comments.

References

1. L. Bachmair and N. Dershowitz. Commutation, transformation, and termination. In J. Siekmann, ed., *Proc. 8th CADE*, LNCS **230**, pp. 5–20. Springer, 1986.
2. N. Dershowitz. Hierarchical termination. In N. Dershowitz and N. Lindenstrauss, eds., *Proc. 4th CTRS (1994)*, LNCS **968**, pp. 89–105. Springer, 1995.
3. N. Dershowitz. Innocuous constructor-sharing combinations. In H. Comon, ed., *Proc. 8th RTA*, LNCS **1232**, pp. 202–216. Springer, 1997.

4. N. Dershowitz and J.-P. Jouannaud. Rewrite systems. In J. van Leeuwen, ed., *Formal models and semantics, Handbook of Theoretical Computer Science*, volume B, chapter 6, pp. 243–320. Elsevier - The MIT Press, 1990.
5. K. Drosten. *Termersetzungssysteme*. Informatik-Fachberichte 210. Springer, 1989.
6. M. Fernández and J.-P. Jouannaud. Modular termination of term rewriting systems revisited. In *Recent Trends in Data Type Specification*, LNCS **906**, pp. 255–272, Springer, 1995.
7. B. Gramlich. Generalized sufficient conditions for modular termination of rewriting. *Applicable Algebra in Engineering, Communication and Computing*, 5:131–158, 1994.
8. B. Gramlich. Abstract relations between restricted termination and confluence properties of rewrite systems. *Fundamenta Informaticae*, 24:3–23, 1995.
9. B. Gramlich. *Termination and Confluence Properties of Structured Rewrite Systems*. PhD thesis, Fachbereich Informatik, Universität Kaiserslautern, Jan. 1996.
10. J. W. Klop. Term rewriting systems. In S. Abramsky, D. Gabbay, and T. Maibaum, eds., *Handbook of Logic in Computer Science*, volume 2, chapter 1, pp. 2–117. Clarendon Press, Oxford, 1992.
11. J. W. Klop, A. Middeldorp, Y. Toyama, and R. Vrijer. Modularity of confluence: A simplified proof. *Information Processing Letters*, 49:101–109, 1994.
12. M.R.K. Krishna Rao. Modular proofs for completeness of hierarchical term rewriting systems. *Theoretical Computer Science*, 151(2):487–512, Nov. 1995.
13. M. Marchiori. Modularity of $UN^{\rightarrow}$ for left-linear term rewriting systems. Technical Report CS-R9433, CWI, Amsterdam, May 1994.
14. M. Marchiori. Modularity of Completeness Revisited . In J. Hsiang, ed., *Proc. 6th RTA*, LNCS **914**, pp. 2–10. Springer, 1995.
15. A. Middeldorp. *Modular Properties of Term Rewriting Systems*. PhD thesis, Free University, Amsterdam, 1990.
16. A. Middeldorp, H.Ohsaki and H. Zantema. Transforming termination by self-labelling. In *Proc. 13th CADE*, LNAI **1104**, pp. 373–387. Springer, 1996.
17. A. Middeldorp and Y. Toyama. Completeness of combinations of constructor systems. *Journal of Symbolic Computation*, 15:331–348, Sept. 1993.
18. E. Ohlebusch. *Modular Properties of Composable Term Rewriting Systems*. PhD thesis, Universität Bielefeld, 1994. Report 94-01.
19. E. Ohlebusch. On the modularity of termination of term rewriting systems. *Theoretical Computer Science*, 136:333–360, 1994.
20. E. Ohlebusch. Modular properties of composable term rewriting systems. *Journal of Symbolic Computation*, 20(1):1–42, 1995.
21. M. Schmidt-Schauß. Unification in a combination of arbitrary disjoint equational theories. *Journal of Symbolic Computation*, 8(1):51–99, 1989.
22. M. Schmidt-Schauß, M. Marchiori, and S. Panitz. Modular termination of *r*-consistent and left-linear term rewriting systems. *Theoretical Computer Science*, 149(2):361–374, 1995.
23. Y. Toyama. Counterexamples to termination for the direct sum of term rewriting systems. *Information Processing Letters*, 25:141–143, 1987.
24. Y. Toyama. On the Church-Rosser property for the direct sum of term rewriting systems. *Journal of the ACM*, 34(1):128–143, 1987.
25. Y. Toyama, J. Klop, and H. Barendregt. Termination for direct sums of left-linear complete term rewriting systems. *Journal of the ACM*, 42(6):1275–1304, 1995.

From Algebra Transformation to Labelled Transition Systems

Martin Große–Rhode *

Università di Roma *La Sapienza*, Dip. Scienze dell'Informazione
Via Salaria 113, I-00198 Roma, Italy, e-mail: mgr@cs.tu-berlin.de

Abstract. The formal specification of multiple viewpoints of a system requires multiple specification formalisms, suitable for the specific concerns of the viewpoints. For rather different viewpoints, such as for instance the information and the computational model of a component of a system, even the underlying paradigms of the specification formalisms may be different. In this paper a general semantical framework for the formal specification of dynamically evolving systems is presented. Its models, algebra transformation systems, have states whose internal data structures are given by partial algebras, which are manipulated by the application of replacement rules. Its paradigm is the descriptive one of general model theory respectively institutions. Partial observations of the internal state structures yield a translation from algebra transformation systems to labelled transition systems, the granularity of which is determined by the specification of the admissible observations. Since labelled transition systems can be considered as the general (operational) models for process calculi, this translation allows comparisons between the descriptive paradigm of the model theoretic approach and the operational one of process calculi. Thus consistency checks of multiple viewpoint specifications are supported.

1 Introduction

In the reference model for open distributed processing RM–ODP ([ODP]) viewpoints are introduced as one of the main architectural features that support the structured specification of components in an open distributed system. Viewpoint specifications allow to separate concerns within a large specification in that only one aspect of a component is specified, independently of the other ones. RM–ODP introduces five viewpoints: enterprise, information, computation, engineering, and technology. In this paper mainly the information and the computational viewpoints are addressed. In the information viewpoint the information model of a component is specified. According to RM–ODP a specification language for this viewpoint must offer concepts for the specification of *static*, *invariant*, and *dynamic* schemata. Static schemata describe single fixed states of information

* This work has partially been supported by the EEC TMR network GETGRATS (General Theory of Graph Transformation Systems).

objects, as for instance initial states. Invariant schemata describe the state invariants, that is, properties that always hold. State transitions, including the creation and deletion of objects, are described by the dynamic schemata. In the computational viewpoint the functional decomposition of the system into objects that interact at interfaces is specified.

ODP also strongly supports the use of formal specification techniques. Z and LOTOS for instance are suggested as formal specification languages for the information and computational viewpoints respectively. Multiple viewpoint modelling immediately prompts the question for the consistency of specifications of the different aspects of one and the same system. In general there are two approaches to this question. In the first one a most general framework is fixed as the common reference, and each viewpoint specification has to be embedded into this framework. The second approach is more local, in that for each pair of specification techniques a translation is defined that allows to interpret one viewpoint specification in terms of the other one. In both cases embedding respectively translation must be defined on the semantical level, such that models of different viewpoint specifications can be compared in order to check their consistency.

The approach presented in this paper lies between these two alternatives. First a very general formal semantical framework is presented that covers information and computation modelling. In particular it meets the requirements of an information viewpoint specification language, but it goes beyond it in that also processual control flow can be modelled. Then a translation of models of this framework into labelled transition systems is introduced, that allows the comparison and consistency check with process models. Labelled transition systems are considered here as the general semantical interface of process calculi, respectively the computational viewpoint.

The semantical models of the specification framework presented here — algebra transformation systems — are transition systems with internally structured states and transitions. More precisely: States are labelled by partial algebras of a common signature as their internal data states, and transitions between states are given by the nondeterministic application of rules that specify how such partial algebras are transformed. The common signature of the states defines the local language and the name space for the component. It is also used to specify the state invariants, and the static and dynamic schemata. Transitions are labelled by a method name and parameters that indicate how the method has been applied to obtain the state transformation.

The consistency of an algebra transformation system with a process model can be checked via a translation of algebra transformation systems to labelled transition systems. Since the states of the transformation systems have an internal structure it does not suffice to simply restrict to its underlying transition system by forgetting the state labels. Tests given by designated observations of the internal structure of the states allow to distinguish different transformation systems according to relevant properties of their information contents.

This translation between algebra transformation systems and labelled transition systems relates semantical models of different paradigms. Algebra transformation systems belong to the general model theoretic paradigm, formalized as *institutions* (see [GB92]). That means, there are *signatures* that determine the local language of a specification and the shape of its models, *model categories* as the admissible interpretations of the syntactical entities given by a signature, and *sentences* that describe the properties of models. In particular, the models are given independently of the sentences, that they may either satisfy or not. Altogether institutions/general model theories are a *descriptive* specification framework. In contrast with that, process specification calculi are used to *construct* models from the description of their properties. That is, given a language and expressions for the behaviour of a process in that language, a *transition system* can be constructed as the operational model of the process concerned. Further abstractions, like bisimulation, then yield the desired model of the relevant behaviour of the process.

One main concern of this paper is the translation between these paradigms: the descriptive paradigm for dynamic systems represented by algebra transformation systems, and the constructive (or operational) paradigm represented by labelled transition systems as operational models of process specifications. Specific specification languages or techniques are not considered here. However, discussing general concepts should enhance the application to concrete instances.

The paper is organized as follows. In the following section syntax and semantics of algebra transformation systems are introduced, defining the semantical framework. As running example a simplified version of the class of linked lists is presented. The logical part, that is, the description of the dynamic behaviour by formulae is discussed in section 3. The formulae are rules that describe the relation between preceding states as pre and post conditions. In section 4 the translation to labelled transition systems via observations is discussed. In the conclusion a summary of the approach is given and a comparison with related works.

2 Algebra Transformation Systems

Algebra transformation systems are transition systems whose states and transitions are labelled, thereby giving an internal structure to states and transitions that carries semantic information. The internal structure of the states is given by partial algebras to a common specification. Partial algebras yield a very general and powerful framework. First, they comprise first order structures since predicates can be expressed as partial functions to a singleton set. But they are much more adequate than first order structures, because partial functions are included as first class citizens. This latter aspect is particularly important in the context of dynamic systems. Partial algebras allow the separation of signatures and their interpretation in a much more flexible way than structures with total functions. The declaration and the instantiation of an entity can be separated as in programming languages. For example, a constant $c :\to s$ in the signature (its declaration) need not be interpreted in all states. Its instantiation is then modelled by the first state in a run in which c can be evaluated, i.e. in which c is bound to a value. This feature, together with the conditional equations as axioms, allows to specify the invariants that include dynamic information about the states. The example below will show how conditions are used to specify within one set of axioms properties with different interpretations in the different states.

Signatures for partial algebras are usual algebraic signatures $SIG = (S, OP)$, given by a set S of sort names and a family $OP = (OP_{w,s})_{w \in S^*, s \in S}$ of operation symbols indexed by their arities. As usual an operation symbol $op \in OP_{w,s}$ is denoted $op : w \to s$. The semantical difference to total algebras is that an operation symbol $op : w \to s$ is interpreted as a *partial* function $op^A : A_{s_1} \times \cdots \times A_{s_n} \twoheadrightarrow A_s$ (if $w = s_1 \ldots s_n$). An equation $t = t'$ is satisfied by a partial algebra A if both terms t and t' can be evaluated (are defined) in A, and yield the same value. Satisfaction of a conditional equation $r = r' \Rightarrow t = t'$ is defined as usual: whenever the premise $r = r'$ is satisfied, then also the conclusion $t = t'$ must be satisfied. The interpretation of equations also yields the definedness predicate for terms t, denoted $t\downarrow$, given by $t\downarrow$ iff $t = t$. A signature SIG with a set CE of conditional equations is called a partial equational specification $SPEC$. Homomorphisms of partial $SPEC$–algebras are given by families of total functions that preserve the domains of definition of the operations, and operation application. The so defined model category is denoted $PAlg(SPEC)$.

The *methods* of the system under consideration, that yield the state changes, are declared in a method signature, as names with arities, similar to operation symbols. However, methods do not have an output sort. Partial equational specifications and method signatures yield the signatures of algebra transformation systems.

Definition 1 (Transformation Signature). A *transformation signature* $T\Sigma = (SPEC, M)$ is given by a partial equational specification $SPEC = (S, OP, CE)$ and a *method signature* $M = (M_w)_{w \in S^*}$. A method name $m \in M_w$ is denoted $m : w$ for short.

As example consider the following transformation signature of a simplified version of the class of linked lists of integers. Linkable cells are introduced as static data type, that is, there is a total constructor $[\_, \_] : int, point(cell) \to cell$ that returns a cell for each integer and cell pointer, and corresponding projections *left* and *right*. The actual list is then determined in each state by the actual contents of the pointers. There are three designated pointers to handle the actual list: *fst* is used to point to its first cell, *act* to the actual cell, representing a list cursor, and *pos* is a pointer to natural numbers that holds the actual position of the cursor.

The dynamic states are determined by partial dereferencing functions $!_{cell} : point(cell) \twoheadrightarrow cell$ and $!_{nat} : point(nat) \twoheadrightarrow nat$ for pointers to cells and natural numbers respectively. That is, $!_{cell}$ and $!_{nat}$ are environments for pointers, defining their contents.

There are further *dependent functions* whose values depend on the actual definitions of the dereferencing functions, that is, on the actual contents of the pointers. They are specified completely in the axioms part. There also the state invariants are specified, as for instance the fixed relation between *fst*, *act*, and *pos*. Thus the invariant schemata of the ODP–information viewpoint are covered.

A method name *move* is introduced with its parameter declaration $i : nat$, for a method that moves the list cursor to the i'th list position. I.e. $move \in M_{nat}$ and $M_w = \emptyset$ for all $w \neq nat$. Partial equational specifications for natural numbers and integers are supposed to be given.

```
linked--list = nat + int +
  sorts  cell, point(cell), point(nat)
  opns   [_,_] : int, point(cell) → cell
         left : cell → int
         right : cell → point(cell)
         fst, act : → point(cell)
         pos : → point(nat)
         ! : point(cell) → cell
         ! : point(nat) → nat
  dependent functions
         get : → int
             get the actual list element

         next_cell : cell → cell
             returns the right neighbour, if it exists

         nth_cell : nat, cell → cell
             returns the n'th right neighbour, if it exists

             ⋮
  axms   invariants
         !fst↓ ∧ !act↓ ∧ !pos↓ ⇒ nth_cell(!pos,!fst) = !act

             if all three pointers are non void then the (!pos)'th right
             neighbour of !fst is !act

             ⋮
  specification of the dependent functions
         !act↓ ⇒ get = left(!act)
         for all p:point(cell), for all n:nat :
         !p↓ ⇒ next_cell([n,p]) = !p
             ⋮
  meths  move(i:nat)
end linked--list
```

The labels of the state transitions in an algebra transformation system are given by method expressions $m(a)$, given by a method name $m \in M_w$ and a list of parameters $a = \langle a_1, \ldots, a_n \rangle \in A_{s_1} \times \cdots \times A_{s_n}$ (if $w = s_1 \ldots s_n$) from a given partial algebra A.

Definition 2 (Method Expressions). Given a transformation signature $T\Sigma = (SPEC, M)$ the set $ME_{T\Sigma}$ of *method expressions* is defined by

$$ME_{T\Sigma} = \bigcup_{A \in |PAlg(SPEC)|} M(A) ,$$

where the components $M(A)$, for $A \in PAlg(SPEC)$, are defined by

$$M(A) = \{m(a) | m \in M_w, a \in A_w\} .$$

The extension of a $SPEC$–homomorphism $h : A \to B$ to method expressions, also denoted $h : M(A) \to M(B)$, is defined by $h(m(a)) = m(h(a))$ for all $m(a) \in M(A)$.

Now an algebra transformation system is given by two layers. The first one is a *transition graph* that models the control flow. It is given by a set $\mathcal{S}$ of control states as nodes and a set $\mathcal{T}$ of state transitions as edges, with functions $src, tar : \mathcal{T} \to \mathcal{S}$ that assign source and target states to the transitions. The second layer is given by the internal data states associated with the control states, given by partial $SPEC$–algebras, and the transformation information associated to the transitions, given by method expressions. Data states and transformation information are formally given as labels of control states and transitions.

Definition 3 (Algebra Transformation System). Let $T\Sigma = (SPEC, M)$ be a transformation signature. A *$T\Sigma$–transformation system* $\mathcal{A} = (TG_{\mathcal{A}}, lab_{\mathcal{A}})$ is given by a transition graph

$$TG_{\mathcal{A}} = (\mathcal{S}, \mathcal{T}, src, tar) \text{ with } src, tar : \mathcal{T} \to \mathcal{S}$$

and a pair of functions

$$lab_{\mathcal{A}} = (lab_{\mathcal{S}} : \mathcal{S} \to |PAlg(SPEC)|,\ lab_{\mathcal{T}} : \mathcal{T} \to ME_{T\Sigma})$$

such that

$$lab_{\mathcal{T}}(l) \in M(lab_{\mathcal{S}}(src(l))) \text{ for all } l \in \mathcal{T} ,$$

i.e. the parameters are always taken from the actual state.

A method expression $m(a)$ that occurs as a label of a transition from a state s to a state t models that fact, that the method m has been applied in state s, and that is has been given the parameters $a = \langle a_1, \ldots, a_n \rangle$. Thus it is clear, that the values a_i must be present in the actual data state of s.

The labelling of the transitions has only been introduced in order to distinguish more clearly between the transition graph, which is independent of the transition signature, and all the syntactic information given in the signature.

Remark 4. As usual, a transition $l \in \mathcal{T}$ with $src(l) = s$ and $tar(l) = t$ will be denoted $l : s \to t$. Moreover both $l \in \mathcal{T}$ and the triple $l : s \to t$ will be called transition. Correspondingly $\mathcal{T}$ is called the *transition relation*. The labels of states and transitions will also be indicated by capital letters, i.e. $lab_{\mathcal{S}}(s) = S$ and $lab_{\mathcal{T}}(l) = L$ for states $s \in \mathcal{S}$ and transitions $l \in \mathcal{T}$. The condition that parameters are always taken from the actual state in the definition above thus reads $L \in M(S)$ for all $l : s \to t \in \mathcal{T}$.

The external transition relation $\mathcal{T}$ on the states induces on each pair (S, T) of data states $S = lab_{\mathcal{S}}(s)$, $T = lab_{\mathcal{S}}(t)$, related by a transition $tr = (l : s \to t)$, an internal relation $track_{tr} \subseteq |S| \times |T|$ on the carrier sets, given by the evaluation of ground terms in both states, i.e.

$$track_{tr} = \{(N^S, N^T) \mid N \in T_{SIG}\}\ .$$

These relations keep track of the development of elements through state changes. Thus the signature yields the global constant identities that are necessary to reason about dynamic state changes. *Partial tracking functions*, not necessarily obtained in this way, are fundamental in the D–oids approach (see [AZ95]).

Since the label functions need not be injective control information need not be encoded in the data states. If for example t and t' are different control states with same data state $T = lab_{\mathcal{S}}(t) = lab_{\mathcal{S}}(t')$, the system behaves differently in the states t and t', although this cannot be detected in the data state alone. In pure rule based approaches, such as graph grammars, control information has to be encoded in the (data) state, unless another technique is added to capture this information. Note furthermore that the state invariants given in the data type specification hold in each state by definition of a transformation model. In the case of a construction of a transformation model from a specification (see [Gro96]) this is assured by the free construction of *SPEC*–algebras for the states. It remains to be shown then, as a proof obligation, that the so constructed states are consistent, for instance, with some contained or predefined data types.

Finally note that the *logical order*, induced by the direction of the labelling function from the transition graph to the state space and method expressions respectively, does not imply the same *methodological order*. In fact, the transition graph is usually obtained by adding the control flow information to representations of the data states. The labelling functions are then projections that remove the additional information. (See the example below.)

To define transformation systems for the *linked–list* signature given above we first fix a partial algebra that models the parts of the system that are considered as unchangeable, that is, natural numbers, integers, and the cells. For that purpose let *static* be the specification given by *linked–list* except the dereferencing functions $!_{cell}$ and $!_{nat}$, the dependent functions *get*, *next_cell*,..., and their axioms. Further let A be a partial *static*–algebra whose *nat* and *int* parts are the natural numbers and the integers respectively, and whose *cell*–part is the cartesian product of A_{int} and $A_{point(cell)}$. Now each pair of partial functions $e = (e_{cell}, e_{nat})$ with $e_{cell} : A_{point(cell)} \rightharpoonup A_{cell}$ and $e_{nat} : A_{point(nat)} \rightharpoonup A_{nat}$ defines a free extension of A to a *linked–list* algebra $S = A[e]$ with $S|_{static} = A$,

$!^S_{cell} = e_{cell}$, and $!^S_{nat} = e_{nat}$. S being a free extension means in this case, that for each partial *linked–list* algebra A and *static*–homomorphism, $h : A \to B|_{static}$ with $!^B_s(h(p)) = h(e_s(p))$ ($p \in A_{point(s)}, s \in \{nat, cell\}$) there is a unique *linked–list* homomorphism $h^* : S \to B$ with $h^*|_{static} = h$. (The universal morphism $\eta : A \to S|_{static}$ is the identity in this example.) For sake of brevity only pairs of partial functions $e = (e_{cell}, e_{nat})$ with $A_{fst}, A_{act} \in dom(e_{cell})$ and $A_{pos} \in dom(e_{nat})$ will be considered, which could also be added as an axiom to the specification. These *environments* e are used to represent control states, whose internal data state is given by $S = A[e]$.

At this point it can be seen how dynamically changing properties are specified with conditional equations for partial functions. Consider the *constant get:*$\to$ *int*. It is defined by the axiom $!act \downarrow \Rightarrow get = left(!act)$. Thus if *act* is void *get* is also undefined. If *act* has contents *get* is the integer entry in the cell *act* points to. Thus in different states the constant term *get* evaluates to different elements of the carrier set A_{int} automatically.

With these preliminaries a *linked–list*–transformation system $\mathcal{A} = (TG_{\mathcal{A}}, lab_{\mathcal{A}})$ can now be defined as follows.

control states $\quad \mathcal{S} = \{ e = (e_{cell}, e_{nat}) \mid$

$$e_{cell} : A_{point(cell)} \nrightarrow A_{cell}; fst^A, act^A \in dom(e_{cell})$$
$$e_{nat} : A_{point(nat)} \nrightarrow A_{nat}; pos^A \in dom(e_{nat})\}$$

data states $\quad lab_{\mathcal{S}}(e) = A[e]$

transitions $\quad \mathcal{T} = \{(i : e \to e') \mid i \in \mathbb{N}, e_{nat}(pos^A) \neq i \}$

where $e' = (e'_{cell}, e'_{nat})$ is given by

$$e'_{cell} = e_{cell}[act^A \mapsto e_{cell}(fst^A)]$$
$$e'_{nat} = e_{nat}[pos^A \mapsto 1]$$

if $e_{nat}(pos^A) > i$, and

$$e'_{cell} = e_{cell}[act^A \mapsto next\_cell^{A[e]}(e_{cell}(act^A))]$$
$$e'_{nat} = e_{nat}[pos^A \mapsto e_{nat}(pos^A) + 1]$$

if $e_{nat}(pos^A) < i$.

method expr's $\quad lab_{\mathcal{T}}(i : e \to e') = move(i)$

Thus with a transition $i : e \to e'$ the transformation of data states $move(i) : A[e] \Rightarrow A[e']$ is associated.

$\mathcal{A}$ models the usual algorithm of the move method. If the actual position is to the right of i it moves to the first position, if it is to the left of i it moves one step to the right, and if i is reached the procedure is finished, i.e. nothing happens any more.

The second model $\mathcal{A}'$ behaves similarly, however, while traversing the list it destroys all its entries. The state space and the labels are the same as the ones of $\mathcal{A}$. Only the transitions are defined differently.

$\mathcal{T}' = \{(i : e \to e') \mid i \in \mathbb{N}, e_{nat}(pos^A) \neq i \}$

where $e' = (e'_{cell}, e'_{nat})$ is given by

$e'_{cell} = e_{cell}[act^A \mapsto e_{cell}(fst^A), p_1 \mapsto [0, right^A(e_{cell}(p_1))]]$
$e'_{nat} = e_{nat}[pos^A \mapsto 1]$

if $e_{nat}(pos^A) > i$, and

$e'_{cell} = e_{cell}[act^A \mapsto next\_cell^{A[e]}(e_{cell}(act^A)), p_1 \mapsto [0, right^A(e_{cell}(p_1))]]$
$e'_{nat} = e_{nat}[pos^A \mapsto e_{nat}(pos^A) + 1]$

if $e_{nat}(pos^A) < i$; where in both cases $p_1 \neq act^A \in A_{point(cell)}$ with $e_{cell}(p_1) = e_{cell}(act^A)$.

The following figure shows a part of this (mis)behaviour.

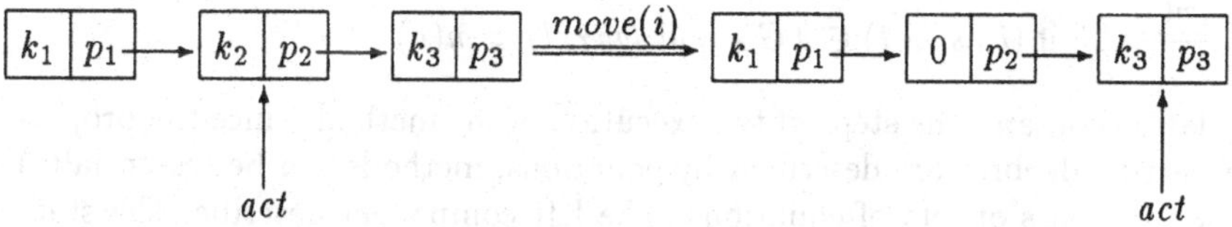

We will see below how this unwanted side effect can be detected.

To conclude this section the complete model categories for transformation signatures are defined. A morphism of algebra transformation systems is given by a graph homomorphism h_{TG} of their transition graphs — i.e. the control flow is preserved — and for each control state s a *SPEC*–homomorphism $h_s : S \rightarrow S'$ of the associated data state S to the data state S' of the image $s' = h_{TG}(s)$. These morphisms must be compatible with the transformation information, i.e. method expressions must be preserved.

Definition 5 (Transformation System Morphism). Let $T\Sigma = (SPEC, M)$ be a transformation signature, and $\mathcal{A} = (TG_{\mathcal{A}}, lab_{\mathcal{A}})$ and $\mathcal{A}' = (TG_{\mathcal{A}'}, lab_{\mathcal{A}'})$ be $T\Sigma$–transformation systems. A $T\Sigma$*–morphism*

$$h = (h_{TG}, (h_s)_{s \in S}) : \mathcal{A} \rightarrow \mathcal{A}'$$

is given by a graph homomorphism

$$h_{TG} : TG_{\mathcal{A}} \rightarrow TG_{\mathcal{A}'},$$

and a family of *SPEC*–homomorphisms

$$(h_s : S \rightarrow S')_{s \in S} \quad \text{with } s' = h_{TG}(s),$$

such that

$$h_s(L) = L' \quad \text{with } l' = h_{TG}(l)$$

for all $l : s \rightarrow t \in T$.

$T\Sigma$–transformation systems and morphisms define the category $\mathbf{ATS(T\Sigma)}$.

Recall that the last condition in the definition means: if $L = m(a)$ then $L' = m(h(a))$.

Model categories for transformation signatures constitute the first part of an institution of algebra transformation systems. However, signature morphisms and the induced forgetful functors on model categories, that would complete the model part of an institution, are not considered in this paper.

3 Specification of Dynamic Properties

In this section the logical part of algebra transformation systems is introduced. That means, the *sentences* for a given transformation signature $T\Sigma$ and *satisfaction* of sentences by $T\Sigma$-transformation systems are defined.

Each method expression $m(a)$ of a given transformation signature $T\Sigma = (SPEC, M)$ induces an interpretation in a $T\Sigma$–transformation system $\mathcal{A} = (TG_{\mathcal{A}}, lab_{\mathcal{A}})$ by a binary relation on the partial $SPEC$–algebras being its data states:

$$S \overset{m(a)}{\Longrightarrow} T \text{ if } (l : s \to t) \in TG_{\mathcal{A}} \text{ and } lab_T(l) = m(a) .$$

This relation contains the steps of the execution of the method. Since the properties of partial algebras are described by equations, methods can be (essentially) described by pairs of sets of equations. The left component describes the state before the actual step, the right one the successor state. Such *transformation rules* are extended by conditions that specify when a method can be applied. In the following definition the notation X_w with $w = s_1 \ldots s_n \in S^*$ is used for a list of variable declarations $X_w = (x_1 : s_1, \ldots, x_n : s_n)$, and analogously P_v $(v = s'_1 \ldots s'_k)$ for a parameter declaration $P_v = (y_1 : s'_1, \ldots, y_k : s'_k)$. Parameters are here variables that are used in method expressions, and thereby bound. Furthermore $Eqns_{SPEC}(Z_u)$ denotes the set of all equations w.r.t. the signature of $SPEC$ and variables or parameters from Z_u.

Definition 6 (Transformation Specification). A *transformation rule* r w.r.t. a partial equational specification $SPEC$ and variables and parameters (X_w, P_v) is defined by

$$r = (\underline{\text{vars}}\ X_w : \underline{\text{replace}}\ \mathit{lft}\ \underline{\text{by}}\ \mathit{rght}\ \underline{\text{if}}\ \mathit{cond})$$

where $\mathit{lft}, \mathit{rght}, \mathit{cond} \subseteq Eqns_{SPEC}(X_w \cup P_v)$. The set of all rules is denoted by $Rules_{SPEC}(X_w, P_v)$.
A *transformation specification* $TS = (T\Sigma, St, Tr)$ is given by

- a transformation signature $T\Sigma = (SPEC, M)$,
- a set St of sets of equations, the *initial state specifications*, and
- a family of functions $Tr = (Tr_v : M_v \to \mathcal{P}(Rules_{SPEC}(X_w, P_v)))_{v \in S^*}$, the *transformation specification.*

For the *linked–list* example we specify one initial state, with actual list (1,2,3) and cursor position 1. That is, St has one element

$$\begin{aligned} st = \{\ & \mathit{left}(!\mathit{fst}) = 1, \mathit{left}(\mathit{nth\_cell}(1, !\mathit{fst})) = 2, \\ & \mathit{left}(\mathit{nth\_cell}(2, !\mathit{fst})) = 3, !\mathit{act} = !\mathit{fst}\} . \end{aligned}$$

The *move* method is specified by two rules, corresponding to whether the actual position is to the left or to the right of the parameter i. The variables that appear in the rules, beyond the method parameter i, are used to relate the states before and after the execution steps.

```
meth  move(i:nat)
rules vars     n : nat, c : cell, c' : cell
      replace  !act = c',    !pos = n
      by       !act = c,     !pos = 1
      if       !pos > i,     !fst = c

      vars     k : int, n : nat, p : point(cell), c : cell
      replace  !act = [k,p], !pos = n
      by       !act = c,     !pos = n + 1
      if       !pos < i,     !p = c
```

Written as conditional parallel assignments these rules would look like

$$act, pos := !fst, 1 \quad \text{if } !pos > i$$
$$act, pos := next\_cell(!act), !pos + 1 \quad \text{if } !pos < i$$

and in a concrete specification language for such pointer models a similar notation would be chosen. The more algebraic notation with equations has been chosen here in order not to restrict the approach to pointer models. In a natural extension of the example for instance, cells would also be modelled as dynamic entities. That is, the constructor $[\_,\_] : int, point(cell) \rightarrow cell$ would be modelled as a partial changeable function. Then individual cells that constitute the actual list in a state can also be created or deleted. Furthermore creation and deletion of pointers could be specified using equations of the form $p = p$ in the right or left hand side of a rule. (Recall that $p = p$, abbreviated $p\downarrow$, means p is defined.) The rule

<u>replace</u> $\emptyset$ <u>by</u> $p\downarrow$

specifies the creation of p, the rule

<u>replace</u> $p\downarrow$ <u>by</u> $\emptyset$

its deletion. Note, however, that in an algebraic framework negative facts can not be forced for all models. Whereas p must exist after the application of the creation rule, the deletion rule only states that p need not exist afterwards. On the other hand, *initial* interpretation of rules, similar to the initial interpretation of equations, guarantees that p does not exist if it need not exist. (See [Gro96] for the existence and construction of initial models.)

In general an algebra transformation system $\mathcal{A}$ satisfies a transformation rule r for a method m with parameters a if the following condition holds. For each state s of $\mathcal{S}$ whose data state S satisfies the condition and the left hand side of r, there is a transition $l : s \rightarrow t$ with method expression is $m(a)$, and each data state T of a state t, that can be reached from s via a transition $l : s \rightarrow t$ labelled with $m(a)$, satisfies the right hand side of r. Thus left and right hand side are pre and post conditions for a method. Note that in this general setting there is no formal semantical difference between application conditions and preconditions. This appears only, when rules are interpreted strictly as in the initial model, where the left hand side is actually deleted, as opposed to the condition. However, the example above also shows, how conditions and left hand sides are used differently in the specification.

To satisfy the whole specification $\mathcal{A}$ must satisfy all rules for all methods and appropriate parameters, and it must contain a state for each set of equations in the initial state specification.

In the following definition instances of rules are required. For a rule r with variable declaration X_w and parameter declaration P_v, $r[t/x, p/y]$ denotes the rule that is obtained by instantiating the variables $x = \langle x_1, \ldots, x_n \rangle$ by the terms $t = \langle t_1, \ldots, t_n \rangle$ and the parameters $y = \langle y_1, \ldots, y_k \rangle$ by terms $p = \langle p_1, \ldots, p_k \rangle$ simultaneously in the left and right hand side and in the condition of r.

Definition 7 (Satisfaction). Let $TS = (T\Sigma, St, Tr)$ be a transformation specification, $\mathcal{A}$ a $T\Sigma$–transformation system. $\mathcal{A}$ satisfies TS, if

- For each initial state specification $st \in St$ there is a state $s \in \mathcal{S}$ such that $S = lab_{\mathcal{S}}(s)$ satisfies st.
- For each method $m : v$, each ground instance $r[t/x, p/y]$ of a rule $r \in Tr(m)$, and each state $s \in \mathcal{S}$ holds:
  If $S = lab_{\mathcal{S}}(s)$ satisfies $cond[t/x, p/y]$ and $lft[t/x, p/y]$,
  then there is a transition $l : s \to t$ such that $lab_{\mathcal{T}}(l) = m(a)$,
  and for each transition $l : s \to t$ with $lab_{\mathcal{T}}(l) = m(a)$ the data state $T = lab_{\mathcal{S}}(t)$ satisfies $rght[t/x, p/y]$.

The full subcategory of $T\Sigma$–transformation systems that satisfy TS is denoted **ATS(TS)** .

Both *linked–list* transformation systems $\mathcal{A}$ and $\mathcal{A}'$ defined above satisfy the given rules for the method *move*. Consider for instance the second rule. The actual position must be to the left of i ($!pos < i$). Then there must be a pointer p which is not void ($!p = c$) and, specified in the left hand side, is the pointer in the actual cell ($!act = [k, p]$). Thus c is the right neighbour of the actual cell. Then *act* is updated to c, and in parallel, *pos* is incremented by one. Both models satisfy this requirement (and there is obviously also a formal proof of this property), the second rule is treated analogously.

4 Observation Systems

The transformation rules for the specification of the *move* method obviously do not suffice to exclude bad behaviours like the model $\mathcal{A}'$ that destroys all the list entries. However, regarding only the underlying labelled transition systems, i.e. not inspecting the states, $\mathcal{A}$ and $\mathcal{A}'$ cannot be distinguished.

In this section observation systems are defined that allow state inspections to distinguish transformation models with different information states. Since properties of algebras are expressed by equations these are also used as labels, and an equation can be observed if and only if it holds in the actual state. To obtain a functorial translation observations are restricted, however, to ground equations. Since checking an equation does not change the state, an observation is always of the form $(t = t') : s \to s$. Admissible inspections can be restricted to parts of the states by a corresponding restriction of the signature, thus implementation aspects can be hidden as well. To obtain the ***observation system*** the observations are added to the underlying transition graph of an algebra transformation system.

Definition 8 (Observation System). Let $TS = (T\Sigma, St, Tr)$ be a transformation specification with $T\Sigma = ((S, OP, CE), M)$, *SIG0* a subsignature of *(S,OP)*, and $\mathcal{A} = (TG_{\mathcal{A}}, lab_{\mathcal{A}})$ a TS–transformation system, with $TG_{\mathcal{A}} = (\mathcal{S}, \mathcal{T}, src, tar)$. The *observation system of* $\mathcal{A}$ w.r.t *SIG0*, denoted by $Obs_{SIG0}(\mathcal{A})$, is the labelled transition system $Obs_{SIG0}(\mathcal{A}) = (St, L, \xrightarrow[Obs]{})$ defined by

- $St = \mathcal{S}$
- $L = ME_{T\Sigma} \cup Eqns_{SIG0}$
- $s \xrightarrow[Obs]{e} t$ if $\begin{cases} l : s \to t \in \mathcal{T} \text{ and } e = lab_{\mathcal{T}}(l), \text{ or} \\ s = t \text{ and } e \in Eqns_{SIG0} \text{ and } \mathcal{S} \models e \end{cases}$

To check whether the actual list in a *linked-list* model is preserved while moving the cursor, equations of the form *left(nth_cell(n,!fst))=k* will be tested in each state looking for the integer entries in the cells to the right of *!fst*. The observation signature is thus

```
test-sig = sig(int) +
  sorts  cell, point(cell)
  opns   left : cell → int
         fst : → point(cell)
         ! : point(cell) → cell
         nth_cell : nat, cell → cell
end test--sig
```

In the observation systems of the *linked-list* transformation models we obtain thus the following two kinds of transitions. The first ones are labelled *move(i)* for some integer i and indicate that the *move*–method is running. The second ones are the observations of the data states that check whether the actual state of the list is still (1,2,3). Now the test equations *left(!fst)=1*, *left(nth_cell(1,!fst))=2*, and *left(nth_cell(2,!fst))=3* can be used to distinguish the observation systems of $\mathcal{A}$ and $\mathcal{A}'$. Since $Obs_{test-sig}(\mathcal{A}')$ fails after the first step of *move(3)* for at least one of the tests, whereas $Obs_{test-sig}(\mathcal{A})$ always passes all tests, the observation models are different. Obviously they are also not bisimilar.

Observation systems are labelled transition systems. However, the label sets are local and have an internal structure. These aspects have to be taken into account when the categories of observation systems are defined. Moreover, up to now only observation systems generated by algebra transformation systems have been considered. To obtain an independent category we also give a complete definition of observation systems.

Definition 9 (Category Obs(TS,SIG0)). Let $TS = (T\Sigma, St, Tr)$ be a transformation specification, with $T\Sigma = (SPEC, M)$, $SIG0$ a subsignature of the signature of $SPEC$. An *observation system* $Obs = (St, L, \xrightarrow[Obs]{})$ w.r.t. TS and $SIG0$ is a labelled transition system whose labels are method expressions w.r.t. $T\Sigma$ or equations w.r.t. $SIG0$, i.e. $L \subseteq ME_{T\Sigma} \cup Eqns_{SIG0}$. A *morphism of observation systems*

$$f = (f_{St}, f_L) : (St, L, \xrightarrow[Obs]{}) \to (St', L', \xrightarrow[Obs']{})$$

is given by functions $f_{St} : St \to St'$ and $f_L : L \to L'$ such that

- $f_L(m(a)) = m(a')$ and $f_L(t = t') = (t = t')$
  for all labels $l = m(a)$ and $l = (t = t')$ respectively, and
- $S \xrightarrow[Obs]{l} T$ implies $f_{St}(S) \xrightarrow[Obs']{f_L(l)} f_{St}(T)$,

i.e. f is compatible with the label structure and the transitions.

The category **Obs(TS,SIG0)** is defined by *(TS,SIG0)*-observation systems and their morphisms.

The morphisms in the categories express the relevant relationships between the objects (systems) under consideration. Any construction, like the construction of observation systems for algebra transformation systems, must be compatible with the morphisms in order to preserve their basic interconnections. Thus the translation should be functorial.

Proposition 10. *Let $TS = (T\Sigma, St, Tr)$ be a transformation specification, with $T\Sigma = (SPEC, M)$, $SIG0$ a subsignature of the signature of $SPEC$. Furthermore let $h = (h_{TG}, (h_s)_{s\in S}) : \mathcal{A} \to \mathcal{A}'$ be a morphism of algebra transformation systems, then $(h_{St}, h_L) : Obs_{SIG0}(\mathcal{A}) \to Obs_{SIG0}(\mathcal{A}')$, defined by $h_{St} = h_{TG}|_S$ and $h_L(m(a)) = m(h_s(a))$ resp. $h_L(t = t') = (t = t')$, is a morphism of observation systems. Moreover, this mapping of morphisms extends Obs_{SIG0} to a functor Obs_{SIG0} : $\mathbf{ATS(TS)} \to \mathbf{Obs(TS, SIG0)}$.*

An important consequence of this proposition is that isomorphisms are preserved. That means, transformation systems that are identical up to bijective renamings of their carrier sets yield observation systems that are also identical up to bijective renaming. Furthermore composition of models in a categorical framework would also be expressed via morphisms, diagrams, and colimits. Compatibility with such compositions requires at least functoriality of the construction. Further conditions, like the preservation of colimits, may be imposed to ensure compatibility. These questions however are outside the scope of this paper.

5 Conclusion

In this paper I have presented a formal semantical framework for the specification of dynamically evolving systems that allows to model both computational and information aspects of a system. In particular it satisfies the requirements of an information viewpoint specification formalism according to RM–ODP. The information states are modelled as partial algebras to a given partial equational specification that contains the invariant schemata, formulated as conditional equations. Static schemata are sets of equations that describe particular partial algebras. The state transitions are given as labelled transitions on algebras that also induce internal relationships between the states. The dynamic schemata for the specification of the transitions are given by rules that consist of sets of equations describing the states before and after one step, and the application conditions of a method.

The presented framework belongs to the *algebras–as–states* approach to the specification of dynamic systems, whose foremost representative are the abstract state machines, formerly called evolving algebras (see [Gur94]). A formalisation of this approach has been presented in [DG94], where however algebraic specifications are considered as algebraic programs, and consistency conditions become part of the definition of the semantics. A very general abstract mathematical model within this approach, D–oids, has been presented in [AZ95]. It introduces a model theory for dynamic systems, parameterized by the underlying static framework for values and state algebras. The main difference of the approach presented here is that a loose semantics is defined for dynamic systems. That means, rules do not (only) specify how to transform states, but are (also) interpreted as pre/post–conditions that may or may not be satisfied in a model. Loose semantics is particularly important, if components and component composition are considered. In this case it cannot be assumed that the specification already contains the whole information, because other components may influence the behaviour of the component. Specification means, i.e. sentences and satisfaction for D–oids are introduced in [Zuc96]. There, however, methods are total functions, which is problematic for the modelling of non–deterministic systems, and identities are modelled by a tracking map, which might be in conflict with the data signature.

For the reasons mentioned in section 2 I have chosen partial algebras as data states. Formally this would not have been necessary, and it is easy to see how *institution independent* the approach is. For the definition of transformation systems (definition 3) data models are only required to have carrier sets to be able to choose the parameters from the actual state, i.e. a *concrete institution* is needed. Transformation rules (definition 6) can be composed of arbitrary sentences, and satisfaction is inherited from the institution in the same way as in definition 7. The only constraint comes up with the translation to observation systems: if it shall be functorial model morphisms are required to preserve satisfaction.

Beyond transformation systems and specifications their functorial translation to labelled transition systems has been discussed. The latter are considered as the general semantical interface of process calculi, which, in principle, cover the computational viewpoint specification requirements according to RM–ODP. The translation is obtained by the *observation* of an admissible part of the internal data structure of states via the side effect free evaluation of equations. This allows to turn the information contents of states into labels. Within labelled transition systems labels represent the only observable parts of a system, thus transformation systems with different (internal) information models can also be distinguished as observation systems.

To compare models of different specifications, for instance to check their consistency, this translation has of course to be combined with the transformation and refinement techniques available within each framework. In general a viewpoint specification is designed independently of the other ones, so the comparison with another viewpoint specification requires to transform the specification in such a way that the relevant parts become explicit. A function within a system for example can be specified as a function within the underlying partial algebra, or as a method (procedure) like the *move*–method in the example presented in this paper. In the first case only the input/output–relation of the function matters, in the latter case its implementation or dynamic behaviour is specified.

For the comparison with processes the other direction of refinement would have to be considered, i.e. the design of an algorithm for a given static function, that would then be realized as a process. Within the level of process calculi other transformation techniques are supported, and subsequent work will be devoted to the study of the relationship between these techniques. For the concrete specification techniques Z and LOTOS such transformations and refinements have been studied in detail in [BBDM96], where also a general framework for ODP–viewpoint consistency is introduced.

References

[AZ95] E. Astesiano and E. Zucca. D-oids: A model for dynamic data types. *Math. Struct. in Comp. Sci.*, 5(2):257–282, 1995.

[BBDM96] H. Bowman, E. Boiten, J. Derrick, and M.Steen. Viewpoint consistency in ODP, a general interpretation. In *Proc. FMOODS 96*, 1996.

[DG94] P. Dauchy and M.C. Gaudel. Algebraic specifications with implicit states. *Tech. Report, Univ. Paris Sud*, 1994.

[GB92] J. A. Goguen and R. M. Burstall. Institutions: Abstract Model Theory for Specification and Programming. *Journals of the ACM*, 39(1):95–146, January 1992.

[Gro96] M. Große-Rhode. First steps towards an institution of algebra replacement systems. Technical Report 96-44, Technische Universität Berlin, 1996.

[Gur94] Y. Gurevich. Evolving algebra 1993. In E. Börger, editor, *Specification and Validation Methods.* Oxford University Press, 1994.

[ODP] RM-ODP. Reference model of open distributed processing, part 1 – 4. ISO/IEC International Standard 10746, ITU-T Recommendation X.901 – X.904.

[Zuc96] E. Zucca. From static to dynamic abstract data–types. In W.Penczek and A. Szałas, editors, *Mathematical Foundations of Computer Science 1996*, volume 1113 of *Lecture Notes in Computer Science*, pages 579–590. Springer Verlag, 1996.

Open Maps as a Bridge between Algebraic Observational Equivalence and Bisimilarity*

Sławomir Lasota

Institute of Informatics, Warsaw University, Banacha 2, 02-097 Warszawa, Poland, phone: +48 22 658-31-65, fax: +48 22 658-31-64, e-mail: sl@mimuw.edu.pl

Abstract. We show that observational equivalences for standard, partial and regular algebras are bisimulation equivalences in the setting of *open maps*, proposed in [JNW93] as an abstract approach to behavioural equivalences of processes. The main advantage of the results is capturing models for sequential and concurrent systems in a uniform framework. In such an abstract setting we formulate the property of *determinism*, shared by all the algebras considered in this paper, and identify some interesting facts about bisimilarity in the deterministic case. All the results for standard, regular and partial algebras are obtained by the applications of general theorems proved in the paper, which we expect to be applicable also in other contexts.

1 Introduction

The concept of *a behavioural equivalence* is a fundamental notion in programming methodology, as it seems to capture appropriately the "black box" character of data abstraction. It plays also a central role in the process of stepwise refinement, where each step embodies some design decisions, under the requirement that behaviour must be preserved. This applies to the observational equivalence of algebras, being models for sequential programs and data types, as well as to bisimulation equivalence (bisimilarity) of processes, modelling concurrent or nondeterministic systems.[1]

Roughly speaking, two algebras are *observationally equivalent* with respect to a distinguished set of *observable sorts* if all computations yielding a result in an observable sort produce the same output in both. This idea goes back to [Mor68], [GGM76], and was widely studied mainly in the case of standard algebras [Rei81], [ST87], but also in some other frameworks, like regular algebras [BT95] and partial algebras [BT96].

Another approach to behavioural equivalence, successful in the case of concurrent processes, is the notion of *bisimilarity* [Mil89,Par81]. Recently a categorical generalisation of bisimulation was proposed, by means of spans of *open maps*

* This work was partially supported by KBN 8 T11C 018 11 grant.

[1] We reserve term *observational equivalence* for algebras, whereas *behavioural equivalence* will be understood more generally, including bisimilarity.

(open morphisms) [JNW93], enabling a uniform definition of bisimulation equivalence across a range of different models for parallel computations (see [CN95], [JNW93] for an overview). Open maps can be understood as arrows witnessing a bisimulation, and are defined abstractly to be those morphisms which satisfy a *computation-lifting property*, where a computation is modelled by a morphisms from any of distinguished *observation objects*.

Both observational equivalence and bisimilarity play a crucial role in formal models of sequential and concurrent systems, respectively. The undeniable analogies between these approaches are well recognised and much work was done in linking the two notions. This paper is a continuation of this stream of research, but the first one to consider abstract data types and processes commonly in the context of open maps. As some examples of related work we should mention [Gor95], where a characterisation of contextual equivalence of PCF terms was given by means of bisimulation, and [AO93], where authors proposed an approach to semantics of lazy λ-calculus, inspired by bisimulation. Moreover, in [Jac95], [Mal96] some links were observed between bisimulation and observational equivalence for hidden-sorted algebras, viewed as coalgebras. We do not consider here any other abstract depictions of bisimulation (like spans of coalgebra maps), leaving this for further research.

The most important corollary from results stated in this paper is that observational equivalences for some classes of algebras (standard, partial and regular algebras are considered) can be captured in the setting of open maps, and thus can be thought of as abstract bisimilarities.

In Section 3.1 we show that the observational equivalence of standard and regular algebras w.r.t. a set OBS of observable sorts is a bisimilarity induced by the class of observation objects consisting of free objects w.r.t. a forgetful functor $|\text{-}|_{OBS}$. More precisely, we show that open maps are precisely *observational morphisms* [BT96], i.e. bijective on observable sorts. To prove this we formulate an observation (Lemma 3.1 in Section 3 and Theorem 4.2 in Section 4) that open maps are transported via *multiadjunction [Die79]*, being a generalisation of adjunction where a whole family of "jointly" free objects is allowed. While for categories of standard and regular algebras usual adjunctions are sufficient, this does not extend to partial algebras, where the class of free objects is too small to induce the observational equivalence. In Section 4.2 the observational equivalence of partial algebras is proved to be a bisimilarity induced by union of all families of free objects of some multiadjunction. This is done in the category of partial algebras with *strong* partial homomorphisms, similarly as in [BT96]. Such a result is not quite satisfactory, as the choice of strong homomorphisms can not be easily justified. To do the same for partial algebras with all (weak) homomorphisms, we introduce the notion of *pre-adjunction*, being a further relaxation of requirements of multiadjunction. It turns out that every pre-adjunction gives rise to some multiadjunction, when certain necessary conditions are satisfied. We prove that bisimilarity induced by pre-free objects of a pre-adjunction and bisimilarity induced by free objects of the related multiadjunction coincide (Theorem 5.7 in Section 5). This allows us to characterise observational equivalence of partial algebras by means of open maps in the category with weak homomorphisms (in Section 5.1). The results concerning open maps in multiadjunction and pre-adjunction are formulated for an arbitrary category, and we expect that they are applicable in other frameworks than considered in this paper.

The surprising observation is that all bisimilarities studied in this paper (i.e. those induced by adjunction, multiadjunction or pre-adjunction) have a common property that each computation in a target of an open map lifts uniquely to a computation in its source. We call such a bisimilarity *deterministic* (cf. Section 4.1) and observe that it corresponds to the class of *etale maps* ([JM94]) in a topos. An interesting conclusion is that etale maps play the same role in models for sequential systems as open maps do for concurrent or nondeterministic ones.

We are also interested in how large the class of observation objects is necessary to induce the observational equivalence. For example, in the case of standard and regular algebras it turns out that free algebras over (at most) two-element sets are sufficient, i.e. observations by means of terms over two variables are powerful enough to distinguish non-equivalent algebras. This gives a useful tool in showing equivalence of algebras, as we only have to examine a small category of observation objects. The detailed study of observation subcategories for standard, regular and partial algebras are summarised in Corollaries 3.5, 3.6, 4.12 and 5.17. Moreover, in the case of deterministic bisimulation, the class of observation objects can be decreased by Lemmas 4.9 and 4.10.

This paper is a short version of [Las97a], where one can find the proofs, omitted here, as well as some minor details, lacking in the sequel.

2 Preliminaries

Standard algebras. Let $\Sigma = \langle S, F\rangle$ be a fixed S-sorted signature throughout this paper, unless stated otherwise. In subsequent sections we need usual definitions of (standard) Σ-algebra, Σ-homomorphism, congruence, quotient and subalgebra [BL70,GTW78]. We will also need notions of Σ-terms, satisfaction of an equation, and the value $t_{A[v]}$ of a term t in algebra A under a valuation $v : X \to |A|$. By $T_\Sigma(\_)$ we denote the left adjoint to the forgetful functor $|\_| : Alg(\Sigma) \to Set^S$, where $Alg(\Sigma)$ denotes the category of Σ-algebras with Σ-homomorphisms.

We will overload the notation $T_\Sigma(X)$ using it also for X from Set^R, $R \subseteq S$; $T_\Sigma(\_)$ will denote the left adjoint to any forgetful functor $|\_|_R : Alg(\Sigma) \to Set^R$, returning the carriers of sorts from R. Moreover, we will use the same notation $|\_|_R$ also for the forgetful functors from categories of partial and regular algebras.

Partial algebras. Partial Σ-algebras [BW82,BT96] are like standard algebras except that their operations can be partial functions. A (weak) partial Σ-homomorphism $h : A \to B$ is a S-sorted function $|h| : |A| \to |B|$ preserving definedness of operations, i.e. for each $f : s_1 \times \ldots \times s_n \to s$ in Σ and $a_1 \in |A|_{s_1}, \ldots,$ $a_n \in |A|_{s_n}$, if $f_A(a_1, \ldots, a_n)$ is defined then $f_B(|h|_{s_1}(a_1), \ldots, |h|_{s_n}(a_n))$ is defined and $|h|_s(f_A(a_1, \ldots, a_n)) = f_B(|h|_{s_1}(a_1), \ldots, |h|_{s_n}(a_n))$. Let $PAlg_w(\Sigma)$ denote the category of partial Σ-algebras with weak partial Σ-homomorphisms.

A partial Σ-homomorphism is *strong* if it preserves and reflects definedness of operations, i.e. $f_A(a_1, \ldots, a_n)$ is defined iff $f_B(|h|_{s_1}(a_1), \ldots, |h|_{s_n}(a_n))$ is defined, for each $f : s_1 \times \ldots \times s_n \to s$ in Σ and $a_1 \in |A|_{s_1}, \ldots, a_n \in |A|_{s_n}$. The subcategory of partial algebras with strong partial Σ-homomorphisms we denote by $PAlg(\Sigma)$. A partial subalgebra of A is any partial algebra B the carrier of which is a subset of A closed under operations and operations of which are equal to operations of A restricted to $|B|$.

A congruence on A is a many sorted equivalence relation $\sim \subseteq |A| \times |A|$, such that for each operation $f : s_1 \times \ldots \times s_n \to s$ in Σ and $a_1, a'_1 \in |A|_{s_1}, \ldots, a_n, a'_n \in |A|_{s_n}$, whenever $a_1 \sim_{s_1} a'_1, \ldots, a_n \sim_{s_n} a'_n$ then $f_A(a_1, \ldots, a_n)$ and $f_A(a'_1, \ldots, a'_n)$ are both undefined or both defined and $f_A(a_1, \ldots, a_n) \sim_s f_A(a'_1, \ldots, a'_n)$. Given a congruence relation $\sim$, the quotient algebra $A/_\sim$ is defined by $|A/_\sim| = |A|/_\sim$ and $f_{A/_\sim}([a_1]_\sim, \ldots, [a_n]_\sim)$ is defined iff $f_A(a_1, \ldots, a_n)$ is defined, and then $f_{A/_\sim}([a_1]_\sim, \ldots, [a_n]_\sim) = [f_A(a_1, \ldots, a_n)]_\sim$.

The quotient projection and the inclusion of subobjects are strong partial homomorphisms, hence definitions above correspond to general definitions from [BT96] (a concrete category) applied to $PAlg(\Sigma)$.

In the case of partial algebras, the value $t_{A[v]}$ of a term under a valuation $v : X \to |A|$ can be undefined, as operations are allowed to be partial functions. The relation of satisfaction of an equation $t = t'$, for $t, t' \in T_\Sigma(X)$, in a Σ-algebra A under a valuation $v : X \to |A|$ is defined as follows: $A[v] \models t = t'$ iff either both $t_{A[v]}$ and $t'_{A[v]}$ are undefined or both are defined and $t_{A[v]} = t'_{A[v]}$.

Regular algebras. Regular Σ-algebras differ from standard algebras in at least two respects: their carriers are partial orders with the least element and lub's of all chains definable by iterations of term-definable functions; morphisms are required to preserve these lub's. This makes them useful for modelling "infinitary" data, expressed by lub's of approximating chains. We omit here a detailed presentation, which can be found in [BT95] and [Tiu78].

Observational equivalence. Throughout the paper let us fix a subset $OBS \subseteq S$ of *observable sorts*. We define here the observational equivalence for standard, partial and regular algebras, called here commonly Σ-algebras. For a Σ-algebra A and $X \subseteq |A|$, by $\langle X \rangle_A$ we denote *the subalgebra of A generated by X*, being the least (standard, partial, regular) subalgebra of A whose carrier includes X.

For a Σ-algebra A we define the partial congruence $\sim_A^{OBS}$ as follows. The domain of $\sim_A^{OBS}$ is the carrier $|A_{OBS}|$ of $A_{OBS} = \langle |A|_{OBS} \rangle_A$. Then we define $\sim_A^{OBS}$ as the greatest congruence on A_{OBS} being identity on observable sorts. The relation $\sim_A^{OBS}$ represents *indistinguishability* of elements of A. There exist explicit characterisations of $\sim_A^{OBS}$ as the relations of indistinguishability by means of observable Σ-contexts (see [BHW94] for standard, [BT96] for partial and [BT95] for regular algebras).

Definition 2.1. An observational equivalence of (standard, partial, regular) algebras is a relation $\equiv_{OBS}$ *factorized* by the family $\{\sim_A^{OBS}\}_{A \in |Alg(\Sigma)|}$ of partial congruences[2], that is to say:

$$A \equiv_{OBS} B \text{ iff } A_{OBS}/_{\sim_A^{OBS}} \simeq B_{OBS}/_{\sim_B^{OBS}}.$$

Equivalence $\equiv_{OBS}$ of standard algebras has a logical characterisation [BHW94]: $A \equiv_{OBS} B$ iff for some set $X \subseteq Set^{OBS}$ of variables and surjective valuations $v_A : X \to |A|_{OBS}$ and $v_B : X \to |B|_{OBS}$, for all terms $t, t' \in |T_\Sigma(X)|_o$ of observable sort $o \in OBS$, $A[v_A] \models t = t'$ iff $B[v_B] \models t = t'$. Similar characterisations, although specific in each case, exist also for regular and partial algebras [BT95,BT96].

In searching for a uniform definition of observational equivalence across different kinds of algebras, the framework of *concrete category* was developed in [BT96], following the lines of Definition 2.1. More interestingly, authors indicated also equivalent formulation of $\equiv_{OBS}$, by means of spans of *observational morphisms*, i.e. bijective on observable sorts. This result will be useful for us in the following sections.

Open maps. Let $\mathcal{N}$ be a category of models, in which we choose a subcategory (not necessarily full) of *observation objects*, denoted by $\mathcal{Q}$. A morphism $p : O \to A$ from an observation object $O \in |\mathcal{Q}|$ can be seen as *a computation* in A. A morphism $h : A \to B$ in $\mathcal{N}$ should be intuitively thought of as a simulation of A in B in a sense that h transforms every computation $p : O \to A$ in A into a computation $(p; h)$ in B. Moreover any morphism $m : O_1 \to O_2$ in $\mathcal{Q}$ making $p = m; q$, for two computations p, q, means intuitively that q is an extension of p (via m).

Definition 2.2 (Open morphisms). A morphism $h : A \to B$ in $\mathcal{N}$ is $\mathcal{Q}$-*open* if for any morphism $m : O \to O'$ in $\mathcal{Q}$, $p : O \to A$, $q : O' \to B$ in $\mathcal{N}$, whenever the diagram

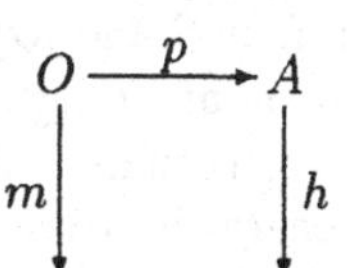

commutes, there exists a *diagonal morphism* $r : O' \to A$ in $\mathcal{N}$ such that the two triangles commute, i.e. $p = m; r$ and $q = r; h$.

Definition 2.3 (Bisimilarity). Two objects A and B in $\mathcal{N}$ are $\mathcal{Q}$-*bisimilar*, denoted $A \sim_\mathcal{Q} B$, if there exists a span of $\mathcal{Q}$-open morphisms $A \longleftarrow Z \longrightarrow B$ in $\mathcal{N}$ from a common object Z.

We often omit the prefix $\mathcal{Q}$-, when it is obvious from a context. The notion of $\mathcal{Q}$-bisimilarity is a generalisation of the strong bisimilarity between transition

[2] By $|\mathcal{C}|$ we denote the class of objects of a category $\mathcal{C}$, deliberately overloading the symbol $|\_|$.

systems: in [JNW93] it was proved that it coincides with the bisimilarity induced by the subcategory $Bran_L$ of finite "linear" transition systems.

Throughout the paper we implicitly rely on the observation that the larger an observation subcategory $\mathcal{Q}$ is, the fewer morphisms are $\mathcal{Q}$-open: if $\mathcal{P}$ is a subcategory of $\mathcal{Q}$ then every $\mathcal{Q}$-open morphism is $\mathcal{P}$-open.

3 Bisimilarity in Adjunction

In this section we state some general result leading to the conclusion that observational equivalence of standard and regular algebras coincide with the bisimilarity defined by means of open morphisms w.r.t. a natural class of observation objects.

We assume to be given a functor $G : \mathcal{M} \to \mathcal{O}$ from the category $\mathcal{M}$ of models to the category $\mathcal{O}$ having the left adjoint $F : \mathcal{O} \to \mathcal{M}$. Throughout this section let us fix the subcategory $\mathcal{P}$ of $\mathcal{O}$. As the subcategory of observation objects in $\mathcal{M}$ we choose the image $F(\mathcal{P})$ of the functor F. We will study especially the case $\mathcal{P} = \mathcal{O}$. Following facts are special cases of Theorem 4.2 and Lemma 4.3, respectively:

Lemma 3.1. *A morphism h in $\mathcal{M}$ is $F(\mathcal{P})$-open iff $G(h)$ is $\mathcal{P}$-open in $\mathcal{O}$.*

Corollary 3.2. *h is $F(\mathcal{O})$-open iff $G(h)$ is an isomorphism in $\mathcal{O}$.*

Lemma 3.1 will be a basic tool in analysing bisimulations for standard and regular algebras. Essentially it states that open bisimulation is transported via adjunction in both directions. Moreover, the $F(\mathcal{O})$-open morphisms in $\mathcal{M}$ are precisely those transported by G to isomorphisms. Some weaker version of Lemma 3.1, concerning only coreflections, was already proved in [JNW93].

For any subcategory $\mathcal{S}$ of $\mathcal{M}$ let $[\mathcal{S}]$ denote the smallest full subcategory of $\mathcal{M}$ containing all the objects of $\mathcal{S}$ (in fact $\mathcal{S}$ and $[\mathcal{S}]$ can differ only on morphisms). The choice of observation objects as an image subcategory $F(\mathcal{O})$ of $\mathcal{M}$ is not the only natural - we could also take (usually different) full subcategory $[F(\mathcal{O})]$, having the same objects but in general more morphisms. It turns out that both choices give the same result, which we prove in Lemma 4.4 in Section 4. The following is a specialisation for an adjunction:

Corollary 3.3. *A morphism h in $\mathcal{M}$ is $F(\mathcal{O})$-open iff it is $[F(\mathcal{O})]$-open.*

As a conclusion from Corollary 3.2, $A \sim_{F(\mathcal{O})} B$ implies $G(A) \simeq G(B)$, but the opposite implication does not hold in general (it holds, in fact, for a coreflection).

3.1 Standard and Regular Algebras

We separated the special case of adjunction from the more general situation since it has some important examples: standard and regular algebras. Recall that by Σ we denote a fixed signature and OBS is a subset of observable sorts.

Standard algebras. As an example consider the category $Alg(\Sigma)$ of standard algebras together with the forgetful functor $|\_|_{OBS} : Alg(\Sigma) \to Set^{OBS}$ returning the carrier sets of sorts in OBS. The functor $|\_|_{OBS}$ has the left adjoint T_Σ. We conclude from Corollary 3.2 that the $T_\Sigma(Set^{OBS})$-open morphisms of standard algebras are precisely the observational morphisms, which gives the characterisation of observational equivalence by means of $T_\Sigma(Set^{OBS})$-open maps:

Corollary 3.4. *The observational equivalence is $T_\Sigma(Set^{OBS})$-bisimilarity.*

The subcategory of observation objects can be reduced while still inducing the same bisimilarity relation. Any subcategory $T_\Sigma(\mathcal{P})$ of $T_\Sigma(Set^{OBS})$ is a candidate when $\mathcal{P}$ is chosen large enough to "distinguish" between non-isomorphic objects in Set^{OBS}. Surprisingly, very small subcategory $\mathbf{2}_{OBS}$ is sufficient, consisting of OBS-sorted sets being empty or two-element, on each sort. We obtain the following fact as a corollary from Lemma 3.1 and Corollaries 3.2 and 3.3:

Corollary 3.5. *The observational equivalence of standard algebras equals $\mathcal{Q}$-bisimilarity, for any subcategory $\mathcal{Q}$ of $[T_\Sigma(Set^{OBS})]$ containing $T_\Sigma(\mathbf{2}_{OBS})$.*

This can be interpreted as follows: terms of observable sorts with only two variables are powerful enough to distinguish non-equivalent algebras. Observational equivalence is thus very robust to changes of subcategory of observation objects.

Regular Algebras. The forgetful functor $|\_|_{OBS} : RAlg(\Sigma) \to Set^{OBS}$ has the left adjoint $T^\mu_\Sigma(\_)$; free regular algebras are algebras of so called regular Σ-trees [Tiu78]. We obtain hence a conclusion, similarly as for standard algebras:

Corollary 3.6. *The observational equivalence of regular algebras coincides with $\mathcal{Q}$-bisimilarity, for any subcategory $\mathcal{Q}$ of $[T^\mu_\Sigma(Set^{OBS})]$ containing $T^\mu_\Sigma(\mathbf{2}_{OBS})$.*

4 Bisimilarity in Multiadjunction

In this section we assume to have a *left multiadjoint* [Die79] to a fixed functor $G : \mathcal{M} \to \mathcal{O}$, where instead of one free object we have a family of free objects. The definition below is adapted from [Die79]:

Definition 4.1 (Left multiadjoint, multiadjunction). Let X be any object of category $\mathcal{O}$. Let $F(X) = \{F_i(X)\}_{i \in \mathcal{I}_X}$ and $\eta_X = \{\eta^i_X : X \to G(F_i(X))\}_{i \in \mathcal{I}_X}$ be families of objects of $\mathcal{M}$ and morphisms of $\mathcal{O}$, respectively, both indexed by the same set $\mathcal{I}_X$. We say that a pair $\langle F(X), \eta_X \rangle$ is *a family of free objects over* X if for any object A in $\mathcal{M}$ and any morphism $f : X \to G(A)$ there exists a unique pair $\langle i, f^\# : F_i(X) \to A \rangle$ where $i \in \mathcal{I}_X$ and $f^\#$ is a morphism in $\mathcal{M}$, such that $\eta^i_X; G(f^\#) = f$.

If for every object X in $\mathcal{O}$ there exists a family of free objects $\langle F(X), \eta_X \rangle$ over X, we say that the pair of families $\langle F = \{F(X)\}_{X \in |\mathcal{O}|}, \eta = \{\eta_X\}_{X \in |\mathcal{O}|} \rangle$ is *the left multiadjoint* to G and that $\langle F, \eta \rangle$ and G form a *multiadjunction*.

An example of multiadjunction for partial algebras is presented in Section 4.2. Assume in this section that G has a left multiadjoint $\langle F, \eta\rangle$ (it is determined uniquely up to isomorphism [Die79]). The main difference between multiadjunction and adjunction is that we have now a bijection

$$(\_)^{\#_{X,A}} : \mathcal{O}(X, G(A)) \to \bigcup_{i \in \mathcal{I}_X} \mathcal{M}(F_i(X), A), \tag{1}$$

for any X in $\mathcal{O}$ and A in $\mathcal{M}$, instead of a bijection $\mathcal{O}(X, G(A)) \to \mathcal{M}(F(X), A)$. Obviously, when sets $\mathcal{I}_X$ are singletons we obtain a usual adjunction. As usual we suppress the index of #, hoping that this should not not lead to an ambiguity.

There is no unique way to define "$F(f)$" for $f : X \to Y$ in $\mathcal{O}$. Instead we can naturally define a family of arrows $\{F_j(f)\}_{j \in \mathcal{I}_Y}$ in $\mathcal{M}$, where for $j \in \mathcal{I}_Y$,

$$F_j(f) := (f; \eta_Y^j)^\# : F_i(X) \to F_j(Y) \quad \text{(for certain } i \in \mathcal{I}_X). \tag{2}$$

Let $\mathcal{P}$ be a fixed subcategory of $\mathcal{O}$ in this section. As a subcategory of observation objects in $\mathcal{M}$ we take now the subcategory $F(\mathcal{P})$ defined as follows: objects are $\{F_i(X)\}_{X \in |\mathcal{P}|, i \in \mathcal{I}_X}$, morphisms are $\{F_j(f)\}_{f:X \to Y \in |\mathcal{P}|, j \in \mathcal{I}_Y}$. This obviously defines a subcategory, as the morphisms are closed under composition and contain all identities. Openness is transported via multiadjunction in both directions; in the following let $h : A \to B$ be an arbitrary map in $\mathcal{M}$.

Theorem 4.2. *h is $F(\mathcal{P})$-open iff $G(h)$ is $\mathcal{P}$-open in $\mathcal{O}$.*

For $\mathcal{P}$ being the whole category $\mathcal{O}$ we obtain the following corollary:

Lemma 4.3. *h is $F(\mathcal{O})$-open iff $G(h)$ is an isomorphism in $\mathcal{O}$.*

Lemma 4.4. *h is $F(\mathcal{O})$-open iff it is $[F(\mathcal{O})]$-open.*

In Lemma 4.9 in Section 4.1, equipped with a notion of determinism, we formulate a result similar to Lemma 4.4 for a wide class of subcategories $F(\mathcal{P})$.

4.1 Determinism

Consider a commuting diagram

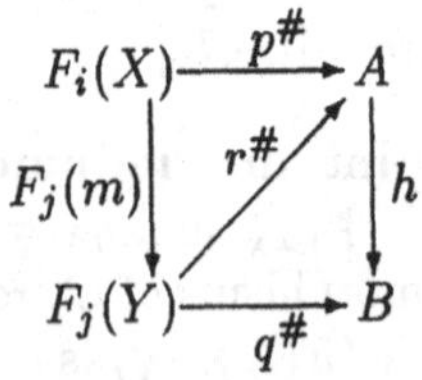

in $\mathcal{M}$ for a $F(\mathcal{O})$-open morphism h and any morphism $m : X \to Y$ in $\mathcal{O}$. It is interesting to note that the diagonal morphism $r^\#$ making both triangles commute is unique. Moreover, this morphism is determined by only h and $q^\#$:

Observation 4.5. For any $F(\mathcal{O})$-open map $h : A \to B$ in $\mathcal{M}$ and $q : Y \to G(B)$ in $\mathcal{O}$ there exists at most one morphism $r : Y \to G(A)$ in $\mathcal{O}$ such that $r^\#; h = q^\#$.

This means that two different computations in A cannot be simulated by the same computation in B, which is not the case in general. The observation above leads to the general notion of determinism of open maps. In the following let $\mathcal{N}$ denote an arbitrary category of models and let $\mathcal{Q}$ be a subcategory of $\mathcal{N}$.

Definition 4.6. We say that $\mathcal{Q}$-bisimilarity in $\mathcal{N}$ is *deterministic* if any $\mathcal{Q}$-open morphism h has exactly one diagonal morphism in every commuting square as in Definition 2.2 in Section 2. Moreover, $\mathcal{Q}$-bisimilarity is *strongly deterministic* if for any $\mathcal{Q}$-open morphism $h : A \to B$ in $\mathcal{N}$ and any computation $q : O' \to B$ in $\mathcal{N}$ there exists at most one computation $r : O' \to A$ in $\mathcal{N}$ satisfying $r; h = q$.

The notion of deterministic bisimilarity corresponds closely to *etale maps* [JM94]. In this paper we concentrate on the strong determinism, as all considered bisimilarities of algebras are strongly deterministic. In Observation 4.5 we noticed that $F(\mathcal{O})$-bisimilarity in $\mathcal{M}$ is strongly deterministic for any left multiadjoint F, which we generalise to:

Lemma 4.7. *Assume $\mathcal{P}$ is a subcategory of $G(\mathcal{M})$. If $\mathcal{P}$-bisimilarity in $G(\mathcal{M})$ is strongly deterministic then $F(\mathcal{P})$-bisimilarity in $\mathcal{M}$ is so as well.*

Definition 4.8. We say that $\mathcal{N}$ is *initialized* (w.r.t. $\mathcal{Q}$) if for any $\mathcal{Q}$-open map $h : A \to B$, any object O' in $\mathcal{Q}$ and a computation $q : O' \to B$ there exists an object O in $\mathcal{Q}$ together with a computation $p : O \to A$ and a morphism $m : O \to O'$ in $\mathcal{Q}$ making the diagram in Definition 2.2 commute.

This is usually the case, for example when $\mathcal{N}$ has the initial object $\mathcal{I}$, belonging to $\mathcal{Q}$ together with all (unique) morphisms from $\mathcal{I}$ to objects of $\mathcal{Q}$. (The unique map from $\mathcal{I}$ to A can be thought of as "the empty computation" in A.)

In the deterministic case some closures of observation subcategory do not change the induced bisimilarity. First "closure" property, stated in Lemma 4.9, is a strengthening of Lemma 4.4.

Lemma 4.9. *If $\mathcal{N}$ is initialized w.r.t. $\mathcal{Q}$ and $\mathcal{Q}$-bisimilarity in $\mathcal{N}$ is strongly deterministic, then every $\mathcal{Q}$-open map in $\mathcal{N}$ is $[\mathcal{Q}]$-open.*

Deterministic bisimilarity has an interesting property, that the closure of observation subcategory under colimits does not change induced bisimilarity. In Lemma 4.10 we show that $\mathcal{Q}$-bisimilarity coincides with $\mathcal{Q}^{\to}$-bisimilarity, where subcategory $\mathcal{Q}^{\to}$ of $\mathcal{M}$ contains colimits of all diagrams in $\mathcal{Q}$, together with all morphisms (between colimits of diagrams of the same shape) induced canonically by the universal property of colimits (for details we refer to [Las97a]).

Lemma 4.10. *If $\mathcal{Q}$-bisimilarity is strongly deterministic, then every $\mathcal{Q}$-open map in $\mathcal{N}$ is $\mathcal{Q}^{\to}$-open and $\mathcal{Q}^{\to}$-bisimilarity is strongly deterministic.*

4.2 Partial Algebras with Strong Homomorphisms

We are going to demonstrate that observational equivalence in the category of partial algebras is a bisimulation equivalence. We will actually characterise the observational equivalence twice: first (in this section) in the category $PAlg(\Sigma)$ of partial algebras with strong homomorphisms and then (in Section 5.1) in the richer category $PAlg_w(\Sigma)$ of weak partial homomorphisms. Restricting to only strong homomorphisms makes the task easier, so we start from that case, following [BT96].

The situation now is not so nice as in the case of standard algebras, where the forgetful functor has the left adjoint. Instead, there exists the left multiadjoint defined as follows. For any X in Set^{OBS}, the set $T(X) = \{T_i(X)\}_{i \in \mathcal{I}_X}$ consists of all partial algebras $T_i(X)$ whose carrier is a subset of the total algebra $T_\Sigma(X)$, which satisfy (for every $i \in \mathcal{I}_X$):

- $X \subseteq |T_i(X)| \subseteq |T_\Sigma(X)|$,
- carrier sets of $T_i(X)$ are closed under subterms, i.e. $t \in |T_i(X)|$ implies that all subterms of t are in $|T_i(X)|$,
- for any function symbol $f : s_1, \ldots, s_n \to s$ in Σ and terms $t_1, \ldots, t_n$ of appropriate sorts in $|T_i(X)|$, the function $f^{T_i(X)}$ is defined on $t_1, \ldots, t_n$ iff the term $f(t_1, \ldots, t_n)$ belongs to $|T_i(X)|$, and if so, then $f^{T_i(X)}(t_1, \ldots, t_n) = f(t_1, \ldots, t_n)$.

For any X in Set^{OBS}, we define the family $\eta_X = \{\eta^i_X : X \to |T_i(X)|_{OBS}\}_{i \in \mathcal{I}_X}$ of inclusions in Set^{OBS}.

Proposition 4.11. *The pair* $\langle T = \{T(X)\}_{X \in |Set^{OBS}|}, \eta = \{\eta_X\}_{X \in |Set^{OBS}|}\rangle$ *is the left multiadjoint to the functor* $|\_|_{OBS}$.

Having defined the multiadjoint, we easily obtain a conclusion from Lemmas 4.3, 4.4 and Theorem 4.2:

Corollary 4.12. *The observational equivalence of partial algebras in category* $PAlg(\Sigma)$ *coincides with the* $\mathcal{Q}$*-bisimilarity in* $PAlg(\Sigma)$*, for any subcategory* $\mathcal{Q}$ *of* $[T(Set^{OBS})]$ *containing* $T(\mathbf{2}_{OBS})$ *(*$\mathbf{2}_{OBS}$ *was introduced in Section 3.1).*

5 Bisimilarity in Pre-adjunction

In this section it is sufficient for our purposes to have a *pre-adjunction* (Definition 5.1) from which we extract a certain multiadjunction by distinguishing the class of *ext-strong* morphisms in $\mathcal{M}$. More precisely, left pre-adjoint turns out to be the left multiadjoint to the functor G restricted to subcategory $\mathcal{M}^{\text{e-s}}$ of ext-strong morphisms of $\mathcal{M}$ (Corollary 5.6). Then we compare the induced bisimilarities in both categories (Theorem 5.7) showing that they coincide for certain classes of observation objects. The results presented in this section are quite general; an illustrative example of application to partial algebras with weak homomorphisms can be found in Section 5.1.

Definition 5.1 (Left pre-adjoint, pre-adjunction). Let $G : \mathcal{M} \to \mathcal{O}$ be a functor and X be any object of category $\mathcal{O}$. Let $F(X) = \{F_i(X)\}_{i \in \mathcal{I}_X}$ and $\eta_X = \{\eta^i_X : X \to G(F_i(X))\}_{i \in \mathcal{I}_X}$ be families of objects of $\mathcal{M}$ and morphisms of $\mathcal{O}$, respectively, both indexed by the same set $\mathcal{I}_X$. We say that a pair $\langle F(X), \eta_X\rangle$ is *a family of pre-free objects over* X (with *units* η_X) if

(i) for any object A in $\mathcal{M}$ and a morphism $f : X \to G(A)$ there exists a (not necessarily unique) pair $\langle i, f' : F_i(X) \to A \rangle$ where $i \in \mathcal{I}_X$ and f' is a morphism in $\mathcal{M}$, such that $\eta^i_X; G(f') = f$ (we call f' *an extension* of f to A),
(ii) $\eta^i_X; G(f') = \eta^i_X; G(f'')$ implies $f' = f''$, for any $f', f'' : F_i(X) \to A$ in $\mathcal{M}$, i.e. no two pairs with the same index i are allowed in (i),
(iii) for any object A in $\mathcal{M}$ and a morphism $f : X \to G(A)$, the set of indexes

$$\mathcal{I}^{f,A}_X = \{ i \in \mathcal{I}_X \mid \text{there exists } f' : F_i(X) \to A \text{ such that } \eta^i_X; G(f') = f \}$$

has the greatest element $\sqcup^{f,A}$ w.r.t. a partial order $\leq_X$ on $\mathcal{I}_X$ defined as:

$$i \leq_X j \text{ iff there exists } w : F_i(X) \to F_j(X) \text{ such that } \eta^i_X; G(w) = \eta^j_X.$$

If for every object X in $\mathcal{O}$ there exists a family of pre-free objects $\langle F(X), \eta_X \rangle$ over X, we say that the pair of families $\langle F = \{F(X)\}_{X \in |\mathcal{O}|}, \eta = \{\eta_X\}_{X \in |\mathcal{O}|} \rangle$ is *a left pre-adjoint* to G and that $\langle F, \eta \rangle$ and G form a *pre-adjunction.*

Note. Relation $\leq_X$ (for any X in $\mathcal{O}$) is a preorder, but from (ii) it follows that $\leq_X$ is a partial order up to isomorphism, namely, for any $i, j \in \mathcal{I}_X$,

$$i \leq_X j \text{ and } j \leq_X i \Leftrightarrow F_i(X) \simeq F_j(X).$$

Thus without loss of generality we can treat $\leq_X$ as a partial order.

From now on we consider a fixed pre-adjunction formed by $\langle F, \eta \rangle$ and a functor $G : \mathcal{M} \to \mathcal{O}$. Let X, A be any objects in $\mathcal{O}$, $\mathcal{M}$, respectively. For any morphism $f : X \to G(A)$ in $\mathcal{O}$, by $f^\#$ we denote the unique morphism $f^\# : F_{\sqcup^{f,A}}(X) \to A$ such that $\eta^{\sqcup^{f,A}}_X; G(f^\#) = f$. We call $f^\#$ *the greatest extension* of f (to A).

Example. As an example consider the category $PAlg_w(\Sigma)$ of partial algebras with weak homomorphisms. It turns out that the left multiadjoint $\langle T, \eta \rangle$, defined in Section 4.2 for category $PAlg(\Sigma)$, is now a left pre-adjoint to the functor $|\text{-}|_{OBS} : PAlg_w(\Sigma) \to Set^{OBS}$ (cf. Section 5.1). The pre-free partial algebra $T_{\sqcup^{v,A}}(X)$ consists of precisely those terms over variables from X, which are defined in A under a valuation $v : X \to |A|_{OBS}$. Moreover, the greatest extension $v^\# : T_{\sqcup^{v,A}}(X) \to A$ is the unique extension of the valuation v to terms.

Now we are going to distinguish a subclass of *ext-strong* morphisms, giving raise to a multiadjunction. An example of partial algebras justifies a claim that ext-strong maps generalise the concept of strong homomorphisms.

Definition 5.2. A morphism $f : A \to B$ in $\mathcal{M}$ is *ext-strong* if whenever composed with the greatest extension of a morphism $p : X \to G(A)$ to A gives the greatest extension of $(p; G(f))$ to B, i.e. if it satisfies the following

$$p^\#; f = (p; G(f))^\#, \quad \text{for any morphism } p : X \to G(A).$$

Note that the opposite implication holds for any morphism $f : A \to B$ in $\mathcal{M}$:

Lemma 5.3. *Let $p' : F_i(X) \to A$ be an extension of $p : X \to G(A)$. Whenever $(p'; f) = (p; G(f))^\#$ then $p' = p^\#$.*

Lemma 5.4. *Every ext-strong extension of a morphism is its greatest extension.*

The greatest extensions need not to be in general ext-strong and a simple counter-example can be constructed. But in many typical situations these two notions coincide (cf. Section 5.1), which justifies the following assumption, needed below for the results in this section:

Assumption 5.5. Every greatest extension is ext-strong, i.e. for any A and $f : X \to G(A)$, $f^{\#}$ is ext-strong.

It follows from Assumption 5.5 that the greatest extensions compose, since ext-strong morphisms compose. Moreover, identities (in fact, all isomorphisms) are ext-strong. We distinguish the subcategory $\mathcal{M}^{\text{e-s}}$ of $\mathcal{M}$, consisting of all objects but only ext-strong morphisms, to formulate the following:

Corollary 5.6. *$\langle F, \eta\rangle$ is the left multiadjoint to G restricted to $\mathcal{M}^{e\text{-}s}$.*

Let $\mathcal{P}$ be arbitrary fixed subcategory of $\mathcal{O}$. As a subcategory of observation objects in $\mathcal{M}$ we take similarly as in Section 4 the subcategory $F(\mathcal{P})$ whose objects are all pre-free objects and morphisms are all extensions of morphisms $(m; \eta^j_Y)$ for any morphism $m : X \to Y$ in $\mathcal{O}$ and any $j \in \mathcal{I}_Y$. We will deliberately use the same notation $F(\mathcal{P})$ for this subcategory of $\mathcal{M}$ as well as for different in general on morphisms subcategory $F(\mathcal{P})$ of $\mathcal{M}^{\text{e-s}}$ determined by the left multiadjoint (this subcategory was defined in Section 4).

Now we relate $F(\mathcal{P})$-openness in $\mathcal{M}$ and $F(\mathcal{P})$-openness in $\mathcal{M}^{\text{e-s}}$. It turns out that if all $F(\mathcal{P})$-open morphisms in $\mathcal{M}$ are ext-strong (i.e. are also in $\mathcal{M}^{\text{e-s}}$; obviously it is not always the case, for example for $\mathcal{P}$ being the empty category), then the notions of $F(\mathcal{P})$-openness coincide in both categories:

Theorem 5.7. *Let $\mathcal{P}$ be any subcategory of $\mathcal{O}$ such that all $F(\mathcal{P})$-open morphisms in $\mathcal{M}$ are ext-strong. Let $h : A \to B$ be a morphism in $\mathcal{M}$. Then h is $F(\mathcal{P})$-open in $\mathcal{M}$ iff h is $F(\mathcal{P})$-open in $\mathcal{M}^{e\text{-}s}$.*

The requirement needed in Theorem 5.7 is satisfied when $\mathcal{P} = \mathcal{O}$ (Lemma 5.8), which gives Corollary 5.9.

Lemma 5.8. *Every $F(\mathcal{O})$-open morphism in $\mathcal{M}$ is ext-strong.*

Corollary 5.9. *An arrow h in $\mathcal{M}$ is $F(\mathcal{O})$-open iff h is $F(\mathcal{O})$-open in $\mathcal{M}^{e\text{-}s}$.*

Determinism. Lemma 4.7 implies that all the bisimilarities considered so far, induced by an adjunction or a multiadjunction, are strongly deterministic. It turns out that pre-adjunction also induces strongly deterministic bisimilarity if the related multiadjunction does so.

Lemma 5.10. *For $\mathcal{P}$ as in Theorem 5.7, whenever $F(\mathcal{P})$-bisimilarity in $\mathcal{M}^{e\text{-}s}$ is strongly deterministic then $F(\mathcal{P})$-bisimilarity in $\mathcal{M}$ is so as well.*

Observation analogous to 4.5 holds also in this case, and can be concluded directly from the above lemma and Lemma 5.8:

Corollary 5.11. *$F(\mathcal{O})$-bisimilarity is strongly deterministic for any left pre-adjoint F.*

5.1 Partial Algebras with Weak Homomorphisms

The forgetful functor $|\text{-}|_{OBS} : PAlg_w(\Sigma) \to Set^{OBS}$ has the left adjoint, but the full subcategory of free objects (in which all operations are totally undefined) is too small to induce the expected observational equivalence of partial algebras. In Section 4.2 we defined the left multiadjoint $\langle T, \eta \rangle$ to the functor $|\text{-}|_{OBS}$. It turns out that the same pair is a left pre-adjoint when we work with weak morphisms:

Proposition 5.12. *Functor $|\text{-}|_{OBS} : PAlg_w(\Sigma) \to Set^{OBS}$ together with $\langle T, \eta \rangle$ form a pre-adjunction.*

The greatest extensions of this pre-adjunction are those homomorphisms which appeared in the multiadjunction as $v^{\#}$, for some $v : X \to |A|_{OBS}$ in Set^{OBS}. In the following we will need some weaker kind of strong morphisms. We say that $f : A \to B$ in $PAlg_w(\Sigma)$ is *R-strong* (for $R \subseteq sorts(\Sigma)$) if it reflects definedness of operations on the carrier of the subalgebra $\langle |A|_R \rangle_A$ of A generated by sorts R. The following lemma gives an abstract characterisation of OBS-strong maps:

Lemma 5.13. *A weak partial homomorphism is OBS-strong iff it is ext-strong.*

Assumption 5.5 is satisfied in this case:

Lemma 5.14. *Every greatest extension in $PAlg_w(\Sigma)$ w.r.t. the forgetful functor $|\text{-}|_{OBS}$ is OBS-strong.*

Category $PAlg_w(\Sigma)^{\text{e-s}}$ contains more morphisms than $PAlg(\Sigma)$, namely all OBS-strong ones. Now, by Corollary 5.9, $T(Set^{OBS})$-open maps in category $PAlg_w(\Sigma)$ are precisely $T(Set^{OBS})$-open morphisms in $PAlg_w(\Sigma)^{\text{e-s}}$. The subtle point here is that the left multiadjoint to $|\text{-}|_{OBS}$ for category $PAlg(\Sigma)$ works also for $PAlg_w(\Sigma)^{\text{e-s}}$, since every OBS-strong extension is a strong homomorphism. This observation implies that $T(Set^{OBS})$-bisimilarity in $PAlg_w(\Sigma)$ coincides with $T(Set^{OBS})$-bisimilarity in $PAlg(\Sigma)$, from which we conclude:

Corollary 5.15. *The observational equivalence $\equiv_{OBS}$ of partial algebras coincides with $T(Set^{OBS})$-bisimilarity in the category $PAlg_w(\Sigma)$.*

The subcategory $T(Set^{OBS})$ can be restricted to $\mathbf{Fin}_{OBS}$, the full subcategory of $T(Set^{OBS})$ of finite algebras. Note that $\mathbf{Fin}_{OBS}$ is not a full subcategory of $PAlg_w(\Sigma)$ in general. The observational equivalence of partial algebras is equal to $\mathbf{Fin}_{OBS}$-bisimilarity in $PAlg_w(\Sigma)$.

Lemma 5.16. *Every $\mathbf{Fin}_{OBS}$-open partial homomorphism is $T(Set^{OBS})$-open.*

The subcategory $\mathbf{Fin}_{OBS}$ of finite partial pre-free algebras plays in the category $PAlg_w(\Sigma)$ a role of finite computations, analogously as the subcategory $Bran_L$ of finite sequences in the category of transition systems. Let us summarize the results concerning category $PAlg_w(\Sigma)$, as a direct conclusion from Corollary 5.15, Lemma 5.16, Corollary 5.11 and Lemma 4.9:

Corollary 5.17. *The observational equivalence of partial algebras equals $\mathcal{Q}$-bisimilarity in $PAlg_w(\Sigma)$, for any subcategory $\mathcal{Q}$ of $[T(Set^{OBS})]$ containing $\mathbf{Fin}_{OBS}$.*

6 Final Remarks

The main results of this paper are characterisations of observational equivalences for a number of classes of algebras as the bisimilarity induced by a subcategory of free, multifree or pre-free objects. We developed the framework of pre-adjunction and stated some facts for multiadjunction and pre-adjunction, which enabled us to show the coincidence of observational equivalence and bisimilarity. We identified small observation subcategories, corresponding to "finitary" observations, which are sufficient to induce the observational equivalences. This can be very useful in proving observational equivalence of algebras. Moreover, we observed that the bisimilarities considered in the paper are deterministic, in contrast to bisimilarities between model for concurrent systems like transition systems or event structures.

As a response to numerous definitions of observational equivalence for different categories of algebras, the idea of behavioural equivalence was generalised in [BT96] to an arbitrary *concrete category*, i.e. assumed to be equipped with a faithful concretization functor to Set^S. The interesting continuation of the work is to try to apply the machinery of the pre-adjunction in such a general case. Moreover, some work is needed to examine applicability of our results to other frameworks than considered in this paper, for example to continuous algebras [TW86].

There arises also another interesting issue: is it possible, for an arbitrary bisimilarity induced by some class of open maps in a concrete category, to find partial congruences factorizing this bisimilarity (cf. Definition 2.1)? We have found a partial answer to this question [Las97b], but elegant necessary and sufficient conditions for such a coincidence are yet to be developed.

Acknowledgements. Many thanks are due to Andrzej Tarlecki for fruitful discussions and helpful comments during this work as well as for introduction into the area of behavioural equivalences and algebraic specifications.

References

[AO93] S. Abramsky, C.-H. Luke Ong. Full Abstraction in the Lazy Lambda Calculus. *Information and Computation* 105(2), 159-267, 1993.

[BHW94] Bidoit, M., Hennicker, R., Wirsing, M. Behavioural and abstractor specifications. *Science of Computer Programming* 25(2-3), 1995.

[BT95] Bidoit, M., Tarlecki, A. Regular algebras: a framework for observational specifications with recursive definitions. Report LIENS-95-12, Ecole Normale Superieure, 1995.

[BT96] Bidoit, M., Tarlecki A. Behavioural satisfaction and equivalence in concrete model categories. *Manuscript*. The short version appeared in *Proc. 20th Coll. on Trees in Algebra and Computing* CAAP'96, Linköping, Springer-Verlag.

[BL70] Birkhoff, G., Lipson, J.D. Heterogeneous algebras. *J. Combinatorial Theory* 8(1970), 115-133.

[BW82] Broy, M., Wirsing, M. Partial abstract data types. *Acta Informatica* 18(1982), 47-64.

[CN95] Cheng, A., Nielsen, M. Open maps (at) work. Research series RS-95-23, BRICS, Department of Computer Science, University of Aarhus, 1995.

[Die79] Diers, Y., Familles universelles de morphismes. *Annales de la Société Scientifique de Bruxelles*, 93, nr 3, 1979, 175-195.

[GGM76] Giarratana, V., Gimona, F., Montanari, U. Observability concepts in abstract data type specification. *Proc. 5th Intl. Symp. Mathematical Foundations of Computer Science*, Gdańsk 1976, LNCS 45, Springer-Verlag 1976, 576-587.

[GTW78] Goguen, J.A., Thatcher, J.W., Wagner, E.G. An initial algebra approach to the specification, correctness and implementation of abstract data types. *Current Trends in Programming Methodology 4, Data Structuring*, Yeh, R.T., ed., 80-149, Prentice-Hall 1978.

[Gor95] Gordon, A. A tutorial on co-induction and functional programming. *Proc. of the 1994 Glasgow Workshop on Functional Programming*, Springer Workshops in Computing, 1995, 78-95.

[Jac95] Jacobs, B.F.P. Objects and classes, coalgebraically. Technical Report CS-R9536, CWI, 1995.

[JM94] Joyal, A., Moerdijk, I. A completeness theorem for open maps. *Annals of Pure and Applied Logic* 70(1994), 51-86.

[JNW93] Joyal, A., Nielsen, G. Winskel, Bisimulation and open maps. *Proc. 8th Annual Symposium on Logic in Computer Science* LICS'93, 1993, 418-427.

[Las97a] Lasota, S. Open Maps as a Bridge between Algebraic Observational Equivalence and Bisimilarity. Technical report 97-12 (249), Institute of Informatics, Warsaw University. Accesible at `http://zls.mimuw.edu.pl/~sl`.

[Las97b] Lasota, S. Partial Congruence Factorization of Bisimilarity Induced by Open Maps. Manuscript, accesible at `http://zls.mimuw.edu.pl/~sl`.

[Mal96] Malcolm, G. Behavioural equivalence, bisimilarity, and minimal realisation. *Recent Trends in Data Type Specification, 11th Workshop on Specification of Abstract Data Types*, Oslo, 1995. LNCS 1130, 359-378.

[Mil89] Milner, R. *Communication and concurrency.* Prentice-Hall International Series in Computer Science, C. A. R. Hoare series editor, 1989.

[Mor68] Morris, J.H. Lambda-Calculus Models of Programming Languages. Ph.D. thesis, MIT, 1968.

[Par81] Park, D.M.R. Concurrency and Automata on Infinite Sequences. *Proc. 5th G.I. Conference*, Lecture Notes in Computer Science 104, Springer-Verlag, 1981.

[Rei81] Reichel, H. Behavioural equivalence – a unifying concept for initial and final specification methods. *Proc. 3rd Hungarian Computer Science Conference, Mathematical Models in Computer Systems*, M. Arato, L. Varga, eds., Budapest, 27-39, 1981.

[ST87] Sannella, D., Tarlecki, A. On observational equivalence and algebraic specification. *J. Computer and System Sciences* 34(1987), 150-178.

[TW86] Tarlecki, A., Wirsing, M. Continuous abstract data types. *Fundamenta Informaticae* 9(1986), 95-126.

[Tiu78] Tiuryn, J. Fixed-points and algebras with infinitely long expressions. Part I. Regular algebras. *Fundamenta Informaticae* 2(1978), 102-128.

A Systematic Study of Mappings Between Institutions

Alfio Martini* Uwe Wolter

Technische Universität Berlin, FB Informatik, Sekr. 6-1
Franklinstr. 28/29, D-10587 Berlin, Germany
{alfio,wolter}@cs.tu-berlin.de

Abstract

Concerning different notions of mappings between institutions, we believe that the current state of the art is somehow unsatisfactory. On the one hand because of the variety of different concepts proposed in the literature. On the other hand because of the apparent lack of a suitable basis to formally discuss about what they mean and how they relate to each other. In this paper we aim at a systematic study of some of the most important notions of these mappings by proposing a methodology based on the concept of power institutions. Firstly, power institutions allow the investigation of the *entire logical structure* of an institution along these mappings, i.e., the satisfaction relation *together* with the satisfaction condition. Secondly, they allow this investigation in a systematic way, i.e., the transformation of the institutional logical structure can be described by means of simpler, more elementary transformations or units which are themselves also power institutions. These units are constructions which denote, e.g., *typing reduction along functors between signatures, borrowing of models, common model theory, semantical restriction, and logical semantical restriction.* The mappings can then be related to each other by showing that they all comprise a particular number of these more fundamental, elementary transformations.

1 Introduction and Preliminaries

Institutions were first introduced in [2] in order to formalize a number of logics used in different specification languages so that abstract model theory (for specification and programming) could be done independently of some pre-established logical system. Briefly, an *institution* $\mathcal{I} = (\mathsf{Sign}, Sen, Mod, \models)$ consists of a category Sign whose objects are called *signatures*; a functor $Sen : \mathsf{Sign} \to \mathsf{Set}$ (indexed category), giving for each signature a set whose elements are called *sentences* over that signature; a functor $Mod : \mathsf{Sign}^{\mathsf{op}} \to \mathsf{Cat}$ (indexed category), giving for each signature Σ a category whose objects are called Σ-models, and whose arrows are called Σ-morphisms; and a function $\models$ associating to each signature Σ a relation $\models_\Sigma \subseteq |Mod(\Sigma)| \times Sen(\Sigma)$, called Σ-*satisfaction relation*,

*Research supported in part by a CNPq-grant 200529/94-3.

such that for each arrow $\phi : \Sigma_1 \to \Sigma_2$ in Sign the *satisfaction condition*

$$\textbf{(SC)} \quad M_2 \models_{\Sigma_2} Sen(\phi)(\varphi_1) \iff Mod(\phi)(M_2) \models_{\Sigma_1} \varphi_1,$$

holds for any $M_2 \in |Mod(\Sigma_2)|$ and any $\varphi_1 \in Sen(\Sigma_1)$.

Concerning mappings between institutions, we believe that the current state of the art is still somehow unsatisfactory. A number of different formal concepts of mappings have been proposed in the literature as, for instance, institution morphisms, plain maps of institutions, institution transformations, simulations, and simple maps of institutions (see, e.g., [1, 6] for a survey and discussion). A key motivation underlying these proposals is based on the fact that heterogenous software specification demands the possibility of soundly moving along different (logical) specification languages (see, e.g., [6, 4]). In the simplest cases, these mappings are comprised by a corresponding pointwise translation of sentences and models (usually in a contravariant way), such that both *satisfaction relations* are preserved.

This paper aims at a systematic study of these mappings by proposing a methodology which allows to systematically investigate the *entire logical structure* of an institution along these mappings. By entire logical structure we mean the satisfaction relation $M \models \varphi$ between models and sentences *together* with the invariant it is supposed to satisfy, namely the one expressed by the satisfaction condition **(SC)**. Surprisingly, every formulation of mapping between institutions has avoided speaking explicitly about the preservation of this "invariant" and concentrated only on the preservation of the model theory as denoted by the satisfaction relation.

Therefore, we are interested in a methodology satisfying two essential requirements: firstly, it should allow to "equivalently" model the notion of an institution in such a way that its entire logical structure could be modeled by a simple and well-known mathematical construction. This is needed to ease the task of *studying the institutional logical structure* along these mappings. Secondly, it should also allow for a procedure which is quite ubiquitous in science: the description and characterization of a particular structure through the combination of some particular canonical set of elementary units or constructions. This is the *systematic study* part of this methodology, i.e, we expect to be able to describe the entire transformation of this logical structure by means of simpler ones, revealing in detail the roles of the translations mentioned above.

Roughly speaking, an institution has an equivalent formulation at the level of specifications and subcategories of models, using only the language of indexed categories as an indexed functor.

Nevertheless, before making the above statement more precise, we introduce in the sequel the basic background machinery needed to follow it, which is fully developed in [8].

In the following, let Σ and $\phi : \Sigma_1 \to \Sigma_2$ be, respectively, an arbitrary signature and an arbitrary signature morphism in Sign. Firstly, we consider the sentence functor Sen : Sign $\to$ Set. Let $\wp_{set}$: Set $\to$ Cat be the powerset functor that associates to each set $A \in |\text{Set}|$, the partial order $\wp_{set}(A) = (\wp(A), \supseteq)$

considered as a category, and to each map $f : A_1 \to A_2$ the image functor $f : (\wp(A_1), \supseteq) \to (\wp(A_2), \supseteq)$. Now we can define the category $Spec(\Sigma)$ of all Σ-specifications where the objects are specifications $\Gamma \subseteq Sen(\Sigma)$, and whose arrows are inverse inclusions $\Gamma_1 \supseteq \Gamma_2$, i.e., $Spec(\Sigma) =_{def} (\wp(Sen(\Sigma)), \supseteq)$. These constructions give, together with the assignment $Spec(\phi) =_{def} Sen(\phi)$, an indexed category $Spec : \mathsf{Sign} \to \mathsf{Cat}$, i.e., we have $Spec =_{def} Sen; \wp_{set}$. Secondly, we consider the model functor $Mod : \mathsf{Sign}^{\mathsf{op}} \to \mathsf{Cat}$. Let $co\wp_{cat} : \mathsf{Cat}^{op} \to \mathsf{Cat}$ be the (contravariant) power category functor that associates to each category $\mathsf{A} \in |\mathsf{Cat}|$, the partial ordering $co\wp_{cat}(\mathsf{A}) = (\wp(\mathsf{A}), \subseteq)$, and to each functor $F : \mathsf{A}_1 \to \mathsf{A}_2$ in Cat, the pre-image functor $F^{-1} : (\wp(\mathsf{A}_2), \subseteq) \to (\wp(\mathsf{A}_1), \subseteq)$, that assigns to any subcategory $\mathcal{B}_2 \in \wp(\mathsf{A}_2)$ the pre-image $F^{-1}(\mathcal{B}_2) \in \wp(\mathsf{A}_1)$. Now we can define the category $Sub(\Sigma)$ of all subcategories, where the objects are all subcategories $\mathcal{M} \subseteq Mod(\Sigma)$, and where the arrows are all inclusion functors $\mathcal{M}_1 \subseteq \mathcal{M}_2$, that is to say, $Sub(\Sigma) =_{def} (\wp(Mod(\Sigma)), \subseteq)$. These constructions give, together with the assignment $Sub(\phi) =_{def} Mod(\phi)^{-1}$, also an indexed category $Sub : \mathsf{Sign} \to \mathsf{Cat}$, i.e., we have $Sub =_{def} Mod^{op}; co\wp_{cat}$. Thirdly, given a specification $\Gamma \in |Spec(\Sigma)|$ we define the category $mod(\Sigma)(\Gamma)$ as the full subcategory of $Mod(\Sigma)$ determined by those models $M \in |Mod(\Sigma)|$ that satisfy all the sentences in Γ, i.e., we have

$$\textbf{(mod)} \quad |mod(\Sigma)(\Gamma)| =_{def} \{M \in |Mod(\Sigma)| \mid \forall \varphi \in \Gamma : M \models_\Sigma \varphi\}.$$

Since $\Gamma_1 \supseteq \Gamma_2$ implies $mod(\Sigma)(\Gamma_1) \subseteq mod(\Sigma)(\Gamma_2)$, these assignments provide a functor $mod(\Sigma) : Spec(\Sigma) \to Sub(\Sigma)$.

The "invariance" of the satisfaction condition is made precise with the help of the above constructions in the following

Theorem 1 (Satisfaction Condition = Indexed Functor) *([8]) Let $\mathcal{I}$ be an institution. Then for any $M_2 \in |Mod(\Sigma_2)|, \varphi_1 \in Sen(\Sigma_1)$ and any $\phi : \Sigma_1 \to \Sigma_2$ in* Sign*,*

$$\textbf{(SC)} \quad M_2 \models_{\Sigma_2} Sen(\phi)(\varphi_1) \iff Mod(\phi)(M_2) \models_{\Sigma_1} \varphi_1$$

iff the family of functors $mod(\Sigma) : Spec(\Sigma) \to Sub(\Sigma)$ constitutes an indexed functor $mod : Spec \Rightarrow Sub : \mathsf{Sign} \to \mathsf{Cat}$. □

In simpler words, given $\phi : \Sigma_1 \to \Sigma_2$ and a Σ_1- specification Γ, the satisfaction condition, as a natural transformation $mod : Spec \Rightarrow Sub : \mathsf{Sign} \to \mathsf{Cat}$, simply says that the expansion of the category of Γ-models along ϕ (or against $Mod(\phi)$) coincides with the category of models of the corresponding translated specification $Sen(\phi)(\Gamma)$. This is the "*semantics is invariant under change of notation*" perspective.

The following concept was first introduced in [7] under the name of institutional frames, but worked out in detail only in [8]. The discussion carried out above concerning the main purposes of this paper, together with the powerset constructions introduced before, justifies a more adequate name for it.

Definition/Proposition 1 (Power Institution) Given an institution $\mathcal{I} = (\mathsf{Sign}, Sen, Mod, \models)$, the canonical *power institution* associated to it is a 4-tuple $\mathcal{P}(\mathcal{I}) = (\mathsf{Sign}, Spec, Sub, mod)$, where $Spec : \mathsf{Sign} \to \mathsf{Cat}$ and $Sub : \mathsf{Sign} \to \mathsf{Cat}$ are indexed categories given by $Spec =_{def} Sen; \wp_{set}$ and $Sub =_{def} Mod^{op}; co\wp_{cat}$ as above, and $mod : Spec \Rightarrow Sub$ is the indexed functor from theorem 1.

The above definition/proposition says that the satisfaction relation of an institution, once freely extended to the level of specifications and subcategories of models, allows one to model the entire institutional logical structure by the simple notion of an indexed functor. This fulfills the first requirement of our methodology. In the sequel we begin to develop its second part, namely, to systematically identify more elementary transformations into this logical structure.

The first *elementary construction* we can identify concerns the translation of signatures. That is, to relate two power institutions $\mathcal{P}(\mathcal{I})$ and $\mathcal{P}(\mathcal{I}')$ we have to define, in any case, firstly a translation of signatures, e.g., from Sign into Sign′. Such a coding of signatures can be seen as defining a reduction together with a "retyping" of the "logical system" represented by $\mathcal{P}(\mathcal{I}')$ as formally described in the following

Proposition 1 (Φ-reduct of power institutions) Let $\mathcal{I}' = (\mathsf{Sign}', Sen', Mod', \models')$ be an institution and $\mathcal{P}(\mathcal{I}') = (\mathsf{Sign}', Spec', Sub', mod')$ be the associated power institution. Then, for any functor $\Phi : \mathsf{Sign} \to \mathsf{Sign}'$, the 4-tuple $\mathcal{P}(\mathcal{I}')_\Phi = (\mathsf{Sign}, \Phi; Spec', \Phi; Sub', \Phi \cdot mod')$ is also a power institution, called the *Φ-reduct* of $\mathcal{P}(\mathcal{I}')$. □

Intuitively, $\mathcal{P}(\mathcal{I}')_\Phi$ is just as $\mathcal{P}(\mathcal{I}')$, but its (natural) model theory, as given by $\Phi \cdot mod' : \Phi; Spec' \Rightarrow \Phi; Sub'$, is only able to take into account specifications out of signatures which can be reached via $\Phi : \mathsf{Sign} \to \mathsf{Sign}'$.

Any of the mappings we are going to analyze includes a Φ-reduction as the *first elementary construction.* That is, each of these mappings can be interpreted as defining actually a relation between a power institution $\mathcal{P}(\mathcal{I})$ and the Φ-reduct of another power institution $\mathcal{P}(\mathcal{I}')$.

Institution transformations reflect a situation where the model theory of $\mathcal{P}(\mathcal{I})$ can be simulated by the model theory of the reduct $\mathcal{P}(\mathcal{I}')_\Phi$. That is, institution transformations incorporate a *second elementary construction* which uses the model theoretic component $\Phi \cdot mod'$ of $\mathcal{P}(\mathcal{I}')_\Phi$, an "encoding" of specifications from $Spec$ to $\Phi; Spec'$ and a "recovering" of models from $\Phi; Sub'$ to Sub to obtain a new power institution thats coincides $\mathcal{P}(\mathcal{I})$.

Institution morphisms and plain maps of institutions are characterized by a new power institution which represents the common model theory of $\mathcal{P}(\mathcal{I})$ and of $\mathcal{P}(\mathcal{I}')_\Phi$ and which is thus related by a transformation with both $\mathcal{P}(\mathcal{I})$ and $\mathcal{P}(\mathcal{I}')_\Phi$. In this way, both concepts can be described as a particular combination of the first and the second elementary construction.

Simulations take into consideration a further power institution which can be constructed by a "semantical restriction" of $\mathcal{P}(\mathcal{I}')_\Phi$, i.e., by a special variant of the second elementary construction. Further simulations reflect the existence of a common model theory of $\mathcal{P}(\mathcal{I})$ and of the "semantical restriction" of $\mathcal{P}(\mathcal{I}')_{\Phi}$. Finally simple maps of institutions can be seen essentially as simulations equipped with a "logical axiomatization" of the "semantical restriction".

2 Institution Transformation

Consider the institutions of many sorted equational logic $\mathcal{I}_{\mathcal{MSEL}}$ and of many sorted equational partial logic $\mathcal{I}_{\mathcal{MSEPL}}$. Firstly we have a functor $\Phi : \mathsf{Sign}_{\mathcal{MSEL}} \to \mathsf{Sign}_{\mathcal{MSEPL}}$ which is identity. Secondly, any equation in context $X \vdash t = u \in Sen_{\mathcal{MSEL}}(\Sigma)$ can be translated into a conditional existence equation $(X \vdash \emptyset \to t \stackrel{e}{=} u) \in Sen_{\mathcal{MSEPL}}(\Sigma)$, i.e., actually we have a natural injection $\alpha : Sen_{\mathcal{MSEL}} \Rightarrow \Phi; Sen_{\mathcal{MSEPL}}$. As a reminder, an assignment of variables X into a partial algebra A is a solution of an existence equation $t \stackrel{e}{=} u$ if both sides can be evaluated in A and provide the same value. A conditional existence equation $(X \vdash t_1 \stackrel{e}{=} u_1, \dots, t_n \stackrel{e}{=} u_n \to t \stackrel{e}{=} u)$ is valid in A if any solution of $t_1 \stackrel{e}{=} u_1, \dots, t_n \stackrel{e}{=} u_n$ becomes a solution of $t \stackrel{e}{=} u$. Thirdly, any total Σ-algebra can be seen as a partial Σ-algebra as well, i.e., actually we have a natural inclusion $\beta : Mod_{\mathcal{MSEL}} \Rightarrow \Phi^{op}; Mod_{\mathcal{MSEPL}}$. This example motivates the following

Definition 1 (Institution transformation) Let $\mathcal{I}$ and $\mathcal{I}'$ be institutions. An *institution transformation* $(\Phi, \alpha, \beta) : \mathcal{I} \to \mathcal{I}'$ is given by a functor $\Phi : \mathsf{Sign} \to \mathsf{Sign}'$, a natural transformation $\alpha : Sen \Rightarrow \Phi; Sen' : \mathsf{Sign} \to \mathsf{Set}$, and a natural transformation $\beta : Mod \Rightarrow \Phi^{op}; Mod' : \mathsf{Sign}^{\mathsf{op}} \to \mathsf{Cat}$ such that the *institution transformation condition*

$$\textbf{(ITC)} \quad M \models_{\Sigma} \varphi \Longleftrightarrow \beta(\Sigma)(M) \models'_{\Phi(\Sigma)} \alpha(\Sigma)(\varphi),$$

holds for any $\Sigma \in |\mathsf{Sign}|, \varphi \in Sen(\Sigma)$, and $M \in |Mod(\Sigma)|$. □

The above definition is to be seen as a "pointwise" variant (for institutions) of the notion of pre-institution transformation [5].

Proposition 2 (ITC) Let $\Sigma \in |\mathsf{Sign}|$, $M \in |Mod(\Sigma)|$, $\varphi \in Sen(\Sigma)$ and $\Gamma \in Spec(\Sigma)$. Then the following conditions are equivalent:

1. **(ITC)** $\quad M \models_{\Sigma} \varphi \Longleftrightarrow \beta(\Sigma)(M) \models'_{\Phi(\Sigma)} \alpha(\Sigma)(\varphi)$.

2. $mod(\Sigma)(\Gamma) = \beta^{-1}(\Sigma)(mod'(\Phi(\Sigma))(\alpha(\Sigma)(\Gamma)))$. □

Using the power constructions introduced before, we can, out of the natural transformations in definition 1, obtain the natural transformations $\alpha \cdot \wp_{set} : Spec \Rightarrow \Phi; Spec'$ and $\beta^{op} \cdot co\wp_{cat} : \Phi; Sub' \Rightarrow Sub$. Note that $\beta^{op} : \Phi; Mod'^{op} \Rightarrow Mod^{op}$. This allows a characterization of institution transformations with the structures of power institutions.

Theorem 2 (Power institution transformation) *Let $\mathcal{I}, \mathcal{I}'$ be institutions, and let $\mathcal{P}(\mathcal{I})$ and $\mathcal{P}(\mathcal{I}')$ be the two associated power institutions. Then, given an institution transformation $(\Phi, \alpha, \beta) : \mathcal{I} \to \mathcal{I}'$, we have that the following equation*

$$mod = (\alpha \cdot \wp_{set}); (\Phi \cdot mod'); (\beta^{op} \cdot co\wp_{cat})$$

holds in Cat, *where $\alpha \cdot \wp_{set} : Spec \Rightarrow \Phi; Spec'$, and $\beta^{op} \cdot co\wp_{cat} : \Phi; Sub' \Rightarrow Sub$.*

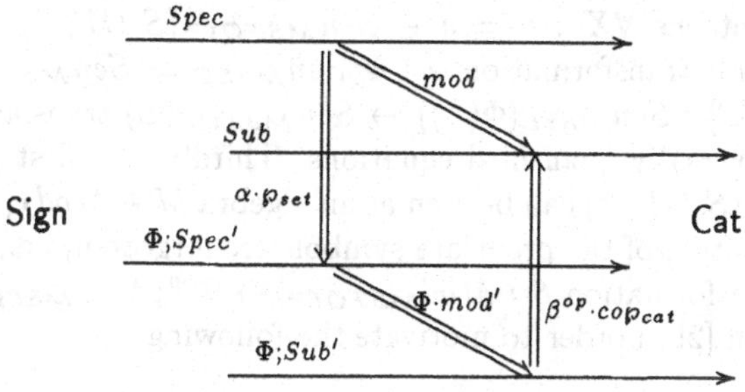

In the following, whenever clear from context, we may use β^{op} for $\beta^{op} \cdot co\wp_{cat}$ and α for $\alpha \cdot \wp_{set}$.

An important observation from the above result is implied by the first step in the proof and is stated objectively in the following

Corollary 1 Given an institution transformation $(\Phi, \alpha, \beta) : \mathcal{I} \to \mathcal{I}'$, we can construct a new power institution

$$\mathcal{P}(\mathcal{I}')_{\Phi,\alpha,\beta} =_{def} (\mathsf{Sign}, Spec, Sub, \alpha; (\Phi \cdot mod'); \beta^{op}),$$

whose model theory coincides with the one from $\mathcal{P}(\mathcal{I})$. □

This construction means that the model theoretic component *mod* of the power institution $\mathcal{P}(\mathcal{I})$ can be simulated via α and β^{op} by the model theoretic component $\Phi \cdot mod'$ of the reduct $\mathcal{P}(\mathcal{I}')_{\Phi}$. That is, we could construct $\mathcal{P}(\mathcal{I})$ by defining the components Sign, *Spec*, *Sub* and then borrowing the model theory from $\mathcal{P}(\mathcal{I}')_{\Phi}$ via α and β^{op}. Roughly speaking, α "encodes" the specifications of $\mathcal{P}(\mathcal{I})$ with the (reduced) logic of $\mathcal{P}(\mathcal{I}')_{\Phi}$, while β^{op} recovers the models of $\mathcal{P}(\mathcal{I})$ which can be represented in $\mathcal{P}(\mathcal{I}')_{\Phi}$.

Example 1 According to the injectivity of the transformation $(\Phi, \alpha, \beta) : \mathcal{I}_{\mathcal{MSEL}} \to \mathcal{I}_{\mathcal{MSEPL}}$ we can interpret $\mathcal{I}_{\mathcal{MSEL}}$ as a *restriction* of $\mathcal{I}_{\mathcal{MSEPL}}$, where α and β describe the restriction of sentences and models, respectively. The models of a specification $\Gamma \in |Spec_{\mathcal{MSEL}}(\Sigma)|$ are obtained by restricting $mod_{\mathcal{MSEPL}}(\Sigma)$ $(\alpha(\Sigma)(\Gamma))$ to total algebras, i.e., by intersecting this category of partial $(\Sigma, \alpha(\Sigma)$ $(\Gamma))$-models with the category of all total Σ-algebras. On the other hand, $\mathcal{I}_{\mathcal{MSEPL}}$ turns out to be a consistent extension of $\mathcal{I}_{\mathcal{MSEL}}$. That is, the validity of existence equations for partial algebras is a proper generalization of the validity of equations for total algebras since $(X \vdash \emptyset \to t \stackrel{e}{=} u)$ is valid in a total algebra A iff $X \vdash t = u$ is valid in A. □

3 Institution Morphism

Consider the institutions of many-sorted first-order predicate logic with equality $\mathcal{I}_{\mathcal{MSFOL}^=}$ and of many sorted equational logic $\mathcal{I}_{\mathcal{MSEL}}$, respectively. Firstly, forgetting predicate symbols, defines a functor $\Phi : \mathsf{Sign}_{\mathcal{MSFOL}^=} \to \mathsf{Sign}_{\mathcal{MSEL}}$ with $\Phi(S, OP, P) = (S, OP)$ for any first order signature (S, OP, P). Secondly, any equation in context $X \vdash t = u \in Sen_{\mathcal{MSEL}}(\Phi(S, OP, P))$ can be regarded as a first order sentence $\forall X : t = u \in Sen_{\mathcal{MSFOL}^=}(S, OP, P)$, i.e., we actually have a natural transformation $\alpha : \Phi; Sen_{\mathcal{MSEL}} \Rightarrow Sen_{\mathcal{MSFOL}^=}$, where the components $\alpha(\Sigma) : Sen_{\mathcal{MSEL}}(\Phi(\Sigma)) \to Sen_{\mathcal{MSFOL}}(\Sigma)$ translate equations in context into universally quantified equations. Thirdly, any first order model $M \in Mod_{\mathcal{MSFOL}^=}(S, OP, P)$ can be seen as an algebra $M \in Mod_{\mathcal{MSEL}}(S, OP)$, where the interpretation of the predicate symbols are forgotten, i.e., we actually have a natural transformation $\beta : Mod_{\mathcal{MSFOL}^=} \to \Phi^{op}; Mod_{\mathcal{MSEL}}$. This was the example used in [2] in order to motivate the following

Definition 2 (Institution morphism) Let $\mathcal{I} = (\mathsf{Sign}, Sen, Mod, \models)$ and $\mathcal{I}' = (\mathsf{Sign}', Sen', Mod', \models')$ be institutions. An *institution morphism* $(\Phi, \alpha, \beta) : \mathcal{I} \to \mathcal{I}'$ is given by a functor $\Phi : \mathsf{Sign} \to \mathsf{Sign}'$, a natural transformation $\alpha : \Phi; Sen' \Rightarrow Sen : \mathsf{Sign} \to \mathsf{Set}$, and a natural transformation $\beta : Mod \Rightarrow \Phi^{op}; Mod' : \mathsf{Sign}^{op} \to \mathsf{Cat}$ such that for each $\Sigma \in |\mathsf{Sign}|$ the *institution morphism condition*

$$\textbf{(IMC)} \quad M \models_\Sigma \alpha(\Sigma)(\varphi') \iff \beta(\Sigma)(M) \models'_{\Phi(\Sigma)} \varphi',$$

holds for any $M \in |Mod(\Sigma)|$ and any $\varphi' \in Sen'(\Phi(\Sigma))$. □

Proposition 3 (IMC) For each $\Sigma \in |\mathsf{Sign}|, M \in |Mod(\Sigma)|, \varphi' \in Sen'(\Phi(\Sigma))$, and $\Gamma' \in Spec'(\Phi(\Sigma))$, the following two conditions are equivalent:

1. **(IMC)** $M \models_\Sigma \alpha(\Sigma)(\varphi') \iff \beta(\Sigma)(M) \models'_{\Phi(\Sigma)} \varphi'$.
2. $mod(\Sigma)\left(\alpha(\Sigma)(\Gamma')\right) = \beta(\Sigma)^{-1}\left(mod'(\Phi(\Sigma))(\Gamma')\right)$. □

Theorem 3 (Institution morphism) *Let $\mathcal{I}, \mathcal{I}'$ be institutions, and let $\mathcal{P}(\mathcal{I})$ and $\mathcal{P}(\mathcal{I}')$ be the two associated power institutions. Then, given an institution morphism $(\Phi, \alpha, \beta) : \mathcal{I} \to \mathcal{I}'$, we have that the following equation*

$$(\alpha \cdot \wp_{set}); mod = (\Phi \cdot mod'); (\beta^{op} \cdot co\wp_{cat})$$

holds in Cat, *where* $\alpha \cdot \wp_{set} : \Phi; Spec' \Rightarrow Spec$ *and* $\beta^{op} \cdot co\wp_{cat} : \Phi; Sub' \Rightarrow Sub$.

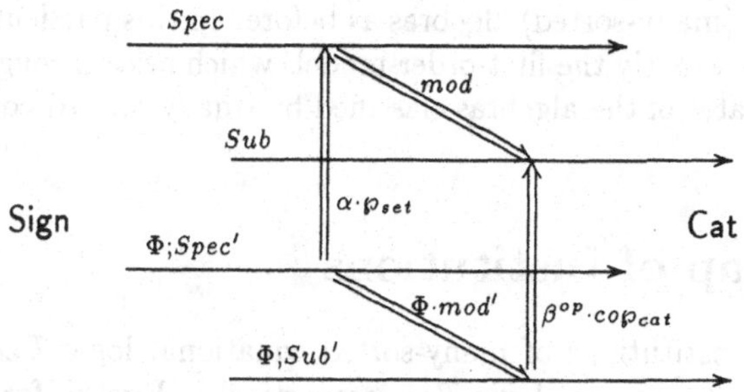

Corollary 2 Given an institution morphism $(\Phi, \alpha, \beta) : \mathcal{I} \to \mathcal{I}'$, there are two ways to construct new power institutions:

1. $\mathcal{P}(\mathcal{I})_\alpha =_{def} (\mathsf{Sign}, \Phi; Spec', Sub, \alpha; mod))$;
2. $\mathcal{P}(\mathcal{I}')_{\Phi,\beta} =_{def} (\mathsf{Sign}, \Phi; Spec', Sub, (\Phi \cdot mod'); \beta^{op})$,

whose model theory coincide. □

The relation of the above construction(s) with the original power institutions $\mathcal{P}(\mathcal{I})$ and $\mathcal{P}(\mathcal{I}')$ can be visualized in the following commutative diagram:

$$\begin{array}{ccccc} Spec & \overset{\alpha}{\Longleftarrow} & \Phi; Spec' & \overset{Id}{\Longrightarrow} & \Phi; Spec' \\ \Downarrow{\scriptstyle mod} & & \Downarrow{\scriptstyle k} & & \Downarrow{\scriptstyle \Phi\cdot mod'} \\ Sub & \underset{Id}{\Longrightarrow} & Sub & \underset{\beta^{op}}{\Longleftarrow} & \Phi; Sub' \end{array}$$

where $k = (\Phi \cdot mod'); \beta^{op} = \alpha; mod$. Firstly, note that $\mathcal{P}(\mathcal{I})_\alpha$ (and $\mathcal{P}(\mathcal{I})_{\Phi,\beta}$) can be viewed as providing a power institution transformation with both $\mathcal{P}(\mathcal{I})$ and $\mathcal{P}(\mathcal{I}')_\Phi$ according to the commutativity of the above squares. This in turn - see corollary 1 - implies that $\mathcal{P}(\mathcal{I})_\alpha$ can be seen as borrower of the model theory from $\mathcal{P}(\mathcal{I})$ via α and $\mathcal{P}(\mathcal{I}')_{\Phi,\beta}$ as borrower of the model theory from $\mathcal{P}(\mathcal{I}')_\Phi$ via β^{op}. The equality $\mathcal{P}(\mathcal{I})_\alpha = \mathcal{P}(\mathcal{I})_{\Phi,\beta}$ means now that the model theories of $\mathcal{P}(\mathcal{I})_\alpha$ and $\mathcal{P}(\mathcal{I}')_{\Phi,\beta}$ coincide, i.e., can be seen as the common model theory of $\mathcal{P}(\mathcal{I})$ and of the reduct $\mathcal{P}(\mathcal{I}')_\Phi$.

Example 2 According to the above corollary, we can construct the power institutions
$\mathcal{P}(\mathcal{I})_{\mathcal{MSFOL}^=,\alpha} = (\mathsf{Sign}_{\mathcal{MSFOL}^=}, \Phi; Spec_{\mathcal{MSEL}}, Sub_{\mathcal{MSFOL}^=}, \alpha; mod_{\mathcal{MSFOL}^=})$,
and
$\mathcal{P}(\mathcal{I})_{\mathcal{MSEL},\Phi,\beta} = (\mathsf{Sign}_{\mathcal{MSFOL}^=}, \Phi; Spec_{\mathcal{MSEL}}, Sub_{\mathcal{MSFOL}^=}, (\Phi \cdot mod_{\mathcal{MSEL}}); \beta^{op})$,
which "happens to be the same". In the particular case discussed in this example, the new power institution has both many-sorted first order signatures and models, and sentences out of many-sorted algebraic signatures. Moreover, specifications formed by equations-in-context classify now (many-sorted) first order

structures, and not (many-sorted) algebras as before. In this particular case, the common models are exactly the first-order models which arise through extension by arbitrary predicates of the algebras classified by (many sorted) equations. □

4 Plain Map of Institutions

Consider now the institutions of many-sorted equational logic $\mathcal{I}_{\mathcal{MSEL}}$ and of unsorted (untyped) equational logic $\mathcal{I}_{\mathcal{EL}}$, respectively. Firstly, forgetting sort symbols defines a functor $\Phi : \mathsf{Sign}_{\mathcal{MSEL}} \to \mathsf{Sign}_{\mathcal{EL}}$ with $\Phi(S, OP) = OP$ for each many-sorted equational signature $(S, OP) \in |\mathsf{Sign}_{\mathcal{MSEL}}|$. Secondly, any equation in context $X \vdash t = u \in Sen_{\mathcal{MSEL}}(S, OP)$ can be considered as a sentence in $Sen_{\mathcal{EL}}(\Phi(S, OP))$ once we omit the sort declaration of variables. This gives a natural transformation $\alpha : Sen_{\mathcal{MSEL}} \to \Phi; Sen_{\mathcal{EL}}$ with $\alpha(S, OP)(X \vdash t = u) =_{def} t = u$, for every equation in context $X \vdash t = u \in Sen_{\mathcal{MSEL}}(S, OP)$ and every signature $(S, OP) \in |\mathsf{Sign}_{\mathcal{MSEL}}|$. Thirdly, any unsorted algebra $M \in |Mod_{\mathcal{EL}}(\Phi(S, OP))|$ gives a many-sorted algebra $\beta(M) \in |Mod_{\mathcal{MSEL}}(S, OP)|$, where $\beta(M)_s =_{def} M$ for every $s \in S$, and $op^{\beta(M)} =_{def} op^M$ for each $op \in OP$. This situation delivers a functor $\beta(S, OP) : Mod_{\mathcal{EL}}(\Phi(S, OP)) \to Mod(S, OP)$, and globally a natural transformation $\beta : \Phi^{op}; Mod_{\mathcal{EL}} \to Mod_{\mathcal{MSEL}}$. This was the example used in [3] in order to motivate the following

Definition 3 (Plain map of institutions) Let $\mathcal{I} = (\mathsf{Sign}, Sen, Mod, \models)$ and $\mathcal{I}' = (\mathsf{Sign}', Sen', Mod', \models')$ be institutions. A *plain map of institutions* $(\Phi, \alpha, \beta) : \mathcal{I} \to \mathcal{I}'$ is given by a functor $\Phi : \mathsf{Sign} \to \mathsf{Sign}'$, a natural transformation $\alpha : Sen \Rightarrow \Phi; Sen' : \mathsf{Sign} \to \mathsf{Set}$, and a natural transformation $\beta : \Phi^{op}; Mod' \Rightarrow Mod : \mathsf{Sign}^{op} \to \mathsf{Cat}$ such that for each $\Sigma \in |\mathsf{Sign}|$ the *plain map of institutions condition*

$$\textbf{(IPC)} \quad M' \models'_{\Phi(\Sigma)} \alpha(\Sigma)(\varphi) \iff \beta(\Sigma)(M') \models_\Sigma \varphi,$$

holds for any $M' \in |Mod'(\Phi(\Sigma))|$ and any $\varphi \in Sen(\Sigma)$. □

Proposition 4 (IPC) For each $\Sigma \in |\mathsf{Sign}|$, $M' \in |Mod'(\Phi(\Sigma))|$, $\varphi \in Sen(\Sigma)$, and $\Gamma \in Spec(\Sigma)$, the following two conditions are equivalent:

1. **(IPC)** $M' \models'_{\Phi(\Sigma)} \alpha(\Sigma)(\varphi) \iff \beta(\Sigma)(M') \models_\Sigma \varphi$.
2. $mod'(\Phi(\Sigma))(\alpha(\Sigma)(\Gamma)) = \beta^{-1}(\Sigma)(mod(\Sigma)(\Gamma))$. □

Theorem 4 (Plain map of institutions) *Let $\mathcal{I}, \mathcal{I}'$ be institutions, and let $\mathcal{P}(\mathcal{I})$ and $\mathcal{P}(\mathcal{I}')$ be the two associated power institutions. Then, given a plain map of institutions $(\Phi, \alpha, \beta) : \mathcal{I} \to \mathcal{I}'$, we have that the following equation*

$$(\alpha \cdot \wp_{set}); (\Phi \cdot mod') = mod; (\beta^{op} \cdot co\wp_{cat}).$$

holds in Cat, *where* $\alpha \cdot \wp_{set} : Spec \Rightarrow \Phi; Spec'$, *and* $\beta^{op} \cdot co\wp_{cat} : Sub \Rightarrow \Phi; Sub'$.

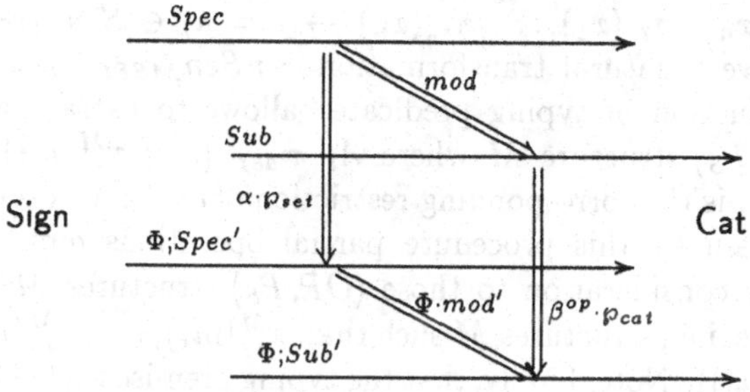

□

Similarly to the situation analysed by institution morphism, the above theorem allow us to construct new power institutions

- $\mathcal{P}(\mathcal{I})_\beta =_{def}$ (Sign, $Spec, \Phi; Sub', mod; \beta^{op}$);
- $\mathcal{P}(\mathcal{I}')_{\Phi,\alpha} =_{def}$ (Sign, $Spec, \Phi; Sub', \alpha; (\Phi \cdot mod')$),

whose model theory coincide.

Note that this abstract characterization is given again by a pair of power institution transformations. The construction is essentially the same as the one presented for institution morphisms. Again, the new constructions describe the common model theory of $\mathcal{P}(\mathcal{I})$ and of the reduct $\mathcal{P}(\mathcal{I}')_\Phi$. Only the roles of α and β^{op} are now interchanged.

Example 3 Consider again the institutions $\mathcal{I}_{\mathcal{MSEL}}$ and $\mathcal{I}_{\mathcal{EL}}$. According to the above collorary we have two power institutions, e.g., $\mathcal{P}(\mathcal{I})_{\mathcal{EL},\Phi,\alpha}$ = (Sign$_{\mathcal{MSEL}}$, $Spec_{\mathcal{MSEL}}, \Phi; Sub_{\mathcal{EL}}, \alpha; (\Phi \cdot mod')$). Summarizing, this power institution has signatures and specifications of many-sorted equational logic, and models of unsorted equational logic. More precisely, specifications formed by equations in context classify now unsorted algebras, and not many-sorted algebras as before. Therefore, this construction summarizes a situation in which we are using a syntax from one logic – in this case of many-sorted equational logic – and "borrowing" models from another one – here unsorted equational logic. Note, that these common models can be seen as many-sorted models which have the same carrier everywhere. □

5 Simulations

Consider the institutions of many-sorted equational logic $\mathcal{I}_{\mathcal{MSEL}}$ and of first order predicate logic $\mathcal{I}_{\mathcal{FOL}^=}$, respectively. Firstly, we can translate any many sorted signature $\Sigma = (S, OP)$ into the first order signature (OP, P_S) with unary typing predicates $P_S =_{def} \{\pi_s \mid s \in S\}$. This defines a functor Φ : Sign$_{\mathcal{MSEL}} \rightarrow$

$\mathsf{Sign}_{\mathcal{FOL}^=}$. Secondly, any equation in context $\varphi = (X \vdash t = u) \in Sen_{\mathcal{MSEL}}(\Sigma)$ where X is a S-sorted family of variables, can be translated into a first order sentence $(\forall x_1, \ldots, x_n : \pi_{s_1}(x_1), \ldots, \pi_{s_n}(x_n) \rightarrow t = u) \in Sen_{\mathcal{FOL}^=}(OP, P_S)$, i.e., we actually have a natural transformation $\alpha : Sen_{\mathcal{MSEL}} \Rightarrow \Phi; Sen_{\mathcal{FOL}^=}$. Thirdly, the introduction of typing predicates allows to extract a Σ-algebra A out of any (OP, P_S)-structure M where $A_s =_{def} \{m \mid \pi_s^M(m)\}$ and $op^A : A_{s_1} \times \cdots \times A_{s_n} \rightarrow A_s$ is the corresponding restriction of $op^M : M \times \cdots \times M \rightarrow M$. In general, we obtain by this procedure partial operations op^A, so that we have to restrict our consideration to those (OP, P_S)-structures M that represent total Σ-algebras, i.e., structures M such that $\pi_{s_1}^M(m_1), \ldots, \pi_{s_n}^M(m_n)$ implies $\pi_s^M(op^M(m_1, \ldots, m_n))$. Note, finally, that the typing premise $\pi_{s_1}(x_1), \ldots, \pi_{s_n}(x_n)$ in $\alpha(\Sigma)(\varphi)$ ensures that all representatives of an algebra A will satisfy $\alpha(\Sigma)(\varphi)$ if A satisfies φ . The implication into the other direction would be valid even if we omit the typing premise. This was the example used in [1] in order to motivate the following

Definition 4 (Simulation) Let $\mathcal{I}$ and $\mathcal{I}'$ be institutions. Further let be given a functor $\Phi : \mathsf{Sign} \rightarrow \mathsf{Sign}'$, a natural transformation $\alpha : Sen \Rightarrow \Phi; Sen' : \mathsf{Sign} \rightarrow \mathsf{Set}$, and a *partial natural transformation* $\beta_p : \Phi^{op}; Mod' \Rightarrow Mod : \mathsf{Sign}^{\mathsf{op}} \rightarrow \mathsf{Cat}$, i.e., a family of functors $\beta(\Sigma) : dom(\beta_p(\Sigma)) \rightarrow Mod(\Sigma)$ with $dom(\beta_p(\Sigma))$ a subcategory of $Mod'(\Phi(\Sigma))$ such that for any $\phi : \Sigma_1 \rightarrow \Sigma_2$ in Sign $Mod'(\Phi(\phi))(dom(\beta_p(\Sigma_2))) \subseteq dom(\beta_p(\Sigma_1))$ and $(Mod'(\Phi(\phi)))_{|\beta}; \beta(\Sigma_1) = \beta(\Sigma_2); Mod(\phi)$ for the corresponding restricted model functor $(Mod'(\Phi(\phi)))_{|\beta} : dom(\beta_p(\Sigma_2)) \rightarrow dom(\beta_p(\Sigma_1))$. Then, $(\Phi, \alpha, \beta_p) : \mathcal{I} \rightarrow \mathcal{I}'$ is a *simulation* if for each $\Sigma \in |\mathsf{Sign}|$ the following conditions hold:

1. The *simulation condition*

$$\textbf{(SIC)} \quad M' \models'_{\Phi(\Sigma)} \alpha(\Sigma)(\varphi) \iff \beta(\Sigma)(M') \models_\Sigma \varphi,$$

   holds for any $M' \in |dom(\beta_p(\Sigma))|$ and any $\varphi \in Sen(\Sigma)$;

2. the functor $\beta(\Sigma) : dom(\beta_p(\Sigma)) \rightarrow Mod(\Sigma)$ is surjective on $|Mod(\Sigma)|$.

□

The partiality of the natural transformation $\beta_p : \Phi^{op}; Mod' \Rightarrow Mod : \mathsf{Sign}^{\mathsf{op}} \rightarrow \mathsf{Cat}$ can be equivalently described as a span of natural transformations, as introduced in the following

Corollary 3 (Partial = span) Let be given a partial natural transformation $\beta_p : \Phi^{op}; Mod' \Rightarrow Mod : \mathsf{Sign}^{\mathsf{op}} \rightarrow \mathsf{Cat}$. Then the inclusions $dom(\beta_p(\Sigma)) \subseteq Mod'(\Phi(\Sigma))$ constitute a natural inclusion $in : (\Phi^{op}; Mod')_{|\beta} \Rightarrow \Phi^{op}; Mod' : \mathsf{Sign}^{\mathsf{op}} \rightarrow \mathsf{Cat}$. Moreover, the partial naturality condition is equivalent to the condition that the functors $\beta(\Sigma) : dom(\beta_p(\Sigma)) \rightarrow Mod(\Sigma)$, $\Sigma \in |\mathsf{Sign}|$ constitutes a natural transformation $\beta : (\Phi^{op}; Mod')_{|\beta} \Rightarrow Mod : \mathsf{Sign}^{\mathsf{op}} \rightarrow \mathsf{Cat}$. □

We first analyse simulations without concern for the surjectivity condition, since it will play no essential rôle for the structural characterization (a discussion which takes into account also this additional requirement will be postponed to the end of this section).

Consider now the Φ-reduct of proposition 1. The partial naturality of β_p induces a semantical restriction on the (natural) model theory of $\mathcal{P}(\mathcal{I}')_\Phi$, which is in fact a further special kind of power institution transformation. This idea is precisely reflected in the next

Proposition 5 (Semantical restriction) Let $\mathcal{P}(\mathcal{I}')_\Phi$ be a reduct as in proposition 1. Then, given a partial natural transformation $\beta_p : \Phi^{op}; Mod' \Rightarrow Mod : \mathsf{Sign}^{\mathsf{op}} \to \mathsf{Cat}$ as in definition 4, we can construct the *semantical restriction* $(\mathcal{P}(\mathcal{I}')_\Phi)_{|\beta}$ of $\mathcal{P}(\mathcal{I}')_\Phi$ as given by

$$(\mathcal{P}(\mathcal{I}')_\Phi)_{|\beta} = (\mathsf{Sign}, \Phi; Spec', (\Phi; Sub')_{|\beta}, (\Phi \cdot mod')_{|\beta}),$$

where $(\Phi; Sub')_{|\beta} =_{def} (\Phi^{op}; Mod')^{op}_{|\beta}; co\wp_{cat}$ and $(\Phi \cdot mod')_{|\beta} =_{def} (\Phi \cdot mod'); (in^{op} \cdot co\wp_{cat})$. □

Intuitively, $(\mathcal{P}(\mathcal{I}')_\Phi)_{|\beta}$ is just as $\mathcal{P}(\mathcal{I}')_\Phi$, but its (natural) model theory is restricted to those subcategories which belong to $\wp(dom(\beta_p(\Sigma)))$, for each $\Sigma \in |\mathsf{Sign}|$.

Up to the semantical restriction introduced above, the concept of simulation presents a striking similarity with the concept of plain maps of institutions, as one can notice by looking at the simulation condition. Therefore, its structural characterization by means of power institutions should also present the same resemblance.

Theorem 5 (Simulation) *Let $\mathcal{I}, \mathcal{I}'$ be institutions, and let $\mathcal{P}(\mathcal{I})$ and $\mathcal{P}(\mathcal{I}')$ be the two associated power institutions. Then, given a simulation $(\Phi, \alpha, \beta_p) : \mathcal{I} \to \mathcal{I}'$, we have that the following equation*

$$(\alpha \cdot \wp_{set}); (\Phi \cdot mod')_{|\beta} \quad = \quad mod; (\beta^{op} \cdot co\wp_{cat})$$

holds in Cat.

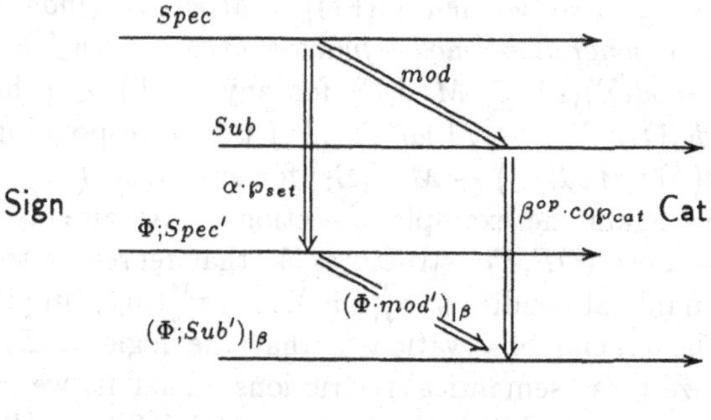

□

Similarly to what we have done so far, the above theorem gives for free two new constructions, $\mathcal{P}(\mathcal{I})_\beta$ and $(\mathcal{P}(\mathcal{I}')_\Phi)|_{\beta,\alpha}$ whose model theory coincide.

Example 4 Consider again the institutions of $\mathcal{I}_{\mathcal{MSEL}}$ and $\mathcal{I}_{\mathcal{FOL}^=}$. Consider the power institution $(\mathcal{P}(\mathcal{I}')_\Phi)_{|\beta,\alpha} = (\mathsf{Sign}_{\mathcal{MSEL}}, Spec_{\mathcal{MSEL}}, (\Phi; Sub')_{|\beta}, \alpha; (\Phi \cdot mod')_{|\beta})$. It has signatures and specifications of many-sorted equational logic. The common model theory of both institutions, as given by $(\mathcal{P}(\mathcal{I}')_\Phi)_{|\beta,\alpha}$ can be seen as consisting of all first-order models which have a straightforward interpretation as total many-sorted algebra, once one makes the construction introduced in the beginning of this section, i.e., simulation of sorted sets by corresponding predicates and use of typing premises. □

To complete the analysis of the notion of simulation we have to clarify the rôle of the surjectivity condition. According to theorem 5, we can show that for any $\Gamma \in |Spec(\Sigma)|$

$$\beta(\Sigma)^{-1}(mod(\Sigma)(\Gamma)) = dom(\beta_p(\Sigma)) \cap mod'(\Phi(\Sigma))(\alpha(\Sigma)(\Gamma)).$$

Now, the surjectivity condition ensures $|\beta(\Sigma)(\beta(\Sigma)^{-1}(mod(\Sigma)(\Gamma)))| = |mod(\Sigma)(\Gamma)|$ and thus that $|mod(\Sigma)(\Gamma)| = |\beta(\Sigma)(dom(\beta_p(\Sigma)) \cap mod'(\Phi(\Sigma))(\alpha(\Sigma)(\Gamma)))|$. This equation could be interpreted in the sense that the model theoretic component of the power institution $\mathcal{P}(\mathcal{I})$ can be simulated via Φ, α, and β_p by the model theoretic component of $\mathcal{P}(\mathcal{I}')$. Even though this kind of simulation is not "natural" it seems to be of relevance for applications.

6 Simple Map of Institutions

In the following we introduce some additional background material, which is needed to follow this section, since we won't introduce Meseguer's concept of simple map of institutions [3] from scratch.

Let $\mathcal{I}$ be an institution . Then, the category Th_0 of *theories* has as objects pairs (Σ, Γ) with $\Sigma \in |\mathsf{Sign}|$ and $\Gamma \in |Spec(\Sigma)|$, and as arrows $(\phi, \supseteq) : (\Sigma_1, \Gamma_1) \to (\Sigma_2, \Gamma_2)$ signature morphisms $\phi : \Sigma_1 \to \Sigma_2$ such that $\Gamma_2 \supseteq Sen(\phi)(\Gamma_1)$. The satisfaction condition ensures according to theorem 1 that $\Gamma_2 \supseteq Sen(\phi)(\Gamma_1)$ implies $mod(\Sigma_2)(\Gamma_2) \subseteq mod(\Sigma_2)(Sen(\phi)(\Gamma_1)) = Mod(\phi)^{-1}(mod(\Sigma_1)(\Gamma_1))$, so that we can define a *generalized model functor* $Mod_{\models} : \mathsf{Th}_0{}^{op} \to \mathsf{Cat}$ with $Mod_{\models}(\Sigma, \Gamma) =_{def} mod(\Sigma)(\Gamma) \subseteq Mod(\Sigma)$ for any $(\Sigma, \Gamma) \in |\mathsf{Th}_0|$ and with $Mod_{\models}(\phi, \supseteq) : Mod_{\models}(\Sigma_2, \Gamma_2) \to Mod_{\models}(\Sigma_1, \Gamma_1)$ the corresponding restriction of the functor $Mod(\phi) : Mod(\Sigma_2) \to Mod(\Sigma_1)$ for any arrow $(\phi, \supseteq)$ in Th_0.

We consider now again the example of section 5. As already pointed out we can characterize those (OP, P_S)-structures M that represent total (S, OP)-algebras by the semantical condition $\pi^M_{s_1}(m_1), \ldots, \pi^M_{s_n}(m_n)$ implies $\pi^M_s(op^M(m_1, \ldots, m_n))$. The crucial observation is that the logic of $\mathcal{I}_{\mathcal{FOL}^=}$ is rich enough to axiomatize these semantical restrictions. That is, we can assign to any many sorted signature (S, OP) a specification $\Psi(S, OP)_{sp} = \{(\forall x_1, \ldots, x_n :$

$\pi_{s_1}(x_1),\ldots,\pi_{s_n}(x_n) \to \pi_s(op(x_1,\ldots,x_n))) \mid op : s_1 \ldots s_n \to s$ in $(S, OP)\}$ such that the subcategory $mod_{\mathcal{FOL}^=}(OP, P_S)(\Psi(S,OP)_{sp})$ of $Mod_{\mathcal{FOL}^=}(OP, P_S)$ describes exactly all representatives of total (S, OP)-algebras.

The above discussion, adapted from [3], motivates the next

Definition/Proposition 2 (α-simple functor) Let $\mathcal{I}$ and $\mathcal{I}'$ be institutions. Let $\Phi : \mathsf{Sign} \to \mathsf{Sign}'$ and $\Psi : Sign \to \mathsf{Th}_0{}'$ be functors, and $\alpha : Sen \Rightarrow \Phi; Sen' : \mathsf{Sign} \to \mathsf{Cat}$ be a natural transformation. Then the functor $\Psi_\alpha : \mathsf{Th}_0 \to \mathsf{Th}_0{}'$ defined by $\Psi_\alpha(\Sigma,\Gamma) =_{def} (\Phi(\Sigma), \alpha(\Sigma)(\Gamma) \cup \Psi(\Sigma)_{sp})$, for each $(\Sigma,\Gamma) \in |\mathsf{Th}_0|$, is is called *$\alpha$-simple.* □

Definition 5 (Simple map of institutions) Let $\mathcal{I}$ and $\mathcal{I}'$ be institutions. A *simple map of institutions* $(\Phi, \Psi_\alpha, \alpha, \beta_{\models}) : \mathcal{I} \to \mathcal{I}'$ is given by a functor Φ : $\mathsf{Sign} \to \mathsf{Sign}'$, a α-simple functor $\Psi_\alpha : \mathsf{Th}_0 \to \mathsf{Th}_0{}'$, a natural transformation $\alpha : Sen \Rightarrow \Phi; Sen'$, and a natural transformation $\beta_{\models} : \Psi_\alpha^{op}; Mod'_{\models'} \Rightarrow Mod_{\models}$: $\mathsf{Th}_0 \to \mathsf{Cat}$, such that the *simple map of institutions condition*

$$\textbf{(SMC)} \quad M' \models'_{\Phi(\Sigma)} \alpha(\Sigma)(\varphi) \iff \beta_{\models}(\Sigma,\emptyset)(M') \models_\Sigma \varphi,$$

holds for each $\Sigma \in |\mathsf{Sign}|$, $M' \in |Mod'_{\models'}(\Psi_\alpha(\Sigma,\emptyset))|$, and $\varphi \in Sen(\Sigma)$. □

Now, the results which put simple maps and simulations under the same perspective are given by the following two propositions.

Proposition 6 Let $\beta_{\models} : \Psi_\alpha^{op}; Mod'_{\models'} \Rightarrow Mod_{\models} : \mathsf{Th}_0 \to \mathsf{Cat}$ be a natural transformation as in definition 5. Then the following hold:

1. There is a restricted natural transformation $\beta : \Psi^{op}; Mod'_{\models'} \Rightarrow Mod$: $\mathsf{Sign}^{op} \to \mathsf{Cat}$ where, for each $\Sigma \in |\mathsf{Sign}|$, $\beta(\Sigma) =_{def} \beta_{\models}(\Sigma,\emptyset)$.
2. There's a natural inclusion $in_\Psi : \Psi^{op}; Mod'_{\models'} \Rightarrow \Phi^{op}; Mod' : \mathsf{Sign}^{op} \to \mathsf{Cat}$.
3. There's a partial natural transformation $\gamma_p : \Phi^{op}; Mod' \Rightarrow Mod : \mathsf{Sign}^{op} \to \mathsf{Cat}$. □

Proposition 7 (Logical semantical restriction) Let $\mathcal{P}(\mathcal{I}')_\Phi$ be a power institution reduct as in proposition 1. Then we can define the *logical semantical restriction* $(\mathcal{P}(\mathcal{I}')_\Phi)_{|\Psi}$ of $\mathcal{P}(\mathcal{I}')_\Phi$ as

$$(\mathcal{P}(\mathcal{I}')_\Phi)_{|\Psi} = (\mathsf{Sign}, \Phi; Spec', (\Phi; Sub')_{|\Psi}, (\Phi \cdot mod')_{|\Psi}),$$

where $(\Phi; Sub')_{|\Psi} =_{def} \Psi; Mod'^{op}_{\models'}; co\wp_{cat}$, and $(\Phi \cdot mod')_{|\Psi} =_{def} (\Phi \cdot mod'); (in_\Psi^{op} \cdot co\wp_{cat})$. □

Note that $(\mathcal{P}(\mathcal{I}')_\Phi)_{|\Psi}$ is just as $(\mathcal{P}(\mathcal{I}')_\Phi)_{|\beta}$, but now $dom(\beta_p(\Sigma))$ can be logically axiomatized.

Theorem 6 (Simple map of institutions) *Let $\mathcal{I}$ and $\mathcal{I}'$ be institutions, and let $\mathcal{P}(\mathcal{I})$ and $\mathcal{P}(\mathcal{I}')$ be the corresponding power institutions. Then, given a simple map of institutions $(\Phi, \Psi_\alpha, \alpha, \beta_\models) : \mathcal{I} \to \mathcal{I}'$, the following equation*

$$(\alpha \cdot \wp_{set}); (\Phi \cdot mod')_{|\Psi} = mod; (\beta^{op} \cdot co\wp_{cat})$$

holds in Cat. □

7 Concluding Remarks

This paper was concerned with a systematic analysis (study) of some particular notions of mappings between institutions, namely institution transformations, institution morphisms, plain maps of institutions, simulations and simple maps of institutions. More precisely, using the concept of power institutions, we have shown that these mappings can be understood as comprising a number of more fundamental, elementary constructions, which we have identified as *typing reduction along functors between signatures, borrowing of models, common model theory, semantical restriction, and logical semantical restriction.* How these mappings are actually reflecting the above mentioned fundamental denotations or constructions, can be seen by considering the following (overview) table:

	Transf.	Morph.	Plain	Simul.	Simple
Reduction	√	√	√	√	√
Borrowing	√	√	√	√	√
Common Model		√	√	√	√
Restriction				√	√
Log.Restriction					√

According to Definition/Proposition 1, as far as information content is concerned, institutions and power institutions are just different descriptional tools. The first as a formalization of the notion of a logical system, and the second one an equivalent (yet more appropriate) one to model the entire logical structure of an institution and hence (actually our claim) to study maps of institutions. Intuitively, the fact that each of the mappings studied in this paper factorize through a number of power institutions, suggest that they are actually reflecting a number of *corresponding construction of institutions* and hence, as motivated in the introduction, a number of elementary transformations of institutional *logical structures.* The only point here is that the overview table shows these transformations at the more abstract level of specifications and subcategory of models. In a forthcoming paper we'll take these results back to the language of institutions with the ultimate goal of providing a very fine grained unifying classification of these mappings based on this systematic transformation of the entire logical structure of the corresponding institutions.

Acknowledgments: We are indebted to Martin Grosse-Rhode for his careful reading of the manuscript and for his many very helpful suggestions to improve the exposition. We cordially thank Harmut Ehrig and Fabio Gaducci for their valuable suggestions that have helped us to significantly improve the presentation as well. Last but not least, we are very grateful to the anonymous referees for their positive suggestions and criticism which, amongst other things, led us to carefully revise the introduction of this paper.

References

[1] Maura Cerioli. *Relationships between Logical Formalisms.* PhD thesis, Università di Pisa–Genova–Udine, 1993. TD-4/93.

[2] J. A. Goguen and R. M. Burstall. Institutions: Abstract Model Theory for Specification and Programming. *Journal of the ACM*, 39(1):95–146, January 1992.

[3] J. Meseguer. General logics. In H.-D. Ebbinghaus et. al., editor, *Logic colloquium '87*, pages 275–329. Elsevier Science Publishers B. V.,North Holland, 1989.

[4] J. Meseguer and M-O. Narciso. From abstract data types to logical frameworks. In *Recent Trends in Data Type Specification*, pages 48–80. Springer, LNCS 906, 1995.

[5] S. Salibra and G. Scollo. Interpolation and compactness in categories of pre-institutions. *Mathematical Structures in Computer Science*, 6:261–286, 1996.

[6] A. Tarlecki. Moving between logical systems. In *Recent Trends in Data Type Specification*, pages 478–502. Springer, LNCS 1130, 1996.

[7] U. Wolter. Institutional frames. In *Recent Trends in Data Type Specification*, pages 469–482. Springer, LNCS 906, 1995.

[8] U. Wolter and A. Martini. Shedding new light in the world of logical systems. In *Category Theory and Computer Science, 7th International Conference, CTCS'97*, pages 159–176. Springer, LNCS 1290, 1997.

Colimits of Order-Sorted Specifications

Till Mossakowski

Department of Computer Science,
University of Bremen, P.O.Box 33 04 40, D-28334 Bremen, Germany,
E-mail till@informatik.uni-bremen.de

Abstract. We prove cocompleteness of the category of CASL signatures, of monotone signatures, of strongly regular signatures and of strongly locally filtered signatures. This shows that using these signature categories is compatible with a pushout or colimit based module system.

1 Introduction

> "Given a species of structure, say widgets, then the result of interconnecting a system of widgets to form a super-widget corresponds to taking the *colimit* of the diagram of widgets in which the morphisms show how they are interconnected." J. Goguen [8]

An important application of this is the slogan "Putting theories together to make specifications" [3]. That is, specifications should be developed in a modular way, using colimits to combine different modules properly.

An orthogonal question is that of the *logic* that is used to specify the individual modules. Order-sorted algebra is a logic that has been proposed as a means to deal with exceptions, partiality and inheritance. See, among others, Goguen and Meseguer's survey paper [10]. The specification languages OBJ3 [11], CafeOBJ [7] and CASL [15, 4] are all based on (extensions of) order-sorted algebra.

Now the combination (or should we say "pushout"?) of the two previous paragraphs would be an approach to order-sorted algebra where signatures and theories indeed can be combined with colimits. Surprisingly, such an approach was proposed only recently by Haxthausen and Nickl [12]. They examine how conditions like regularity or local filtration proposed by [10] interact with pushouts. They prove the existence of pushouts of signatures which are constrained by these conditions. Unfortunately, the existence of pushouts is guaranteed only under some severe restrictions. In particular, all signature morphisms have to be embeddings.

In this work, we propose an alternative way to deal with colimits of order-sorted specifications. Rather than trying to find restrictive conditions under which pushouts of order-sorted signatures preserve properties like monotonicity, regularity or local filtration, we change the perspective and allow the pushout of regular signatures to differ from the pushout of ordinary signatures. Technically, we drop the implicit assumption of [12] that the functor forgetting regularity (or other conditions) should preserve colimits.

The main result states that the categories of monotone, of (strongly) regular and of (strongly) locally filtered signatures, resp., together with morphisms which preserve in some natural way the monotone, (strongly) regular and (strongly) locally filtered structure, resp., all are cocomplete. Thus we can combine order-sorted signatures in an unrestricted way.

In this paper, we choose the subsorting approach of CASL, since CASL does not impose any condition on order-sorted signatures and thus seems to be the most general approach. The paper is organized as follows: We introduce some preliminaries about subsorting in CASL and some category theory results in Sect. 2. The following sections deal with pure subsorting (without any conditions), monotonicity, partial orderedness, regularity and local filtration, resp. In particular, in each section a cocompleteness theorem is proved. Sect. 8 contains the conclusions.

2 Preliminaries

In this section we first recall the CASL approach to order-sorted algebra, and then some category theory that will be useful for dealing with colimits.

2.1 Subsorting in CASL

CASL stands for "Common Algebraic Specification Language" and is the central language within the family of CoFI languages. CoFI[1] is an initiative to design *a Common Framework for Algebraic Specification and Development* [15].

We are interested in CASL here since it has a new and very general approach to subsorting, which is combined with overloading and partiality.

We first recall the notion of CASL-signature and signature morphism from [5, 4], using a slightly different but equivalent formulation.

Definition 2.1. A *CASL subsorted signature* $(S, F, TF, P, \leq_S)$ consists of

- a set S of *sorts*
- an $S^* \times S$-sorted family $F = (F_{w,s})_{w \in S^*, s \in S}$ of *function symbols,*
- an $S^* \times S$-sorted family $TF = (TF_{w,s} \subseteq F_{w,s})_{w \in S^*, s \in S}$ of subsets indicating the *total function symbols,*
- an S^*-sorted family $P = (P_w)_{w \in S^*}$ of *predicate symbols* and
- a pre-order (i.e. a reflexive transitive relation) $\leq_S$ of *subsort embeddings* on the set S of sorts. □

Notation: The relation $\leq_S$ naturally extends to sequences of sorts. We drop the subscript S when obvious from the context. We write $f: w \to s \in F$ for $f \in F_{w,s}$ (even if f is a partial function symbol, which is denoted by $f : w \to ? \; s$

[1] CoFI is an acronym for *Common Framework Initiative* and is pronounced like 'coffee'.

in CASL specifications) and $p : w$ for $p \in P_w$. $f\colon w \to s \in F$ and $p : P_w$ are called *profiles*, w is called the *arity* and s the *coarity*.

Note that the signatures are not required to be monotonic, regular or locally filtered as in [10]. This decision was taken in order both to allow arbitrary overloading and to get a cocompleteness theorem (see Theorem 3.1 below).

For a subsorted signature $\Sigma = (S, F, TF, P, \leq_S)$, we define *overloading relations* (also called *monotonicity orderings*), $\sim_F$ and $\sim_P$, for function and predicate symbols, respectively:

Definition 2.2. For $f : w_1 \to s_1, f : w_2 \to s_2 \in F, f : w_1 \to s_1 \sim_F f : w_2 \to s_2$ iff there exist $w \in S^*$ with $w \leq w_1, w_2$ and $s \in S$ with $s \geq s_1, s_2$.

For $p : w_1,\ p : w_2 \in P$, $p : w_1 \sim_P\ p : w_2$ iff there exists $w \in S^*$ with $w \leq w_1, w_2$.

These overloading relations lead to semantical identifications of different profiles in the models.

Definition 2.3. Given signatures $\Sigma = (S, F, TF, P, \leq)$ and $\Sigma' = (S', F', TF', P', \leq')$, a signature morphism $\sigma\colon \Sigma \to \Sigma'$ consists of

- a map $\sigma^S\colon S \to S'$,
- a map $\sigma^F_{w,s}\colon F_{w,s} \to F'_{\sigma^{S^*}(w), \sigma^S(s)}$[2] for each $w \in S^*, s \in S$ and
- a map $\sigma^P_w\colon P_w \to P'_{\sigma^{S^*}(w)}$ for each $w \in S^*$

such that

- subsorting is preserved, i. e. $s_1 \leq s_2$ implies $\sigma^S(s_1) \leq \sigma^S(s_2)$ for $s_1, s_2 \in S$,
- totality is preserved, i. e. $\sigma^F_{w,s}(TF_{w,s}) \subseteq TF'_{\sigma^{S^*}(w), \sigma^S(s)}$,
- the overloading relations are preserved[3], i. e. $f : w_1 \to s_1 \sim_F f : w_2 \to s_2$ implies $\sigma^F_{w_1,s_1}(f) : \sigma^{S^*}(w_1) \to \sigma^S(s_1) \sim_F \sigma^F_{w_2,s_2}(f) : \sigma^{S^*}(w_2) \to \sigma^S(s_2)$ and $p : w_1 \sim_P\ p : w_2$ implies $\sigma^P_{w_1}(p) : \sigma^{S^*}(w_1) \sim_P\ \sigma^P_{w_2}(p) : \sigma^{S^*}(w_2)$.

Note that, due to preservation of subsorting, the last condition can be simplified to: $f : w_1 \to s_1 \sim_F f : w_2 \to s_2$ implies $\sigma^F_{w_1,s_1}(f) = \sigma^F_{w_2,s_2}(f)$ (and similarly for the predicate symbols).

This gives us a category **CASLSig** of subsorted CASL signatures and signature morphisms. □

Since we will later on specify the CASL signature category in CASL itself, we now give a brief overview over basic specifications in CASL. A basic specification in CASL consists of a list of basic items. A basic item may either be a declaration of some sorts, subsorts, partial or total functions or predicates, or be an axiom. Axioms are usual first-order formulae over the following atomic formulae:

---

[2] σ^{S^*} is the extension of σ^S to finite strings

[3] The purpose of preservation of overloading is the following: the overloading relations lead to semantical identifications in the models. These are preserved by taking reducts only if the overloading relations are preserved by signature morphisms.

- Existential ($\stackrel{e}{=}$) or strong ($=$) equalities between terms,
- applications of predicates to a list of terms,
- definedness tests (*defined* t), which check if a term is defined, or
- membership tests ($t \in s$), which check if a term belongs to a subsort.

As an example, consider the specification of **CASLSig** within CASL itself given in Fig. 1.

spec ORDEREDSTRINGS =
    sorts S, S^*
    preds $\_ \leq \_ : S \times S$
        $\_ \leq \_ : S^* \times S^*$
    ops $\lambda : S^*$
        $\_\,\_ : S \times S^* \rightarrow S^*$
    vars $s, s_1, s_2, s_3 : S;\ w_1, w_2 : S^*$
    axioms $s \leq s$
        $s_1 \leq s_2 \wedge s_2 \leq s_3 \Rightarrow s_1 \leq s_3$
        $\lambda \leq \lambda$
        $w_1 \leq w_2 \wedge s_1 \leq s_2 \Leftrightarrow s_1 w_1 \leq s_2 w_2$
        $s_1 w_1 = s_2 w_2 \Rightarrow (s_1 = s_2 \wedge w_1 = w_2)$
        $\lambda = sw \Rightarrow (s_1 = s_2 \wedge w_1 = w_2)$

spec CASLSIG =
    ORDEREDSTRINGS **then**
    sorts *FunProfiles*, *PredProfiles*
    ops *arity*: $FunProfiles \rightarrow S^*$
        coarity: $FunProfiles \rightarrow S$
        arity: $PredProfiles \rightarrow S^*$
    preds *istotal* : *FunProfiles*
        $\_ \sim_F \_ : FunProfiles \times FunProfiles$
        $\_ \sim_P \_ : PredProfiles \times PredProfiles$
    vars $fp_1, fp_2 : FunProfiles;\ pp_1, pp_2 : PredProfiles;\ s : S;\ w : S^*$
    axioms $(fp_1 \sim_F fp_2 \wedge arity(fp_1) = arity(fp_2)$
            $\wedge\ coarity(fp_1) = coarity(fp_2))$
          $\Rightarrow fp_1 = fp_2$
        $(pp_1 \sim_F pp_2 \wedge arity(pp_1) = arity(pp_2) \Rightarrow fp_1 = fp_2$
        $\lambda = sw \Rightarrow (fp_1 = fp_2 \wedge pp_1 = pp2 \wedge istotal(fp1))$

Figure 1. A specification of the CASL signature category within CASL

Subsorted partial-first order logic (*SubPFOL*) is the underlying institution of CASL. It is described in [6, 4]. For our cocompleteness proofs, we will need the following proposition, which follows from Corollaries 15 and 17 in [13].

Proposition 2.4. *Let* $\theta: T \rightarrow T1$ *be a theory morphism in SubPFOL between universal Horn theories which do not use subsorting and which do use strong*

equalities only in the conclusions of Horn formulae.[4]

Then both **Mod**(T) *and* **Mod**$(T1)$ *are cocomplete, and the forgetful functor* $\_|_\theta$: **Mod**$(T1) \to$ **Mod**(T) *has a left adjoint.*[5] □

2.2 Some categorical tools for proving cocompleteness

In this section, we recall some results from category theory that comprise a useful toolkit for proving cocompleteness theorems.

Definition 2.5 ([1], 13.17). A functor $F: \underline{A} \to \underline{B}$ is said to *lift colimits*, if for every diagram $D: \underline{I} \to \underline{A}$ and every colimit C of $F \circ D$ there exists a colimit C' of D with $F(C') = C$. □

Definition 2.6 ([1], 4.16). Let $\underline{A}$ be a subcategory of $\underline{B}$, and let B be a $\underline{B}$-object. An $\underline{A}$*-reflection arrow* for B is a morphism $r: B \to A$ into an $\underline{A}$-object A with the following universal property:
For any morphism $f: B \to A'$ from B into some $\underline{A}$-object A', there exists a unique $\underline{A}$-morphism $f': A \to A'$ such that $f = f' \circ r$.

$\underline{A}$ is called a *reflective subcategory* of $\underline{B}$ provided that each $\underline{B}$-object has an $\underline{A}$-reflection. [6]

The dual notion is that of *co-reflective* subcategory. □

Proposition 2.7 ([2], 3.5). *If* $\underline{A}$ *is a full, reflective or coreflective, subcategory of* $\underline{B}$, *and* $\underline{B}$ *is cocomplete, then* $\underline{A}$ *is cocomplete as well.* □

Proposition 2.8. *Consider a commuting diagram of four subcategories*

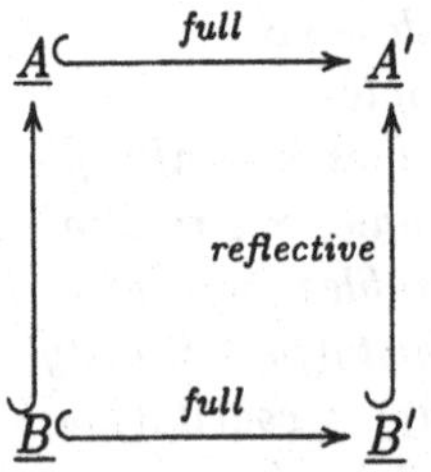

where $\underline{A}$ *is a full subcategory of* $\underline{A}'$, $\underline{B}$ *is a full subcategory of* $\underline{B}'$ *and* $\underline{B}'$ *is a reflective subcategory of* $\underline{A}'$.

If the $\underline{B}'$*-reflection of an arbitrary* $\underline{A}'$*-object coming from* $\underline{A}$ *is already a* $\underline{B}$*-object, then* $\underline{B}$ *is a reflective subcategory of* $\underline{A}$. □

---

[4] The proposition can also be proved without the restriction that subsorting is not used, but forbidding strong equations in the premises of the Horn formulae is essential.

[5] For the purpose of this paper, we assume that empty carriers are allowed in the models of a CASL theory.

[6] Note that this is equivalent to the condition that the inclusion functor from $\underline{A}$ to $\underline{B}$ has a left adjoint, which then produces the reflection.

3 Cocompleteness of the CASL signature category

For modularity aspects, it is important that signatures can be combined. In particular, CASL has the construct of generic extensions, which have a formal parameter part that can be later instantiated with different actual parameters. Instantiation of generic extensions is done via a construction which is a pushout. The existence of pushouts in CASL is guaranteed by the following:

Theorem 3.1. *The category* **CASLSig** *of subsorted CASL signatures and signature morphisms is cocomplete.*

Proof. Consider the universal Horn CASL specification CASLSɪɢ given in Fig. 1. It might look a bit strange that we do not introduce a sort for names of functions and predicates, but rather for profiles. This is because we want to capture CASL signature morphisms by CASLSɪɢ-homomorphisms. Now CASL signature morphisms can map different profiles for one and the same symbol name to profiles with different symbol names, while homomorphisms of course have only one choice for mapping elements of carriers. That's why profiles are appropriate as carriers, and not symbol names. Identity of names can be recovered by using the overloading relations. The axioms starting with $\lambda = sw$ ensure that any confusion of elements in S^* enforces the model to be terminal.

By Prop. 2.4 we know that **Mod**(CASLSɪɢ), the model-category of CASLSɪɢ, is cocomplete. The intention is now that CASLSɪɢ specifies **CASLSig** somehow and thus **CASLSig** inherits cocompleteness.

Unfortunately, **Mod**(CASLSɪɢ) is not equivalent to **CASLSig**: it is not guaranteed that there are no junk strings in the interpretation of S^*. Moreover, the profiles do not exactly capture function and predicate symbols. To repair this, we need the extension of CASLSɪɢ given in Fig. 2.

spec CASLSɪɢGᴇɴ =
  CASLSɪɢ **then**
  generated { **sort** S^*
    ops $\lambda : S^*$
      $\_\_ \_\_ : S \times S^* \to S^*$ }
  vars $s : S;\ w : S^*$
  axioms $\neg\ (\lambda = sw)$
    $\neg\ (\lambda \leq sw)$
    $\neg\ (sw \leq \lambda)$
  vars $fp_1, fp_2 : \mathit{FunProfiles};\ pp_1, pp_2 : \mathit{PredProfiles}$
  axioms $fp_1 \sim_F fp_2 \Rightarrow \exists w : S^*; s : S.$
    $(w \leq arity(fp_1) \wedge w \leq arity(fp_2)$
    $\wedge\ coarity(fp_1) \leq s \wedge coarity(fp_2) \leq s)$
    $pp_1 \sim_F pp_2 \Rightarrow \exists w : S^* \bullet (w \leq arity(pp_1) \wedge w \leq arity(pp_2))$

Figure 2. A specification of the CASL signature category within CASL

Lemma 3.2. **Mod**(CASLSIGGEN) *is equivalent to* **CASLSig**.

Proof. A CASL-signature is mapped to a CASLSIGGEN-model by taking the obvious interpretations of the symbols. A CASLSIGGEN-model M is mapped to is the CASL-signature $(M_S, F, TF, P, \leq_M)$ which consists of

- a set of sorts M_S (without loss of generality, $(M_S)^*$ can be identified with M_{S^*}),
- the pre-order $\leq_M$,
- let $\mathit{FunNames} = M_{\mathit{FunProfiles}}/(\sim_F)_M$,
- for each $w \in (M_S)^*$ and $s \in M_S$, $F_{w,s} = \{c \in \mathit{FunNames} \mid$ there is some $\mathit{prof} \in c$ with $\mathit{arity}_M(\mathit{prof}) = w, \mathit{coarity}_M(\mathit{prof}) = s\}$,
- for each $w \in (M_S)^*$ and $s \in M_S$, $c \in TF_{w,s}$ iff there is some $\mathit{prof} \in c$ with $\mathit{arity}_M(\mathit{prof}) = w, \mathit{coarity}_M(\mathit{prof}) = s$ and $\mathit{istotal}_M(\mathit{prof})$,
- let $\mathit{PredNames} = M_{\mathit{FunProfiles}}/(\sim_P)_M$,
- for each $w \in (M_S)^*$, $P_{w,s} = \{c \in \mathit{PredNames} \mid$ there is some $\mathit{prof} \in c$ with $\mathit{arity}_M(\mathit{prof}) = w\}$. □

To prove cocompleteness of **Mod**(CASLSIGGEN), we cannot apply Prop. 2.4, since CASLSIGGEN is not in Universal Horn form. But we can apply the following lemma:

Lemma 3.3. **Mod**(CASLSIGGEN) *is a coreflective subcategory of* **Mod**(CASLSIG).

Proof. The coreflection of a CASLSIG-model M that is not terminal is the following submodel M' of M:

- $M'_S = M_S$,
- M'_{S^*} is the subset of M_{S^*} generated by λ_M and $\_\_ \_\__M$,
- $\leq$ on M'_{S^*} is $\leq$ on M_{S^*} restricted to arguments of same length (otherwise, it yields false),
- $\mathit{FunProfiles}_{M'} = \{\mathit{prof} \in \mathit{FunProfiles}_M \mid \mathit{arity}_M(\mathit{prof}) \in M'_{S^*}\}$,
- $\mathit{prof}_1 (\sim_F)_{M'} \mathit{prof}_2$ iff $\mathit{prof}_1 (\sim_F)_M \mathit{prof}_2$ and there exist some $w \in M'_{S^*}$ and $s \in M'_S$ with $w \leq_M \mathit{arity}_{M'}(\mathit{prof}_1)$, $w \leq_M \mathit{arity}_{M'}(\mathit{prof}_2)$, $\mathit{coarity}_{M'}(\mathit{prof}_1) \leq_M s$ and $\mathit{coarity}_{M'}(\mathit{prof}_2) \leq_M s$,
- $\mathit{PredProfiles}_{M'} = \{\mathit{prof} \in \mathit{PredProfiles}_M \mid \mathit{arity}_M(\mathit{prof}) \in M'_{S^*}\}$,
- $\mathit{prof}_1 (\sim_P)_{M'} \mathit{prof}_2$ iff $\mathit{prof}_1 (\sim_P)_M \mathit{prof}_2$ and there exists some $w \in M'_{S^*}$ with $w \leq_M \mathit{arity}_{M'}(\mathit{prof}_1)$ and $w \leq_M \mathit{arity}_{M'}(\mathit{prof}_2)$ and
- the other operations and relations are inherited from M.

The coreflection property can be seen as follows: Given a CASLSIGGEN-model MGen, a CASLSIG-model M and a CASLSIG-homomorphism $\sigma: \mathit{MGen} \rightarrow M$, σ is already a CASLSIG-homomorphism into the coreflection of M, since σ preserves both generatedness of elements of M_{S^*} and common super- and subsorts of arities and coarities, resp.

The coreflection of the terminal CASLSIG-model is the terminal CASLSIGGEN-model consisting of one sort and, for each number of arguments, one total function and one predicate profile. □

Thus we can apply Prop. 2.7 to conclude that **CASLSig** is cocomplete as well. □

Corollary 3.4. *The category of SubPFOL-theories is cocomplete.*

Proof. Cocompleteness of the signature category implies cocompleteness of the theory category for arbitrary institutions [9]. □

4 Monotonicity

Monotonicity is a condition that is assumed throughout the most popular survey paper about order-sorted algebra by Goguen and Meseguer [10]. A consequence of monotonicity is that overloading of constants is forbidden. More importantly, monotonicity is a preparing condition for regularity, which is discussed below.

Definition 4.1. A signature $\Sigma = (S, F, TF, P, \leq_S)$ is called *monotone*, if for $f : w_1 \to s_1$ and $f : w_2 \to s_2$, $w_1 \leq w_2$ implies $s_1 \leq s_2$. Let **MonSig** be the full subcategory of **CASLSig** consisting of monotone signatures. □

Proposition 4.2 (Haxthausen and Nickl[12]). *The inclusion functor from* **MonSig** *to* **CASLSig**[7] *does not lift colimits (in particular, it does not lift pushouts).* □

This is illustrated by the following example:

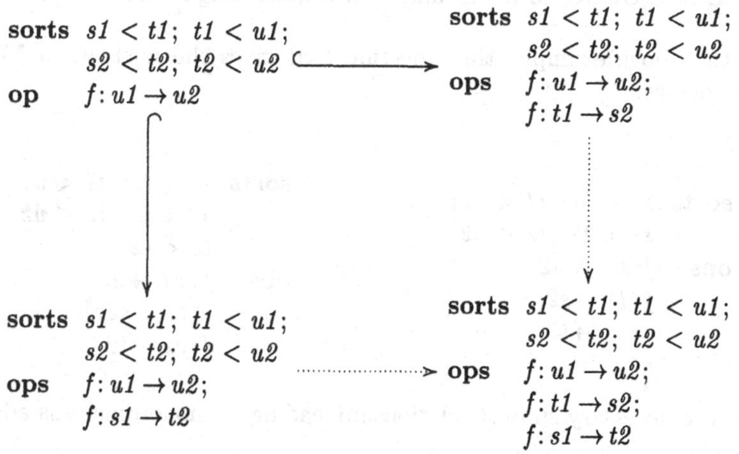

The resulting signature is no longer monotone. That is, we cannot use the construction of colimits in **CASLSig** for constructing colimits in **MonSig** in

[7] Haxthausen and Nickl use a category of signatures slightly different from **CASLSig**: they do not allow partial function and predicate symbols, they assume antisymmetry of the subsorting relation, and their preservation of overloading is stronger. But these differences do not matter here.

general. Now Haxthausen and Nickl study conditions under which pushouts are lifted from **CASLSig** to **MonSig**. These conditions are rather technical and restrictive. The signature morphisms used in the pushout have to be embeddings, which rules out instantiations of generics where different parts of the formal parameter are instantiated in the same way.

But there is a different way to get pushouts in **MonSig**. Rather than trying to lift colimits from **CASLSig** to **MonSig**, we can try to modify **CASLSig**-colimits in order to get **MonSig**-colimits.

Proposition 4.3. **MonSig** *is a full reflective subcategory of* **CASLSig**.

Corollary 4.4. **MonSig** *is cocomplete.*

Proof. Given a CASL signature $\Sigma \in$ **CASLSig**, we construct its reflection $r: \Sigma \rightarrow \mathit{Monotonify}(\Sigma)$ as follows:

Let θ be the inclusion of CASLSIG into the theory

spec MONSIG =
  CASLSIG **then**
  vars $pr_1, pr_2 : \mathit{FunProfile}$
  axiom $pr_1 \sim_F pr_2 \wedge \mathit{arity}(pr_1) \leq \mathit{arity}(pr_2) \Rightarrow \mathit{coarity}(pr_1) \leq \mathit{coarity}(pr_2)$

By Prop. 2.4, **Mod**(MONSIG) is a full reflective subcategory of **Mod**(CASLSIG). Moreover, the reflection of a model induced by a CASL signature again is a model induced by a CASL signature, which is monotone. Thus by Prop. 2.8, **MonSig** is a reflective subcategory of **CASLSig**, and *Monotonify*(Σ) is obtained by first viewing Σ as a CASLSIG-model and then constructing its free θ-extension. □

For the above example, the reflection that gives the pushout in **MonSig** looks as follows:

sorts $s1 < t1;\ t1 < u1;$
  $s2 < t2;\ t2 < u2$
ops $f: u1 \rightarrow u2;$
  $f: t1 \rightarrow s2;$
  $f: s1 \rightarrow t2$

$\cdots\cdots\cdots\cdots>$

sorts $s1 < t1;\ t1 < u1;$
  $s2 < t2;\ t2 < u2;$
  $t2 < s2$
ops $f: u1 \rightarrow u2;$
  $f: t1 \rightarrow s2;$
  $f: s1 \rightarrow t2$

That is, the necessary subsort relationship leading to monotonicity is added.

5 Pre-ordered subsorts versus partially ordered subsorts

In CASL, the subsort relation on the set of sorts is a *pre-order*, that is, it may have sort cycles and thus fail to satisfy the law of antisymmetry:

$$s \leq s' \wedge s' \leq s \Rightarrow s = s'$$

In the models, a sort cycle leads to an isomorphism of all the carrier sets that correspond to sorts in the cycle. This possibility was consciously included into the design of CASL.

Now there are approaches to subsorting where the subsorting relation is required to be a *partial order*, and thus satisfy the law of antisymmetry. If one wants to feed a CASL specification into a tool which expect a partially-ordered set of sorts, one first has to quotient out the sort cycles. This can be done with the following:

Proposition 5.1. *Let* **PoSig** *be the full subcategory of* **CASLSig** *determined by the signatures whose subsorting relation satisfy the law of antisymmetry. Then* **PoSig** *is a reflective subcategory of* **CASLSig**.

Corollary 5.2. **PoSig** *is cocomplete.*

Proof. Let θ be the inclusion of CASLSIG into the theory
spec POSIG =
  CASLSIG **then**
  vars $s1, s2 : S$
  axiom $s_1 \leq s_2 \wedge s_2 \leq s_1 \Rightarrow s_1 = s_2$

The rest of the proof parallels the proof of Prop. 4.3. □

PoMonSig denotes the intersection of **PoSig** and **MonSig** consisting of all partially ordered monotone signatures. Cocompleteness of this category follows with analogous arguments.

6 Regularity

Regularity is a property of subsorted signatures which leads to an extremely simple parsing algorithm which does the overload resolution.

Definition 6.1. A signature is *regular* if

- it is monotone and
- for every $f : w \to s$ and $w_0 <= w$, there is a least profile for f with arity $>= w_0$, and
- for every $p : w$ and $w_0 <= w$, there is a least profile for p with arity $>= w_0$.
- Moreover, we assume that the subsort relation is a partial order[8].

Let **PoRegSig** be the (non-full) subcategory of **PoSig** consisting of regular signatures and signature morphisms preserving the least profiles that are guaranteed by regularity. □

[8] Throughout this section, we assume that all signatures have a subsort relation that is a partial order. In principle, the definitions and results of this section can be generalized to arbitrary pre-orders. However, the construction *Regularify* defined below gives satisfactory results only for partial orders (otherwise, too many new sorts are introduced). Note that a pre-ordered subsort relation can always be turned into a partially-ordered one, using the result of the previous section.

The main thing we gain from regularity is the existence of a simple bottom-up least sort parse algorithm due to Goguen and Meseguer [10]. To parse a term $f(t_1, \dots, t_n)$, recursively parse the t_i, giving a least sort parse $u_i : s_i$, and let $w_0 = s_1 \dots s_n$. If there is no $f\colon w \to s$ satisfying $w_0 \leq w$ in the signature, the term is ill-sorted. Otherwise, take the least such $f\colon w \to s$, which exists by regularity. The least sort parse of $f(t_1, \dots, t_n)$ is then $(f\colon w \to s)(u_1 : s_1, \dots, u_n : s_n) : s$.

Without regularity, we have to consider all possible sorts and parses of a given unparsed term or use a more complicated overload resolution algorithm [14].

A drawback of regularity is that is does not behave so well w. r. t. modularity:

Proposition 6.2 (Haxthausen and Nickl[12]). *The inclusion functor from* **RegSig** *to* **CASLSig** *does not lift colimits (in particular, it does not lift pushouts).* □

This can be illustrated with the following example:

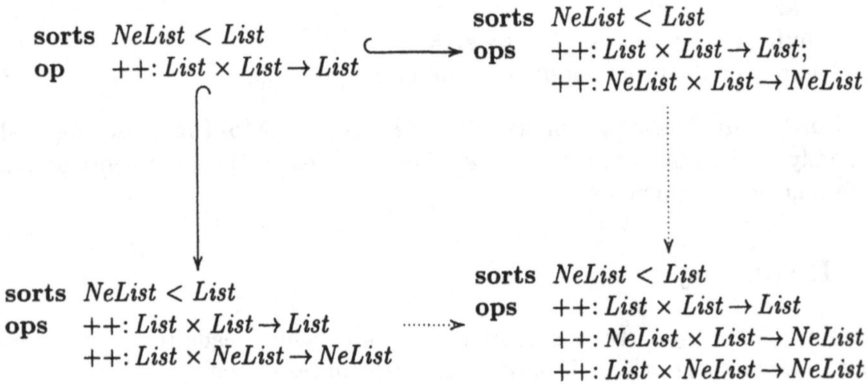

The three signatures forming the diagram are regular, but the pushout signature is not regular: if we take $w_0 = \mathit{NeList} \times \mathit{NeList}$, there is no least profile for ++ with arity $\geq w_0$.

Again, Haxthausen and Nickl examine rather technical restrictions of pushout diagrams which guarantee that the pushout is lifted to **PoRegSig**. In particular, they require the signature morphisms to be embeddings. But as in the case of monotonicity, instead of trying to lift colimits, we can try to modify them by using a full reflective embedding.

Unfortunately, **PoRegSig** is not a full subcategory of **PoSig** (even though it is reflective). Therefore, we introduce the following restriction of regularity:

Definition 6.3. A monotone CASL-signature in **PoMonSig** is called *strongly regular*, if

- for every $f : w \to s$ and $w_0 <= w$, w_0 is already an arity for f, that is, there is some s_0 with $f\colon w_0 \to s_0$, and

– for every $p : w$ and $w_0 <= w$, w_0 is already an arity for p, that is, $p : w_0$.

This gives us a subcategory **PoStrRegSig** of **PoMonSig**. □

Corollary 6.4. *Every strongly regular signature is regular.* □

The restriction to strongly regular signatures is not as harmful as it may look in the first place. If we want to make a regular signature strongly regular, we just have to add some new profiles which can be interpreted by restricting interpretations of old profiles. Moreover, **PoStrRegSig** is a full[9] subcategory of **PoRegSig** by the following proposition:

Proposition 6.5. *A signature morphism in* **PoMonSig** *starting from a strongly regular signature preserves the least profiles that are required to exist by regularity.*

Proof. Let $\sigma: \Sigma \to \Sigma'$ be a signature morphism in **PoMonSig** and Σ strongly regular. Let $f: w \to s$ and $w_0 \leq w$ in Σ. By strong regularity, the least profile for f with arity $\geq w_0$ has the form $f: w_0 \to s_0$. By preservation of overloading, $\sigma^F_{w,s}(f) = \sigma^F_{w_0,s_0}(f) = f'$. So by monotonicity, $f': \sigma^S(w_0) \to \sigma^S(s_0)$ is the least profile for f' with arity $\geq \sigma^S(w_0)$ in Σ'. □

Proposition 6.6. *There is a functor Regularify:* **PoSig** → **PoStrRegSig** *making* **PoStrRegSig** *a reflective subcategory of* **PoSig**.

Corollary 6.7. **PoStrRegSig** *is cocomplete.*

Proof. Let θ be the inclusion of PoSig into the theory

spec PoStrRegSig =
  PoMonSig **then**
  op *new_profile* : *FunProfiles* × S^* →? *FunProfiles*
  vars $w_0 : S^*$; f : *FunProfile*
  axioms *defined new_profile*$(f, w_0) \Leftrightarrow w_0 \leq$ *arity*(f)
    defined new_profile$(f, w_0) \Rightarrow$ *arity*(*new_profile*$(f, w_0)) = w_0$
    defined new_profile$(f, w_0) \Rightarrow f \sim_F$ *new_profile*(f, w_0)
  op *new_profilePredProfiles* × S^* →? *PredProfiles*
  vars p : *PreProfile*
    defined new_profile$(p, w_0) \Leftrightarrow w_0 \leq$ *arity*(p)
    defined new_profile$(p, w_0) \Rightarrow$ *arity*(*new_profile*$(p, w_0)) = w_0$
    defined new_profile$(p, w_0) \Rightarrow p \sim_P$ *new_profile*(p, w_0)

The partial function *new_profile* generates, for each existing profile and each string of sorts w_0 less than the arity of the profile, a profiles with arity w_0, in order to make the signature strongly regular. Note that the coarity of the new profile is left unspecified; this may lead to the generation of new sorts.

The rest of the proof parallels the proof of Prop. 4.3. □

[9] even reflective

Now let us come back to the example. *Regularify* adds a new sort $L \leq NeList$ and a profile $++: NeList \times NeList \to L$ to the above pushout signature, giving a strongly regular pushout signature in *PoRegSig*.

sorts $NeList < List$
ops $++: List \times List \to List$
$++: NeList \times List \to NeList$
$++: List \times NeList \to NeList$

⋯⋯▸

sorts $NeList < List$;
$L < NeList$
ops $++: List \times List \to List$
$++: NeList \times List \to NeList$
$++: List \times NeList \to NeList$
$++: NeList \times NeList \to L$

For practical purposes, one would rename L by *NeList* after forming the pushout.

7 Local Filtration

Locally filtered signatures have the property that satisfaction is closed under isomorphism in the approach of Goguen and Meseguer [10]. Also, the reducibilitiy to many-sorted algebra is guaranteed only in case of local filtration. In CASL, we have both properties without local filtration. However, in CASL well-formedness of $t_1 = t_2$ and $t_2 = t_3$ do not necessarily ensure well-formedness of $t_1 = t_3$, but in case of local filtration, this implication holds[10].

Definition 7.1. A signature $\Sigma \in$ **PoSig** is called *locally filtered*, if each pair of sorts that is connected (i. e. that is in the symmetric transitive closure of $\leq$) has a common upper bound. This gives us a full subcategory **LFiltSig** of **PoSig**. □

Now like the other properties, local filtration does not behave so well w. r. t. modularity:

Proposition 7.2 (Haxthausen and Nickl[12]). *The inclusion functor from* **LFiltSig** *to* **CASLSig** *does not lift colimits (in particular, it does not lift pushouts).* □

This can be illustrated with the following example:

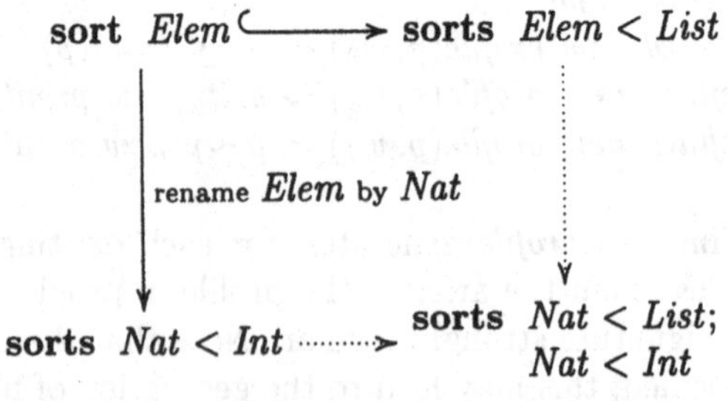

[10] In CASL, the implication holds also without local filtration, if we use *sorted* equations.

The three signatures forming the diagram are locally filtered, but the pushout signature is not: the sorts *List* and *Int* are connected, but do not have a common upper bound.

Again, Haxthausen and Nickl study rather restrictive conditions which guarantee that colimits are lifted to **LFiltSig**.

We cannot here apply our reflection technique directly, but need the following sharpening of local filtration:

Definition 7.3. A partially ordered signature in **PoSig** is called *strongly locally filtered*, if each pair of connected sorts has a *least* common upper bound. Let **SLFiltSig** be the (non-full) subcategory of **LFiltSig** consisting of strongly locally filtered signatures and morphisms preserving least common upper bounds of connected pairs. □

Unfortunately, we cannot use a reflection technique here to prove cocompleteness: **SLFiltSig** is a reflective subcategory of **PoSig**, but not full! But we can prove cocompleteness more directly:

Proposition 7.4. SLFiltSig *is cocomplete.*

Proof. Consider the CASL specification

spec SLFILTSIG =
  POSIG **then**
  op $\_\wedge\_ : S \times S \rightarrow ? S$
  vars $s, s_1, s_2 : S$
  axioms $s_1 \leq s_2 \Leftrightarrow s_1 \wedge s_2 = s_2$
  $s \wedge s = s$[11]
  $s_1 \wedge s_2 = s_2 \wedge s_1$
  $(s_1 \wedge s_2) \wedge s_3 = s_1 \wedge (s_2 \wedge s_3)$
  defined $s_1 \wedge s_2 \ \wedge$ *defined* $s_2 \wedge s_3 \Rightarrow$ *defined* $s_1 \wedge s_3$

$\wedge$ is an operation that satisfies the axioms for *sup*-semilattices, but it is a *partial* operation. Of course, any pair of sorts for which $\wedge$ is defined is connected. The last axiom ensures that the converse is true as well (note that by the first axiom, $s_1 \leq s_2$ implies *defined* $s_1 \wedge s_2$). Thus the set of sorts consists of a number of connected components, each of which is a *sup*-semilattice. This is nothing else than local filtration.

The rest of the proof parallels that of Theorem 3.1. □

How do pushouts in **SLFiltSig** look? A pushout in **SLFiltSig** is constructed by taking the pushout in **PoSig** and then adding, for each (finite) set of connected sorts without least upper bound, a new sort which now serves as (the set's) least upper bound. In particular, for the above example, we get

[11] Note that = denotes strong equality: both sides are undefined or both defined and equal.

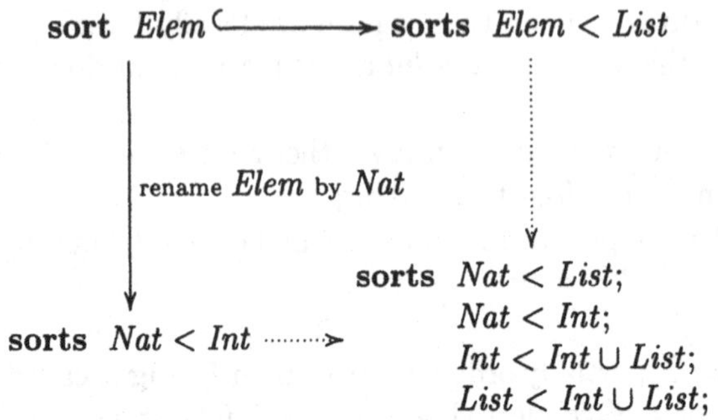

We can also use the left adjoint (=free construction) along the theory inclusion from PoSig to SLFiltSig to get a reflection which makes a partially ordered signature strongly locally filtered. By a general categorical theorem, this reflection also preserves colimits. But note that, different from the monotone and regular case, **SLFiltSig** is a non-full subcategory, which implies that the reflection of strongly locally filtered signatures may not be the identity. Thus a reflection of a colimit in **PoSig** of a diagram of signatures in **SLFiltSig** is a colimit in **SLFiltSig** but a colimit of the *reflected* diagram, which need not be the original one.

8 Conclusion and Comparison With Related Work

We have shown that the CASL specification language (which includes a new and general approach to subsorting) has a cocomplete signature category. This is important to be able to form instantiations of generic extension as pushouts.

For restrictions to monotone, regular or locally filtered signatures, Haxthausen and Nickl [12] have shown that colimits (in particular, pushouts) do not lift to the restricted signature categories. We argue that the approach of Haxthausen and Nickl of imposing severe restrictions on signatures and morphisms is not appropriate in all practical cases: they require morphisms to be embeddings, but there are instantiations of generics with non-injective fitting morphisms.

Therefore, we have developed a different approach to pushouts (and colimits in general) for monotone, regular and locally filtered signatures. We have shown that appropriate subcategories of such signatures are reflective, which means that colimits in the subcategories can be obtained by taking the reflection of the colimit in the original category.

A short comparison of the advantages and disadvantages of the CASL approach (without any monotonicity, regularity and local filtration restriction) and of Haxthausen and Nickl's and our approach to colimits for regular signatures is given in the following table:

Approach	advantages	disadvantages
CASL[15]	conceptually very simple and general	overload resolution is involved
Haxthausen and Nickl[12]	easy parsing, literal pushouts	restricted to embeddings, technically involved
Reflections (our approach)	easy parsing, conceptually simple	New sorts and relations are added, amalgamation property may fail

The crucial feature of our approach is the generation of new profiles, subsort relationships and even sorts. The former are not so harmful (and even may be consider as helpful), and the new sorts generated for locally filtered signatures may be interpreted as sort unions. Only for regular signatures, pushouts may contain new sorts that cannot be interpreted in a useful way and rather should be renamed into existing sorts after forming the pushout.

The main advantage of our approach is the combination of the conceptual simplicity of the CASL approach with the possibility to use monotonicity, regularity and local filtration, e.g. in order to ease parsing.

Owe and Dahl [16] also describe an algorithm that makes a signature regular. It works with intersections and unions of types, assuming that there is a family of basic types with disjoint interpretation. To make the algorithm work, they have to assume that signatures are consistent, which means that some interpretation with non-empty carriers exists (in CASL, such an interpretation always exists, but in their approach, there is a disjointness requirement that may lead to the non-existence of such an interpretation). Now unfortunately consistency is a property that is not stable under pushouts. So Owe and Dahl's regularifying algorithm, while interesting in general, does not help in the present setting.

Acknowledgements I would like to thank Anne Haxthausen and Olaf Owe for fruitful discussions and the anonymous referee for some useful hints.

References

1. J. Adámek, H. Herrlich, and G. Strecker. *Abstract and Concrete Categories.* Wiley, New York, 1990.
2. F. Borceux. *Handbook of Categorical Algebra I – III.* Cambridge University Press, 1994.
3. R. M. Burstall and J. A. Goguen. Putting theories together to make specifications. In *Proceedings of the 5th International Joint Conference on Artificial Intelligence*, pages 1045–1058. Cambridge, 1977.
4. M. Cerioli, A. Haxthausen, B. Krieg-Brückner, and T. Mossakowski. Permissive subsorted partial logic in CASL. In M. Johnson, editor, *Algebraic methodology and software technology: 6th international conference, AMAST 97*, volume 1349 of *Lecture Notes in Computer Science.* Springer-Verlag, 1997.

5. CoFI Task Group on Language Design. CASL – The CoFI Algebraic Specification Language – Summary. CoFI Document: CASL/Summary. WWW[12], FTP[13], September 1997.
6. CoFI Task Group on Semantics. CASL – The CoFI Algebraic Specification Language (version 0.97) – Semantics. CoFI Note: S-6. WWW[14], FTP[15], July 1997.
7. R. Diaconescu and K. Futatsugi. Logical semantics of CafeOBJ. Technical report IS-RR-96-0024S, JAIST, 1996.
8. J. A. Goguen. A categorical manifesto. *Mathematical Structures in Computer Science*, 1:49–67, 1991.
9. J. A. Goguen and R. M. Burstall. Institutions: Abstract model theory for specification and programming. *Journal of the Association for Computing Machinery*, 39:95–146, 1992. Predecessor in: LNCS 164, 221–256, 1984.
10. J. A. Goguen and J. Meseguer. Order-sorted algebra I: equational deduction for multiple inheritance, overloading, exceptions and partial operations. *Theoretical Computer Science*, 105:217–273, 1992.
11. J. A. Goguen and T. Winkler. Introducing OBJ3. Research report SRI-CSL-88-9, SRI International, 1988.
12. A. Haxthausen and F. Nickl. Pushouts of order-sorted algebraic specifications. In *Proceedings of AMAST'96*, volume 1101 of *Lecture Notes in Computer Science*, pages 132–?? Springer-Verlag, 1996.
13. T. Mossakowski. Equivalences among various logical frameworks of partial algebras. In H. K. Büning, editor, *Computer Science Logic. 9th Workshop, CSL'95. Paderborn, Germany, September 1995, Selected Papers*, volume 1092 of *Lecture Notes in Computer Science*, pages 403–433. Springer Verlag, 1996.
14. T. Mossakowski, Kolyang, and B. Krieg-Brückner. Static semantic analysis of CASL. 12th Workshop on Algebraic Development Techniques, Tarquinia. This volume, 1997.
15. P. D. Mosses. CoFI: The common framework initiative for algebraic specification and development. In M. Bidoit and M. Dauchet, editors, *Theory and Practice of Software Development*, LNCS 1214, pages 115– 137. Springer Verlag, 1997.
16. O. Owe and O.-J. Dahl. Generator induction in order sorted algebras. *Formal Aspects of Computing*, 3:2–20, 1991.

[12] http://www.brics.dk/Projects/CoFI/Documents/CASL/Summary/
[13] ftp://ftp.brics.dk/Projects/CoFI/Documents/CASL/Summary/
[14] http://www.brics.dk/Projects/CoFI/Notes/S-6/
[15] ftp://ftp.brics.dk/Projects/CoFI/Notes/S-6/

Static Semantic Analysis and Theorem Proving for CASL

Till Mossakowski, Kolyang and Bernd Krieg-Brückner

Bremen Institute for Safe Systems, Universität Bremen, P.O. Box 330440, D-28334 Bremen. Email: {till, kol, bkb}@informatik.uni-bremen.de

Abstract. This paper presents a static semantic analysis for CASL, the Common Algebraic Specification Language. Abstract syntax trees are generated including subsorts and overloaded functions and predicates. The static semantic analysis, through the implementation of an overload resolution algorithm, checks and qualifies these abstract syntax trees. The result is a fully qualified CASL abstract syntax tree where the overloading has been resolved. This abstract syntax tree corresponds to a theory in the institution underlying CASL, subsorted partial first-order logic (*SubPFOL*).
Two ways of embedding *SubPFOL* in higher-order logic (*HOL*) of the logical framework Isabelle are discussed: the first one from *SubPFOL* to *HOL* via *PFOL* (partial first-order logic) first drops subsorting and then partiality, and the second one is the counterpart via *SubFOL* (subsorted first-order logic). Finally, we sketch an integration of the embedding of CASL into the UniForM Workbench.

1 Introduction

During the past decades a large number of algebraic specification languages have been developed, based on a diversity of basic algebraic specification concepts. The goal of CoFI, the *Common Framework Initiative for Algebraic Specification and Development* [17] is to get a common agreement in the algebraic specification community about basic concepts, and to provide a family of specification languages at different levels, a development methodology and tool support. The family of specification languages comprises a central, common language, called CASL[1] and various restrictions and extensions of CASL. A design proposal [6] of CASL has already been completed by representatives of most algebraic specification language groups and has been tentatively approved by the IFIP working group 1.3 (Foundations of Systems Specification). The final version is expected shortly.

A variety of tools and experience with them is available, and it is neither sensible nor feasible to build CASL tools anew from scratch. Rather, a more promising strategy is to build *bridges* to existing tools, which transform a CASL

[1] CASL is an acronym for *Common Algebraic Specification Language* and is pronounced like 'castle'.

specification to some format recognizable by an existing tool. An important property of such bridges is of course that they are *sound* and *complete* in some way. Moreover, they should not be too complicated, but rather preserve as much structure as possible, and be, if possible, bi-directional in order to be able to re-transform intermediate steps that appear in the application of the target tool and visualize them as CASL-formulae or theories.

In this paper, we provide a first set of tools for CASL, which can be used for theorem-proving and transformation development. A bridge from CASL to an existing theorem proving tool has to perform syntactic analysis, static semantic analysis and an encoding of CASL's logic in the tool's logic. We want to use a tool that allows to integrate all these steps. Isabelle [18] is a generic tool well-suited for doing this. We represent the institution underlying CASL in Isabelle/HOL, very much in the spirit of a shallow embedding like the encoding of Z in Isabelle/HOL [13]. This comprises a minimal set of tools that make the "in-the-small" part of CASL amenable to machine support.

The outline of this paper is as follows: After some preliminaries about CASL and Isabelle, we describe the embedding of CASL in Isabelle by presenting the parsing techniques, the static semantic analysis (including an overload resolution algorithm) and the encoding in Isabelle/HOL, which enables us to borrow some theorem prover facilities. The paper concludes by sketching future extensions of the encoding and also integration possibilities in a development tool environment as pursued by the UniForM project [14].

2 Preliminaries

In the following section, we present some preliminaries about CASL and Isabelle. Isabelle is the logical framework used for embedding the underlying logic.

2.1 CASL

CASL (The Common Algebraic Specification Language) is the language of the Common Framework Initiative (CoFI) [17], whose major novelty lies in the combination of existing concepts and constructs. CASL encompasses basic specifications, structured specifications, architectural specifications and libraries.

In the focus of this work, we concentrate on basic specifications (i.e. CASL-in-the-small) and the following features in particular:

- *Partiality.* Functions may be partial, the value of a function application in a term being possibly undefined. Total functions may be declared as such. The underlying logic is 2-valued.
- *Subsorts and Overloading.* Subsort inclusions are represented by injective embedding functions. These injections (from a subsort to a supersort) may be left implicit, allowing a concise notation, while projections (from a supersort to a subsort) have to be inserted explicitly, since they may be partial. Furthermore, functions (and predicates) may be overloaded, the same symbol being declared with more than one profile of argument and result sorts.

- *Formulae.* The usual first-order quantification and logical connectives are provided, as well as sort generation constraints.

A basic specification in CASL consists of a list of basic items. A basic item may either be a declaration of some sorts, subsorts, partial or total functions or predicates, or be an axiom.[2]

Axioms are usual first-order formulae over the following atomic formulae:

- Existential ($\stackrel{e}{=}$) or strong ($=$) equalities between terms,
- applications of predicates to a list of terms,
- definedness tests (*defined* t), which check if a term is defined, or
- membership tests ($t \in s$), which check if a term belongs to a subsort.

Terms are built from variables and function symbols in the usual way, where additionally a term may be forced to have a certain sort in two ways: a *sorted term* ($t : s$) forces a term to have a given sort by implicitly adding injections to supersorts, if the term does not already have the specified sort, while a *cast* (t *as* s) projects a term to a subsort (note that this is undefined if the value of the term does not belong to the subsort). Moreover, function and predicate symbols may be *qualified* with their profile (i.e. argument and result sorts).

There is another kind of axioms: sort generation constraints, specifying generatedness of some set of sorts w.r.t. some set of functions.

A list of CASL signature declarations leads to a signature of *SubPFOL* (subsorted partial first-order logic, see [7, 3] for a formal definition), the institution underlying CASL[3]. Such a signature $\Sigma = (S, TF, PF, P, \leq_S)$ consists of a set S of *sorts*, two $S^* \times S$-indexed sets TF and PF of *total* and of *partial function symbols*, an S^*-indexed set P of *predicate symbols* and a *subsorting relation* $\leq$ that is a pre-order (i.e. reflexive, transitive relation) over S.

Notation: We write $f : w \to s \in TF$ for $f \in TF_{w,s}$, $f : w \to ?\, s \in PF$ for $f \in PF_{w,s}$, $p : w \in P$ for $p \in P_w$. Moreover, we write $f : w \longrightarrow s \in TF \cup PF$ or even just $f : w \longrightarrow s$ for $f \in TF_{w,s} \cup PF_{w,s}$.

For each signature, there are two associated *overloading relations* $\sim_F$ and $\sim_P$. They indicate which pairs of profiles for one and the same function (or predicate) symbol have to be interpreted in the same way in the models. Their definition is guided by the principle that arguments valid for two different profiles, say $f : w_1 \to s_1$ and $f : w_2 \to s_2$, should be interpreted in the same way by the interpretations of both profiles.

More specifically, for a model M, those arguments that are in the intersection of w_1^M and w_2^M should lead to the same result in the union of s_1^M and s_2^M. In the subsorts-as-injections approach taken by CASL, we approximate the intersection by considering any argument in any common subsort w of w_1 and w_2, and the union by considering the injection of the result into any common supersort s of s_1 and s_2.

---

[2] There are some other forms of basic items. However, these can easily be translated to signature declations and axioms.

[3] More precisely, the logic is *SubPCFOL*, which is *SubPFOL* with sort generation constraints. But we omit these here and will come back to them in the sequel.

Formally, given a *SubPFOL*-signature $\Sigma = (S, TF, PF, P, \leq_S)$, consider two profiles $f\colon w_1 \longrightarrow s_1, f\colon w_2 \longrightarrow s_2 \in TF \cup PF$ of a function symbol f. Then

$$f\colon w_1 \longrightarrow s_1 \sim_F f\colon w_2 \longrightarrow s_2$$

iff there exist both $w \in S^*$ with $w \leq w_1, w_2$ and $s \in S$ with $s \geq s_1, s_2$. Let $p : w_1, p : w_2 \in P$ be two profiles of a predicate symbol p. Then

$$p : w_1 \sim_P \; p : w_2$$

iff there exists $w \in S^*$ with $w \leq w_1, w_2$.

The overloading relations play a role for well-formedness of terms and formulae: The ambiguity due to overloading in a well-formed formula must be harmless in the sense that all its possible fully qualified *expansions* (with profiles and explicit injections inserted) are interpreted in the same way. A first-order formula is well-formed if it is built from well-formed atomic formulae in the usual way. An atomic formula (in a given context of variables) is well-formed if it is well-sorted and expands to an atomic formula for constructing sentences of *SubPFOL* that is unique up to an equivalence on atomic formulae and terms. Equivalence between fully-qualified terms t, t' is the least congruence including the following cases:

- t' is the application of an embedding to t;
- t and t' are of the same sort, and are applications of different compositions of embedding functions to the same term;
- t' is the same as t up to profiles of function symbols in the overloading relation $\sim_F$ and embedding.

Equivalence between fully-qualified atomic formulae is the least congruence including the following cases:

- the formulae are identical, up to replacement of equivalent terms; or
- the formulae are the same up to the profiles of predicate symbols in the overloading relation $\sim_P$ and embedding.

This finishes the overview over CASL in-the-small. CASL also incorporates in-the-large concepts such as structured and architectural specifications and libraries. This is not subject of this paper, but the hope is that the experiences gained here can be extended to CASL in-the-large as well.

2.2 Isabelle

Isabelle [18] is an interactive tactic-based generic theorem prover that supports a family of logics, e.g. first-order logic (FOL), Zermelo-Fränkel set theory (ZF), constructive type theory (CTT), the Logic of Computable Functions (LCF), higher-order logic (HOL), and others. Logics can be encoded using Isabelle's metalogic. Isabelle supports a natural deduction style. Its principal inference techniques are resolution (based on higher-order unification) and term-rewriting.

In our approach, we will encode CASL into Isabelle/HOL, much in the style of other HOL-encodings [13]. This allows us to re-use a rich set of HOL-theorems and tactics for CASL. Furthermore, we aim at using Isabelle as a basis for a transformational development framework as proposed by [12]. Isabelle then is used as a host for the development of transformation rules, basic tactics, proof scripts etc.

3 Representing CASL in Isabelle/HOL

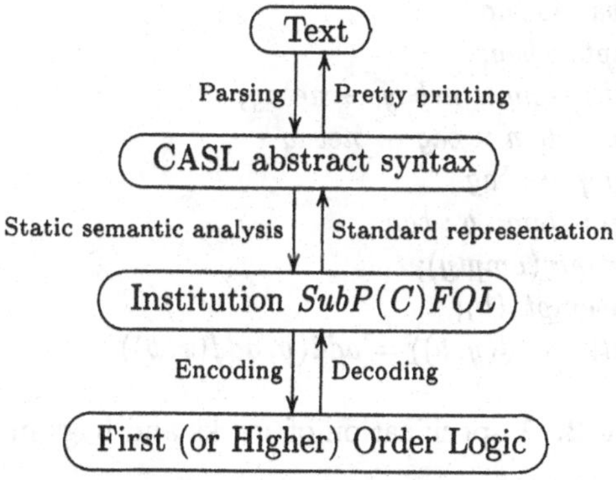

Figure 1. Overall architecture for representing CASL in Isabelle/HOL

The representation of CASL in Isabelle/HOL consists of three steps (cf. Fig. 1):

1. *Parsing* of the input text, yielding an abstract syntax tree (which is represented as an ML-data structure).
2. *Static semantic analysis*, including overload resolution, which checks if an abstract syntax tree corresponds to a well-formed CASL specification and in parallel constructs a theory in *SubPCFOL*, the underlying institution of CASL. The static semantic analysis has to follow the static semantic rules of the CASL language semantics [7].
3. *Encoding SubPCFOL* into Isabelle/HOL. This encoding is achieved via a map between the underlying *institutions* of the logics.

In the following subsection, we will concentrate on each of these steps and also discuss how the steps can be reversed.

3.1 Parsing

As an example, consider the specification of stacks and bags in CASL in Fig. 2.

The concrete syntax of CASL (see [1]) can easily be represented in the parsing machinery of Isabelle. To keep the static analysis (which is written in ML) simple, an ML-function converts the abstract syntax tree from the Isabelle format

spec STACKS_AND_BAGS =
    sort *stack*
    ops *empty* : *stack*;
        push : *elem* × *stack* → *stack*;
        pop : *stack* →? *elem*
    pred *isempty* : *stack*
    vars *x* : *elem*; *s* : *stack*
    axioms *isempty*(*empty*);
        ¬ *isempty*(*push*(*x*, *s*));
        pop(*push*(*x*, *s*)) = *x*
    sorts *elem*, *bag*, *nebag*;
        nebag < *bag*
    ops *empty* : *bag*;
        add : *elem* × *nebag* → *nebag*;
        add : *elem* × *bag* → *nebag*
    pred *isempty* : *bag*
    vars *x*, *y* : *elem*; *b* : *bag*
    axioms *isempty*(*empty*);
        ¬ *isempty*(*b*);
        add(*x*, *add*(*y*, *b*)) = *add*(*y*, *add*(*x*, *b*))

Figure 2. A specification of stacks and bags in CASL

to a more concise ML-format. Furthermore, in Isabelle one can easily implement so-called *print-translations* for pretty printing CASL abstract syntax trees.

3.2 Static Semantic Analysis

The static semantic analysis has to follow the static semantic rules of the CASL semantics [7]. These rules can be encoded in a rather straightforward way, with the exception of the overload resolution described below.

The outcome of the static analysis is a theory (that is, a signature and a set of sentences) of subsorted partial first-order logic with sort constraints (*SubPCFOL*), the underlying institution of CASL. Sentences in *SubPCFOL* are fully qualified, which means that each function and predicate symbol is qualified with its profile. As a consequence, any ambiguities caused by overloading are completely resolved.

The task of overload resolution is to generate a fully qualified *SubPCFOL*-sentence from an unqualified or only partially qualified CASL formula (this sentence is also called a fully qualified *expansion* of the CASL formula). Overload resolution for an arbitrary closed CASL formula proceeds inductively over the structure of the formula, while an environment with the sorts of the variables in the current scope is carried around. Thus it suffices to design an overload resolution algorithm for atomic formulae, which receives an atomic formula and an environment as input and which returns either an expansion of the atomic formula or the message "not resolvable".

For an atomic formula, overload resolution has to check that there is just one fully qualified expansion of the atomic formula up to the equivalences $\sim_F$ and $\sim_P$ defined in Sect. 2.1. There is a simple, naive overload resolution algorithm: collect the set of all expansions of a given atomic formula, and check that they are all semantically equivalent, using the definitions of $\sim_F$ and $\sim_P$; but this bears a lot of redundancy. The "least sort parse" algorithm for regular signatures described in [8], p. 252, just picks *one* expansion (the one with minimal profiles) and forgets about all the other ones. Of course, regularity is defined precisely in such a way that this algorithm works. Since CASL does not impose any condition on signatures, we cannot expect to get such a simple algorithm here. But we can expect that our algorithm behaves as simple as the least sort parse if the signature *happens* to be regular.

Given an atomic formula φ over a signature Σ and a context X of sorted variables, our overload resolution algorithm inductively computes the set of *minimal expansions* of φ as follows:

- For a constant or variable term c[4], the set $MinExpTerm_{\Sigma,X}(c)$ is computed as follows: Consider the set $\{c : s \mid c : s \in X\} \cup \{c : s \mid c : s \in PF \cup TF\}$. Take the subset of all $c : s$ with s minimal. Divide this subset into equivalence classes w.r.t. to the equivalence relation generated by $\sim_F$.
- Given a function application $f(t_1, \ldots, t_n)$[5], for $i = 1, \ldots, n$, inductively compute $MinExpTerm_{\Sigma,X}(t_i)$ (which is a set of equivalence classes). For each $(C_1, \ldots, C_n) \in MinExpTerm_{\Sigma,X}(t_1) \times \cdots \times MinExpTerm_{\Sigma,X}(t_n)$, consider the set of possible profiles w.r.t. f:[6]

$$P(C_1, \ldots, C_n) = \{(\mathbf{op}\ f{:}\, w \longrightarrow s)(t_1 : s_1, \ldots, t_n : s_n) : s \mid$$

$$t_i : s_i \in C_i, s_1, \ldots, s_n \leq w, f{:}\, w \longrightarrow s \in TF \cup PF\}$$

Define an equivalence relation $\sim$ on $P(C_1, \ldots, C_n)$ generated by

$$\frac{\begin{array}{c} s_1, \ldots, s_n \leq w_1,\ s_1, \ldots, s_n \leq w_2 \\ f{:}\, w_1 \longrightarrow u_1 \sim_F f{:}\, w_2 \longrightarrow u_2 \\ t_i : s_i \in C_i (i = 1, \ldots, n) \end{array}}{(\mathbf{op}\, f{:}w_1 \longrightarrow u_1)(t_1 : s_1, \ldots, t_n : s_n) : u_1 \sim (\mathbf{op}\, f{:}w_2 \longrightarrow u_2)(t_1 : s_1, \ldots, t_n : s_n) : u_2}$$

$$\frac{\begin{array}{c} (t_1 : s_1, \ldots, t_n : s_n), (t'_1 : s'_1, \ldots, t'_n : s'_n) \in C_1 \times \cdots \times C_n \\ s_1, \ldots, s_n \leq w \\ s'_1, \ldots, s'_n \leq w \\ f{:}\, w \longrightarrow s \in TF \cup PF \end{array}}{(\mathbf{op}\, f{:}\, w \longrightarrow s)(t_1 : s_1, \ldots, t_n : s_n) : s \sim (\mathbf{op}\, f{:}\, w \longrightarrow s)(t'_1 : s'_1, \ldots, t'_n : s'_n) : s}$$

[4] Note that we cannot distinguish constants from variables before overload resolution.

[5] For simplicity, projection casts are considered as special partial function symbols.

[6] To get a *SubPCFOL*-formula, the algorithm has to be refined by inserting an injection $inj_{s \rightarrow s'}$ whenever a term $t : s$ is used in a context where a term of sort s' (with $s' \geq s$) is required. Due to full qualification, this insertion of injections is unambiguous.

Now take

$$P = \bigcup\{P(C_1,\ldots,C_n)/\sim_{P(C_1,\ldots,C_n)}|$$

$$(C_1,\ldots,C_n) \in MinExpTerm_{\Sigma,X}(t_1) \times \cdots \times MinExpTerm_{\Sigma,X}(t_n)\}$$

and replace each $C \in P$ by a subset C' which, for each s minimal such that there is some $t : s \in C$, chooses one such $t : s \in C$. The set of these subsets C' is $MinExpTerm_{\Sigma,X}(f(t_1,\ldots,t_n))$.[7]

- Given a sorted term $t : s$, inductively compute $MinExpTerm_{\Sigma,X}(t)$. For each $C \in MinExpTerm_{\Sigma,X}(t)$, $MinExpTerm_{\Sigma,X}(t : s)$ contains the set

$$\{t : s \mid t : s' \in C, s' \leq s\}$$

- Given a predicate application $\varphi = p(t_1,\ldots,t_n)$[8], compute the set P in a way that parallels the computation of P for function applications, and output P as $MinExp_{\Sigma,X}(\varphi)$.

Now if $MinExp_{\Sigma,X}(\varphi)$ contains exactly one equivalence class, each element of this equivalence class can serve as a fully qualified expansion of φ. Otherwise, φ is ambiguous, and we can output one element of each equivalence class to show the set of possible disambiguations to the user.

In the regular case, all the *MinExpTerm*-sets are just singletons and consist of the least sort parse. Thus only in (rather exceptional) non-regular cases the algorithm becomes more complicated. By computing minimality and equivalence at each inductive step, the above algorithm avoids exponential time traps typical for such disambiguations, and always runs in polynomial time.

The example given in Fig. 2 can be analyzed by the overload resolution algorithm and returns the following error:

`Cannot disambiguate:`

$$isempty : bag(\mathbf{op}\ empty : bag), isempty : stack(\mathbf{op}\ empty : stack)$$

This error can be repaired by replacing the unqualified axiom $isempty(empty)$ with one of the two qualified expansions given above, respectively. With this correction, all sorts can be inferred correctly by the overload resolution algorithm. For example,

$$add(x, add(y, b)) = add(y, add(x, b))$$

is analyzed as

$$(\mathbf{op}\ add : elem \times nebag \to nebag)(x, (\mathbf{op}\ add : elem \times bag \to nebag)(y, b)) =$$

$$(\mathbf{op}\ add : elem \times nebag \to nebag)(y, (\mathbf{op}\ add : elem \times bag \to nebag)(x, b))$$

---

[7] Note that, different from the case of constant or variable terms, we have to first factor w.r.t. $\sim$ and then select the minimal expansions.

[8] For simplicity, equality, definedness and subsort membership are considered as overloaded predicates.

which is a minimal expansion equivalent to all other possible expansions.

To reverse the process of static semantic analysis, given a *SubPCFOL*-theory, we have to construct a CASL-specification out of it. This can be done as follows: first declare all sort symbols, then all subsorting relations, then all total and partial function symbols, and then all predicate symbols. Finally, represent each *SubPCFOL*-sentence as a CASL-axiom. Here, we have two options:

1. Represent a *SubPCFOL*-sentence directly (this is possible since it is allowed to fully qualify each function and predicate symbol with its profile in CASL).
2. Remove all qualifications of function and predicate symbols where they are not necessary for disambiguation. This has the advantage that axioms are not cluttered with all the qualifying profiles and become much more readable.

For our example, the second option yields the original input.

3.3 Encoding in HOL

The central point of the bridge from CASL to existing tools is the encoding of *SubPCFOL*, the underlying institution of CASL, into higher-order logic (or of *SubPFOL* into first-order logic, if sort generation constraints are omitted). For these logics, a number of theorem proving tools are available.

Technically, this is achieved by a simple map of institutions in the sense of [15] with a surjective model translation. In this setting an application of the *borrowing* technique proposed in [4] allows to re-use theorem provers in a sound and complete way.

For the moment, we will ignore sort generation constraints and concentrate on *SubPFOL*. There are basically two ways to encode *SubPFOL* within *FOL*: one via partial first-order logic (*PFOL*) and the other one via subsorted first-order logic (*SubFOL*), where *SubFOL* is the subinstitution of *SubPFOL* that singles out signatures with no partial function symbols and sentences with no partial projection symbols.

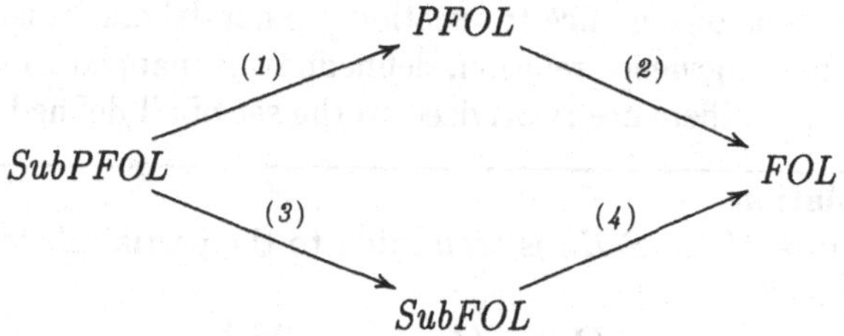

(1) Encoding of *SubPFOL* into *PFOL*. A translation of *SubPFOL* to *PFOL* is already built into the definition of *SubPFOL*: in [3], first *PFOL* is defined, and then *SubPFOL*-signatures a translated to *PFOL*-signatures by adding explicit total injection and partial projection functions between subsorts, and membership predicates for testing membership in the image of an injection; *SubPFOL*-sentences and models are even *defined* to be certain *PFOL*-sentences and models, respectively.

(2) Encoding of *PFOL* into *FOL*. A translation of *PFOL* into *FOL* is described in [5]. We refine this translation here. The main idea is to add to each carrier set a new element that represents "undefined", so that partial functions can be totalized.

Signature translation
A *PFOL*-signature $\Sigma = (S, TF, PF, P)$ is translated to a *FOL*-theory having the signature

$$sign(\Phi(\Sigma)) = (S, TF \uplus PF, P \uplus (\{\stackrel{e}{=}_s : s \times s \mid s \in S\} \cup \{D_s : s \mid s \in S\})$$

and the set of axioms $ax(\Phi(\Sigma))$:

$\exists ! x : s \bullet \neg D_s(x)$	$s \in S$
$x \stackrel{e}{=}_s y \Leftrightarrow (x = y \wedge D_s(x))$	$s \in S$
$D_s(f(x_1, \ldots, x_n)) \Leftrightarrow \bigwedge_{i=1..n} D_{s_i}(x_i)$	$f : s_1, \ldots, s_n \rightarrow s \in TF$
$D_s(g(x_1, \ldots, x_n)) \Rightarrow \bigwedge_{i=1..n} D_{s_i}(x_i)$	$g : s_1 \ldots s_n \rightarrow ? s \in PF$
$p(x_1, \ldots, x_n) \Rightarrow \bigwedge_{i=1..n} D_{s_i}(x_i)$	$p : s_1 \ldots s_n \in P$

D_s plays the role of a definedness predicate, and $\stackrel{e}{=}_s$ the role of existential equality. The axioms in the signature translation state that there is exactly one undefined element, total operations are indeed total and all operations and predicates are strict.

Sentence translation
A Σ-sentence φ (in *PFOL*) is translated to the $\Phi(\Sigma)$-axiom $\alpha_\Sigma(\varphi)$:

$\alpha_\Sigma(defined\ (t)) = D_s(t)$	$\alpha_\Sigma(\neg\varphi) = \neg\alpha_\Sigma(\varphi)$
$\alpha_\Sigma(t_1 = t_2) = (t_1 = t_2)$	$\alpha_\Sigma(\varphi \wedge \psi) = \alpha_\Sigma(\varphi) \wedge \alpha_\Sigma(\psi)$
$\alpha_\Sigma(t_1 \stackrel{e}{=} t_2) = t_1 \stackrel{e}{=}_s t_2$	$\alpha_\Sigma(\varphi \vee \psi) = \alpha_\Sigma(\varphi) \vee \alpha_\Sigma(\psi)$
$\alpha_\Sigma(P(t_1, \ldots, t_n)) = P(t_1, \ldots, t_n)$	$\alpha_\Sigma(false) = false$
$\alpha_\Sigma(\varphi \Rightarrow \psi) = \alpha_\Sigma(\varphi) \Rightarrow \alpha_\Sigma(\psi)$	$\alpha_\Sigma(true) = true$
$\alpha_\Sigma(\forall x : s \bullet \varphi) = \forall x : s \bullet D_s(x) \Rightarrow \alpha_\Sigma(\varphi)$	$\alpha_\Sigma(\exists x : s \bullet \varphi) = \exists x : s \bullet D_s(x) \wedge \alpha_\Sigma(\varphi)$

Only three kinds of sentence translation are non-trivial: existential equality is mapped to the congruence relation, definedness is mapped to the definedness predicate, and quantifiers are relativized to the set of all defined elements.

Model translation
A $\Phi(\Sigma)$-structure M (in *FOL*) is translated to the partial Σ-structure $\beta_\Sigma(M)$ (in *PFOL*) with

$$\beta_\Sigma(M) = (M|_{\Sigma \longrightarrow \Phi(\Sigma)}) / \stackrel{e}{=}_M$$

where $M|_{\Sigma \longrightarrow \Phi(\Sigma)}$ is the reduct of M along the obvious map $\Sigma \longrightarrow \Phi(\Sigma)$

That is, M is interpreted as a partial Σ-structure. The quotient is taken as in [2]. Note that since $\stackrel{e}{=}_M$ is a subset of the equality relation, taking the quotient here only has the effect to take the subset of defined elements and to define the operations as the partial restrictions of the original operations.

The model translation can be easily shown to be surjective: For a partial

Σ-structure M, just add one element, $\bot$, to all carriers, let all functions map $\bot$ to itself and behave as in M otherwise, where undefinedness of partial functions is mapped to $\bot$. Predicates are false on the new element.

The representation condition is proved as in [5].

(3) Encoding of *SubPFOL* into *SubFOL*

Signature translation

Let $\Sigma = (S, TF, PF, P, \leq_S)$ be a *SubPFOL*-signature. Let

- $?S = \{s \in S \mid$ there exists a variable system X and a $\Sigma(X)$-term of sort s built using at least one partial function symbol or projection cast$\}$,
- $S' = S \cup \{?s \mid s \in ?S\}$[9] and $Q(s) = \begin{cases} ?s, & \text{if } s \in ?S \\ s, & \text{otherwise} \end{cases}$

Then Σ is translated to the *SubFOL*-signature $sign(\Phi(\Sigma))$ consisting of the sort set S', ordered by $s \leq' s'$ if $s \leq s'$; $s \leq' ?s'$ if $s \leq s'$; $?s \not\leq' s'$ and $?s \leq' ?s'$ if $s \leq s'$, augmented by the following function symbols:

$f: s_1, \ldots, s_n \to s$ for $f: s_1, \ldots, s_n \to s \in TF$

$f: s_1, \ldots, s_n \to (?s)$ for $f: s_1, \ldots, s_n \to ?\, s \in PF$,

$f: Q(s_1), \ldots, Q(s_n) \to (?s)$ for $f: s_1, \ldots, s_n \longrightarrow s \in TF \cup PF$, $\{s_1, \ldots, s_n\} \cap ?S \neq \emptyset$

$\mathrm{pr}_{(s,s')}: s \to (?s')$ for $s' \leq s$

$\mathrm{pr}_{(s,s')}: ?s \to (?s')$ for $s' \leq s$, if $x \in ?S$

the following predicate symbols:

$p: s_1, \ldots, s_n$ for $p: s_1, \ldots, s_n \in P$,

$p: Q(s_1), \ldots, Q(s_n)$ for $p: s_1, \ldots, s_n \in P, \{s_1, \ldots, s_n\} \cap ?S \neq \emptyset$

$\stackrel{e}{=}_s: ?s \times ?s$ for $s \in ?S$,

$D_s: ?s$ for $s \in ?S$

and the following axioms $ax(\Phi(\Sigma))$:

$\forall x : ?s \bullet D_s(x : ?s)$ $\quad s \in ?S$

$\forall x : ?s \bullet D_s(x) \Leftrightarrow x \in s$ $\quad s \in ?S$

$\exists! x : ?s \bullet \neg D_s(x)$ $\quad s \in ?S$

$\forall x : s' \bullet D_s(\mathrm{pr}_{(s',s)}(x)) \Leftrightarrow x \in s$ $\quad s \leq s'$

$\forall x, y : ?s \bullet x \stackrel{e}{=}_s y \Leftrightarrow (x = y \wedge D_s(x) \wedge D_s(y))$ $\quad s \in ?S$

$\forall x : s \bullet \mathrm{pr}_{(s',s)}(x : s') = x$ $\quad s \leq s'$

$\forall x, y : s' \bullet \mathrm{pr}_{(s',s)}(x) = \mathrm{pr}_{(s',s)}(y) \Rightarrow x = y$ $\quad s \leq s'$

$\forall x_1 : Q(s_1), \ldots, x_n : Q(s_n) \bullet D(f(x_1, \ldots, x_n)) \Leftrightarrow \bigwedge_{s_i \in ?S} D(x_i)$ $\quad f: s_1, \ldots, s_n \to s \in TF$

$\forall x_1 : Q(s_1), \ldots, x_n : Q(s_n) \bullet D(f(x_1, \ldots, x_n)) \Rightarrow \bigwedge_{s_i \in ?S} D(x_i)$ $\quad f: s_1, \ldots, s_n \to ?\, s \in PF$

$\forall x_1 : Q(s_1), \ldots, x_n : Q(s_n) \bullet p(x_1, \ldots, x_n) \Rightarrow \bigwedge_{s_i \in ?S} D(x_i)$ $\quad p: s_1, \ldots, s_n \in P$

The idea here is again to add a new element that represents "undefined", but to retain the old sorts without the undefined element. That is, the sort set is doubled: each sort s now gets a companion supersort $?s$, which contains all values of s plus the undefined element. Since we keep the old sorts, the definedness predicate is just membership in the old sorts. Moreover, we add the companion supersort only to those sorts for which it is necessary, i.e. for those where a term involving partial operation symbols exists.

In the signature translation, we introduce two versions for each function and predicate symbol: the original one and its (overloaded) strict extension that just propagates the undefined element. The strict extension is necessary if we want to apply a function or predicate symbol to a term involving partial operation symbols, since this term can be translated only to a term of sort $?s$.

For the partial projections, we introduce new special function symbols, and for definedness and existential equality, we introduce new predicate symbols.

Sentence translation

A Σ-sentence in *SubPFOL* is translated to a $\Phi(\Sigma)$-sentence in *SubFOL* inductively as follows:

- $\alpha_\Sigma(x : s) = x : s$
- $\alpha_\Sigma((\mathbf{op}\ f : s_1, \dots, s_n \to s)(t_1, \dots, t_n)) = \qquad (f : s_1, \dots, s_n \to s \in TF)$
  $\begin{cases} (\mathbf{op}\ f : s_1, \dots, s_n \to s)(\alpha_\Sigma(t_1), \dots, \alpha_\Sigma(t_n)) \text{ if for } i = 1..n\ \alpha_\Sigma(t_i) : s_i \\ (\mathbf{op}\ f : Q(s_1), \dots, Q(s_n) \to (?s))(\alpha_\Sigma(t_1), \dots, \alpha_\Sigma(t_n)), \text{ otherwise} \end{cases}$
- $\alpha_\Sigma((\mathbf{op}\ f : s_1, \dots, s_n \to ?\ s)(t_1, \dots, t_n)) = \qquad (f : s_1, \dots, s_n \to ?\ s \in PF)$
  $\begin{cases} (\mathbf{op}\ f : s_1, \dots, s_n \to (?s))(\alpha_\Sigma(t_1), \dots, \alpha_\Sigma(t_n)) \text{ if for } i = 1..n\ \alpha_\Sigma(t_i) : s_i \\ (\mathbf{op}\ f : Q(s_1), \dots, Q(s_n) \to (?s))(\alpha_\Sigma(t_1), \dots, \alpha_\Sigma(t_n)), \text{ otherwise} \end{cases}$
- $\alpha_\Sigma((\mathbf{pred} p : s_1, \dots, s_n)(t_1, \dots, t_n)) = \qquad (p : s_1, \dots, s_\in P)$
  $\begin{cases} (\mathbf{pred} p : s_1, \dots, s_n)(\alpha_\Sigma(t_1), \dots, \alpha_\Sigma(t_n)) \text{ if for } i = 1, \dots, n\ \alpha_\Sigma(t_i) : s_i \\ (\mathbf{pred} p : Q(s_1), \dots, Q(s_n))(\alpha_\Sigma(t_1), \dots, \alpha_\Sigma(t_n)), \text{ otherwise} \end{cases}$
- $\alpha_\Sigma(t\ as\ s) = \mathbf{pr}_{(s',s)}(\alpha_\Sigma(t))\ (t : s', s \leq s')$
- $\alpha_\Sigma(\mathit{defined}\ (t)) = \begin{cases} D_s(\alpha_\Sigma(t)), & \text{if } \alpha_\Sigma(t) : ?s, s \in ?S \\ \mathit{true}, & \text{otherwise} \end{cases}$
- $\alpha_\Sigma(t \in s') = t \in s'$
- $\alpha_\Sigma(t_1 = t_2) = \alpha_\Sigma(t_1) = \alpha_\Sigma(t_2)$
- $\alpha_\Sigma((t_1 \stackrel{e}{=} t_2) = \begin{cases} \alpha_\Sigma(t_1) \stackrel{e}{=}_s \alpha_\Sigma(t_2), & \text{if } \alpha_\Sigma(t_1), \alpha_\Sigma(t_2) : ?s \\ \alpha_\Sigma(t_1) = \alpha_\Sigma(t_2), & \text{otherwise} \end{cases}$

This can easily be extended from atomic to all formulae.

Apart from mapping definedness and projection to the special symbols introduced above, the translation mainly checks whether the original version of a function or predicate symbol can be used, or if we have to use its strict extension that propagates the undefined element. The strict extension is necessary

[9] We assume that S does not contain sorts of the form $?s$. In a higher-order extension, there could be a type constructor "?" behaving like our "?".

whenever an argument term involves a partial operation symbol.

Model translation
A $\Phi(\Sigma)$-structure M (in *SubFOL*) is translated to the partial Σ-structure $\beta_\Sigma(M)$ (in *SubPFOL*) having carriers sets M_s for $s \in S$ and total functions and predicates inherited from M, while partial functions $g_{\beta_\Sigma(M)}$ are the corestrictions of g_M to M_s (where s is the result sort of g).

Again, we have a straightforward inverse: a Σ-structure M is extended by adding new carriers $M_{?s} = M_s \uplus \{\bot\}$ and adding the strict extensions that map $\bot$ to itself.

The representation condition can be proved by an easy induction over the structure of the sentences.

(4) Encoding of *SubFOL* into *FOL*. Here we can use a restriction of the translation of *SubPFOL* to *PFOL* described above, which leaves out the partial projection symbols. There is one problem connected with this: the membership predicate was originally axiomatized using partial projections and definedness:

$$\forall x : s' \bullet \in^s_{s'} (x) \Leftrightarrow D_s(\mathtt{pr}_{(s',s)}(x))$$

for $s \le s'$. This axiom has to be replaced by

$$\forall x : s' \bullet \in^s_{s'} (x) \Leftrightarrow \exists y \bullet \mathtt{inj}_{(s,s')}(y) = x$$

since we do not have partial projections in *SubFOL*[10].

Comparison of (2)∘(1) with (4)∘(3) Comparing both ways of encoding *SubPFOL* into *FOL*, we can see that the way via *SubFOL* has the disadvantage that many new sorts are introduced, but the advantage than the original sort system and its interpretation are kept. Moreover, many-sorted total specifications are left unchanged (but the encoding via *PFOL* may also be adapted to have this property).

Encoding of *SubPCFOL* into *HOL*. *SubPCFOL* adds sort generation constraints to *SubPFOL*. These cannot be expressed within *FOL*, but can be translated to the usual induction schemes, which can be expressed in *HOL* by second-order quantification over predicates.

3.4 The Way Back

In the previous section, two ways of encoding *SubPFOL* (*SubPCFOL*) into *FOL* (*HOL*) were described, which allow to re-use *HOL*-theorem provers for

[10] This axiom is not in universal Horn form, but is still in the form of a limit theory that admits initial models, see [16].

SubPCFOL, the underlying institution of CASL. Intermediate steps within a *HOL* proof can be visualized to the user as a *SubPCFOL* or even CASL formulae by interpreting the special operation and predicate symbols introduced in the translation with their original meaning. This decoding is used for displaying pretty-printed *HOL*-formulae that occur as goals in Isabelle theorem proving sessions in an CASL-like syntax.

However, the target of this decoding is not the set of sentences over the original signature, but the signature enriched with sorts of form $?s$. A value of sort $?s$ then may either be undefined or a value of sort s. A variable of sort $?s$ may be interpreted as a potentially undefined variable over sort s, which resembles the partial logic of Scott [19]. Standard proof calculi for partial logics do not have this notion of possibly undefined variable. This means that each application of the substitution rule has to be preceded by a proof of the definedness of the term being substituted. The advantage of working with these possibly undefined variables is that the usual substitution rule of first- and higher-order logic remains valid (and it is just the substitution rule of *HOL* inherited via the encoding).

4 Conclusion

We have shown how to implement a static semantic analysis for CASL, including an overload resolution algorithm, which we have described and implemented. The result of overload resolution is a fully qualified CASL abstract tree involving subsorts and partiality, which can be considered as a theory of subsorted partial first-order logic with sort generation constraints (*SubPCFOL*), the underlying institution of CASL.

In the second part of the paper, we have described two translations, each going either from *SubPFOL* (CASL without sort generation constraints) to *FOL* (first-order logic) or from *SubPCFOL* (the full CASL logic) to *HOL* (higher-order logic) The first translation first drops subsorts and then partiality, the other one first partiality and then subsorts. This allows one to use (among others) the usual substitution- and resolution-based theorem provers for *FOL*/*HOL* when doing proofs in CASL. We have also shown that intermediate proof steps in the *FOL*/*HOL*-proof can be translated back to a slight extension of *SubP*(*C*)*FOL*. All this has been realized in a prototype implementation, which also uses the IsaWin system providing an interactive window based user interface [11].

The results of this paper do not depend on specific features of Isabelle. We have designed an encoding of CASL which can also be used in connection with any other theorem prover for first or higher-order logic.

4.1 Future Work

An important task for the future is to prove derived rules and to design tactics within the Isabelle/HOL-encoding that are particularly well-suited for theorem proving in CASL. For example, one can think of simplifier sets (an Isabelle term rewriting machinery) that automatically deal with injections and projections.

Another direction of future work is the restriction of the CASL encoding to suitable subsets of CASL, e. g. to axioms that are universal Horn formulae or just equations. With slight modifications, the embedding should then have as target universal Horn theories as well, which means that paramodulation and (conditional) term rewriting become applicable.

A third direction is the extension of the current encoding to extensions of CASL, such as higher-order CASL.

One important and challenging field of research remains: CASL in-the-large. One can think of at least two ways of extending the embedding: either working on the abstract syntax of a structured specification at a syntactical level or encoding the in-the-large model theory in HOL.

For transformational program development, the embedding of CASL in Isabelle/HOL will be integrated into the Universal Formal Methods Workbench of the UniForM project [14]. This integration will be just an instantiation of a generic graphical user interface [10], which generates a transformational program development environment, using a Transformation Application System (TAS). The theorem prover Isabelle is used to prove the correctness of a particular development as well as the correctness of the transformation rules (see [12]).

A main topic to be tackled in the future is the development of suitable transformations for CASL. In PROSPECTRA [9], a whole bunch of transformations has been developed. They range from simple transformations to complex ones such as split of postcondition, finite differencing etc. These should be adapted to the algebraic approach to program development and proved correct in the CASL embedding itself.

Acknowledgements We would like to thank all the participants of CoFI.

References

1. M. Bidoit, C. Choppy, B. Krieg-Brückner, P. D. Mosses, and F. Voisin. Concrete syntax for CASL basic and structured specifications. Postscript[11], December 1997.
2. P. Burmeister. *A model theoretic approach to partial algebras.* Akademie Verlag, Berlin, 1986.
3. M. Cerioli, A. Haxthausen, B. Krieg-Brückner, and T. Mossakowski. Permissive subsorted partial logic in CASL. In M. Johnson, editor, *Algebraic methodology and software technology: 6th international conference, AMAST 97*, volume 1349 of *Lecture Notes in Computer Science*. Springer-Verlag, 1997.
4. M. Cerioli and J. Meseguer. May I borrow your logic? (transporting logical structures along maps). *Theoretical Computer Science*, 173:311–347, 1997.
5. M. Cerioli, T. Mossakowski, and H. Reichel. From total equational to partial first order. In E. Astesiano, H.-J. Kreowski, and B. Krieg-Brückner, editors, *Algebraic Foundations of Systems Specifications*. 1997. To appear[12].

[11] http://www.brics.dk/Projects/CoFI/Documents/CASL/SyntaxIssues
[12] http://www.informatik.uni-bremen.de/~kreo/ifip_chapters/chapters.html

6. CoFI task group on Language Design. CASL – The CoFI Algebraic Specification Language – Summary. CoFI Document: CASL/Summary. WWW[13], FTP[14], September 1997.
7. CoFI task group on Semantics. CASL – The CoFI Algebraic Specification Language (version 0.97) – Semantics. CoFI Note: S-6. WWW[15], FTP[16], July 1997.
8. J. A. Goguen and J. Meseguer. Order-sorted algebra I: equational deduction for multiple inheritance, overloading, exceptions and partial operations. *Theoretical Computer Science*, 105:217–273, 1992.
9. B. Hoffmann and B. Krieg-Brückner. *Program Development by Specification and Transformation.* LNCS 690. Springer Verlag, 1993.
10. Kolyang, C. Lüth, T. Meier, and B. Wolff. TAS and IsaWin: Generic interfaces for transformational program development and theorem proving. In M. Bidoit and M. Dauchet, editors, *Proceedings of the Seventh International Joint Conference on the Theory and Practice of Software Development (TAPSOFT'97)*, pages 855–858. Springer-Verlag LNCS 1214, 1997.
11. Kolyang and T. Mossakowski. Implementation of CASL into Isabelle/HOL. Bremen University[17], 1997.
12. Kolyang, T. Santen, and B. Wolff. Correct and user-friendly implementations of transformation systems. In M. C. Gaudel and J. Woodcock, editors, *Formal Methods Europe*, LNCS 1051, pages 629– 648. Springer Verlag, 1996.
13. Kolyang, T. Santen, and B. Wolff. A structure preserving encoding of Z in Isabelle. In J. von. Wright, J. Grundy, and J. Harrison, editors, *Theorem Proving in Higher Order Logics*, LNCS 1125, pages 283 – 298. Springer Verlag, 1996.
14. B. Krieg-Brückner, J. Peleska, E.-R. Olderog, D. Balzer, and A. Baer. UniForM Workbench — Universelle Entwicklungsumgebung für formale Methoden. Technischer Bericht 8/95, Universität Bremen, 1995. English version in: Statusseminar Softwaretechnologie BMBF[18].
15. J. Meseguer. General logics. In *Logic Colloquium 87*, pages 275–329. North Holland, 1989.
16. T. Mossakowski. Equivalences among various logical frameworks of partial algebras. In H. K. Büning, editor, *Computer Science Logic. 9th Workshop, CSL'95. Paderborn, Germany, September 1995, Selected Papers*, volume 1092 of *Lecture Notes in Computer Science*, pages 403–433. Springer Verlag, 1996.
17. P. D. Mosses. CoFI: The common framework initiative for algebraic specification and development. In M. Bidoit and M. Dauchet, editors, *TAPSOFT'97*, volume 1214 of *LNCS*, pages 115–137. Springer–Verlag, 1997.
18. L. C. Paulson. *Isabelle - A Generic Theorem Prover.* Number 828 in LNCS. Springer Verlag, 1994.
19. D. S. Scott. Identity and existence in intuitionistic logic. In M. Fourman, C. Mulvey, and D. Scott, editors, *Application of Sheaves*, volume 753 of *Lecture Notes in Mathematics*, pages 660–696. Springer Verlag, 1979.

[13] http://www.brics.dk/Projects/CoFI/Documents/CASL/Summary/
[14] ftp://ftp.brics.dk/Projects/CoFI/Documents/CASL/Summary/
[15] http://www.brics.dk/Projects/CoFI/Notes/S-6/
[16] ftp://ftp.brics.dk/Projects/CoFI/Notes/S-6/
[17] http://www.informatik.uni-bremen.de/~casl
[18] http://www.informatik.uni-bremen.de/~uniform/

Combining and Representing Logical Systems Using Model-Theoretic Parchments*

Till Mossakowski[1], Andrzej Tarlecki**[2] and Wiesław Pawłowski[3]

[1] Department of Computer Science, University of Bremen
[2] Institute of Informatics, Warsaw University and Institute of Computer Science, Polish Academy of Sciences, Warsaw.
[3] Institute of Computer Science, Polish Academy of Sciences, Gdańsk

Abstract. The paper addresses important problems of building complex logical systems and their representations in universal logics in a systematic way. We adopt the model-theoretic view of logic as captured in the notions of *institution* and of *parchment* (an algebraic way of presenting institutions). We propose a new, modified notion of parchment together with parchment morphisms and representations. In contrast to the original parchment definition and our earlier work, in *model-theoretic parchments* introduced here the universal semantic structure is distributed over individual signatures and models. We lift formal properties of the categories of institutions and their representations to this level: the category of model-theoretic parchments is complete, and their representations may be put together using categorical limits as well. However, model-theoretic parchments provide a more adequate framework for systematic combination of logical systems than institutions. We indicate how the necessary invention for combination of various logical features may be introduced either on an ad hoc basis or via representations in a universal logic.

1 Introduction

The evident and needed multitude of logical systems used for software specification and development led Goguen and Burstall to propose an ambitious goal: as much of the area as possible should be developed for an arbitrary logical system rather than for a number of specific systems separately. A part of this program must address the issues of relating logical systems with each other, for instance to provide a framework for systematic construction of new, gradually more and more complex logical systems, and to present the possibilities of encoding practical logical systems in a powerful, in some sense universal logic. This paper provides a further step in our foundational work on these issues.

We take a model-theoretic view of logic, built around the semantic satisfaction, as captured in the notion of *institution* [6]. The two general goals mentioned above are reflected in the framework of institutions by two different notions of

* A full version of this paper is available as [11]. It contains all the proofs, complete technicalities and more discussion and examples, omitted here due to space limitations.

** This work has been partially supported by KBN grant 8T11C 018 11.

arrow: institution *morphisms* [6] that capture how one institution is built over another and institution *representations* [15] (or *simple maps* of institutions [8]) that capture how one institution is encoded in another.

In [15] the role of these notions and the interplay between them have been studied. The category of institutions and institution morphisms was discussed as a rudimentary framework for systematic construction of logical systems via limits (which are shown to exist in [14], [16]). Maps between institution representations, consisting of an institution morphism and an extra "fitting" component have been introduced. Representations related by such maps can be combined via category-theoretic limits, so that the combined institution is represented in some "universal" institution provided that all the components are represented in it.

One deficiency of combinations of institutions via limits is that sets of sentences are simply united in the combination rather then being properly combined, and the features of the combined logics do not really interact in the result. A possible solution is to move to *parchments* [5], certain algebraic presentations of institutions providing an abstract syntax and evaluator-based semantics for sentences and therefore potentially more useful for logic combination [9].

In [10] we have introduced λ*-parchments*, a slightly modified notion of parchment, and lifted the main notions of arrow and their properties from institutions to λ-parchments. One justification for working with λ-parchments rather than with the original parchments was that the latter inherently rely on a concept of an internal "universal" signature and a similarly "universal" semantic structure — which seems rather dubious from the foundational point of view, as visible even in the simplest examples [5]. In λ-parchments we have separated models from signatures and thus moved the "universal" signature and structure from the internal level of the logic to its presentation. We have also gained more flexibility to represent concepts that manipulate signatures and models separately.

In the present work we go further along this line: we introduce *model-theoretic parchments*, modifying λ-parchments so that the universal semantic structure is distributed over all signatures and models of the logic. This removes entirely any foundational doubt one may have as no "large" signatures or structures need to be dealt with anymore. Moreover, linking the semantic structures that describe evaluation of the linguistic constructs directly with individual signatures and models makes model-theoretic parchments and the related notions more flexible. While the required degree of uniformity is ensured by the appropriate naturality conditions, we can capture now examples of logical systems where the semantics may be fine-tuned to the need of individual signatures and models.

2 Preliminaries

Basic knowledge of category theory is assumed throughout the paper; for the standard notions and facts we refer for instance to [1]. We briefly recall some concepts and facts on relational structures and sources in their categories (cf. [1]).

Relational algebraic signatures $\Sigma = (S, OP, REL)$, consist of a set of sort symbols $s \in S$, an $(S^* \times S)$-indexed set OP of total operation symbols and an

S^*-indexed set REL of relation symbols. Signature morphisms are defined in the standard way — the resulting category is denoted by **AlgSig**.

Let *Logic* $\in$ |**AlgSig**| be the relational algebraic signature consisting of a sort $*$ and a relation symbol $D : *$. Let **AlgSig**$_*$ be the category of many-sorted relational signatures from **AlgSig** having *Logic* as a subsignature and signature morphisms being the identity on *Logic*.

For each signature $\Sigma \in |\mathbf{AlgSig}|$, we have a category $\mathbf{Str}(\Sigma)$ of *Σ-structures* and *Σ-homomorphisms*, where homomorphisms only preserve, but not necessarily reflect the relations, see [12]. A Σ-homomorphism $h: A \longrightarrow B$ is *closed* if it also reflects the relations, that is, for all $R : w \in REL$, $h_w^{-1}[R_B] = R_A$. $\mathbf{Str}(\Sigma)$ is cocomplete. Each signature morphism $\sigma: \Sigma \longrightarrow \Sigma' \in \mathbf{AlgSig}$ determines the *forgetful functor* $\mathbf{Str}(\sigma): \mathbf{Str}(\Sigma') \longrightarrow \mathbf{Str}(\Sigma)$. As usual, given $\sigma: \Sigma \longrightarrow \Sigma'$ and $A' \in |\mathbf{Str}(\Sigma')|$, we will write $A|_\sigma$ rather than $\mathbf{Str}(\sigma)(A')$.

Let $\mathcal{S} = \langle A, \langle A \xrightarrow{g_i} B_i \rangle_{i \in I} \rangle$ be a *source* (that is, a class of arrows with a common domain). The composition $\mathcal{S} \circ f$ of $\mathcal{S}$ with a morphism $f: A' \longrightarrow A$ is defined as expected. $\mathcal{S}$ is a *mono-source* if for any morphisms $f, g: A' \longrightarrow A$, $\mathcal{S} \circ f = \mathcal{S} \circ g$ implies $f = g$. A mono-source $\mathcal{S}$ is *extremal* provided that if $\mathcal{S} = \mathcal{T} \circ e$ for some epimorphism e then e is an isomorphism.

Proposition 1. *In* $\mathbf{Str}(\Sigma)$, $\mathcal{S} = \langle A, \langle A \xrightarrow{g_i} B_i \rangle_{i \in I} \rangle$ *is an extremal mono-source iff* $\mathcal{S}$ *is point-separating (that is, for* $s \in S$, $a_1, a_2 \in A_s$, *if for all* $i \in I$, $(g_i)_s(a_1) = (g_i)_s(a_2)$, *then* $a_1 = a_2$*) and* $\mathcal{S}$ *is closed (that is, for* $R : w \in REL$ *and* $\bar{a} \in A_w$, *if for all* $i \in I$, $(g_i)_w(\bar{a}) \in R_{B_i}$ *then* $\bar{a} \in R_A$*). In particular,* $g: A \longrightarrow B$ *is an extremal monomorphism iff it is a closed injection.*

3 Institutions and institution morphisms

Any specification formalism is usually based on some notion of signature, model, sentence and satisfaction. These are the usual ingredients of abstract model theory [2] and are the essence of Goguen and Burstall's notion of institution [6].

An *institution* $I = (\mathbf{Sign}, \mathbf{Sen}, \mathbf{Mod}, \models)$ consists of

- a category **Sign** of *signatures*,
- a functor $\mathbf{Sen}: \mathbf{Sign} \longrightarrow \mathbf{Set}$ giving the set of *sentences* $\mathbf{Sen}(\Sigma)$ over each signature Σ, and for each signature morphism $\sigma: \Sigma \longrightarrow \Sigma'$, the sentence translation map $\mathbf{Sen}(\sigma): \mathbf{Sen}(\Sigma) \longrightarrow \mathbf{Sen}(\Sigma')$,
- a functor $\mathbf{Mod}: \mathbf{Sign}^{op} \longrightarrow \mathbf{Class}$ giving the class[4] of *models* over a given signature, and for each signature morphism $\sigma: \Sigma \longrightarrow \Sigma'$, the *reduct functor* $\mathbf{Mod}(\sigma): \mathbf{Mod}(\Sigma') \longrightarrow \mathbf{Mod}(\Sigma)$ ($\mathbf{Mod}(\sigma)(M')$ is written as $M'|_\sigma$),
- a satisfaction relation $\models_\Sigma \subseteq \mathbf{Mod}(\Sigma) \times \mathbf{Sen}(\Sigma)$ for each $\Sigma \in |\mathbf{Sign}|$

such that for each $\sigma: \Sigma \longrightarrow \Sigma'$ in **Sign** the following *Satisfaction Condition* holds:

$$M' \models_{\Sigma'} \mathbf{Sen}(\sigma)(\varphi) \iff \mathbf{Mod}(\sigma)(M') \models_\Sigma \varphi$$

[4] There is no problem in allowing *categories* of models here, we leave model morphisms out for simplicity of presentation only. We will identify classes with discrete categories, so that **Class** is a subcategory of the category **CAT** of all categories.

for each $M' \in \mathbf{Mod}(\Sigma')$ and $\varphi \in \mathbf{Sen}(\Sigma)$.

Given institutions $I = (\mathbf{Sign}, \mathbf{Sen}, \mathbf{Mod}, \models)$ and $I' = (\mathbf{Sign}', \mathbf{Sen}', \mathbf{Mod}', \models')$, an *institution morphism* $\mu = (\Phi, \alpha, \beta)\colon I \longrightarrow I'$ consists of a functor $\Phi\colon \mathbf{Sign} \longrightarrow \mathbf{Sign}'$, a natural transformation $\alpha\colon \mathbf{Sen}' \circ \Phi \longrightarrow \mathbf{Sen}$, and a natural transformation $\beta\colon \mathbf{Mod} \longrightarrow \mathbf{Mod}' \circ \Phi^{op}$ such that the following satisfaction invariant holds:

$$M \models_{\Sigma} \alpha_{\Sigma}(\varphi) \iff \beta_{\Sigma}(M) \models'_{\Phi(\Sigma)} \varphi'$$

for each $\Sigma \in |\mathbf{Sign}|$, $M \in \mathbf{Mod}(\Sigma)$ and $\varphi' \in \mathbf{Sen}'(\Phi(\Sigma))$. With the obvious composition this yields the category **Ins** of institutions and institution morphisms.

Theorem 2 [14]. *The category* **Ins** *of institutions and institution morphisms is complete.*

The idea of combining things via colimits [4] applies to institutions as objects in **Ins** — since following [6] we write institution morphisms from a "richer" institution to "poorer" one, limits rather than colimits provide a tool for combination of institutions. Most roughly, limits in **Ins** are constructed by taking limits of the categories of signatures and of classes of models, and colimits of sets of sentences. We refer to [15] for simple examples. To show how this works and to indicate some problems here, let us introduce some well-known institutions and morphisms between them.

Example 1. The institution ALG of many-sorted algebras without sentences. □

Example 2. The institution $ALG(=)$ of many-sorted algebras with equalities between ground terms. Its detailed definition and the proof of the satisfaction condition may be extracted for instance from [6]. □

Example 3. The institution $PALG$ of partial many-sorted algebras (without sentences) has signatures of the form $\Sigma = (S, OP, POP)$, consisting of sort symbols $s \in S$, total operation symbols $op\colon w \longrightarrow s \in OP$ and partial operation symbols $pop\colon w \longrightarrow s \in POP$, with signature morphisms defined as expected. A Σ-model A consists of an (S, OP)-model in ALG together with a family of partial operations $(pop_A\colon A_w \rightharpoonup A_s)_{pop:w \rightharpoonup s \in POP}$. Reducts are defined as in ALG. □

Example 4. The institution morphism $\mu^{=} = (Id, \alpha^{=}, Id)\colon ALG(=) \longrightarrow ALG$ is the identity on signatures and models, and for sentences, $\alpha^{=}_{\Sigma}$ is the inclusion. □

Example 5. The institution morphism $\mu^{\mathrm{P}} = (\Phi^{\mathrm{P}}, Id, \beta^{\mathrm{P}})\colon PALG \longrightarrow ALG$ maps a signature $\Sigma = (S, OP, POP)$ to $\Phi^{\mathrm{P}}(\Sigma) = (S, OP)$. Σ-model consists of a $\Phi^{\mathrm{P}}(\Sigma)$-model plus some partial operations, and $\beta^{\mathrm{P}}_{\Sigma}$ forgets the partial operations. □

Consider a combination of $ALG(=)$ and $PALG$ via a pullback in **Ins**:

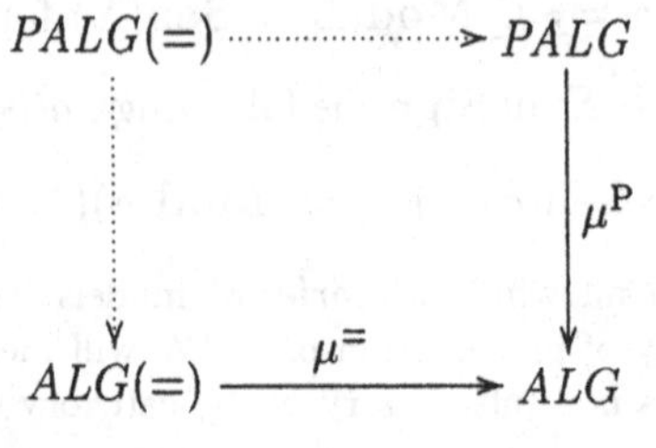

Signatures and models in $PALG(=)$ are those of $PALG$, while (S, OP, POP)-sentences are (S, OP)-sentences of $ALG(=)$. Thus all equations in $PALG(=)$ are built up from total operation symbols, while the expected interaction of equality (as added to ALG in $ALG(=)$) and partiality (as added to ALG in $PALG$) should introduce equations containing partial operation symbols as well.

This indicates a major problem with combination of institutions via limits in **Ins**: the sentences of the institutions involved are just put side by side, which cannot result in a proper interaction of the underlying concepts. A possible solution is to reveal the internal structure of sentences and their semantics.

Notational convention. We will always use symbols **Sign**, **Sen**, **Mod** and $\models$ for the corresponding components of an institution. We will omit expansions of institutions to such components if some natural decorations are used to make the relationship evident. We will tacitly adopt a similar convention for other concepts introduced in this paper hoping that this will never lead to any real confusion.

4 Model-theoretic parchments

When combining logical systems we would like to unite *constructions*, or *operators* on sentences (and their semantics) rather than sets of sentences. This requires that such constructions are captured in a formal way in a logic presentation — which was first attempted for institutions in the notion of *parchment* introduced by Goguen and Burstall [5]. The idea is to present Σ-sentences as terms over an algebraic signature $\mathbf{L}(\Sigma)$ giving their abstract syntax, and interpret them semantically using initiality of the term algebra. A combination of abstract syntaxes then causes a true interaction of different syntactic operators in the abstract syntax terms, which also has to be reflected in the semantics. The semantics is given by a special signature acting as a semantic universe — models are signature morphisms into it — and a "universal" semantic structure of this signature specifying semantic evaluation.

Parchments have a rather syntactic flavour, mixing up signatures and models in a signature category. This works fine as technical means to present institutions, but conceptually, putting signatures and models into one category need not be a good idea. This causes troubles for instance when presenting institution morphisms and representations, where signatures and models may be mapped differently, even in the opposite directions. In [10] we have introduced λ-*parchments*, a variant of parchments with a clean separation between signatures and models.

Each λ-parchment still involves, however, a large algebraic signature in which syntactic constructs for all signatures of the presented logic can be embedded and a corresponding Procrustean semantic structure, universal for the constructs over all signatures. The following new notion spreads such a "very large" structure over individual components suited for each signature separately, thus allowing them to be "small". Moreover, such a distribution of semantic concepts allows for more flexibility in the exact definitions, which may slightly differ for various signatures. The expected uniformity of concepts and their semantics is reflected by the appropriate naturality conditions which imply that the semantics of the concepts involved changes smoothly when we move from one signature to another.

A *model-theoretic parchment* $P = (\mathbf{Sign}, \mathbf{Mod}, \mathbf{L}, \mathbb{G})$ consists of

- a category **Sign** of signatures,
- a functor $\mathbf{Mod}\colon \mathbf{Sign}^{op} \longrightarrow \mathbf{Class}$ (as for institutions),
- a functor $\mathbf{L}\colon \mathbf{Sign} \longrightarrow \mathbf{AlgSig}_*$[5] giving the abstract syntax (or the language) of sentences,
- for each signature $\Sigma \in |\mathbf{Sign}|$ and Σ-model $M \in \mathbf{Mod}(\Sigma)$, an $\mathbf{L}(\Sigma)$-structure $\mathbb{G}_\Sigma(M) \in |\mathbf{Str}(\mathbf{L}(\Sigma))|$ which determines semantic evaluation of Σ-sentences in M,
- for each signature morphism $\sigma\colon \Sigma \longrightarrow \Sigma'$ and model $M' \in \mathbf{Mod}(\Sigma')$, an $\mathbf{L}(\Sigma)$-homomorphism $\mathbb{G}_\sigma(M')\colon \mathbb{G}_\Sigma(M'|_\sigma) \longrightarrow \mathbb{G}_{\Sigma'}(M')|_{\mathbf{L}(\sigma)}$ such that for all $\Sigma \xrightarrow{\sigma''} \Sigma'' \xrightarrow{\sigma'} \Sigma'$, $\mathbb{G}_{\sigma' \circ \sigma''}(M') = \mathbb{G}_{\sigma'}(M')|_{\mathbf{L}(\sigma'')} \circ \mathbb{G}_{\sigma''}(M'|_{\sigma'})$.

For each signature Σ and Σ-model $M \in \mathbf{Mod}(\Sigma)$, the carrier $|\mathbb{G}_\Sigma(M)|_*$ is the *space of logical values* of P for M. Typically we have $|\mathbb{G}_\Sigma(M)|_* = \mathbf{Bool}$ (with $\mathbf{Bool} = \{\mathit{true}, \mathit{false}\}$) as originally required in [5]. Without this restriction we naturally capture multi-valued logics. To recover the ultimatively two-valued notion of logical satisfaction, the unary predicate $D_{\mathbb{G}_\Sigma(M)} \subseteq |\mathbb{G}_\Sigma(M)|_*$ singles out the *designated logical values*, which are interpreted as truth, while the other values are interpreted as falsity. By the homomorphism condition, the designation of logical values is preserved when the signature is changed along a signature morphism; it is also reflected if the following additional requirement is imposed.

We call a model-theoretic parchment *logical* if for all $\sigma\colon \Sigma \longrightarrow \Sigma'$ in **Sign** and $M' \in \mathbf{Mod}(\Sigma')$, the $\mathbf{L}(\Sigma)$-homomorphism $\mathbb{G}_\sigma(M')\colon G_\Sigma(M'|_\sigma) \longrightarrow \mathbb{G}_{\Sigma'}(M')|_{\mathbf{L}(\sigma)}$ is an extremal monomorphism (i.e., a closed injection).

A logical model-theoretic parchment $P = (\mathbf{Sign}, \mathbf{Mod}, \mathbf{L}, \mathbb{G})$ *presents* an institution $\mathbf{I}(P) = (\mathbf{Sign}, \mathbf{Sen}, \mathbf{Mod}, \models)$, where

- $\mathbf{Sen}(\Sigma) = |T_{\mathbf{L}(\Sigma)}|_*$, where $T_{\mathbf{L}(\Sigma)}$ is the initial $\mathbf{L}(\Sigma)$-algebra,
- $\mathbf{Sen}(\sigma\colon \Sigma \longrightarrow \Sigma') = (T_{\mathbf{L}(\sigma)})_*$, where $T_{\mathbf{L}(\sigma)}\colon T_{\mathbf{L}(\Sigma)} \longrightarrow T_{\mathbf{L}(\Sigma')}|_{\mathbf{L}(\sigma)}$ is the initial homomorphism,
- $M \models_\Sigma \varphi$ iff $M^\heartsuit_*(\varphi) \in D_{\mathbb{G}_\Sigma(M)}$, where $M^\heartsuit\colon T_{\mathbf{L}(\Sigma)} \longrightarrow \mathbb{G}_\Sigma(M)$ is the initial homomorphism.

Example 6. The institution *ALG* is presented by the following (logical) model-theoretic parchment $(\mathbf{Sign}, \mathbf{L}, \mathbf{Mod}, \mathbb{G})$, also denoted by *ALG*:

- **Sign** and **Mod** are taken from the institution *ALG*,
- **L** takes an *ALG*-signature (S, OP) to the $\mathbf{AlgSig}_*$-signature $\mathit{Logic} + (S, OP)$,
- **L** extends a signature morphism $\sigma\colon \Sigma \longrightarrow \Sigma'$ to the signature morphism $\mathbf{L}(\sigma)\colon \mathbf{L}(\Sigma) \longrightarrow \mathbf{L}(\Sigma')$ that is the identity on *Logic*.
- For $\Sigma = (S, OP)$ and $M \in \mathbf{Mod}(\Sigma)$, $\mathbb{G}_\Sigma(M) \in |\mathbf{Str}(\mathbf{L}(\Sigma))|$ is defined by:

[5] Recall that signatures in $\mathbf{AlgSig}_*$ always contain *Logic* as a subsignature preserved by the morphisms considered.

- $|\text{₵}_\Sigma(M)|_* = \mathbf{Bool}$, $D_{\text{₵}_\Sigma(M)} = \{true\}$,
- $|\text{₵}_\Sigma(M)|_s = |M|_s$ for $s \in S$, and
- $op_{\text{₵}_\Sigma(M)} = op_M$ for $op \in OP$.

– For $\sigma: \Sigma \longrightarrow \Sigma'$ and $M' \in \mathbf{Mod}(\Sigma')$, $\text{₵}_\sigma(M')$ is the identity. □

Example 7. The institution $ALG(=)$ is presented by the following (logical) model-theoretic parchment $(\mathbf{Sign}, \mathbf{L}, \mathbf{Mod}, \text{₵})$, also denoted by $ALG(=)$:

– $\mathbf{Sign} = \mathbf{Sign}^{ALG}$, and $\mathbf{Mod} = \mathbf{Mod}^{ALG}$ are taken from ALG,
– $\mathbf{L}(S, OP)$ is $\mathbf{L}^{ALG}(S, OP)$ together with $=: s \times s \longrightarrow *$ for $s \in S$.
– For $\Sigma = (S, OP)$ and $M \in \mathbf{Mod}(\Sigma)$, $\text{₵}_\Sigma(M) \in |\mathbf{Str}(\mathbf{L}(\Sigma))|$ is defined by:
  - $\text{₵}_\Sigma(M)|_{\mathbf{L}^{ALG}(\Sigma)} = \text{₵}^{ALG}_\Sigma(M)$,
  - $=_{\text{₵}_\Sigma(M)} (a, b) = \begin{cases} true, & \text{if } a = b \\ false, & \text{if } a \neq b \end{cases}$.

– For $\sigma: \Sigma \longrightarrow \Sigma'$ and $M' \in \mathbf{Mod}(\Sigma')$, $\text{₵}_\sigma(M')$ is the identity. □

Example 8. The institution $PALG$ is presented by the following (logical) model-theoretic parchment $(\mathbf{Sign}, \mathbf{L}, \mathbf{Mod}, \text{₵})$, also denoted by $PALG$:

– $\mathbf{Sign}$ and $\mathbf{Mod}$ are taken from the institution $PALG$,
– $\mathbf{L}(S, OP, POP) = \mathbf{L}^{ALG}(S, OP \cup POP)$.
– For $\Sigma = (S, OP, POP)$ and $M \in \mathbf{Mod}(\Sigma)$, $\text{₵}_\Sigma(M) \in |\mathbf{Str}(\mathbf{L}(\Sigma))|$ has:
  - $|\text{₵}_\Sigma(M)|_* = \mathbf{Bool}$, $D_{\text{₵}_\Sigma(M)} = \{true\}$,
  - $|\text{₵}_\Sigma(M)|_s = |M|_s \uplus \{\perp\}$, where $\perp$ is a special new element, for $s \in S$,
  - for $op \in OP$ and $pop \in POP$, $op_{\text{₵}_\Sigma(M)}$ and $pop_{\text{₵}_\Sigma(M)}$ are the usual extensions of op_M and pop_M, respectively, to total functions on the carriers with the additional "undefined" element $\perp$.

– For $\sigma: \Sigma \longrightarrow \Sigma'$ and $M' \in \mathbf{Mod}(\Sigma')$, $\text{₵}_\sigma(M')$ is the identity. □

Model-theoretic parchments may also be viewed as functors into a category **MPRoom** of *rooms for model-theoretic parchments*[6], where **MPRoom** has:

– objects $(L, \mathcal{M}, G)$ where $L \in \mathbf{AlgSig}_*$, $\mathcal{M} \in \mathbf{Class}$ and $G: \mathcal{M} \longrightarrow \mathbf{Str}(L)$,
– morphisms $(\alpha, \beta, g): (L, \mathcal{M}, G) \longrightarrow (L', \mathcal{M}', G')$ where $\alpha: L \longrightarrow L' \in \mathbf{AlgSig}_*$, $\beta: \mathcal{M}' \longrightarrow \mathcal{M} \in \mathbf{Class}$ and a natural transformation $g: G \circ \beta \longrightarrow \mathbf{Str}(\alpha) \circ G'$ between functors in $\mathcal{M}' \longrightarrow \mathbf{Str}(L)$ (because the source category is discrete, the naturality condition is void here), and
– composition $(\alpha', \beta', g') \circ (\alpha, \beta, g) = (\alpha' \circ \alpha, \beta \circ \beta', (\mathbf{Str}(\alpha) * g') \circ (g * \beta'))$.[7]

Then a model-theoretic parchment P can be identified with a functor from **Sign** to **MPRoom** which, by abuse of notation, we also denote by P. Namely, given a functor $P: \mathbf{Sign} \longrightarrow \mathbf{MPRoom}$ we can construct the corresponding model-theoretic parchment $(\mathbf{Sign}, \mathbf{Mod}, \mathbf{L}, \text{₵})$ as follows:

– for $\Sigma \in |\mathbf{Sign}|$, let $P(\Sigma) = (L, \mathcal{M}, G)$; then $\mathbf{Mod}(\Sigma) = \mathcal{M}$, $\mathbf{L}(\Sigma) = L$ and $\text{₵}_\Sigma = G$, and

---

[6] The terminology comes from a similar presentation of institutions in [7], cf. also [16].
[7] By $*$ we denote the usual multiplication of natural transformations by functors.

- for $\sigma: \Sigma \longrightarrow \Sigma' \in \mathbf{Sign}$, let $P(\sigma) = (\alpha, \beta, g): (\mathbf{L}(\Sigma), \mathbf{Mod}(\Sigma), ⊈_\Sigma) \longrightarrow (\mathbf{L}(\Sigma'), \mathbf{Mod}(\Sigma'), ⊈_{\Sigma'})$; then $\mathbf{Mod}(\sigma) = \beta$, $\mathbf{L}(\sigma) = \alpha$ and $⊈_\sigma(M') = g_{M'}: ⊈_\Sigma(\beta(M')) \longrightarrow ⊈_{\Sigma'}(M')|_\alpha$.

The reader is encouraged to provide the inverse construction. We will switch between the two views of model-theoretic parchments whenever convenient.

4.1 Example

The model-theoretic parchments in Examples 6, 7 and 8 present logics that can also be easily presented by ordinary parchments of [5] or λ-parchments of [10]. To show the extra flexibility of model-theoretic parchments, we sketch a model-theoretic parchment for a simple ground (no variables) modal logic [3], which cannot be naturally presented as either a parchment or λ-parchment.

The model theoretic parchment $GML = (\mathbf{Sign}, \mathbf{Mod}, \mathbf{L}, ⊈)$ of *ground modal logic* is defined as follows:

- The category **Sign** of signatures for *GML* is the category **AlgSig** of relational algebraic signatures.
- $\mathbf{Mod}: \mathbf{AlgSig}^{op} \longrightarrow \mathbf{Class}$ assigns to every every *GML*-signature Σ the class of all models over that signature. A Σ-model MM is a triple $(W, \rightsquigarrow, \mathcal{M})$, where W is a set of *possible worlds*, $\rightsquigarrow$ is the *accessibility relation*, and $\mathcal{M} = \{M_w \mid w \in W\}$ is a W-indexed family of Σ-structures, such that for all $w_1, w_2 \in W$, $|M_{w_1}| = |M_{w_2}|$ (we shall denote the "common carrier" of $\mathcal{M}$ by $|\mathcal{M}|$). Models satisfying these requirements are called *constant domain models* in [3].
  For a signature morphism $\sigma: \Sigma \longrightarrow \Sigma'$, $\mathbf{Mod}(\sigma): \mathbf{Mod}(\Sigma') \longrightarrow \mathbf{Mod}(\Sigma)$ is the obvious reduct.
- The functor $\mathbf{L}: \mathbf{AlgSig} \longrightarrow \mathbf{AlgSig}_*$ takes a *GML*-signature (S, OP, REL), to the $\mathbf{AlgSig}_*$-signature which puts together *Logic* and (S, OP, REL), and extends this by a sort *WPred* as well as by the following operations:
  - $R: s_1, \ldots, s_n \longrightarrow WPred$, for all relational symbols $R: s_1 \ldots s_n \in REL$,
  - $\Box: WPred \longrightarrow WPred$, and
  - bind: $WPred \longrightarrow *$.

  For any signature morphism $\sigma: \Sigma \longrightarrow \Sigma'$, $\mathbf{L}(\sigma)$ is defined in the obvious way.
- For any signature $\Sigma = (S, OP, REL)$ and Σ-model $MM = (W, \rightsquigarrow, \mathcal{M})$, the $\mathbf{L}(\Sigma)$-algebra $⊈_\Sigma(MM)$ is defined as follows:
  - $|⊈_\Sigma(MM)|_s = [W \to |\mathcal{M}|_s]$[8] for every $s \in S$,
  - $|⊈_\Sigma(MM)|_{WPred} = [W \to \mathbf{Bool}\,]$,
  - $|⊈_\Sigma(MM)|_* = \mathbf{Bool}$, $D_{⊈_\Sigma(MM)} = \{true\}$.
  - for $op \in OP$, $op_{⊈_\Sigma(MM)}(t_1, \ldots, t_n)(w) = op_{M_w}(t_1(w), \ldots, t_n(w))$,
  - for $R \in REL$, $R_{⊈_\Sigma(MM)}(t_1, \ldots, t_n)(w) = \begin{cases} true, & \text{if } (t_1(w), \ldots, t_n(w)) \in R_{M_w} \\ false, & \text{otherwise} \end{cases}$
  - $\Box_{⊈_\Sigma(MM)}(\phi)(w) = \begin{cases} true, & \text{if } \phi(w') = true \quad \text{for all} \quad w \rightsquigarrow w' \\ false, & \text{otherwise} \end{cases}$

---

[8] For any sets X, Y, $[X \to Y]$ denotes the set of all functions from X to Y.

- $\text{bind}_{\mathcal{G}_\Sigma(MM)}(\phi) = \begin{cases} \mathit{true}, \text{ if } \phi(w) = \mathit{true} & \text{for all } \quad w \in W \\ \mathit{false}, \text{ otherwise} \end{cases}$

– For each signature morphism $\sigma : \Sigma \to \Sigma'$ in **AlgSig**, and every Σ'-model MM', $\mathcal{G}_\sigma(MM')$: $\mathcal{G}_\Sigma(MM'|_\sigma) \longrightarrow \mathcal{G}_{\Sigma'}(MM')|_{\mathbf{L}(\sigma)}$ is the identity.

There is also an obvious generalisation of the above parchment to capture first-order modal logic, with variables, quantification etc. This would multiply the sorts in $\mathbf{L}(S, OP, REL)$ by the sets of variables and would additionally parameterise values in $\mathcal{G}_\Sigma(MM)$ by environments (valuations of variables). When these are considered, the homomorphisms $\mathcal{G}_\sigma(MM')$ are no longer just identities.

4.2 Model-theoretic parchment morphisms

We lift the concept of institution morphism to the level of model-theoretic parchments to capture how one model-theoretic parchment is built over another.

A *model-theoretic parchment morphism* between model-theoretic parchments P: **Sign** $\longrightarrow$ **MPRoom** and P': **Sign**$'$ $\longrightarrow$ **MPRoom** consists of a functor Φ: **Sign** $\longrightarrow$ **Sign**$'$ and a natural transformation $\mu: P' \circ \Phi \longrightarrow P$. **MPar** is the category of model-theoretic parchments and their morphisms (with the obvious composition).

At a more concrete level, a morphism between model-theoretic parchments $P = (\mathbf{Sign}, \mathbf{Mod}, \mathbf{L}, \mathcal{G})$ and $P' = (\mathbf{Sign}', \mathbf{Mod}', \mathbf{L}', \mathcal{G}')$ consists of:

– a functor Φ: **Sign** $\longrightarrow$ **Sign**$'$,
– a natural transformation $\alpha: \mathbf{L}' \circ \Phi \longrightarrow \mathbf{L}$,
– a natural transformation $\beta: \mathbf{Mod} \longrightarrow \mathbf{Mod}' \circ \Phi^{op}$,
– a 2-natural transformation $g: (\mathcal{G}' * \Phi^{op}) \circ \beta \longrightarrow (\mathbf{Str} * \alpha^{op}) \circ \mathcal{G}$, that is, for each $\Sigma \in |\mathbf{Sign}|$, a natural transformation $g_\Sigma: \mathcal{G}'_{\Phi(\Sigma)} \circ \beta_\Sigma \longrightarrow \mathbf{Str}(\alpha_\Sigma) \circ \mathcal{G}_\Sigma$, which in turn is, for each $M \in \mathbf{Mod}(\Sigma)$, a $\mathbf{L}'(\Phi(\Sigma))$-homomorphism

$$g_{\Sigma,M}: \mathcal{G}'_{\Phi(\Sigma)}(\beta_\Sigma(M)) \longrightarrow \mathcal{G}_\Sigma(M)|_{\alpha_\Sigma}$$

such that for $\sigma: \Sigma_1 \longrightarrow \Sigma_2$ and $M_2 \in \mathbf{Mod}(\Sigma_2)$, $\mathcal{G}_\sigma(M_2)|_{\alpha_{\Sigma_1}} \circ g_{\Sigma_1, M_2|_\sigma} = g_{\Sigma_2, M_2}|_{\mathbf{L}'(\Phi(\sigma))} \circ \mathcal{G}'_{\Phi(\sigma)}(\beta_{\Sigma_2}(M_2))$.

Each $g_{\Sigma,M}$ preserves, but not necessarily reflects, the designated truth values.

We call a model-theoretic parchment morphism $\mu = (\Phi, \alpha, \beta, g): P \longrightarrow P'$ *logical* if for all signatures $\Sigma \in |\mathbf{Sign}|$ and Σ-models $M \in \mathbf{Mod}(\Sigma)$ in P, $g_{\Sigma,M}$ is an extremal monomorphism (i.e., a closed injection). This determines a subcategory **LogMPar** of **MPar**, consisting of logical model-theoretic parchments and their logical morphisms.

Proposition 3. *The construction of institutions out of logical model-theoretic parchments can be extended to a functor* **I**: **LogMPar** $\longrightarrow$ **Ins**.

Example 9. Define a model-theoretic parchment morphism $\mu^=: ALG(=) \longrightarrow ALG$ to be (Id, α, Id, Id), where α_Σ is the inclusion. Then $\mathbf{I}(\mu^=)$ is the institution morphism presented in Example 4. □

Example 10. Define a model-theoretic parchment morphism $\mu^{\mathrm{P}}: PALG \longrightarrow ALG$ to be $(\Phi^{\mathrm{P}}, \alpha^{\mathrm{P}}, Id, g^{\mathrm{P}})$, where $\Phi^{\mathrm{P}}(S, OP, POP) = (S, OP)$, $\alpha^{\mathrm{P}}_{\Sigma}$ is the inclusion, $(g^{\mathrm{P}}_{\Sigma,M})_s$ is the inclusion of $|M|_s$ into $|M|_s \uplus \{\bot\}$, and $(g^{\mathrm{P}}_{\Sigma,M})_*$ is the identity. Then $\mathbf{I}(\mu^{\mathrm{P}})$ is the institution morphism presented in Example 5. □

4.3 Putting model-theoretic parchments together using limits

When combing model-theoretic parchments, a natural condition to require would be that the space of truth values is preserved, so that the combination determines a combination of the corresponding institutions. That is, we would like to take limits in the category **LogMPar** of logical model-theoretic parchments and their logical morphisms. Unfortunately, this is not always possible: **LogMPar** is *not* complete! Therefore, we have to start by considering limits in **MPar**.

Theorem 4. *The category* **MPar** *of model-theoretic parchments and their morphisms is complete.*

Example 11. Consider the combination of $ALG(=)$ and $PALG$ via a pullback in **MPar**:

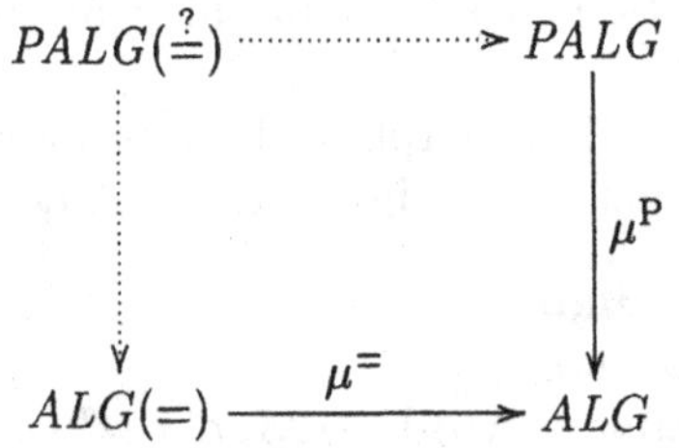

Then, within the model-theoretic parchment $PALG(\stackrel{?}{=})$:

- the signatures and models are those of $PALG$, as expected;
- the language on a signature (S, OP, POP) coincides with the language of $ALG(=)$ on $(S, OP \cup POP)$ — and so the sentences generated include equalities between arbitrary (S, OP, POP)-terms, as expected;
- the semantic structures coincide with those of $PALG$ except for their reducts to the subsignature *Logic*. For each $\Sigma = (S, OP, POP)$ and $A \in \mathbf{Mod}(\Sigma)$, the interpretation of the extra operations $=_{\mathcal{G}_\Sigma(A)}$ on arguments from $|A|_s$ is inherited from $ALG(=)$, as expected; however, $|\mathcal{G}_\Sigma(A)|_*$ in $PALG(\stackrel{?}{=})$ extends **Bool** by new values that are freely generated as the results of $=_{\mathcal{G}_\Sigma(A)}(a, \bot)$ and $=_{\mathcal{G}_\Sigma(A)}(\bot, a)$. □

The new "logical values" emerging in the above parchment combination are certainly not what we want — but at least they clearly identify where some further work on the combination is necessary. To produce the expected result we can proceed in two ways. One possibility, discussed in [9] (for the original parchments) is to further "massage" the resulting parchment and quotient its semantic structures by a congruence that would identify the new values with either *true* or *false*. In the rest of the paper we will pursue another path.

5 Representations

While institution and model-theoretic parchment morphisms capture the intuition that a logical system is built over another one (and the morphism is the corresponding projection), representations serve to *encode* a logical system into another one. We begin with weak institution representations — similar to institution representations (or *simple maps of institutions*, cf. [8]) but with a somewhat modified representation invariant, which turns out to be sufficient and at the same time less restrictive in practice.

A *weak institution representation* [15] $\mu = (\Phi, \alpha, \beta): I \longrightarrow I'$ between institutions I and I' consists of a functor Φ: **Sign** $\longrightarrow$ **Sign**$'$, a natural transformation α: **Sen** $\longrightarrow$ **Sen**$' \circ \Phi$, and a natural transformation β: **Mod**$' \circ \Phi^{op} \longrightarrow$ **Mod** such that the following *weak representation condition* is satisfied for $\Sigma \in |\mathbf{Sign}|$, $M \in \mathbf{Mod}(\Sigma)$ and $\varphi \in \mathbf{Sen}(\Sigma)$:[9]

$$M \models_\Sigma \varphi \iff \beta_\Sigma^{-1}(M) \models'_{\Phi(\Sigma)} \alpha_\Sigma(\varphi)$$

With the straightforward composition, this defines a category **WInsRep** of institutions and their weak representations.

One important application of ordinary institution representations is the reuse of theorem provers. With weak representations, this is only slightly restricted [15]: given a weak institution representation $\rho: I \longrightarrow I'$, any proof system for I' may be soundly reused for I, but its completeness is preserved only if entailment may be internalised in sentences (as it is the case for example in first-order logic).

Given model-theoretic parchments P: **Sign** $\longrightarrow$ **MPRoom** and P': **Sign**$' \longrightarrow$ **MPRoom**, a *model-theoretic parchment representation* of P in P' consists of a functor Φ: **Sign** $\longrightarrow$ **Sign**$'$ and a natural transformation $\mu: P \longrightarrow P' \circ \Phi$. With the straightforward composition, this defines a category **MParRep** of model-theoretic parchments and their representations.

As with model-theoretic parchment morphisms, at a more concrete level, given model-theoretic parchments P and P', a representation of P in P' consists of

- a functor Φ: **Sign** $\longrightarrow$ **Sign**$'$,
- a natural transformation α: **L** $\longrightarrow$ **L**$' \circ \Phi$,
- a natural transformation β: **Mod**$' \circ \Phi^{op} \longrightarrow$ **Mod**,
- a 2-natural transformation $g: \mathcal{G} \circ \beta \longrightarrow (\mathbf{Str} * \alpha^{op}) \circ (\mathcal{G}' * \Phi^{op})$, that is, for each $\Sigma \in |\mathbf{Sign}|$, a natural transformation $g_\Sigma: \mathcal{G}_\Sigma \circ \beta_\Sigma \longrightarrow \mathbf{Str}(\alpha_\Sigma) \circ \mathcal{G}'_{\Phi(\Sigma)}$.

Each $g_{\Sigma,M'}$ preserves, but not necessarily reflects, the truth values.

We call a model-theoretic parchment representation $\mu = (\Phi, \alpha, \beta, g): P \longrightarrow P'$ *weakly logical* if for all signatures $\Sigma \in |\mathbf{Sign}|$ and all Σ-models $M \in \mathbf{Mod}(\Sigma)$ in P, the source $\langle g_{\Sigma,M'}: \mathcal{G}_\Sigma(M) \longrightarrow \mathcal{G}_{\Phi(\Sigma)}(M')|_{\alpha_\Sigma}\rangle_{M' \in \beta_\Sigma^{-1}(M)}$ is an extremal mono-source. This defines a subcategory **WLogMParRep** of **MParRep**, consisting of logical model-theoretic parchments and weakly logical representations.

Proposition 5. *The construction of institutions out of logical model-theoretic parchments can be extended to a functor* **I**: **WLogMParRep** $\longrightarrow$ **WInsRep**.

[9] $\beta_\Sigma^{-1}(M) = \{M' \in \mathbf{Mod}'(\Phi(\Sigma)) \mid \beta_\Sigma(M') = M\}$ is the coimage of M under β_Σ.

5.1 Combinations of representations

Can we combine representations in a fixed "universal" model-theoretic parchment (which can be assumed to have a well-developed proof theory, support tools etc.)? Let us consider a model-theoretic parchment $UP\colon \mathbf{USign} \longrightarrow \mathbf{MPRoom}$, very informally viewed as such a rich "universal" logic.

The idea of a *representation map* is to capture the situation where one model-theoretic parchment is built over another one, and, moreover, the representation in UP of the former is also built over the representation of the latter.

Given model-theoretic parchments $P\colon \mathbf{Sign} \longrightarrow \mathbf{MPRoom}$ and $P'\colon \mathbf{Sign}' \longrightarrow \mathbf{MPRoom}$, and their representations $\langle \Phi, \rho\rangle\colon P \longrightarrow UP$ and $\langle \Phi', \rho'\rangle\colon P' \longrightarrow UP$ in UP, a *representation map* from $\langle \Phi, \rho\rangle$ to $\langle \Phi', \rho'\rangle$ consists of a model-theoretic parchment morphism $\langle \tilde{\Phi}, \tilde{\mu}\rangle\colon P' \longrightarrow P$, and a natural transformation $\theta\colon \Phi \circ \tilde{\Phi} \longrightarrow \Phi'$ such that the following diagram commutes:

$$\begin{array}{ccc} P \circ \tilde{\Phi} & \xrightarrow{\ \rho * \tilde{\Phi}\ } & UP \circ \Phi \circ \tilde{\Phi} \\ \big\downarrow{\scriptstyle \tilde{\mu}} & & \big\downarrow{\scriptstyle UP * \theta} \\ P' & \xrightarrow{\ \rho'\ } & UP \circ \Phi' \end{array}$$

Consider model theoretic parchments $P\colon \mathbf{Sign} \longrightarrow \mathbf{MPRoom}$, $P'\colon \mathbf{Sign}' \longrightarrow \mathbf{MPRoom}$ $P''\colon \mathbf{Sign}'' \longrightarrow \mathbf{MPRoom}$, and their representations $\rho\colon P \longrightarrow UP$, $\rho'\colon P' \longrightarrow UP$ and $\rho''\colon P'' \longrightarrow UP$. Given representation maps $\langle \tilde{\mu}_1, \theta_1\rangle\colon \rho \longrightarrow \rho'$ and $\langle \tilde{\mu}_2, \theta_2\rangle\colon \rho' \longrightarrow \rho''$, their *composition* is $\langle \tilde{\mu}_1 \circ \tilde{\mu}_2, \theta_2 \circ (\theta_1 * \Phi_2)\rangle$.

This yields a category $\mathbf{MParRep}_{UP}$ of model-theoretic parchment representations in UP and their maps, with the obvious (contravariant) projection functor $\Pi\colon \mathbf{MParRep}_{UP}^{op} \longrightarrow \mathbf{MPar}$.

Theorem 6. *Suppose that the category* **USign** *of signatures of the "universal" model-theoretic parchments is cocomplete. Then the category* $\mathbf{MParRep}_{UP}$ *of model-theoretic parchment representations in UP and their maps is cocomplete, and moreover, the projection functor* $\Pi\colon \mathbf{MParRep}_{UP}^{op} \longrightarrow \mathbf{MPar}$ *is continuous.*

The continuity of the projection functor $\Pi\colon \mathbf{MParRep}_{UP}^{op} \longrightarrow \mathbf{MPar}$ shows that colimits in $\mathbf{MParRep}_{UP}$ are as inadequate for proper combination of logics (and their representations) as limits in **MPar**. Moreover, in general we cannot even map the colimits to the category of institutions. Fortunately, a "logical variant" of the above theorem holds as well.

Let $\mathbf{WLogMParRep}_{UP}$ be the full subcategory of $\mathbf{MParRep}_{UP}$ determined by the weakly logical representations in UP of logical model-theoretic parchments.

Theorem 7. *If* $UP\colon \mathbf{USign} \longrightarrow \mathbf{MPRoom}$ *is a logical model-theoretic parchment and* **USign** *is cocomplete then* $\mathbf{WLogMParRep}_{UP}$ *is cocomplete as well.*

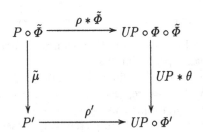

The colimits in $\mathbf{WLogMParRep}_{UP}$ are built by taking colimits in $\mathbf{MParRep}_{UP}$ (which in turn involves taking limits in **MPar**) and then quotienting the semantic structures in the limit model-theoretic parchment so that its representation in *UP* becomes logical.

5.2 Example

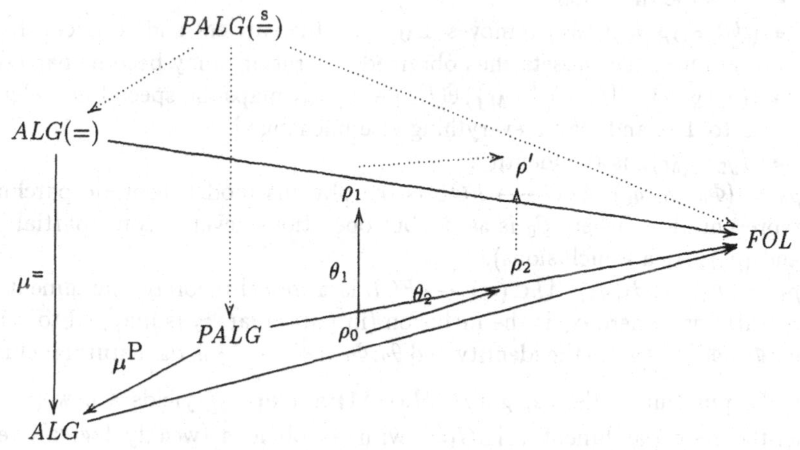

Fig. 1. A pushout in $\mathbf{WLogMParRep}_{FOL}$.

Without cluttering the example with the details of any more practical "universal model-theoretic parchment", let us consider as a universal parchment the model-theoretic parchment *FOL* for first-order logic (with equality), roughly following the presentation in [13]. In particular (other details are irrelevant here):

- $\mathbf{Sign}^{FOL}$ has as objects theories (Σ, Ax) constisting of an **AlgSig**-signature Σ plus some first-order axioms Ax,
- $\mathbf{Mod}^{FOL}(\Sigma, Ax) = \{M \in |\mathbf{Str}(\Sigma)| \mid M \models Ax\}$,
- $\mathbf{L}^{FOL}(S, OP, REL)$ includes sorts S, operations OP and $R{:}\, s_1, \ldots, s_n \longrightarrow *$ for $R : s_1, \ldots, s_n \in REL$, as well as at least operations $=: s \times s \longrightarrow *$ for $s \in S$. These are augmented with propositional connectives and operations to form terms with variables, open formulae, etc.
- for $\Sigma = (S, OP, REL)$ and $M \in \mathbf{Mod}^{FOL}(\Sigma)$, $\Phi^{FOL}_{(\Sigma,Ax)}(M)$ is defined as expected; in particular
  - $|\Phi^{FOL}_{(\Sigma,Ax)}(M)|_* = \mathbf{Bool}$ and $D_{\Phi^{FOL}_{(\Sigma,Ax)}(M)} = \{true\}$,
  - $=_{\Phi^{FOL}_{(\Sigma,Ax)}(M)}(a,b) = \begin{cases} true, & \text{if } a = b \\ false, & \text{otherwise} \end{cases}$

Then we can consider the following morphisms and representations of model-theoretic parchments (see diagram in Fig. 1):

- $\mu^{=}: ALG(=) \longrightarrow ALG$ and $\mu^{\mathrm{P}}: PALG \longrightarrow ALG$ are model-theoretic parchment morphisms defined as in Examples 9 and 10, respectively.
- $\rho_2 = (\Phi_2, \alpha_2, \beta_2, g_2): PALG \longrightarrow FOL$ is a model-theoretic parchment representation defined as follows:
  - $\Phi_2(S, OP, POP)$ is the theory $(S, OP \cup POP, \emptyset)$ extended by constants $\bot: \longrightarrow s$, for $s \in S$, and axioms, for $op \in OP$ and $pop \in POP$:
    $\forall x_1 : s_1, \ldots, x_n : s_n . op(x_1, \ldots, x_n) \neq \bot \Leftrightarrow (x_1 \neq \bot \wedge \cdots \wedge x_n \neq \bot)$
    $\forall x_1 : s_1, \ldots, x_n : s_n . pop(x_1, \ldots, x_n) \neq \bot \Rightarrow (x_1 \neq \bot \wedge \cdots \wedge x_n \neq \bot)$
  - α_2 is the inclusion
  - $(\beta_2)_{(S,OP,POP)}(M)$ removes $\bot_M$ from the carriers and restricts the operation to the subsets thus obtained (operations may become partial),
  - $((g_2)_{\Sigma,M})_s: (|M|_s \setminus \{\bot_M\}) \uplus \{\bot\} \longrightarrow |M|_s$ maps the special new element $\bot$ to $\bot_M$ and leaves everything else unchanged,
  - $((g_2)_{\Sigma,M})_*$ is the identity.
- $\rho_0 = (\Phi_0, \alpha_0, g_0): ALG \longrightarrow FOL$ is the obvious model-theoretic parchment representation (where β_0 is as β_2, but operations never become partial here, and $(g_0)_{\Sigma,M}$ are inclusions).
- $\rho_1 = (\Phi_0, \alpha_1, \beta_0, g_0): ALG(=) \longrightarrow FOL$ is a model-theoretic parchment representation, where α_1 is the inclusion (in particular, $=$ is mapped to $=$),
- $\theta_1: \Phi_0 \circ \Phi^{=} \longrightarrow \Phi_1$ is the identity and $\theta_2: \Phi_0 \circ \Phi^{\mathrm{P}} \longrightarrow \Phi_2$ is the natural inclusion.

Then the pushout in the category $\mathbf{WLogMParRep}_{FOL}$ yields a new (logical) model-theoretic parchment $PALG(\stackrel{s}{=})$ with its obvious (weakly logical) representation into FOL. This model-theoretic parchment has two truth values and presents the institution of partial algebras with ground strong equalities.

This should be contrasted with Example 11: the combination of $ALG(=)$ and $PALG$ via a pullback in **MPar** leads to a model-theoretic parchment with many new logical values. By Theorem 7, we can generate automatically a congruence that identifies these new logical values with either *true* or *false* (as the least congruence such that the representation of the combined model-theoretic parchment in FOL is weakly logical) leading to the desired model-theoretic parchment $PALG(\stackrel{s}{=})$. The information to determine this congruence is given by the representation which maps total equality $(=)$ to the equality in FOL, which means strong equality $(\stackrel{s}{=})$ there. The colimit in $\mathbf{WLogMParRep}_{FOL}$ uses "combinations" that are wired into the universal model-theoretic parchment FOL. By using a different representation here, we can get other congruences providing other semantical identifications — and hence different combinations of equality with partiality of operations. For instance, we could instead consider a representation mapping $=$ to a derived operation symbol $\lambda xy.x = y \wedge x \neq \bot$ in FOL. Then the pushout construction as above would yield a parchment $PALG(\stackrel{e}{=})$ for existential equalities between ground terms (with its representation in FOL).

6 Conclusion

We have modified Goguen and Burstall's notion of parchment [5] and our earlier notion of λ-parchment [10] and introduced model-theoretic parchments (Sect. 4), their morphisms (Sect. 4.2), representations (Sect. 5) and representation maps (Sect. 5.1). We propose these as a framework for combining logical systems presented as model-theoretic parchments using categorical (co)limits in the appropriate categories. The main results show that this is indeed possible, at least in principle: the category of model-theoretic parchments is complete (Theorem 4), the category of model-theoretic parchment representations in a "universal" parchment is cocomplete (under some technical condition on the universal parchment — see Theorem 6), and the same is true for the subcategory of weakly logical representations of logical model-theoretic parchments (Theorem 7).

None of these results provides a miraculous tool for automatic combination of various independent features without any human intervention. Most significantly, the limits in the category of model-theoretic parchments typically yield results which have to be further manipulated "by hand" to describe the interaction of features coming from various logical systems. In fact, this result may be viewed as an exact description of where such a manipulation is essential. With the combination of representations the human intervention happens even before we combine the logical systems: the logical features are combined then exactly as prescribed by the combination of the features used to encode them in the universal logic. The exact way a particular logical concept is represented, even if immaterial for the particular logic at hand, may become crucial when some new features are added.

We have illustrated these ideas using an extremely simple example: a combination of ground equational logic over total algebras with partiality of operations. We hope that the intended simplicity allowed the user to grasp the essential ideas without cluttering them with unnecessary details. However, we believe that more complex examples may be presented in a similar way as well. This would use more realistic "universal parchments", modeling for instance some well-known general logical frameworks. Two new technical features of the framework proposed in this paper should make this possible. One is that we distributed the universal semantic structure over individual signatures and models, thus allowing for more flexibility in parchment representation. Another crucial change is that we propose here to work with "weak representation invariant" where a model is represented not so much by each individual model of the universal parchment from which it may be extracted but by the whole class of such models. This seems sufficient for practical needs of logic representation at the same time better reflecting what happens when a logic is represented for instance in a general logical framework based on some rich type theory. We still need another technical modification, allowing for more complex notions of signature morphism and related translations of models and sentences, for instance by introducing appropriate monads on the category of model-theoretic parchment representations. The details of this are yet to be worked out.

References

[1] J. Adámek, H. Herrlich, G. Strecker. *Abstract and Concrete Categories.* Wiley, New York, 1990.

[2] Jon Barwise. Axioms for abstract model theory. *Annals of Mathematical Logic* **7**, 221–265, 1974.

[3] M. Fitting. Basic Modal Logic. In D.M. Gabbay, C.J. Hogger, J.A. Robinson, and J. Siekmann, eds., *Handbook of Logic in Artificial Intelligence and Logic Programming*, Volume 1: *Logical Foundations.* Clarendon Press, Oxford, 1993.

[4] J. A. Goguen. A categorical manifesto. *Mathematical Structures in Computer Science* **1**, 49–67, 1991.

[5] J. A. Goguen, R. M. Burstall. A study in the foundations of programming methodology: Specifications, institutions, charters and parchments. In D. Pitt et al., ed., *Category Theory and Computer Programming, LNCS* **240**, 313–333. Springer 1985.

[6] J. A. Goguen, R. M. Burstall. Institutions: Abstract model theory for specification and programming. *Journal of the Association for Computing Machinery* **39**, 95–146, 1992. Predecessor in: LNCS 164, 221–256, Springer 1984.

[7] B. Mayoh. *Galleries and Institutions.* Report DAIMI PB-191, Aarhus University, 1985.

[8] J. Meseguer. General logic. In H.-D. Ebbinghaus et al., eds., *Logic Colloquium'87*, 279–329. North-Holland, 1989.

[9] T. Mossakowski. Using limits of parchments to systematically construct institutions of partial algebras. In M. Haveraaen, O. Owe, O.-J. Dahl, eds., *Recent Trends in Data Type Specifications. 11th Workshop on Specification of Abstract Data Types, LNCS* **1130**, 379–393. Springer 1996.

[10] T. Mossakowski, A. Tarlecki, W. Pawłowski. Combining and representing logical systems. In E. Moggi, ed., *Category Theory and Computer Science, LNCS* **1290**, 177–198, Springer 1997.

[11] T. Mossakowski, A. Tarlecki, W. Pawłowski. Combining and Representing Logical Systems Using Model-Theoretic Parchments. Technical report, Warsaw University, 1997. See: `http://wwwat.mimuw.edu.pl/ tarlecki/drafts/adt97.ps`.

[12] P. Padawitz. *Computing in Horn Clause Theories.* Springer 1988.

[13] P. Stefaneas. The first order parchment. Report PRG-TR-16-92, Oxford University Computing Laboratory, 1992.

[14] A. Tarlecki. Bits and pieces of the theory of institutions. In D. Pitt, S. Abramsky, A. Poigné, D. Rydeheard, eds., *Proc. Intl. Workshop on Category Theory and Computer Programming, Guildford 1985, LNCS* **240**, 334–363. Springer 1986.

[15] A. Tarlecki. Moving between logical systems. In M. Haveraaen, O. Owe, O.-J. Dahl, eds., *Recent Trends in Data Type Specifications. 11th Workshop on Specification of Abstract Data Types, LNCS* **1130**, 478–502. Springer 1996.

[16] A. Tarlecki, J. A. Goguen, R. M. Burstall. Some fundamental algebraic tools for the semantics of computation. Part III: Indexed categories. *Theoretical Computer Science* **91**, 239–264, 1991.

Towards the One-Tiered Design of Data Types and Transition Systems

Peter Padawitz
padawitz@cs.uni-dortmund.de
http://ls5.cs.uni-dortmund.de/~peter
University of Dortmund

Abstract. States build up hidden sorts. Consequently, labelled transition systems can be specified as ternary predicates within many-sorted specifications with visible (static, constructor-based) as well as hidden (dynamic, state-based) components. In modal logic, transition systems determine the equality of states, usually called *bisimilarity.* In many-sorted type logic, the equality of hidden data usually comes as contextual or behavioural equivalence. We integrate both concepts and specify transition systems in terms of *functional, relational* and/or *transitional actions.*

We introduce *standard specifications*, which, on the one hand, specialize many-sorted specifications insofar as visible domains are axiomatized by particular Horn clauses that can be regarded as constructor-based functional-logic programs for *defined functions* and *safety predicates.* On the other hand, hidden domains are axiomatized by *co-Horn clauses* for *liveness predicates* such as bisimilarity, fairness and properties concerned with infinite state sequences.

The formal reasoning about a standard specification is based on a hierarchical model construction that starts out from the initial model (for inductive reasoning), interprets liveness predicates (including bisimilarity) as the greatest solutions of their co-Horn axioms and becomes a final model by factoring the initial one through bisimilarity. This model construction, which is quite natural because it combines standard semantics of logic programs, data types and modal logic, relies upon certain constraints, namely *functionality* (existence and uniqueness of visible normal forms), *behavioural congruence* (compatibility of bisimilarity with defined functions and safety predicates) and *behavioural consistency* (closedness of the model class under factoring through bisimilarity). We introduce the notions of a *factorizable* goal and a *coinductive* Horn clause for obtaining syntactical criteria for behavioural congruence and consistency, respectively.

1 Introduction

Systems with both static and dynamic components are usually specified in a two-tiered way (cf., e.g., [9]). Constructor-based data domains are associated with, say, equational logic, while state-based transition systems refer to, say, modal logic. The data domains constitute the states of a Kripke structure, which interprets a transition system independently of the state structure. The concept of a data type with *hidden* and *visible* sorts (cf. [11], [13] or [25][1]) allows us to integrate both tiers and to reason about static and dynamic components within the same logic.

We start out from **standard specifications**, which have the following main ingredients: **constructors** for building up visible as well as hidden domains; **defined functions** and **safety predicates**, which are specified by Horn axioms that reflect the abstract syntax of functional-logic programs; **liveness predicates**, which are specified by **co-Horn axioms** that are complementary to Horn axioms both syntactically and semantically. Transition relations and definedness predicates come as particular safety predicates, while behavioural equivalence or **bisimilarity**, which defines the equality of hidden data, is a liveness predicate. The **final model** of a standard specification is the quotient by bisimilarity of the initial model, enriched by the greatest interpretation of all liveness predicates. The quotient construction preserves the validity of axioms because these are required to have **factorizable** premises.

Bisimilarity is specified in terms of **actions**, which are particular defined functions or safety predicates that determine the behaviour and thus the identity of hidden objects. Whereas hidden data type approaches such as [13] deal with *functional* actions, process logics and λ-calculi employ *transitional* actions and – for coping with partiality – *relational* actions. Only transitional actions admit the specification of *non*-deterministic transition systems. [4] and [7] also regard transition systems as predicates resp. Boolean functions of many-sorted specifications, but these papers are not concerned neither with the *axiomatization* of bisimilarity in terms of transition systems nor with criteria for **behavioural congruence**, i.e. compatibility of bisimilarity with functions and (non-transitional) safety predicates. We introduce **coinductivity** as a syntactical condition for the latter. With respect to functional actions, this condition was inspired by *coinductive coalgebra definitions* (cf. [30], [18]). With respect to transitional actions, it generalizes typical formats of transition system specifications used in process logics (cf. [29], [16]).

SP is **functional** if all visible data have unique normal forms, i.e. terms built up of constructors of SP. Hence visible data are identified by the *structure* of their normal forms so that functions and predicates are specified in terms of this structure and conjectures about them are proved by structural induction. Functionality also allows us to exchange SP for the **relational version** of SP where all defined functions of SP are replaced by their corresponding graphs or input-output relations. Though hidden data are also supposed to have normal forms, these need not represent them uniquely. A hidden normal form should be regarded as a mere *name* of an object. The identity of a hidden object is only defined indirectly in terms of bisimilarity.

While visible functions and predicates are specified inductively in terms of on constructors, this dependency relation is inverted in the case of hidden symbols: hidden constructors are defined in terms of hidden functions and predicates.

[1] where visible and hidden sorts are called constructor resp. action sorts

For instance, take the visible type of finite lists and the hidden type of infinite streams. Both contain a constructor $\& : elem \times list \to list$ (append an element to the left), defined functions $head : list \to elem$ and $tail : list \to list$ and axioms $head(x\&s) \equiv x$ and $tail(x\&s) \equiv s$. Together with $head(nil) \equiv \bot$ and $tail(nil) \equiv \bot$, these axioms define *head* and *tail* *inductively* on *finite* lists in terms of their constructors $nil :\to list$ and &. Conversely, the same axioms define the stream constructor & *inductively* in terms of *head* and *tail*. Moreover, bisimilarity on streams is the greatest solution in $\sim$ of the co-Horn axiom

$$s \sim s' \;\Rightarrow\; head(s) \equiv head(s') \wedge tail(s) \sim tail(s'),$$

which identifies *head* and *tail* as functional actions. In terms of a transitional action, $\sim$ could be specified as follows:

$$s \sim s' \;\Rightarrow\; (s \xrightarrow{x} s_1 \;\Rightarrow\; \exists s_2 : (s' \xrightarrow{x} s_2 \wedge s_1 \sim s_2)).$$

In this case & would be specified coinductively by the axiom $x\&s \xrightarrow{x} s$. In category-theoretic approaches, visible domains come as algebras, hidden ones as coalgebras, constructors and defined functions are morphisms, but the former are *inverses* of the latter and thus reflect the duality of induction and coinduction (cf., e.g., [30], [18]).

Data types with visible and hidden components involve a further duality. Safety predicates are interpreted as *least* solutions of Horn clauses. Liveness predicates are interpreted as **greatest** solutions of co-Horn clauses. The former is well-known from logic programming (cf., e.g., [2]) as well as from algebraic data types (cf. [23]). If SP is functional, the least fixpoint property also applies to the graphs of defined functions of SP. While greatest fixpoints are used in modal logic and relational semantics since a long time (cf., e.g., [33], [15]), they have entered data type theory only recently when behavioural equivalence was identified as bisimilarity. A first approach regarding other liveness predicates as greatest fixpoints was presented in [25].

Proof methods for standard specifications benefit enormously from the combination of fixpoint semantics with initial/final semantics. For instance, the complement $\overline{r}$ of a safety or liveness predicate r can often be specified simply by turning the Horn resp. co-Horn axioms for r into co-Horn resp. Horn axioms for $\overline{r}$ (cf. [26]). Moreover, least and greatest fixpoints provide for the main inference rules for proving conjectures about a standard specification (cf. Section 3). Both facts are particularly useful for proving hidden equations (= bisimilarities) and inequations.

As to liveness and safety predicates, modal logic usually adopts a different view. The former are called safety properties insofar as they are *invariance* conditions. The latter are called liveness properties because they describe *reachability* conditions (cf., e.g., [21]). This view is reasonable in two-tiered approaches where liveness and safety tell us how the validity of predicates – that refer to *individual* states – changes in time. But here a (hidden) predicate always refers to the *set* of states that are reachable from a single one. Roughly said, the validity of a safety predicate for a state set S can be deduced *inductively* from considering finite subsets of S, while the validity of a liveness predicate for S can only be "decided" if the entire - maybe infinite - set is taken into account. Nevertheless, despite the different liveness-versus-safety view both modal logic and (co-)Horn logic formalize invariance and reachability conditions as least and greatest fixpoints, respectively.

Section 2 introduces standard specifications and their initial semantics and links them with least fixpoints. Section 3 focuses on the hidden part of a standard specification, extends its initial model via a greatest fixpoint to the final model and establishes coinductivity as a syntactical criterion for behavioural congruence.

2 Standard specifications and initial semantics

We assume familiarity with the basic notions of many-sorted logic with equality (cf., e.g., [14], [8], [35]). It has been shown that this logic admits presenting not only basic data types with first-order functions, but also generic types, relational and object-oriented programming, higher-order functions and local function definitions, which simulate λ-expressions (cf., e.g., [12], [22], [23], [24], [26]).

Given a set S of sorts, $s_1, \dots, s_n \in S$ and an S-sorted set A, $A_{s_1 \dots s_n}$ denotes the product $A_{s_1} \times \dots \times A_{s_n}$. An **$S$-sorted relation R on A** is an S-sorted set such that for all $s \in S$, $R_s \subseteq A_s \times A_s$. R extends to an S^+-sorted relation on A^+ as usual.

For any expression (term or formula) e, **var(e)** denotes the set of variables occurring in e. e is **linear** if each variable occurs in e at most once. e is **ground** if $var(e)$ is empty. Both **e(t)** and **e[t/u]** stand for expressions that include the subexpression t. $e[t/u]$ indicates that the subexpression u of e has been replaced by t. Set brackets $\{$ and $\}$ in a formula enclose either quantified variables or optional subexpressions. For instance, $t\{\equiv u\}$ stands for t or $t \equiv u$.

We introduce special many-sorted signatures that separate visible from hidden sorts and data constructors from defined functions and include "undefined elements" as well as particular predicates for specifying definedness and equality relations for visible resp. hidden sorts.

Definition 2.1 A **standard signature** $\Sigma = (S, CO, DF, PR)$ consists of a set $S = visS \uplus hidS$ of **sorts** and S^+-sorted sets CO of **constructors**, DF of **defined functions** and PR of **predicates**. The sorts of $visS$ resp. $hidS$ are called **visible** resp. **hidden**. Given $w \in S^+$, a symbol $f \in (CO \cup DF \cup PR)_w$ is **visible** resp. **hidden** if $w \in visS^+$ resp. $w \notin visS^+$. Both constructors and defined functions are called **function symbols**.

As usual, a function symbol $f \in (CO \cup DF)_{ws}$ is written as $f : w \to s$. If $w = \varepsilon$, f is called a **constant**. A predicate $r \in PR_w$ is written as $r : w$. For each $s \in S$, Σ implicitly contains an **equality predicate** $\equiv: ss$ and a **bisimulation predicate** $\sim: ss$. For all $s \in visS$, $\equiv: ss$ agrees with $\sim: ss$.

Non-equality predicates are called **logical predicates**. Given $r : w \in PR$ and $t \in T_\Sigma(X)_w$, the formula $r(t)$ is an **atom**. If r is a logical predicate, then $r(t)$ is a **logical atom**, otherwise $r(t)$ is an **equation**. A Σ-term t is **visible** resp. **hidden** if the sort of t is visible resp. hidden. An equation $t \equiv t'$ is **visible** resp. **hidden** if t and t' are visible resp. hidden. For all $s \in visS$, $\equiv: ss$ and $\sim: ss$ are the same predicates and Σ implicitly contains a **bottom constant** $\bot :\rightarrow s$ and a **definedness predicate** $Def : s$. For all hidden constructors $f : w \rightarrow s$, s is hidden.

Given an S-sorted set X of variables, the S-sorted set $NF_\Sigma(X)$ of **Σ-normal forms** is inductively defined as follows:

- $X \cup \{\bot :\rightarrow s \mid s \in visS\} \subseteq NF_\Sigma(X)$.
- For all visible constructors $f : w \rightarrow s$ and $t \in (NF_\Sigma(X) \setminus \{\bot\})_w$, $f(t) \in NF_\Sigma(X)_s$.
- For all hidden constructors $f : w \rightarrow s$ and $t \in NF_\Sigma(X)_w$, $f(t) \in NF_\Sigma(X)_s$.

T_Σ and NF_Σ denote the sets of all ground Σ-terms and Σ-normal forms, respectively. Σ is **inhabited** if for all sorts s of Σ, $T_{\Sigma,s}$ is not empty. Σ is **visible** if $hidS$ is empty. The greatest visible subsignature of Σ is denoted by **vis**Σ. ❑

The purpose of ground normal forms is to represent data. Visible data should have unique normal form representations. Hidden data also have normal forms, but here the representation need not be unique. Instead, the identity of hidden objects is defined in terms of the bisimulation predicates of Σ. The same conceptual difference between visible and hidden data also applies to definedness: visible sorts have bottom constants in order to make undefined values visible; hidden normal forms, however, are mere names for objects, so that hidden bottom constants would not make sense.

Let Σ be a standard signature with sort set S and X be an S-sorted set of variables.

An S-sorted function $\sigma : X \rightarrow T_\Sigma(X)$ is called a **substitution** over Σ. The **domain of** σ, $dom(\sigma)$, is the set of all variables x with $x\sigma \neq x$. σ_Y denotes the restriction of σ with $dom(\sigma) \subseteq Y$. σ is **ground** if $dom(\sigma)\sigma \subseteq T_\Sigma$. T_Σ^X denotes the set of ground substitutions over Σ. The **instance** $t\sigma$ of a term or atom t **by** σ is obtained from t by replacing each variable x by $x\sigma$. Let $Y, Z \subseteq X$. Given a further substitution τ, the composition $\sigma\tau$ is defined by $x(\sigma\tau) = (x\sigma)\tau$ for all $x \in X$.

Definition 2.2 (Gentzen clauses) Let $Y \subseteq X$. Given a finite conjunction (or set) G of atoms, the formula $\forall Y G$ is a **goal**. $\emptyset$ denotes the empty goal and stands for "always true". A goal $G \wedge \emptyset$ is identified with G. Given a finite disjunction (or set) GS of goals, the formula $\exists Y GS$ is a **goal set**. $FALSE$ denotes the empty goal set and stands for "always false". A goal set $GS \vee FALSE$ is identified with GS.

Given two goal sets GS and HS, the formula $GS \Leftarrow HS$ is a **Gentzen clause**. We write $TRUE$ for $FALSE \Leftarrow \emptyset$ and GS for $GS \Leftarrow \emptyset$ if $GS \neq FALSE$. The free variables of a Gentzen clause are implicitly universally quantified over the entire clause.

Given a logical atom $r(t)$, an equation $f(u) \equiv v$ and a goal H, the Gentzen clauses $r(t) \Leftarrow H$ and $f(u) \equiv v \Leftarrow H$ are called **Horn clauses for** r resp. f. Several Horn clauses $p \Leftarrow H_1, \ldots, p \Leftarrow H_n$ with the same conclusion p are sometimes combined to the Gentzen clause $p \Leftarrow H_1 \vee \cdots \vee H_n$, which is also called a Horn clause.

Given a logical atom $r(x)$ with $x \in X^+$, a goal H and a goal set GS, the Gentzen clause $GS \Leftarrow (r(x) \wedge H)$ is called a **co-Horn clause for** r and written as $r(x) \Rightarrow (H \Rightarrow GS)$. ❑

Gentzen clauses have a number of advantages. First of all, they are as expressive as arbitrary first-order formulas. Secondly, functional-logic programs as well as conjectures about them, such as such as pre/postconditions, invariants, etc., are usually *given* as Gentzen clauses. Thirdly, Gentzen clauses can be proved (or refuted) directly without subjecting them to a normalization process. They are amenable to automatic simplification steps as well as to interactive applications of unfold, induction and coinduction rules.

Conjectures may be presented as arbitrary Gentzen clauses. The axioms of a data type, however, will be restricted Horn and co-Horn clauses. This complies with usual syntactic schemas adopted by functional, relational and even state- or object-oriented programs. Semantically, the restriction to Horn and co-Horn clauses guarantees the existence of initial and final models and thus of concrete implementations. Moreover, these models satisfy a couple of "meta-theorems", which give rise to powerful proof rules (cf. [26]).

Definition 2.3 A **standard specification** $SP = (\Sigma, AX)$ consists of a standard signature Σ and a set AX of Gentzen clauses, the **axioms** of SP, such that PR splits into a set **SPR** of **safety predicates** and a set **LPR** of **liveness predicates** and the following conditions hold true.

- All axioms for a function or a safety predicate are Horn clauses (cf. 2.2).
- All axioms for a liveness predicate are co-Horn clauses.
- Equality predicates are safety predicates. Bisimulation predicates for hidden sorts are liveness predicates.
- AX implicitly contains the congruence axioms for Σ, i.e. reflexivity, symmetry and Σ-compatibility axioms for $\equiv$.
- AX implicitly contains the **partiality axioms** for each visible constructor $c : s_1 \ldots s_n \to s$ of Σ:

$$Def(c(x_1, \ldots, x_n)) \Leftarrow Def(x_1) \wedge \cdots \wedge Def(x_n),$$
$$c(x_1, \ldots, x_{i-1}, \bot, x_{i+1}, \ldots, x_n) \equiv \bot \quad \text{for all } 1 \leq i \leq n.$$

- For all other Horn axioms $C = (f(t_1, \ldots, t_n)\{\equiv u\} \Leftarrow H)$,

(1) f is a defined function or a logical predicate $\neq Def$, $(t_1, \ldots, t_n)$ is a linear tuple of normal forms, $var(u) \subseteq var(t_1, \ldots, t_n, H)$,

(2) for all $1 \leq i \leq n$, $t_i \in X \cup \{\bot\}$ or $Def(t_i) \in H$,[2]
(3) C consists of visible symbols if f is visible,
(4) H is factorizable (see below).

(5) For all co-Horn axioms $p \Rightarrow (H \Rightarrow GS)$, H is factorizable (see below).

(6) AX implicitly contains the axiom $x \sim y \Rightarrow y \sim x$ and each other axiom for $\sim$ has the form

$$x \sim y \;\Rightarrow\; f(x,z) \sim f(y,z) \quad \text{or}$$
$$x \sim y \;\Rightarrow\; (r(x,z) \;\Rightarrow\; r(y,z)) \quad \text{or}$$
$$x \sim y \;\Rightarrow\; (\delta(x,z,x') \;\Rightarrow\; \exists y'(\delta(y,z,y') \wedge x' \sim y'))$$

where $f : sv \to s'$ is a defined function and $r : sv$ and $\delta : svs''$ are safety predicates with $s, s'' \in hidS$ and $v \in visS^*$. f, r and δ are called **functional**, **relational** and **transitional** ***s*-actions**, respectively.

TA denotes the set of transitional actions of SP. Atoms $\delta(t,u,v)$ with $\delta \in TA$ are called **transition atoms**. A goal H is **factorizable** if all predicates of H are safety predicates, all equations of H are visible and there is a partition $\{G_0, G_1, \ldots, G_n\}$ of H such that G_0 has no transition atoms and for all $1 \leq i \leq n$, G_i contains a single transition atom $\delta(t,u,v)$ and v is a variable that does not occur anywhere else in $G_0 \wedge \cdots \wedge G_i$.

SP is **visible** if Σ is visible (cf. 2.1). By (3), SP has a visible subspecification **visSP** that consists of $vis\Sigma$ and the (Horn) axioms for $vis\Sigma$. ❑

Conditions (3)-(5) ensure that the class of SP-models is closed under quotients (see Lemma 2.6) and thus SP is **behaviourally consistent** in the sense of [5]). In particular, (3) implies that Horn axiom premises do not contain hidden equations. This condition is already present in previous approaches to behavioural specifications (cf., e.g., [6], Cor. 4; [35], Thm. 5.4.5; [5], Ex. 3.24). Since we will reason about SP with respect to the final model, which is the quotient by bisimilarity of the initial model (see Section 3), behavioural consistency is crucial.

The factorizability conditions in (4) and (5) also ensure that liveness predicates can be defined in terms of safety predicates, but not vice versa. This allows us to construct the final model in a hierarchical way: the initial model of the Horn axioms is first extended by the greatest solutions of the co-Horn axioms and then factored through bisimilarity. (5) guarantees that the corresponding *consequence operator* Ψ is monotonic and thus greatest solutions do exist (cf. Thm. 3.1). The hierarchy argument used for establishing the greatest fixpoint of Ψ is quite similar to the one that accounts for minimal models of stratified logic programs (cf. [3]).

Conditions (1)-(4) allow us to specify all partial functions that are continuous in the sense of recursive function theory. Moreover, defined functions need not be *strict*, i.e. preserve bottom constants. By (2), a defined function may map $\bot$ to other "error values". However, the partiality axioms imply that all visible constructors are strict.

[2] For the sake of brevity, Def-atoms in the premise of an axiom are usually omitted.

Example 2.4 Let NAT and LIST be visible specifications of natural number arithmetic and finite lists of elements of a generic sort *entry*. We extend LIST + NAT by a specification of **finite and infinite streams**:

STREAM = NAT + LIST +

`hidsorts`	$stream$
`consts`	$\_\&\_ : entry \times stream \to stream$
	$stop, blink :\to streamstream$
	$nats : nat \to stream$
	$odds : stream \to stream$
	$zip : stream \times stream \to stream$
	$map : (entry \to entry) \times stream \to stream$
	$filter : (entry \to bool) \times stream \to stream$
`defuns`	$\_\#\_ : list \times stream \to stream$
	$evens : stream \to stream$
`preds`	$\_ \overset{\_}{\longrightarrow} \_ : stream \times entry \times stream$
	$\_ \sim \_ : stream \times stream$
	$empty, finite, infinite, fair : stream$
	$exists, forall :: entry \to bool) \times stream$
`vars`	$n : nat \quad x, y : entry \quad L : list \quad s, s', t, t' : stream$
	$f : entry \to entry \quad g : entry \to bool$
`Horn axioms`	$x\&s \overset{x}{\longrightarrow} s$
	$blink \overset{0}{\longrightarrow} 1\&blink$
	$nats(n) \overset{n}{\longrightarrow} nats(n+1)$
	$odd(s) \overset{x}{\longrightarrow} odd(t) \Leftarrow s \overset{x}{\longrightarrow} s' \wedge s' \overset{y}{\longrightarrow} t$
	$zip(s, s') \overset{x}{\longrightarrow} zip(s', t) \Leftarrow s \overset{x}{\longrightarrow} t$
	$zip(s, s') \overset{x}{\longrightarrow} zip(s, t') \Leftarrow empty(s) \wedge s' \overset{x}{\longrightarrow} t'$
	$map(f, s) \overset{f(x)}{\longrightarrow} map(f, t)) \Leftarrow s \overset{x}{\longrightarrow} t$
	$filter(g, s) \overset{x}{\longrightarrow} filter(g, t) \Leftarrow s \overset{x}{\longrightarrow} t \wedge g(x) \equiv true$
	$filter(g, s) \overset{y}{\longrightarrow} t' \Leftarrow s \overset{x}{\longrightarrow} t \wedge g(x) \equiv false \wedge filter(g, t) \overset{y}{\longrightarrow} t'$
	$empty(stop)$
	$empty(odds(s)) \Leftarrow empty(s)$
	$empty(zip(s, s')) \Leftarrow empty(s) \wedge empty(s')$
	$empty(map(f, s)) \Leftarrow empty(s)$
	$empty(filter(g, s)) \Leftarrow empty(s)$
	$empty(filter(g, s)) \Leftarrow s \overset{x}{\longrightarrow} t \wedge g(x) \equiv false \wedge empty(filter(g, t))$
	$finite(s) \Leftarrow empty(s)$
	$finite(s) \Leftarrow s \overset{x}{\longrightarrow} t \wedge finite(t)$
	$exists(g, s) \Leftarrow s \overset{x}{\longrightarrow} t \wedge g(x) \equiv true$
	$exists(g, s) \Leftarrow s \overset{x}{\longrightarrow} t \wedge exists(g, t)$
	$nil\#s \equiv s$
	$(x :: L)\#s \equiv x\&t \Leftarrow L\#s \equiv t$
	$evens(s) \equiv stop \Leftarrow empty(s)$
	$evens(s) \equiv odds(t) \Leftarrow s \overset{x}{\longrightarrow} t$
`co-Horn axioms`	$s \sim s' \Rightarrow (s \overset{x}{\longrightarrow} t \Rightarrow \exists t'(s' \overset{x}{\longrightarrow} t' \wedge t \sim t'))$
	$infinite(s) \Rightarrow \exists\{x, t\}(s \overset{x}{\longrightarrow} t \wedge infinite(t))$
	$forall(g, s) \Rightarrow (s \overset{x}{\longrightarrow} t \Rightarrow (g(x) \equiv true \wedge forall(g, t)))$

$$fair(s) \Rightarrow \exists\{L,s'\}(s \equiv L\#(0\&s') \wedge fair(s'))$$

The axioms of STREAM are inspired by transition system specifications given in [31] and [18]. We expect a model that interprets STREAM as follows. $s \xrightarrow{x} t$ holds true if x is the first entry and t is the rest of s, *stop* denotes the empty stream. & appends an entry to a stream. 01 and 10 denote the streams whose elements alternate between zeros and ones. $nats(n)$ generates the stream of all numbers starting from n. $odds(s)$ returns the stream of all elements of s that have odd-numbered positions in s. *zip* merges two streams into a single stream by alternatively appending an element of one stream to an element of the other stream. # concatenates a list and a stream into a stream. *map*, *filter*, *exists* and *forall* have the same meaning as stream functions as they have as list functions. $\sim$ is stream equality. *finite* and *infinite* distinguish finite from infinite streams. $fair(s)$ holds true iff s contains infinitely many zeros. ❑

Definition 2.5 (semantics) Given a signature Σ with sort set S, a **Σ-structure** A consists of an S-sorted set, the **carrier** of A, also denoted by A, a function $f^A : A_w \to A_s$ for each function symbol $f : w \to s$ and a relation $r^A \subseteq A_w$ for each predicate $r : w$ of Σ. Given two Σ-structures A and B, a **Σ-homomorphism** $h : A \to B$ is a homomorphism in the algebraic sense such that $h(r^A) \subseteq r^B$ for all predicates r of Σ. A and B are **Σ-isomorphic**, written: $A \cong B$, if $g \circ h = id_A$ and $h \circ g = id_B$ hold true for some Σ-homomorphism $g : B \to A$.

Let $\approx$ be an S-sorted equivalence relation on A. $\approx$ is **compatible** with a function symbol $f : w \to s \in \Sigma$ if for all $a, b \in A_w$, $a \approx b$ implies $f^A(a) \approx f^A(b)$. $\approx$ is a **Σ-congruence** if $\approx$ is compatible with all function symbols of Σ. $\approx$ is **compatible** with a predicate $r \in \Sigma$ if for all $a \in r^A$, $a \approx b$ implies $b \in r^A$.

If $\approx$ is a Σ-congruence, then the **quotient $A/\!\approx$ of A by $\approx$** is the Σ-structure that interprets sorts and function symbols as usual. The equivalence class of all $b \in A$ with $a \approx b$ for some $a \in A$ is denoted by $[a]$. This notation extends to tuples: $[(a_1,\dots,a_n)]$ is an abbreviation of $([a_1],\dots,[a_n])$. $A/\!\approx$ interprets each predicate $r \in \Sigma$ as the set of equivalence classes $[a]$ with $b \in r^A$ for some $b \approx a$.

The interpretation of Σ-terms in a Σ-structure A depends on a **valuation** of variables in A, i.e. an S-sorted function $b : X \to A$. The unique Σ-homomorphism extending b to a function from $T_\Sigma(X)$ to A is denoted by b^*. If t is a ground term, then for all $b : X \to A$, $b^*(t)$ has the same value and we write t^A instead of $b^*(t)$. A is **reachable** or **finitely generated** if for all $a \in A$ there is a ground term t with $t^A = a$. Each Σ-structure A has a unique reachable substructure, denoted by **gen(A)**.

A valuation $b : X \to A$ **solves** an equation $t \equiv t'$ **in** A if $b^*(t) = b^*(t')$. b **solves** a logical atom $r(t)$ in A if $b^*(t) \in r^A$. This notion extends to goals, goal sets and Gentzen clauses as in first-order logic. If b solves a clause C in A, we write $A \models_b C$. A **satisfies** C or C is **valid in** A, written $A \models C$, if all valuations in A solve C in A.

Let $SP = (\Sigma, AX)$ be a standard specification. A Σ-structure A is an SP-

model if A satisfies all axioms of SP. The class of all SP-models is denoted by **Mod(SP)**. ❑

Lemma 2.6 [27] If A be an SP-model such that $\sim^A$ is compatible with $\Sigma \setminus TA \setminus LPR$, then $B =_{def} A/{\sim^A}$ is also an SP-model. ❑

Definition 2.7 Given a class $\mathcal{C}$ of Σ-structures, $I \in \mathcal{C}$ is **initial in** $\mathcal{C}$ if for all $A \in \mathcal{C}$ there is a unique Σ-homomorphism $ini^A : I \to A$. $T \in \mathcal{C}$ is **final** or **terminal** in $\mathcal{C}$ if for all $A \in \mathcal{C}$ there is a unique Σ-homomorphism $fin^A : A \to T$. ❑

Each two initial (resp. final) Σ-structures are Σ-isomorphic. The initial model of SP is usually presented as the quotient of T_Σ by the equivalence relation consisting of all ground equations that are derivable from the axioms of SP via the *cut calculus*:

Definition 2.8 Let SP be a standard specification with signature Σ and Horn axiom set AX. The **cut calculus for** SP consists of AX and the following rules for deriving goals. Let p be an atom and G, H be goals.

$$\begin{array}{llll} p \Leftarrow H & \vdash_{cut} & p\sigma \Leftarrow H\sigma & \text{for all substitutions } \sigma \text{ over } \Sigma \\ \{p \Leftarrow G \wedge H,\ H\} & \vdash_{cut} & p \Leftarrow G & \text{(cut)} \\ \{G,\ H\} & \vdash_{cut} & G \wedge H & \\ \bigwedge\{G\sigma_Y \mid \sigma \in T_\Sigma^X\} & \vdash_{cut} & \forall Y G & \text{for all } Y \subseteq X \end{array}$$

The set of derivable equations $t \equiv t'$ induces the SP**-equivalence** relation $\equiv_{SP}$, defined by: $t \equiv_{SP} t'$ iff $SP \vdash_{cut} t \equiv t'$. A normal form u is a **normal form of** a term t if t and u are SP-equivalent. SP is **complete** if each ground Σ-term has a normal form. SP is **consistent** if each two SP-equivalent ground normal forms are equal. SP is **functional** if SP is complete and consistent.

SP-equivalence yields an interpretation of $\equiv$ in T_Σ. If one interprets all logical safety predicates r of SP accordingly, i.e. $r_{SP}(t) \Longleftrightarrow_{def} SP \vdash_{cut} r(t)$, then T_Σ becomes a $(\Sigma \setminus LPR)$-structure. The quotient of T_Σ by $\equiv_{SP}$ is called the **initial** SP**-model** and denoted by **Ini(SP)**. ❑

Theorem 2.9 (initial semantics) Let SP be a standard specification with signature Σ. $Ini(SP)$ is an initial Σ-structure in $Mod(SP)$. ❑

Definition 2.10 Let $SP = (\Sigma, AX)$ be a standard specification. A Gentzen clause $GS \Leftarrow HS$ is an **inductive theorem** of SP if for all goals H of HS and $\rho \in T_\Sigma^X$, $SP \vdash_{cut} H\rho$ implies $SP \vdash_{cut} G\tau$ for some goal G of GS and $\tau \in T_\Sigma^X$ such that $\tau_V = \rho_V$ where V is the set of free variables of $GS \Leftarrow HS$. **ITh(SP)** denotes the set of inductive theorems of SP. ❑

Theorem 2.11 Let SP be an inhabited standard specification (cf. 2.1) and C be a Gentzen clause. $Ini(SP) \models C$ iff $C \in ITh(SP)$. ❑

From now on we assume that standard signatures are inhabited. The initial SP-model interprets safety predicates as the **least relations** satisfying their axioms. This follows from *Kleene's fixpoint theorem*, which is also used for defining the semantics of pure logic programs. *Tarski's fixpoint theorem* forbids the use of negation, while Kleene's theorem enforces more restrictive conditions on the axioms of a specification (see [27]). When applying fixpoint theorems we refer to the *relational version* of SP:

Definition 2.12 (relational version) Let $SP = (\Sigma, AX)$ be a standard specification and $f : w \rightarrow s$ be a defined function of SP. The **graph** $r_f : ws$ **of** f is an implicit predicate of SP and specified by the Horn axiom $r_f(x, y) \Leftarrow f(x) \equiv y$.

An equation $f(t) \equiv u$ such that f is a defined function and t and u are normal forms is called **flat**. SP is **flat** if defined functions do only occur in flat equations of AX. If SP is flat, then the **relational version** of SP is the specification $rel(SP)$ obtained from SP by replacing all defined functions of Σ by their graphs and each flat equation $f(t) \equiv u$ of AX by the logical atom $r_f(t, u)$. ❑

Theorem 2.13 Let SP be a functional specification (cf. 2.8) with signature Σ and $rel(SP)$ be the relational version of SP. $Ini(SP)$ and $Ini(rel(SP))$ are isomorphic. More precisely, for all defined functions $f : w \rightarrow s \in \Sigma$, logical safety predicates $r : w \in \Sigma$, $t \in T_{\Sigma,w}$ and $u \in T_{\Sigma,s}$,

$$f(t) \equiv_{SP} u \iff rel(SP) \vdash_{cut} r_f(nf(t), nf(u)),$$
$$SP \vdash_{cut} r(t) \iff rel(SP) \vdash_{cut} r(nf(t))$$

where $nf(t)$ and $nf(u)$ are the unique normal forms of t and u, respectively. ❑

The carrier elements of $Ini(rel(SP))$ are equivalence classes consisting of normal forms. Since SP is functional, each equivalence class $[t]$ is either a singleton or all terms of $[t]$ contain $\perp$ (see 2.3).

Theorem 2.14 (least fixpoint semantics) Let SP be a functional specification, $rel(SP) = (\Sigma, AX)$ be the relational version of SP, PR be the set of logical safety predicates of $rel(SP)$ and $\mathcal{C}$ be the class of Σ-structures that interpret all sorts and constructors of SP the same as $Ini(SP)$ does. The PR-sorted **consequence operator** $\Phi : \mathcal{C} \rightarrow \mathcal{C}$, defined by

$$b^*(t) \in r^{\Phi(A)} \iff_{def} \exists (r(t) \Leftarrow H) \in AX : A \models_b H$$

for all $r \in PR$, $b : X \rightarrow Ini(SP)$ and $A \in \mathcal{C}$ is monotonic with respect to PR-sorted set inclusion on $\mathcal{C}$ where $A \subseteq B \iff_{def} \forall r \in PR : r^A \subseteq r^B$. Hence by Tarski's Theorem, Φ has the least fixpoint $\mu\Phi = \bigcap\{A \in \mathcal{C} \mid \Phi(A) \subseteq A\}$.

$A \in \mathcal{C}$ satisfies AX iff $\Phi(A) \subseteq A$. Hence by Kleene's Theorem, $\mu\Phi$ agrees with $\Phi^\infty = \bigcup_{i \in \mathbb{N}} \Phi^i(\perp)$ if Φ^∞ satisfies AX or Φ is upward continuous where $\perp$ is the least element of $\mathcal{C}$. Moreover, by Thm. 2.13 and Tarski's Theorem, $\mu\Phi = Ini(SP)$ because $Ini(SP)$ is the least SP-model.[3] ❑

3 Final semantics and coinductivity

Final semantics was introduced for modelling **permutative types** such as finite sets, finite bags (multisets) and functions with a finite domain (stores, arrays,

[3] proof by induction on cut calculus derivations

indexed lists) (cf., e.g., [10], [34], [20]). These types are still constructor-based, but *constructor equations* are needed to axiomatize data equality. Hence specifications of permutative types are complete, but not consistent (cf. 2.8). From a model-theoretic viewpoint, initial semantics is sufficient for handling permutative types. Constructor equations are Horn axioms and thus we obtain an initial model as in the case of standard specifications where the only constructor equations are the partiality axioms (cf. 2.3). From a proof-theoretic viewpoint, however, initial semantics is less appropriate. Efficient resolution- and rewriting-oriented proof methods treat constructor equations CE on a lower level than other axioms (cf., e.g., [28], [32], [19]). Normal forms are replaced by equivalence classes of normal forms modulo the equivalence relation $\equiv_{CE}$ induced by CE. Resolution and rewriting modulo $\equiv_{CE}$ work well if CE is restricted to particular axioms such as associativity, commutativity, idempotence, etc. Otherwise corresponding proof rules are difficult to handle.

With respect to final semantics, constructor equations become *theorems* that can be derived, e.g., by **context induction** (cf. [17]). Studies in category theory and modal logic dealing with coalgebras, coinduction and greatest fixpoints suggested both subsuming permutative types under hidden types and adopting final semantics as well for **object types** and **infinite types** such as streams and processes (cf., e.g., [4], [11], [30], [31], [13], [18]). In order to keep the proof-theoretic benefit of initial semantics we present even infinite types as *functional* specifications (cf. Def. 2.8). Consistency is not restrictive here because hidden data equality is defined in terms of bisimulation and not equality predicates. Completeness, however, seems to be restrictive. How can uncountably many streams be represented by countably many normal forms? The answer is that not all elements of a carrier set need to be covered by normal forms as long as the theory of a specification SP is preserved whenever further normal forms are added to SP. Since a theorem about a hidden carrier cannot be proved by structural induction (because the structure is hidden), it will mostly be valid in extensions of SP by further normal forms. For instance, many streams are not represented as normal forms of STREAM, while a single stream may have several normal forms (cf. Ex. 2.4). Whereas *visible* data are considered to be equal if they have the same normal forms, bisimulation predicates take over the rôle of identifying terms representing the same *hidden* objects.

Besides separating visible from hidden sorts we have distinguished safety from liveness predicates. By Theorem 2.14, the initial model interprets safety predicates as least relations satisfying their (Horn) axioms. Dually, liveness predicates are interpreted as greatest relations satisfying their (co-Horn) axioms. Horn axioms for a visible symbol f usually entail an inductive definition of f in the initial model. Category theory set up the duality between inductively defined algebras and coinductively defined coalgebras (cf., e.g., [1], [15], [30], [18]). Initial algebras and final coalgebras are isomorphisms built up from constructors in one direction and actions ("destructors") in the other. For instance, the final coalgebra of infinite streams (cf. Section 1) splits into the constructor & and the actions *head* and *tail*. We keep the duality in mind, but follow more algebraic lines by constructing final models as quotients of initial ones. Let us first "dualize" Thm. 2.14 for characterizing the initial model not only as a least, but also as a greatest fixpoint:

Theorem 3.1 (greatest fixpoint semantics) Let $SP = (\Sigma, AX)$ be a standard specification and $\mathcal{D}$ be the class of Σ-structures that interpret $\Sigma \setminus LPR$ the same as $Ini(SP)$ does. The LPR-sorted **consequence operator** $\Psi : \mathcal{D} \to \mathcal{D}$, defined by

$$b(x) \in r^{\Psi(A)} \iff_{def} \forall (r(x) \Rightarrow (H \Rightarrow GS)) \in AX : A \models_b H \Rightarrow GS$$

for all $r \in LPR$, $b : X \to Ini(SP)$ and $A \in \mathcal{D}$ is monotonic with respect to LPR-sorted set inclusion (cf. 2.14). Hence by Tarski's Theorem, Ψ has the greatest fixpoint $\nu\Psi = \cup\{A \in \mathcal{D} \mid A \subseteq \Psi(A)\}$.

$A \in \mathcal{D}$ satisfies AX iff $A \subseteq \Psi(A)$. Hence by Kleene's Theorem, $\nu\Psi$ agrees with $\Psi_\infty = \cap_{i \in \mathbb{N}} \Psi^i(\top)$ if Ψ_∞ satisfies AX or Ψ is downward continuous where $\top$ is the greatest element of $\mathcal{D}$.

The interpretation of $\sim$ in $\nu\Psi$ is called SP-**bisimilarity** and denoted by $\sim_{SP}$. Since the equality in $\nu\Psi$ reduces to SP-equivalence (cf. 2.8), we write $SP \vdash_{cut} t \sim t'$ for $[t] \sim_{SP} [t']$. SP is **behaviourally congruent** if $\sim_{SP}$ is compatible with $\Sigma \setminus TA \setminus LPR$. ❑

Definition 3.2 (final semantics) Let SP be a behaviourally congruent specification and $\nu\Psi$ be defined as in Thm. 3.1. The **final** SP**-model Fin(SP)** is the quotient of $\nu\Psi$ by $\sim_{SP}$. A **coinductive theorem** of SP is a (Gentzen) clause satisfied by $Fin(SP)$. ❑

By Lemma 2.6, the final SP-model is an SP-model. It can be represented as the quotient $T_\Sigma/\approx$ where $t \approx t' \iff_{def} [t] \sim_{SP} [t']$. $Fin(SP)$ coincides with $Ini(SP)$ if SP has neither hidden sorts nor liveness predicates.

Definition 3.3 (hierarchical models) Let Σ and Σ' be standard signatures with $\Sigma' \subseteq \Sigma$. Let A and B be reachable Σ'- resp. Σ-structures (cf. 2.5).

B is **complete wrt** A if $B_{\Sigma'}$ and $gen(B_{\Sigma'})$ are isomorphic or, equivalently, if for all sorts $s \in \Sigma'$ and $t \in T_{\Sigma,s}$ there is $t \in T_{\Sigma'}$ such that $B \models t \equiv t'$. B is **consistent wrt** A if A and $gen(B_{\Sigma'})$ are isomorphic or, equivalently, if for all ground Σ'-atoms p, $B \models p$ implies $A \models p$. Given a standard specification SP with signature Σ, an SP-model B is **hierarchical over** A if B is complete and consistent wrt A and for all safety predicates $r \in \Sigma \setminus \Sigma'$ and $t \in T_\Sigma$, $B \models r(t)$ implies $Ini(SP) \models r(t)$. ❑

Note that SP is functional (cf. 2.8) iff $Ini(SP)$ is hierarchical over $Ini(NF(SP))$ where $NF(SP)$ is the subspecification of SP that consists of all constructors, bottom constants, definedness predicates and partiality axioms of SP (cf. 2.3).

Theorem 3.4 [Pad 97a,b] Let SP be a behaviourally congruent specification.

(1) For all safety predicates $r : w$ of SP, $r^{Fin(SP)}$ is the least subset of $Fin(SP)_w$ that satisfies the Horn axioms of SP.

(2) For all liveness predicates $r : w$ of SP, $r^{Fin(SP)}$ is the greatest subset of $Fin(SP)_w$ that satisfies the co-Horn axioms of SP.

Let SP be functional.

(3) For all defined functions $f : w \to s$ of SP, the graph $r_f : ws$ of f is the least subset of $Fin(SP)_{ws}$ that satisfies the Horn axioms of the relational version of SP.

(4) $Fin(SP)$ is final in the class of hierarchical SP-models over $Ini(visSP)$ (cf. 2.3). ❑

An important proof-theoretic consequence of (1)-(3) are the fixpoint induction rules of [26]. Let us now present (almost) syntactical conditions on the Horn axioms for hidden symbols of SP that guarantee behavioural congruence.

Definition 3.5 (coinductivity) Let Σ be a standard signature. Given hidden (empty or singleton) term tuples t, u and visible term tuples a, b, an atom $p(t, a, b, u)$ is **oriented** if t and a are normal forms and $p(t, a, b, u)$ has one of the following forms:

(1) $\delta(t, ab, u)$ for a transitional action δ,

(2) $r(t, a)$ for another hidden safety predicate r,

(3) $f(t, a) \equiv b$ for a hidden defined function f with visible range sort,

(4) $f(t, a) \equiv u$ for a hidden defined function f with hidden range sort.

A Horn clause $p_0(t_0, a_0, b_0, u_0) \Leftarrow G_0 \wedge \bigwedge_{i=1}^{n}(p_i(t_i, a_i, b_i, u_i) \wedge G_i)$ is **deterministic** if for all $0 \leq i \leq n$, G_i is a goal over $vis\Sigma$, $p_i(t_i, a_i, b_i, u_i)$ is oriented, $b_i, u_i \in NF_{vis\Sigma}(X)$,[4] $var(t_i, a_i) \subseteq V_{i-1}$ and $var(b_0, u_0) \subseteq V_n$ where $V_0 = var(t_0, a_0, G_0)$ and $V_i = V_{i-1} \uplus var(b_i, u_i) \cup var(G_i)$.

Let SP be a standard specification with signature Σ. p is **bisimilarity compatible** if for all $t, a, b, u \in NF_\Sigma$,

$$SP \vdash_{cut} p(t, a, b, u) \wedge (t, a) \sim (t', a') \text{ implies } SP \vdash_{cut} p(t', a', b', u') \wedge (b, u) \sim (b', u')$$

for some b', u' (see 3.1). The above Horn clause is **coinductive** if it is deterministic and

(5) $t_0, a_0 \in NF_{vis\Sigma}(X)$ and $b_0, u_0 \in NF_\Sigma(X)$ **or**

(6) p_0 is bisimilarity compatible, $a_0 \in NF_{vis\Sigma}(X)$ and there are $t \in NF_{vis\Sigma}(X)$, $c, d \in NF_\Sigma(X)$ and a term v over $vis\Sigma \cup DF$ such that $t_0 = c(t)$, $c \notin X$ and $(b_0, u_0) = d(v(t))$.

SP is **coinductive** if all axioms for hidden defined functions or safety predicates of SP are coinductive. ❑

Deterministic clauses have factorizable premises (see 2.3). In Cases (2)-(4), p is bisimilarity compatible iff $\sim_{SP}$ is compatible with r resp. f.

a and b comprise all visible arguments of an oriented atom $p(t, a, b, u)$. a collects the "input", b the "output" arguments. In Cases (2)-(4), the separation of a from b is fixed. In Case (1), it is usually determined by the requirement that the axioms for δ are coinductive. Intuitively, the above Horn clause is deterministic

[4] Note that the only hidden normal forms of $NF_{vis\Sigma}(X)$ are variables.

if all variables "flow" only left to right, i.e. from t_0, a_0, G_0 to t_i, a_i, from b_j, u_j, G_j to t_{j+k}, a_{j+k}, $k > 0$, or right to left from b_m, u_m, G_m to b_0, u_0.

We show that STREAM is coinductive (cf. Ex. 2.4). All axioms for $\longrightarrow$ satisfy Condition (6) because $\longrightarrow$ is a transitional action and thus bisimilarity compatible by definition. The axioms for *empty* satisfy (6) because *Ini*(STREAM) interprets *empty* as the complement of $\lambda s.\exists\{x,t\}s \xrightarrow{x} t$[5] and thus *empty* is bisimilarity compatible. All other Horn axioms of STREAM satisfy (5) because all hidden normal forms occurring in their conclusions resp. left-hand sides are variables and thus $vis\Sigma$-normal forms. Since this does not hold for the axioms of $\longrightarrow$ and *empty*, we had to employ (6) for showing that they are coinductive.

Theorem 3.6 [27] A functional and coinductive specification is behaviourally congruent. ❑

Since functionality criteria presented in [26] apply to STREAM, we conclude from Thm. 3.6 that STREAM has all the – proof-theoretically significant – properties stated in Thm. 3.4. Similarly, CCS-like process types can be specified coinductively in a quite straightforward way.

References

1. M.A.. Arbib, E.G. Manes, *Parametrized Data Types Do Not Need Highly Constrained Parameters*, Information and Control 52 (1982) 139-158
2. K.R. Apt, *Logic Programming*, in: J. van Leeuwen, ed., Handbook of Theoretical Computer Science, Elsevier (1990) 493-574
3. K.R. Apt, H.A. Blair, A. Walker, *Towards a Theory of Declarative Knowledge*, in: J. Minker, ed., Deductive Databases and Logic Programming, Morgan Kaufmann (1988) 89-148
4. E. Astesiano, M. Wirsing, *Bisimulation in Algebraic Specifications*, in: H. Ait-Kaci, M. Nivat, eds., Resolution of Equations in Algebraic Structures 1, Academic Press (1989) 1-31
5. M. Bidoit, R. Hennicker, M. Wirsing, *Behavioural and Abstractor Specifications*, Science of Computer Programming 25 (1995) 149-186
6. M. Broy, M. Wirsing, *Partial Abstract Types*, Acta Informatica 18 (1982) 47-64
7. G. Costa, G. Reggio, *Specification of Abstract Dynamic Data Types: A Temporal Logic Approach*, Theoretical Computer Science 173 (1997) 513-554
8. H. Ehrig, B. Mahr, *Fundamentals of Algebraic Specification 1*, Springer 1985
9. H. Ehrig, F. Orejas, *Dynamic Abstract Data Types: An Informal Proposal*, EATCS Bulletin 53 (June 1994) 162-169
10. V. Giarratana, F. Gimona, U. Montanari, *Observability Concepts in Abstract Data Type Specifications*, Proc. MFCS '76, Springer LNCS 45 (1976) 576-587
11. J.A. Goguen, R. Diaconescu, *Towards an Algebraic Semantics for the Object Paradigm*, Proc. 9th ADT Workshop, Springer LNCS 785 (1994) 1-29
12. J.A. Goguen, J. Meseguer, *Unifying Functional, Object-Oriented and Relational Programming with Logical Semantics*, in: B. Shriver, P. Wegner, eds., Research Directions in Object-Oriented Programming, MIT Press (1987) 417-477

[5] proof by induction on cut calculus derivations

13. J.A. Goguen, G. Malcolm, *A Hidden Agenda*, UCSD Technical Report CS97-538, 1997, http://www-cse.ucsd.edu/users/goguen/ps/ha.ps.gz
14. J.A. Goguen, J.W. Thatcher, E.G. Wagner, *An Initial Algebra Approach to the Specification, Correctness and Implementation of Abstract Data Types*, in: R. Yeh, ed., Current Trends in Programming Methodology 4, Prentice-Hall (1978) 80-149
15. A.D. Gordon, *A Tutorial on Co-induction and Functional Programming*, Proc. Functional Programming Glasgow 1994, Springer (1995) 78-95
16. J.F. Groote, F. Vaandrager, *Structured Operational Semantics and Bisimulation as a Congruence*, Information and Computation 100 (1992) 202-260
17. R. Hennicker, *Context Induction: A Proof Principle for Behavioural Abstractions*, Formal Aspects of Computing (1991) 326-345
18. B. Jacobs, J. Rutten, *A Tutorial on (Co)Algebras and (Co)Induction*, EATCS Bulletin 62 (June 1997) 222-259
19. J.-P. Jouannaud, H. Kirchner, *Completion of a Set of Rules Modulo a Set of Equations*, SIAM J. Computing 15 (1986) 1155-1194
20. S. Kamin, *Final Data Type Specifications: A New Data Type Specification Method*, ACM TOPLAS 5 (1983) 97-123
21. K.G. Larsen, *Proof Systems for Hennessy-Milner Logic with Recursion*, Proc. CAAP '88, Springer LNCS 299 (1988) 215-230
22. B. Möller, A. Tarlecki, M. Wirsing, *Algebraic Specifications of Reachable Higher-Order Algebras*, Proc. 5th ADT Workshop, Springer LNCS 332 (1988) 154-169
23. P. Padawitz, *Computing in Horn Clause Theories*, Springer 1988
24. P. Padawitz, *Deduction and Declarative Programming*, Cambridge University Press 1992
25. P. Padawitz, *Swinging Data Types: Syntax, Semantics, and Theory*, Proc. WADT '95, Springer LNCS 1130 (1996) 409-435
26. P. Padawitz, *Proof in Flat Specifications*, in: E. Astesiano, H.-J. Kreowski, B. Krieg-Brückner, eds., IFIP WG 1.3 State-of-the-Art Report, *Algebraic Foundations of System Specification*, Springer, to appear, http://ls5.cs.uni-dortmund.de/~peter/Deduct.ps.gz
27. P. Padawitz, *Towards the One-Tiered Design of Data Types and Transition Systems*, full version of this paper, FB Informatik, University of Dortmund 1997, http://ls5.cs.uni-dortmund.de/~peter/Rome.ps.gz
28. G.D. Plotkin, *Building-in Equational Theories*, in: B. Meltzer, D. Michie, eds., Machine Intelligence 7, Elsevier (1972) 73-90
29. G.D. Plotkin, *An Operational Semantics for CSP*, in: D. Bjørner, ed., Proc. IFIP TC-2 Working Conf. Formal Description of Programming Concepts II, North-Holland (1983) 199-225
30. H. Reichel, *An Approach to Object Semantics based on Terminal Coalgebras*, Math. Structures in Comp. Sci. 5 (1995) 129-152
31. J.J.M.M. Rutten, *Universal Coalgebra: A Theory of Systems*, Report CS-R9652, CWI, SMC Amsterdam 1996
32. M. Stickel, *Automated Deduction by Theory Resolution*, J. Automated Reasoning 1 (1985) 333-356
33. C. Stirling, *Modal and Temporal Logics*, in: S. Abramsky et al., eds., Handbook of Logic in Computer Science, Clarendon Press (1992) 477-563
34. M. Wand, *Final Algebra Semantics and Data Type Extensions*, J. Computer and System Sciences 19 (1979) 27-44
35. M. Wirsing, *Algebraic Specification*, in: J. van Leeuwen, ed., Handbook of Theoretical Computer Science, Elsevier (1990) 675-788

Context Parchments

Wiesław Pawłowski

Institute of Computer Science, Polish Academy of Sciences,
ul. Abrahama 18, 81-825 Sopot, Poland.
w.pawlowski@ipipan.gda.pl.

Abstract. The paper introduces a notion of *context parchment*. The notion is illustrated by several examples. It is shown, that every *logical* context parchment generates a context institution. Morphisms between context parchments are introduced, thus yielding a category of context parchments. The use of universal constructions in the category of context parchments, for *modular construction of logics* is discussed and illustrated by examples.

1 Introduction

Institutions, were introduced to provide an "abstract model theory for specification and programming"—quoting from the title of [8]. The *model-theoretic* view of logic, advocated by institutions, seems to be very natural in computer science applications, considering the fact, that our main concern is to specify, create, and reason about *concrete objects*—such as programs or VLSI chips.

Context institutions (cf. [13]), enrich the structure of institutions by adding notions such as contexts, and substitutions, retaining at the same time the *model-theoretic flavour* of institutions.

One of the most important features of *any* system (software or hardware) is a *modularity* of its structure. Modular structure makes the task of understanding and using the system much easier. The same principle applies to *formal systems*, such as *specification formalisms* and *logics* on which such formalisms are based. Therefore, the ability to *construct logics* in a modular fashion, or to *combine* them, is very important.

Although (context) institutions provide a very useful, and general framework for *describing* logical systems, they are not suitable for the task of their modular construction. The reason is quite simple—(context) institutions do not provide any information about the *inner structure* of their components, such as sentences (formulae), and models.

In a not that well known paper [7], Goguen and Burstall introduced structures, which they called *parchments*. The idea was, to make the task of proving *satisfaction condition* for institutions easier. Some years later, parchments were "rediscovered" as a tool for combining logics (cf. [10]). The original notion of parchment has further been refined in [11]. A notion of λ-*parchment*, introduced there, makes the distinction between "object" and "meta" theory much more clear, by moving the *internal* "universal signature", and "universal semantic structure", present in the original definition of parchment, to the *meta-framework*. In [12], yet another step has been made, in the same direction. *Model-theoretic parchments* introduced there, eliminate the need for *universal signature* and *universal semantic structure* altogether.

In the present paper, we introduce *context parchments*, i.e., "parchments (or rather λ-parchments), for context institutions". Although context institutions have much more complex structure than the "ordinary" ones, it turns out that context parchments (or rather concrete examples of them), are actually simpler. The reason for this apparent "contradiction" comes from the "uniform" way, in which context institutions are built from context parchments. For example, in the case of ordinary parchments, the "denotations" for quantifiers present in the logic, have to be "hard-wired" into the parchment, in the form of functions on "predicates", while in the case of context parchments, only their "logical meaning", and the information about the "range of quantification" is needed.

The structure of the paper is as follows. After presenting some categorical, and algebraic preliminaries, in Sect. 3, we recall the notion of *context institution*. In Sect. 4, a *meta-framework*, on which context parchments will be based is introduced. *Context parchments* are introduced in Sect. 5, and illustrated by several examples. In Sect. 5, we show also that *logical* context parchments generate context institutions. The notion of *morphism* between context parchments is defined in Sect. 6, together with several examples. Then, the use of universal constructions in the category of context parchments, for *modular construction of logics* is discussed.

2 Preliminaries

Categories with Inclusions. The definition of *context institution* (see Def. 1 in Sect. 3), uses a notion of a *category with inclusions*. A category $\mathbf{C}$, is a category with inclusions, if there is a distinguished class of morphisms in $\mathbf{C}$, having properties very similar to the properties of the set-theoretic inclusions. More formally, we require that $\mathbf{C}$, has an *inclusion system*, i.e., there is a class $\mathcal{I}_\mathbf{C}$ of morphisms and a class of epimorphisms $\mathcal{E}_\mathbf{C}$ in $\mathbf{C}$, such that both $\mathcal{I}_\mathbf{C}$ and $\mathcal{E}_\mathbf{C}$ are subcategories of $\mathbf{C}$ such that $|\mathcal{I}_\mathbf{C}| = |\mathcal{E}_\mathbf{C}| = |\mathbf{C}|$; every morphism f in $\mathbf{C}$ has a unique factorisation as $e;i$ with $e \in \mathcal{E}_\mathbf{C}$ and $i \in \mathcal{I}_\mathbf{C}$; and $\mathcal{I}_\mathbf{C}$ is a partial order. The morphisms in $\mathcal{I}_\mathbf{C}$ are called *inclusions*.

Categories with inclusions and inclusion-preserving functors form a category which we shall denote by $\mathbf{ICat}$. By a "discretization" functor for categories with inclusions, we shall mean a functor $\mathbf{Inc} : \mathbf{ICat} \to \mathbf{ICat}$, mapping every $\mathbf{C}$, to $\mathcal{I}_\mathbf{C}$, and every inclusion preserving F, to its appropriate restriction.

We shall also need a category of $\mathbf{Set}$-*diagrams with variable shape*. It will be denoted by $\mathbf{Func}_{\mathbf{ICat}}(\mathbf{Set})$. Its objects are inclusion-preserving functors $F : \mathbf{C} \to \mathbf{Set}$ ($\mathbf{Set}$ is an obvious example of a category with inclusions). A morphism from a functor $F : \mathbf{C} \to \mathbf{Set}$ to a functor $F' : \mathbf{C}' \to \mathbf{Set}$ is a pair $\langle P, \alpha \rangle$ such that $P : \mathbf{C} \to \mathbf{C}'$ is an inclusion-preserving functor, and $\alpha : F \dot{\to} P; F'$ is a natural transformation. Let $\langle P, \alpha\rangle : F \to G$ and $\langle Q, \beta\rangle : G \to H$ be two morphisms in $\mathbf{Func_{ICat}}(\mathbf{Set})$. Their composition is defined as $\langle P; Q, \alpha; (P \cdot \beta)\rangle$. There is an obvious projection functor from $\mathbf{Func_{ICat}}(\mathbf{Set})$ to $\mathbf{ICat}$, which we shall denote by π_{dom}.

Indexed and Sorted Sets. Let S be a set. It is easy to see that the category of *S-indexed sets* $\mathbf{ISet}[S]$, with S-indexed families of sets—$\langle X_s\rangle_{s \in S}$—as objects, and S-indexed functions, i.e., families $\langle h_s : X_s \to Y_s\rangle_{s \in S}$—as morphisms, is a category with inclusions. Moreover, indexed sets form an *indexed category* $\mathbf{ISet} : \mathbf{Set}^{op} \to \mathbf{ICat}$ (cf. [16]).

In what follows, we shall also need a "discrete version" of $\mathbf{ISet}$, which we shall denote by $\mathbf{iset}$. For any set S, $\mathbf{iset}(S)$, gives the *class* of all S-indexed sets. The reduct functor $\mathbf{ISet}_f$, induced by $f : S \to S'$, by just forgetting its action on arrows (i.e., S'-indexed functions), becomes a *reduct function*—$\mathbf{iset}_f : \mathbf{iset}(S') \to \mathbf{iset}(S)$.

For every $S \in |\mathbf{Set}|$, the full subcategory of $\mathbf{ISet}[S]$ determined by the following condition:

$$\forall X \in |\mathbf{SSet}[S]| \ \ \forall s_1, s_2 \in S \ \ s_1 \neq s_2 \Rightarrow X_{s_1} \cap X_{s_2} = \emptyset$$

will be called the category of *S-sorted sets*, and denoted by $\mathbf{SSet}[S]$.

The correspondence $S \mapsto \mathbf{SSet}[S]$, in an obvious way, extends to a functor $\mathbf{SSet} : \mathbf{Set} \to \mathbf{ICat}$.

Universal Algebra. Throughout the paper we shall use the standard notions and notation for (many-sorted) *algebraic signatures*, (many-sorted) *algebras*, and their morphisms. The category of algebraic signatures will be denoted by $\mathbf{AlgSig}$. The indexed functor "constructing" categories of algebras will be denoted by $\mathbf{Alg} : \mathbf{AlgSig}^{op} \to \mathbf{Cat}$.

For any signature $\Sigma = \langle S, \Omega\rangle$, and every S-indexed set X, the *free Σ-algebra* over X, will be denoted by $T_\Sigma(X)$.

Let *Voc* be an infinite set—the *vocabulary of variable names*, and let $\Sigma = \langle S, \Omega\rangle$ be an algebraic signature. By a *category of Σ-substitutions* we shall mean a category $\mathbf{T}_\Sigma$, having S-sorted sets of elements of *Voc* as objects, and functions $f : X \to |T_\Sigma(Y)|$, as morphisms from X to Y. Composition in $\mathbf{T}_\Sigma$, for $f : X \to Y$ and $g : Y \to Z$, is given by $f; g \stackrel{df}{=} f; |g^\sharp|$, where $g^\sharp$ is the *free extension* of g, and $|\_|$ is the *forgetful functor* from $\mathbf{Alg}(\Sigma)$, to $\mathbf{ISet}(S)$.

It is not difficult to check that for every signature Σ, $\mathbf{T}_\Sigma$ is a category with inclusions. Using the idea of "corresponding assignments", introduced in [17], one can show as well, that the construction of $\mathbf{T}_\Sigma$ can be extended to a functor $\mathbf{T} : \mathbf{AlgSig} \to \mathbf{ICat}$.

3 Context Institutions

Institutions, introduced by Goguen and Burstall (cf. [6], [8]), represent an abstract *model-theoretic* view of logics. The idea, inspired by the work of Barwise—[1], turned out to be a very useful basis for considering specification formalisms.

Context institutions, introduced in [13], are institution-like structures, with notions of *context*, and *substitution* built-in. Additionally, for every signature, the class of models for this signature is concrete over certain class of *indexed sets*, and consequently, there is a notion of a *carrier* for each model. Contexts are abstract objects and all what is assumed about them, is that they in some technical sense "contain" *sorted sets* of variables. Having variables and carriers it is quite natural to talk about *valuations*. The *satisfaction relation* in the case of context institutions takes these valuations into account. Also *open formulae* arise in a natural way—they are just formulae built *over a context*. Substitutions are modelled by *context morphisms*. They induce appropriate translations of sets of formulae, which "perform" the substitution. All the components are tight together by three conditions, which ensure that the whole structure behaves in a "smooth" way.

In this paper, for simplicity, we shall consider context institutions with "discrete model functors", i.e., giving *classes of models* instead of *categories of models*, for each signature. Everything we are going to present however, can be rephrased in the general setting as well. As a consequence of the simplification, some parts of the definition of context institution simplify as well—e.g., some naturality conditions become trivial, and disappear.

Contexts and Formulae. In a context institution $\mathfrak{C}$, the vocabulary of a logic in question will be modelled by a suitable category of *signatures* $\mathbf{Sig}^{\mathfrak{C}}$.

For every signature Σ, we want to consider (possibly *open*) *formulae* over Σ. Therefore we shall assume that for every Σ, there is a category of Σ*-contexts* $\mathbf{Ctxt}^{\mathfrak{C}}_{\Sigma}$. Context morphisms are meant to model *substitutions*.

The fact that for every Σ-context we have a corresponding set of formulae (built "over" that context), will be modelled by a functor $\mathbf{Frm}^{\mathfrak{C}}_{\Sigma} : \mathbf{Ctxt}_{\Sigma} \to \mathbf{Set}$. For any context morphism (substitution), its image under $\mathbf{Frm}^{\mathfrak{C}}_{\Sigma}$, is a function "performing" the substitution. It seems natural to require that if a certain context is "included" in some other one, then the corresponding sets of formulae are also related via (set-theoretic) inclusion. In other words—$\mathbf{Ctxt}^{\mathfrak{C}}_{\Sigma}$, has to be a *category with inclusions*, and the functor $\mathbf{Frm}^{\mathfrak{C}}_{\Sigma}$, has to *preserve* them.

To take the change of notation into account, we shall eventually define the *formula functor* as a functor $\mathbf{Frm}^{\mathfrak{C}} : \mathbf{Sig}^{\mathfrak{C}} \to \mathbf{Func}_{\mathbf{ICat}}(\mathbf{Set})$.

Carriers, Variables and Valuations. The notion of a *model functor* for context institutions is exactly the same as for ordinary ones. Additionally, we assume that models are equipped with *carriers*. This is done by assuming that with every signature Σ, we can associate a set $\mathbf{Srt}^{\mathfrak{C}}_{\Sigma}$, called its *set of sorts*. We require this correspondence to be functorial. The association of carriers to models is given by a function $\mathbf{Carr}^{\mathfrak{C}}_{\Sigma} : \mathbf{Mod}^{\mathfrak{C}}_{\Sigma} \to \mathbf{iset}(\mathbf{Srt}^{\mathfrak{C}}_{\Sigma})$. Signature morphisms induce translations of model classes (going in the opposite direction). Via the sort functor $\mathbf{Srt}^{\mathfrak{C}}$, they also induce appropriate reduct functions between classes of indexed sets. It seems reasonable to require that both these semantic translations agree, i.e., that "the carrier of the reduct of a model is the reduct of its carrier". In other words $\mathbf{Carr}^{\mathfrak{C}}$ has to be a natural transformation.

The Reader may think of contexts as "properly decorated" sorted sets of variables. This idea is formalized by requiring, that for every signature Σ, there is an inclusion preserving functor from (the discretization of) the category of Σ-contexts $\mathbf{Ctxt}^{\mathfrak{C}}_{\Sigma}$, to the category of $\mathbf{Srt}^{\mathfrak{C}}_{\Sigma}$-sorted sets. We shall further require this interpretation to be natural with respect to change of notation (i.e., signature morphisms). The corresponding natural transformation will be denoted by $\mathbf{Var}^{\mathfrak{C}}$.

For every signature Σ, every Σ-context Γ, and any Σ-model M, we define the set of *valuations of* Γ *in* M—denoted by $\mathbf{Val}^{\mathfrak{C}}_{\Sigma}(\Gamma, M)$, to be the set of all $\mathbf{Srt}^{\mathfrak{C}}_{\Sigma}$-indexed functions from $\mathbf{Var}^{\mathfrak{C}}_{\Sigma}(\Gamma)$ to $\mathbf{Carr}^{\mathfrak{C}}_{\Sigma}(M)$.

We assume, that for every substitution (i.e., context morphism) $f : \Gamma \to \Delta$, there is a corresponding (partial) translation of valuations, going in the opposite direction—from valuations of Δ to valuations of Γ. In practical examples the resulting valuation is obtained by "evaluating" the "terms" from the codomain of the substitution.

In [17], a notion of *corresponding assignments* has been introduced as a tool for proving satisfaction condition in several examples of (ordinary) institutions. In the framework of context institutions, it turns out that for any signature morphism $\sigma : \Sigma \to \Sigma'$, any Σ-context Γ, and any Σ'-model M', there is a bijection:

$$\sigma^{\mathrm{Val}}_{\Gamma,M'} : \mathbf{Val}^{\mathfrak{C}}_{\Sigma'}(\mathbf{Ctxt}^{\mathfrak{C}}_{\sigma}(\Gamma), M') \to \mathbf{Val}^{\mathfrak{C}}_{\Sigma}(\Gamma, \mathbf{Mod}^{\mathfrak{C}}_{\sigma}(M')).$$

Valuations: $v \in \mathbf{Val}^{\mathfrak{C}}_{\Sigma'}(\mathbf{Ctxt}^{\mathfrak{C}}_{\sigma}(\Gamma), M')$, and $\bar{v} \in \mathbf{Val}^{\mathfrak{C}}_{\Sigma}(\Gamma, \mathbf{Mod}^{\mathfrak{C}}_{\sigma}(M'))$, such that $\bar{v} = \sigma^{\mathrm{Val}}_{\Gamma,M'}(v)$, shall be called *$\sigma$-adjoint.*

Assuming the meaning of $\mathbf{Ctxt}^{\mathfrak{C}}$ and $\mathbf{Val}^{\mathfrak{C}}$ as introduced above, we have:

Definition 1. A *context institution* $\mathfrak{C}$ consists of the following data:

- a category $\mathbf{Sig}^{\mathfrak{C}}$ of *signatures*
- a functor $\mathbf{Frm}^{\mathfrak{C}} : \mathbf{Sig}^{\mathfrak{C}} \to \mathbf{Func}_{\mathbf{ICat}}(\mathbf{Set})$ called a *formula functor.*
- a functor $\mathbf{Mod}^{\mathfrak{C}} : (\mathbf{Sig}^{\mathfrak{C}})^{op} \to \mathbf{Class}$ called a *model functor.*
- a functor $\mathbf{Srt}^{\mathfrak{C}} : \mathbf{Sig}^{\mathfrak{C}} \to \mathbf{Set}$ called a *sort functor.*
- a natural transformation: $\mathbf{Var}^{\mathfrak{C}} : \mathbf{Ctxt}^{\mathfrak{C}}; \mathbf{Inc} \to \mathbf{Srt}^{\mathfrak{C}}; \mathbf{SSet}$,
- a natural transformation $\mathbf{Carr}^{\mathfrak{C}} : \mathbf{Mod}^{\mathfrak{C}} \to (\mathbf{Srt}^{\mathfrak{C}})^{op}; \mathbf{iset}$,
- for every $\Sigma \in |\mathbf{Sig}^{\mathfrak{C}}|$, $M \in \mathbf{Mod}^{\mathfrak{C}}_{\Sigma}$, $\Gamma \in |\mathbf{Ctxt}^{\mathfrak{C}}_{\Sigma}|$ a *satisfaction relation:*

$$M[\,\text{-}\,] \models^{\mathfrak{C}}_{\Sigma,\Gamma} \text{-} \;\subseteq\; \mathbf{Val}^{\mathfrak{C}}_{\Sigma}(\Gamma, M) \times \mathbf{Frm}^{\mathfrak{C}}_{\Sigma}(\Gamma)$$

- for every $\Sigma \in |\mathbf{Sig}^{\mathfrak{C}}|$, any Σ-model M, and every Σ-context morphism $f : \Gamma \to \Delta$, a partial function[1], $f^{\mathrm{Val}}_{M} : \mathbf{Val}^{\mathfrak{C}}_{\Sigma}(\Delta, M) \rightharpoonup \mathbf{Val}^{\mathfrak{C}}_{\Sigma}(\Gamma, M)$,

such that the following three conditions hold:

[1] In [13], f^{Val}_{M} (which was a component of a natural transformation $f^{\mathrm{Val}}_{\_}$, there), was *total*, what turned out to be too restrictive in some cases.

- *Substitution condition:* for any Σ-context morphism $f : \Gamma \to \Delta$, any Σ-formula $\phi \in \mathbf{Frm}^{\mathfrak{C}}_{\Sigma}(\Gamma)$, every model $M \in \mathbf{Mod}^{\mathfrak{C}}_{\Sigma}$, and every valuation $v \in \mathbf{Val}^{\mathfrak{C}}_{\Sigma}(\Delta, M)$, such that $v \in \mathbf{dom}(f^{\mathbf{Val}}_{M})$:

$$M[v] \models^{\mathfrak{C}}_{\Sigma,\Delta} \mathbf{Frm}^{\mathfrak{C}}_{\Sigma}(f)(\phi) \;\Leftrightarrow\; M[f^{\mathbf{Val}}_{M}(v)] \models^{\mathfrak{C}}_{\Sigma,\Gamma} \phi.$$

- *Satisfaction condition:* for every signature morphism $\sigma : \Sigma \to \Sigma'$, every model $M' \in \mathbf{Mod}^{\mathfrak{C}}_{\Sigma'}$, any context $\Gamma \in |\mathbf{Ctxt}^{\mathfrak{C}}_{\Sigma}|$, any σ-adjoint valuations v and $\bar{v}$, and every formula $\phi \in \mathbf{Frm}^{\mathfrak{C}}_{\Sigma}(\Gamma)$:

$$M'[v] \models^{\mathfrak{C}}_{\Sigma',\mathbf{Ctxt}^{\mathfrak{C}}_{\sigma}(\Gamma)} \mathbf{Frm}^{\mathfrak{C}}_{\sigma}(\Gamma)(\phi) \;\Leftrightarrow\; \mathbf{Mod}^{\mathfrak{C}}_{\sigma}(M')[\bar{v}] \models^{\mathfrak{C}}_{\Sigma,\Gamma} \phi$$

- *Coherence condition:* for any signature Σ, any Σ-context morphism $f : \Gamma \to \Delta$, any signature morphism $\sigma : \Sigma \to \Sigma'$, and any Σ'-model M', the following diagram commutes in **PSet** (the category of sets and partial functions):

$$\begin{array}{ccc}
\mathbf{Val}^{\mathfrak{C}}_{\Sigma'}(\mathbf{Ctxt}^{\mathfrak{C}}_{\sigma}(\Delta), M') & \xrightarrow{\mathbf{Ctxt}^{\mathfrak{C}}_{\sigma}(f)^{\mathbf{Val}}_{M'}} & \mathbf{Val}^{\mathfrak{C}}_{\Sigma'}(\mathbf{Ctxt}^{\mathfrak{C}}_{\sigma}(\Gamma), M') \\
\Big\downarrow{\scriptstyle \sigma^{\mathbf{Val}}_{\Delta,M'}} & & \Big\downarrow{\scriptstyle \sigma^{\mathbf{Val}}_{\Gamma,M'}} \\
\mathbf{Val}^{\mathfrak{C}}_{\Sigma}(\Delta, \mathbf{Mod}^{\mathfrak{C}}_{\sigma}(M')) & \xrightarrow[f^{\mathbf{Val}}_{\mathbf{Mod}^{\mathfrak{C}}_{\sigma}(M')}]{} & \mathbf{Val}^{\mathfrak{C}}_{\Sigma}(\Gamma, \mathbf{Mod}^{\mathfrak{C}}_{\sigma}(M'))
\end{array}$$

4 Meta-Framework

In the present section we shall introduce structures, which will play the role of a *meta-framework* for representing both syntax and semantics of logics presented by context parchments.

Meta-Signatures

Let us start by defining a category of *pointed algebraic signatures*, $\mathbf{AlgSig}^{\star}$. Objects in this category will be algebraic signatures $\langle S, \Omega \rangle$, such that S is a *pointed set*, i.e., a set with a distinguished element, which we shall always denote by $\star$. A morphism $\sigma : \langle S, \Omega \rangle \to \langle S', \Omega' \rangle$, in $\mathbf{AlgSig}^{\star}$, is an (ordinary) algebraic signature morphism, such that $\sigma_{\mathsf{srts}}(s) = \star$, iff $s = \star$.

Definition 2. A *meta-signature* Π is a quadruple $\langle S, \Omega, V, \mathbb{Q} \rangle$, such that $\langle S, \Omega \rangle$ is a pointed algebraic signature, $V \subseteq S \setminus \{\star\}$, and $\mathbb{Q}$ is a set.

Definition 3. A *meta-signature morphism* from $\langle S, \Omega, V, \mathbb{Q} \rangle$ to $\langle S', \Omega', V', \mathbb{Q}' \rangle$, is a pair $\langle \sigma, f \rangle$, such that $\sigma : \langle S, \Omega \rangle \to \langle S', \Omega' \rangle$, is a pointed signature morphism, such that $\sigma_{\mathsf{srts}}[V] \subseteq V'$, and $f : \mathbb{Q} \to \mathbb{Q}'$ is a function.

Meta-signatures and their morphisms in an obvious way constitute a category, which we shall denote by **MSig**.

Similarly, as in the case of algebraic, and relational signatures we can show that:

Proposition 4. *The category of meta-signatures* **MSig** *is cocomplete.* □

Meta-Structures

Definition 5. A *meta-structure* over a signature $\Pi = \langle S, \Omega, V, \mathbb{Q} \rangle$, is a quadruple $\langle \mathcal{A}, V^{\mathcal{A}}, D^{\mathcal{A}}, \mathbb{Q}^{\mathcal{A}} \rangle$, such that:

- $\mathcal{A}$ is a pointed $\langle S, \Omega \rangle$-algebra,
- $V^{\mathcal{A}}$ is a V-indexed set, such that for every $s \in V$, $|V^{\mathcal{A}}|_s \subseteq |\mathcal{A}|_s$,
- $D^{\mathcal{A}} \subseteq |\mathcal{A}|_\star$,
- $\mathbb{Q}^{\mathcal{A}} = \{Q^{\mathcal{A}} : \mathcal{P}(|\mathcal{A}|_\star) \rightharpoonup |\mathcal{A}|_\star \mid Q \in \mathbb{Q}\}$.

We shall quite often simply write $\mathcal{A}$, instead of $\langle \mathcal{A}, V^{\mathcal{A}}, D^{\mathcal{A}}, \mathbb{Q}^{\mathcal{A}} \rangle$, where it will not lead to confusion. The set $|\mathcal{A}|$ will be called the *carrier* of $\mathcal{A}$, the subset $V^{\mathcal{A}}$—the set of *assignable values* of $\mathcal{A}$, and the set $D^{\mathcal{A}}$—the set of *designated values.* Instead of saying "a meta-structure over a signature Π" we shall say simply Π-structure.

Note, that apart from the interpretation of the $\mathbb{Q}$-symbols and the distinguished subsets of *assignable* and *designated values*, Π-structures are just (pointed) algebras. We intend to use Π-structures for semantic interpretation of logical syntax. The distinguished sort $\star$, will correspond to the space of *truth values.* The functions interpreting symbols from $\mathbb{Q}$ (sometimes called *generalized operations*), shall be used for interpreting quantifier symbols. The idea of using generalized operations for interpreting quantifiers is inspired by the work on algebraization of logic, by Rasiowa and Sikorski—[15], [14] (although we use it in a slightly different context).

Definition 6. Let $\mathcal{A}$ and $\mathcal{A}'$ be two structures over a signature $\Pi = \langle S, \Omega, V, \mathbb{Q} \rangle$. A *$\Pi$-structure morphism* $g : \mathcal{A} \to \mathcal{A}'$ is a pointed $\langle S, \Omega \rangle$-homomorphism $g : \mathcal{A} \to \mathcal{A}'$ such that the following three conditions are satisfied:

- for every symbol $Q \in \mathbb{Q}$:
  $O \in \mathbf{dom}(Q^{\mathcal{A}}) \Rightarrow g[O] \in \mathbf{dom}(Q^{\mathcal{A}'}) \wedge g(Q^{\mathcal{A}}(O)) = Q^{\mathcal{A}'}(g[O])$,
- $g[V^{\mathcal{A}}] \subseteq V^{\mathcal{A}'}$,
- $g[D^{\mathcal{A}}] \subseteq D^{\mathcal{A}'}$.

The notion of a Π-structure morphism, as defined, corresponds to a notion of homomorphism between partial algebras, as described in [2]. Therefore instead of Π-structure morphism (or Π meta-structure morphism), we shall simply say Π-homomorphism.

Similarly to the case of ordinary (partial) algebras, we can introduce notions of *congruence* and *quotient structure.*

For every meta-signature Π, the class of all Π-structures together with their morphisms constitute a category which we shall denote by $\mathbf{MStr}_\Pi$. Similarly as in the algebraic case, for any meta-signature morphism $\pi = \langle\sigma, f\rangle : \Pi \to \Pi'$, there is an obvious *reduct functor* $\_|_\pi : \mathbf{MStr}_{\Pi'} \to \mathbf{MStr}_\Pi$.

Using the notion of quotient structure, in a way very similar to the (partial) algebraic case, we can show the following two "standard propositions":

Proposition 7. *For every signature $\Pi = \langle S, \Omega, V, \mathbb{Q}\rangle$, the category $\mathbf{MStr}_\Pi$ is cocomplete.* □

Proposition 8. *For any meta-signature morphism $\pi : \Pi \to \Pi'$, the reduct functor $\_|_\pi : \mathbf{MStr}_{\Pi'} \to \mathbf{MStr}_\Pi$, has a left adjoint.* □

5 Context Parchments

5.1 Definition

Originally, parchments were introduced merely, as a tool for "getting the satisfaction condition for free" (cf. [7]). Later on, it became clear, that for certain purposes, such as "modular construction" of logics, or "combining" logics, institutions are not adequate, and that parchments may be used instead (cf. [10], [11]). The main advantage of parchments over institutions is that they describe certain aspects of logical systems in a *generative* way.

The notion of *context parchment* is meant to play the same role for *context institutions* (cf. [13]), as the notion of *parchment* plays for ordinary institutions. Actually, context parchments take what is called λ-*parchments* in [11] as a starting point. λ-parchments shift some foundational difficulties present in the original parchment definition, allowing for a better separation of the "object" and "meta" frameworks.

Definition 9. A *context parchment* $\mathfrak{P}$ consists of the following data:

- a category of (abstract) signatures $\mathbf{Sig}^{\mathfrak{P}}$,
- a *language* functor $\mathbf{Lan}^{\mathfrak{P}} : \mathbf{Sig}^{\mathfrak{P}} \to \mathbf{MSig}$,
- a *universal language* signature $\mathbb{L}^{\mathfrak{P}} \in |\mathbf{MSig}|$,
- a *model* functor $\mathbf{Mod}^{\mathfrak{P}} : (\mathbf{Sig}^{\mathfrak{P}})^{op} \to \mathbf{Class}$,
- a natural transformation $\mathbf{mod}^{\mathfrak{P}} : \mathbf{Mod}^{\mathfrak{P}} \dot{\to} \hom(\mathbf{Lan}^{\mathfrak{P}}(\_), \mathbb{L}^{\mathfrak{P}})$,
- a *semantic universe* $\mathbb{G}^{\mathfrak{P}} \in |\mathbf{MStr}(\mathbb{L}^{\mathfrak{P}})|$.

We shall call $\mathfrak{P}$, *logical*, if all the generalized operations in $\mathbb{G}^{\mathfrak{P}}$ are total.

Observe that the definition looks almost identical to the definition of λ-parchment in [11]. The only difference is that we use meta-signatures and meta-structures instead of ordinary algebraic signatures and algebras for semantic interpretation.

It is easy to see, that every context parchment $\mathfrak{P}$, defines a "natural inclusion" $\iota : \mathbf{ValSrt}^{\mathfrak{P}} \dot{\to} \mathbf{AllSrt}^{\mathfrak{P}}$, in $[\mathbf{Sig}^{\mathfrak{P}} \to \mathbf{Set}]$, where for $\langle S, \Omega, V, \mathbb{Q}\rangle = \mathbf{Lan}^{\mathfrak{P}}_\Sigma$, $\mathbf{ValSrt}^{\mathfrak{P}}_\Sigma \stackrel{df}{=} V$, and $\mathbf{AllSrt}^{\mathfrak{P}}_\Sigma \stackrel{df}{=} S$. Functors $\mathbf{ValSrt}^{\mathfrak{P}}$, and $\mathbf{AllSrt}^{\mathfrak{P}}$ will be called the *value sorts* functor, and *all sorts* functor for $\mathfrak{P}$, respectively. The natural transformation ι consists of the appropriate inclusions $V \hookrightarrow S$.

Intuitively speaking, for the logic presented by $\mathfrak{P}$, the natural inclusion ι, determines which sorts may be ranged over by variables.

5.2 From Context Parchments to Context Institutions

In this section we shall describe a construction, which for any *logical context parchment* $\mathfrak{P}$, yields a *context institution* $\mathcal{I}(\mathfrak{P})$. As some parts of the construction are rather tedious, and slightly complicated notationally (although not difficult), we shall present here only the main points, without going into the details.

Signatures, Models, Sorts, and Carriers.

- As a category of signatures of $\mathcal{I}(\mathfrak{P})$, we take the category of signatures of $\mathfrak{P}$—$\mathbf{Sig}^{\mathfrak{P}}$.
- Situation is equally simple with the model functor for $\mathcal{I}(\mathfrak{P})$—we can simply take $\mathbf{Mod}^{\mathfrak{P}}$.
- As the sort functor for $\mathcal{I}(\mathfrak{P})$—$\mathbf{Srt}^{\mathcal{I}(\mathfrak{P})}$, we take $\mathbf{ValSrt}^{\mathfrak{P}}$, as defined above.
- The $\mathbf{Carr}^{\mathcal{I}(\mathfrak{P})}$ natural transformation, for any signature Σ, and every Σ-model M, gives the assignable values of $\mathbf{mod}^{\mathfrak{P}}_{\Sigma}(M)$-reduct of the semantic universe $\mathbb{G}^{\mathfrak{P}}$. More precisely:

$$\mathbf{Carr}^{\mathcal{I}(\mathfrak{P})}_{\Sigma}(M) \stackrel{df}{=} \{\, |\mathbb{G}^{\mathfrak{P}}|_{\mathbf{mod}^{\mathfrak{P}}_{\Sigma}(M)}|_s \mid s \in \mathbf{ValSrt}^{\mathfrak{P}}_{\Sigma} \}.$$

Variables, Contexts and Substitutions. For every (abstract) signature Σ, let us consider all finite, $\mathbf{ValSrt}^{\mathfrak{P}}_{\Sigma}$-sorted sets of elements of a fixed vocabulary of variable names *Voc*. Every such a set X, can be "extended" to $\mathbf{AllSrt}^{\mathfrak{P}}_{\Sigma}$-sorted $\overline{X}$, by taking $|\overline{X}|_s = |X|_s$ provided that $s \in \mathbf{ValSrt}^{\mathfrak{P}}_{\Sigma}$, and $|\overline{X}|_s = \emptyset$ otherwise. Let $\mathbf{Lan}^{\mathfrak{P}}_{\Sigma} = \langle S, \Omega, V, \mathbb{Q} \rangle$. As the category $\mathbf{Ctxt}^{\mathfrak{P}}_{\Sigma}$ we shall take the full subcategory of the "usual" category of substitutions—$\mathbf{T}_{\langle S,\Omega \rangle}$, generated by the above class of objects. It means that Σ-context morphisms will be $\mathbf{AllSrt}^{\mathfrak{P}}_{\Sigma}$-indexed functions of the form: $f : \overline{X} \to |T_{\langle S,\Omega \rangle}(\overline{Y})|$. It is not difficult to see that the construction of $\mathbf{Ctxt}^{\mathfrak{P}}$ is indeed functorial.

The natural transformation $\mathbf{Var}^{\mathcal{I}(\mathfrak{P})}$, is determined by putting for any signature Σ, and every Σ-context $\overline{X}$—$\mathbf{Var}^{\mathcal{I}(\mathfrak{P})}_{\Sigma}(\overline{X}) \stackrel{df}{=} X$.

Valuations. For any signature Σ, any Σ-context $\overline{X}$, and any Σ-model M, $\mathbf{Val}^{\mathcal{I}(\mathfrak{P})}_{\Sigma}(\overline{X}, M) = [X \to \mathbf{Carr}^{\mathcal{I}(\mathfrak{P})}_{\Sigma}(M)]$. Now, for an arbitrary Σ-substitution (context morphism) $f : \overline{X} \to |T_{\langle S,\Omega \rangle}(\overline{Y})|$, let us define $f^{\mathsf{Val}}_M : \mathbf{Val}^{\mathcal{I}(\mathfrak{P})}_{\Sigma}(\overline{Y}, M) \rightharpoonup \mathbf{Val}^{\mathcal{I}(\mathfrak{P})}_{\Sigma}(\overline{X}, M)$. Value of $f^{\mathsf{Val}}_M(v)$ is undefined iff for some $s \in \mathbf{Srt}^{\mathcal{I}(\mathfrak{P})}_{\Sigma}$, and $x \in |X|_s$, $v^{\sharp}(f(x)) \notin |\mathbf{Carr}^{\mathcal{I}(\mathfrak{P})}_{\Sigma}(M)|_s$. Otherwise—$f^{\mathsf{Val}}_M(v)$ is defined, and equals $v^{\sharp}(f(x))$, for any $x \in X$.

Algebras of Σ-Evaluators. Let Σ be an arbitrary signature from $\mathbf{Sig}^{\mathfrak{P}}$, and $\mathbf{Lan}^{\mathfrak{P}}_{\Sigma} = \langle S, \Omega, V, \mathbb{Q}\rangle$. Let $\mathbf{EvSig}_{\Sigma} = \langle S^{\mathrm{Ev}(\Sigma)}, \Omega^{\mathrm{Ev}(\Sigma)}\rangle$, be an algebraic signature defined as follows:

- *Sorts.* $S^{\mathrm{Ev}(\Sigma)} \stackrel{df}{=} |\mathbf{Ctxt}^{\mathcal{I}(\mathfrak{P})}_{\Sigma}| \times S$,
- *Operations symbols.* Let $\Omega^{\mathrm{Ev}(\Sigma)}$, be the smallest (wrt. $\subseteq$), $S^{\mathrm{Ev}(\Sigma)}$-indexed set, satisfying the following conditions:
  - $\Omega_{s_1 \ldots s_n \to s} \subseteq \Omega^{\mathrm{Ev}(\Sigma)}_{\langle \overline{X}, s_1\rangle \ldots \langle \overline{X}, s_n\rangle \to \langle \overline{X}, s\rangle}$, for any Σ-context $\overline{X}$,
  - $\{\, Qx{:}s \mid Q \in \mathbb{Q}\} \subseteq \Omega^{\mathrm{Ev}(\Sigma)}_{\langle \overline{X}, \star\rangle \to \langle \overline{Y}, \star\rangle}$, for any Σ-contexts $\overline{X}$, and $\overline{Y}$, and every x, such that $X \setminus Y = \{x\}_s$.

Using the functoriality of $\mathbf{Lan}^{\mathfrak{P}}$, and $\mathbf{Ctxt}^{\mathcal{I}(\mathfrak{P})}$, it is easy to see that the above construction, actually yields a functor $\mathbf{EvSig} : \mathbf{Sig}^{\mathfrak{P}} \to \mathbf{AlgSig}$.

For any Σ-model M, we shall now construct an algebra $\mathrm{Eval}_{\Sigma}(M)$, over $\mathbf{EvSig}_{\Sigma}$, which we shall call the *algebra of Σ-evaluators for M*.

- *Carriers.* For any $\langle \overline{X}, s\rangle \in S^{\mathrm{Ev}(\Sigma)}$.
  $|\mathrm{Eval}_{\Sigma}(M)|_{\langle \overline{X}, s\rangle} \stackrel{df}{=} [\, \mathbf{Val}^{\mathcal{I}(\mathfrak{P})}_{\Sigma}(\overline{X}, M) \to |\mathbb{G}(M)|_s \,]$, where $\mathbb{G}(M)$, denotes $\mathbb{G}|_{\mathbf{mod}^{\mathfrak{P}}_{\Sigma}(M)}$.
- *Operations.*
  - For any $\mathrm{op} : \langle \overline{X}, s_1\rangle \ldots \langle \overline{X}, s_n\rangle \to \langle \overline{X}, s\rangle$,
    $\mathrm{op}^{\mathrm{Eval}_{\Sigma}(M)}(f_1, \ldots, f_n)(v) \stackrel{df}{=} \mathrm{op}^{\mathbb{G}(M)}(f_1(v), \ldots, f_n(v))$.
  - For any $(Qx{:}s) : \langle \overline{X}, \star\rangle \to \langle \overline{Y}, \star\rangle$,
    $(Qx{:}s)^{\mathrm{Eval}_{\Sigma}(M)}(f)(v) \stackrel{df}{=} Q^{\mathbb{G}(M)}(\{\, f(v^x) \mid v^x \in \mathbf{Val}^{\mathcal{I}(\mathfrak{P})}_{\Sigma}(\overline{X}, M),\ \text{s.t. } v^x\!\restriction_Y = v\,\})$

Σ-Formulae and Satisfaction. Let us define a *universe of variables* for Σ, which we denote by $\mathcal{V}_{\Sigma}$. It is an $S^{\mathrm{Ev}(\Sigma)}$-indexed set, such that for any $\langle \overline{X}, s\rangle \in S^{\mathrm{Ev}(\Sigma)}$, $|\mathcal{V}_{\Sigma}|_{\langle \overline{X}, s\rangle} \stackrel{df}{=} |X|_s$.

In the case of context institutions, the set of all Σ-formulae is stratified into "layers", indexed by contexts. One might think, that for each Σ-context $\overline{X}$, as the set $\mathbf{Frm}^{\mathcal{I}(\mathfrak{P})}_{\Sigma}(\overline{X})$, we could take the set of terms $|T_{\mathbf{EvSig}_{\Sigma}}(\mathcal{V}_{\Sigma})|_{\langle \overline{X}, \star\rangle}$. However, our "pre-syntax" includes *variable binding operations*—all the operations of the form "$Qx{:}s$". Therefore, to be able to actually "perform" *substitutions* on formulae, we need to consider elements of $|T_{\mathbf{EvSig}_{\Sigma}}(\mathcal{V}_{\Sigma})|$, up to an appropriate notion of *α-conversion* (i.e., renaming of *bound variables*). As it turns out, the α-convertibility relation $\sim_{\alpha}$, is a congruence on the algebra $T_{\mathbf{EvSig}_{\Sigma}}(\mathcal{V}_{\Sigma})$. Moreover, for any Σ-model M, this congruence is included in the kernel of the unique homomorphism $(\_)^{\heartsuit} : T_{\mathbf{EvSig}_{\Sigma}}(\mathcal{V}_{\Sigma}) \to \mathrm{Eval}_{\Sigma}(M)$, extending the *canonical interpretation* of $\mathcal{V}_{\Sigma}$ in $\mathrm{Eval}_{\Sigma}(M)$, given by $x \mapsto \lambda \mathrm{env}.\mathrm{env}(x)$.. Hence, as the set of Σ-formulae over $\overline{X}$, we take $|T_{\mathbf{EvSig}_{\Sigma}}(\mathcal{V}_{\Sigma})/\!\sim_{\alpha}|_{\langle \overline{X}, \star\rangle}$. Observe, that for any signature morphism $\sigma : \Sigma \to \Sigma'$, the universe of variables for Σ, is simply included in the corresponding reduct of $\mathcal{V}_{\Sigma'}$. Therefore, we can translate Σ-formulae *along σ*. We shall skip the details of the above construction—both because of the lack of space, and because they are fairly "standard". Let us instead, proceed directly to the definition of the satisfaction relation for $\mathcal{I}(\mathfrak{P})$.

For an arbitrary Σ-model M, the satisfaction $M[\text{-}] \models^{\mathcal{I}(\mathfrak{P})}_{\Sigma,\overline{X}}$ -, is a binary relation between the set of *valuations* $\mathbf{Val}^{\mathcal{I}(\mathfrak{P})}(\overline{X}, M)$, and the set of formulae $\mathbf{Frm}^{\mathcal{I}(\mathfrak{P})}_{\Sigma}(\overline{X})$. It is defined as follows:

$$M[v] \models^{\mathfrak{C}}_{\Sigma,\Gamma} [\phi]_{\sim_\alpha} \text{ iff } (\phi)^{\heartsuit}(v) \in D^{\mathbb{G}(M)}$$

i.e., valuation v satisfies ϕ in $\mathcal{I}(\mathfrak{P})$, if and only if, the semantic interpretation of ϕ under v, belongs to the set of *designated values* of $\mathbb{G}^{\mathfrak{P}}$.

It is relatively easy, although quite tedious, to check that the components described above indeed define a context institution. Again, due to the lack of space we shall not do it here.

Let us observe instead, that the whole construction "mimics" quite closely the usual construction of an institution out of a (λ-)parchment (cf. [10], [11]). What we gain by using context parchments, apart from the richer structure of a context institution, is certain *uniformity* and *economy* of presentation of the logic in question. For example, a typical (λ-)parchment, already "contains" all possible *evaluator algebras* $\text{Eval}_{\Sigma}(M)$, as a "part" of the semantic universe. In the case of context parchments, evaluator algebras are uniformly generated in the process of constructing (context) institution.

5.3 Examples of Context Parchments

In this section we would like to present some examples of context parchments. All (well—almost all) of them will be based on the same "relational base"—relational signatures as a category of (abstract) signatures, and relational structures as a model functor. To set up the ground let us start with some auxiliary definitions.

Signatures and model functor. By a *relational signature* we shall mean a triple $\langle S, \mathrm{OP}, \mathrm{REL}\rangle$, such that $\langle S, \mathrm{OP}\rangle$ is an algebraic signature, and REL is an S^+-indexed family of *relational symbols*. Morphisms of relational signatures are similar to algebraic signatures morphisms, with the only difference being an additional component translating relational symbols. The translation of relational symbols has to respect the sort mapping of course. Relational signatures with their morphisms constitute a category, which we shall denote by **RelSig**.

A relational structure over a signature $\langle S, \mathrm{OP}, \mathrm{REL}\rangle$, is an algebra A over $\langle S, \mathrm{OP}\rangle$, together with for every relational symbol $R \in \mathrm{REL}_w$, a relation $R^A \subseteq |A|_w$. Similarly to the algebraic case, for every relational signature morphism there is a corresponding *reduct* function on classes of relational structures. In other words, the relationship between relational signatures and relational structures can be described as a functor ("discrete" indexed category): **RelStr** : $\mathbf{RelSig}^{op} \to \mathbf{Class}$.

In order to introduce the examples let us, in a parametric way, define some further components.

Language Functor. Let VSrt, be any **Set**-endofunctor (i.e., VSrt : **Set** $\rightarrow$ **Set**), "naturally included" in $\mathrm{ASrt}(\_) \stackrel{df}{=} ((\_)^* \times \_) + (\_)^+$. By "natural inclusion" we mean a natural transformation $\iota : \mathrm{VSrt} \dot{\rightarrow} \mathrm{ASrt}$, consisting of set-theoretic inclusions.

We shall define a family of functors $\mathbf{Lan}^{\mathrm{VSrt}} : \mathbf{RelSig} \rightarrow \mathbf{MSig}$, parameterized by VSrt. Let us define:

$$\mathbf{Lan}^{\mathrm{VSrt}}(\langle S, \mathrm{OP}, \mathrm{REL}\rangle) \stackrel{df}{=} \langle \mathrm{ASrt}(S) \cup \{\star\}, \mathrm{VSrt}(S), \mathcal{O}, \{\mathit{forall}\}\rangle,$$

where $\mathcal{O}$ is the least (wrt. $\subseteq$), suitably indexed set, satisfying the following conditions:[2]

- $\mathrm{OP}_{w,s} \subseteq \mathcal{O}_{\rightarrow F(w,s)}$, for any $w \in S^*$, and $s \in S$
- $\mathrm{REL}_w \subseteq \mathcal{O}_{\rightarrow R(w)}$, for any $w \in S^+$,
- "*app*" $\in \mathcal{O}_{F(w,s),F(\lambda,s_1),\ldots,F(\lambda,s_n) \rightarrow F(\lambda,s)}$, for any sequence $w = s_1 \ldots s_n \in S^*$, and $s \in S$,
- "ε" $\in \mathcal{O}_{R(w),F(\lambda,s_1),\ldots,F(\lambda,s_n) \rightarrow \star}$, for any sequence $w = s_1 \ldots s_n \in S^+$,
- "$\neg$" $\in \mathcal{O}_{\star \rightarrow \star}$, "$\wedge$" $\in \mathcal{O}_{\star,\star \rightarrow \star}$.

Extending the above definition to relational signature morphisms is rather s-traightforward.

Universal Language. As a next step, we shall define (a schema of) the *universal language* signature $\mathbb{L}$. Its construction (as well as the construction of the *semantic universe* structure—$\mathbb{G}$), may look rather unorthodox from set-theoretical point of view. There are several ways of justifying it—the construction may be explained using either Grothendieck universes, or assumption about existence of several inaccessible cardinals, or using some stratified version of ZFC (as described for example in Chapter II of [9]). The same "foundational difficulties" arise in the framework of ordinary parchments (cf. [7], [10]), as well as λ-parchments–[11]. In this paper, we shall stay at a "naive" level, and will not even try to be more precise in this respect.[3]

Let us now consider the functor ASrt, as an endofunctor on the category of classes—**Class**. As previously, we shall define the signature $\mathbb{L}$ in a "parametric" way—with an arbitrary endofunctor VSrt on **Class**, naturally included in ASrt, as a parameter.

Let $\mathbb{L}$(VSrt) be a meta-signature defined as follows:

- *Sorts names.* For any set s and any finite sequence of sets $s_1, \ldots, s_n$, $n \geq 0$, $F(s_1 \ldots s_n, s)$, and $R(s_1 \ldots s_n)$, $(n \geq 0)$, (i.e., all elements of $\mathrm{ASrt}(|\mathbf{Set}|)$) are sort names. Additionally, $\star$ is also a sort name in $\mathbb{L}$(VSrt).

[2] We shall use "$F(w,s)$", and "$R(w)$" for denoting elements of the first, and second component of $S^* \times S \uplus S^+$, respectively (purists should think of $F(w,s)$ and $R(w)$, as "syntactic sugar" for $\langle 0, \langle w, s\rangle\rangle$, and $\langle 1, w\rangle$, for example).

[3] For an interesting discussion of the rôle of "big sets" in set-theory, see [5], §6, and §7.

- *Value sorts.* As the sorts of values, let us take VSrt(|**Set**|). Remember that VSrt(|**Set**|) is a *subclass* of the class of all sort names—ASrt(|**Set**|).
- *Operation symbols.* The collection of operation names of $\mathbb{L}$(VSrt) is the smallest collection satisfying the following conditions:
  - for every function $f : s_1 \times ... \times s_n \to s$, $\ulcorner f \urcorner : \to F(s_1 ... s_n, s)$,
  - for every relation $r \subseteq s_1 \times ... \times s_n$, $\ulcorner r \urcorner : \to R(s_1 ... s_n)$,
  - "*app*": $F(s_1 ... s_n, s), F(\lambda, s_1), ..., F(\lambda, s_n) \to F(\lambda, s)$, for any finite sequence of sets $s_1 ... s_n$, and any set s,
  - "ε": $R(w), F(\lambda, s_1), ..., F(\lambda, s_n) \to \star$, for any finite sequence of sets $w = s_1 ... s_n$,
  - "$\neg$": $\star \to \star$, "$\wedge$": $\star, \star \to \star$.
- *Generalized operation names.* There is only one generalized operation name in $\mathbb{L}$(VSrt): "$\forall$".

Language Interpretation for Models. For any functor VSrt, naturally included in ASrt, let us define a natural transformation $\mathbf{mod}^{\mathrm{VSrt}}$, from **RelStr**, to hom($\mathbf{Lan}^{\mathrm{VSrt}}(\_), \mathbb{L}(\mathrm{VSrt})$). For any relational signature $\Sigma = \langle S, \mathrm{OP}, \mathrm{REL} \rangle$, $\mathbf{mod}_{\Sigma}^{\mathrm{VSrt}}$ assigns to every relational structure $A = \langle |A|, \mathrm{OP}^A, \mathrm{REL}^A \rangle$, over Σ, a meta-signature morphism $\mathbf{mod}_{\Sigma}^{\mathrm{VSrt}}(A) : \mathbf{Lan}_{\Sigma}^{\mathrm{VSrt}} \to \mathbb{L}(\mathrm{VSrt})$, defined as follows:

- *Sort names.* The carrier $|A|$ of A, may be thought of, as a function $C_A : S \to |\mathbf{Set}|$. The sort-part of $\mathbf{mod}_{\Sigma}^{\mathrm{VSrt}}(A)$ is then defined as:

$$(\mathbf{mod}_{\Sigma}^{\mathrm{VSrt}}(A))_{\mathbf{srts}} \stackrel{df}{=} \mathrm{ASrt}(C_A) \cup \mathbf{id}_{\star}.$$

  Observe that $\mathrm{ASrt}(C_A)$ preserves value sorts, i.e., that

$$\mathrm{ASrt}(C_A)[\mathrm{VSrt}(S)] \subseteq \mathrm{VSrt}(|\mathbf{Set}|).$$

- *Operation names.*
  - for any $f : \lambda \to F(w, s)$, $(\mathbf{mod}_{\Sigma}^{\mathrm{VSrt}}(A))_{\mathbf{ops}}(f) \stackrel{df}{=} \ulcorner f^A \urcorner$,
  - for any $r : \lambda \to R(w)$, $(\mathbf{mod}_{\Sigma}^{\mathrm{VSrt}}(A))_{\mathbf{ops}}(r) \stackrel{df}{=} \ulcorner r^A \urcorner$,
  - for $op \in \{app, \varepsilon, \neg, \wedge\}$, $(\mathbf{mod}_{\Sigma}^{\mathrm{VSrt}}(A))_{\mathbf{ops}}(op) \stackrel{df}{=} op$.
- *Generalized operation names.* For the only generalized operation symbol "*forall*", we define $(\mathbf{mod}_{\Sigma}^{\mathrm{VSrt}}(A))(\mathit{forall}) \stackrel{df}{=} \forall$.

Naturality of $\mathbf{mod}^{\mathrm{VSrt}}$, easily follows from the above construction—it formalizes the idea, that taking reducts corresponds to a "meta-level renaming".

Semantic Universe. For an arbitrary universal language signature $\mathbb{L}$(VSrt), let $\mathbb{G}$(VSrt) be a meta-structure defined as follows:

- *carrier*
  - $|\mathbb{G}|_{F(s_1 ... s_n, s)} \stackrel{df}{=} [s_1 \times ... \times s_n \to s]$,
  - $|\mathbb{G}|_{R(s_1 ... s_n)} \stackrel{df}{=} \mathcal{P}(s_1 \times ... \times s_n)$
  - $|\mathbb{G}|_{\star} \stackrel{df}{=} \{\mathrm{tt}, \mathrm{ff}\}$

- *operations*
  - for any $\ulcorner f \urcorner : \langle\rangle \to F(s_1...s_n, s)$, $\ulcorner f \urcorner^{\mathbb{G}} \stackrel{df}{=} f$,
  - for every $\ulcorner r \urcorner : \langle\rangle \to \mathcal{R}(s_1...s_n)$, $\ulcorner r \urcorner^{\mathbb{G}} \stackrel{df}{=} r$,
  - $app^{\mathbb{G}}(f, e_1, ..., e_n)() \stackrel{df}{=} f(e_1(), ..., e_n())$,
  - $\varepsilon^{\mathbb{G}}(r, e_1, ..., e_n) \stackrel{df}{=}$ tt, whenever $\langle e_1(), ..., e_n()\rangle \in r$, and equals ff otherwise,
  - $\neg^{\mathbb{G}}$, and $\wedge^{\mathbb{G}}$ are the usual negation and conjunction on {tt, ff},
- *assignable values*: for any $s \in \mathrm{VSrt}(|\mathbf{Set}|)$, $|V^{\mathbb{G}(\mathrm{VSrt})}|_s \stackrel{df}{=} |\mathbb{G}(\mathrm{VSrt})|_s$, i.e., *all* the values in the value sorts are assignable,
- *designated values*: $D^{\mathbb{G}} \stackrel{df}{=} \{\mathrm{tt}\}$,
- *generalized operation*: $\forall^{\mathbb{G}} : \mathcal{P}(\{\mathrm{tt}, \mathrm{ff}\}) \to \{\mathrm{tt}, \mathrm{ff}\}$, such that $\forall^{\mathbb{G}}(B) = \mathrm{ff}$, if and only if $\mathrm{ff} \in B$.

Examples. The first three examples illustrate (a well known fact), that using a common "relational base", by just changing the allowed variable range, we can obtain different logics. By "common relational base", as we mentioned at the beginning of this section, we mean: a common category of signatures—**RelSig**, and a common model functor—**RelStr**.

Let ISrt, and SSrt be endofunctors on **Set** (**Class**) such that for an arbitrary set (class) S, $\mathrm{ISrt}(S) \stackrel{df}{=} \{\, F(\lambda, s) \mid s \in S \,\}$, and $\mathrm{SSrt}(S) \stackrel{df}{=} \{\, R(s) \mid s \in S \,\}$ (with an obvious extension to morphisms). It is not difficult to see, both ISrt, and $\mathrm{MSrt} \stackrel{df}{=} \mathrm{ISrt} \cup \mathrm{SSrt}$ are naturally included in ASrt.

- *First-order logic.* To get a context parchment $\mathfrak{P}^{\mathrm{FOL}}$, for first-order logic (without equality), we should take $\mathbf{Lan}^{\mathrm{ISrt}}$—as the language functor, a meta-signature $\mathbb{L}(\mathrm{ISrt})$—as the universal language, $\mathbf{mod}^{\mathrm{ISrt}}$—as the language interpretation for models, and $\mathbb{G}(\mathrm{ISrt})$—as the semantic universe. "Instantiating" VSrt with ISrt, expresses the fact, that in first-order logic, variables range over individual elements of the model carrier only.
- *Monadic second-order logic.* In monadic second-order logic, variables may range over individuals as well as over arbitrary subsets of individuals. Therefore in order to get a context parchment $\mathfrak{P}^{\mathrm{MSOL}}$, for monadic second-order logic, we have to "instantiate" VSrt with MSrt, and take: $\mathbf{Lan}^{\mathrm{MSrt}}$—as the language functor, $\mathbb{L}(\mathrm{MSrt})$—as the universal language, $\mathbf{mod}^{\mathrm{MSrt}}$—as the language interpretation for models, and $\mathbb{G}(\mathrm{MSrt})$—as the semantic universe.
- *Second-order logic.* To obtain the full second-order logic, where variables may range over not only individuals, and sets of individuals, but over arbitrary functions and relations as well, we have to instantiate VSrt with ASrt. Hence, to obtain a context parchment $\mathfrak{P}^{\mathrm{SOL}}$, for (full) second-order logic we take: $\mathbf{Lan}^{\mathrm{ASrt}}$—as the language functor, $\mathbb{L}(\mathrm{ASrt})$—as the universal language signature, $\mathbf{mod}^{\mathrm{ASrt}}$—as the language interpretation for models, and $\mathbb{G}(\mathrm{ASrt})$—as the semantic universe.

By looking at the three examples above, one might get an impression that meta-structures do not really need to contain the distinguished subset of *assignable values*. It seems always possible to determine it purely syntactically—"*assignable values* = subcarrier of the structure, indexed by the *value sorts* of its signature". Unfortunately it is not always the case. Let us illustrate it with two different examples— weak second-order logic, and partial first-order logic.

- *Weak second-order logic.* In the weak second-order logic (cf. [4], for example), variables may range over individuals, and *finite* sets of individuals. It is however possible to have constants, which denote infinite sets of individuals. Therefore, in order to obtain a context parchment $\mathfrak{P}^{\mathrm{WSOL}}$, it is not sufficient to redefine the semantic universe $\mathbb{G}$ of $\mathfrak{P}^{\mathrm{MSOL}}$, by imposing a restriction that $|\mathbb{G}(\mathrm{MSrt})|_{R(s)} = \mathcal{F}(s)$, for any set s (where $\mathcal{F}(\_)$ is the finite powerset operator). We need to change the distinguished subclass of assignable values in $\mathbb{G}(\mathrm{MSrt})$ from $\mathcal{P}(s)$, for any value sort $R(s)$, to $\mathcal{F}(s)$. All the other components of $\mathfrak{P}^{\mathrm{WSOL}}$ remain exactly the same as in $\mathfrak{P}^{\mathrm{MSOL}}$.
- *Partial first-order logic.* With logics of *partial functions*, the problem is, that our meta-framework is essentially based on *total* algebras (only the generalized operations may be partial). Therefore we need to represent partiality somehow. A standard approach is to introduce a special element for the *undefined "value"*.
  To define a context parchment $\mathfrak{P}^{\mathrm{PFOL}}$, for partial first-order logic (without equality), we first have to define *partial relational signatures*, and *partial relational structures*. A partial relational signature is a quadruple $\langle S, \mathrm{OP}, \mathrm{POP}, \mathrm{REL}\rangle$, such that $\langle S, \mathrm{OP}, \mathrm{REL}\rangle$, is a (total) relational signature, and POP is an $S^* \times S$-indexed set of *partial operation names* (for purely technical reasons we assume that sets OP and POP are disjoint). Morphisms of partial relational signatures are defined in an obvious way. A partial relational structure A, over $\langle S, \mathrm{OP}, \mathrm{POP}, \mathrm{REL}\rangle$, is a relational structure over $\langle S, \mathrm{OP}, \mathrm{REL}\rangle$, plus for every $\mathrm{pop}\mathrm{POP}_{w,s}$, a partial function $\mathrm{pop}^A : |A|_w \rightharpoonup |A|$. Similarly, as in the total case, every partial relational signature morphism, induces an appropriate reduct functor on structures. In other words, partial relational structures constitute an indexed category $\mathbf{PRelStr} : \mathbf{PRelSig}^{op} \to \mathbf{Class}$, where **PRelSig**, denotes the category of partial relational signatures.
  PRelSig, and **PRelStr**, are the category of signatures, and the model functor of $\mathfrak{P}^{\mathrm{PFOL}}$, respectively. Other components of $\mathfrak{P}^{\mathrm{PFOL}}$—the language functor $\mathbf{Lan}^{\mathrm{PFOL}}$, the universal language $\mathbb{L}^{\mathrm{PFOL}}$, and the language interpretation for models—$\mathbf{mod}^{\mathrm{PFOL}}$ can be obtained from their total counterparts in $\mathfrak{P}^{\mathrm{FOL}}$, by adding information about partial operations. The only nontrivial part is the semantic universe $\mathbb{G}^{\mathrm{PFOL}}$. Below, we shall define only those "parts" of $\mathbb{G}^{\mathrm{PFOL}}$, which are different from the corresponding parts of $\mathbb{G}^{\mathrm{FOL}}$ ($= \mathbb{G}(\mathrm{ISrt})$):
  - *Carrier*: $|\mathbb{G}^{\mathrm{PFOL}}|_{F(s_1 \ldots s_n, s)} \stackrel{df}{=} [s_1^{\perp} \times \ldots \times s_n^{\perp} \xrightarrow{s} s^{\perp}]$, where for any set s, $s^{\perp} \stackrel{df}{=} s \cup \{\perp\}$, and $[\_ \xrightarrow{s} \_]$ denotes the *strict function space* operator.
  - *Operations*: for any $\ulcorner f \urcorner : \langle\rangle \to F(w, s)$, $\ulcorner f \urcorner^{\mathbb{G}^{\mathrm{PFOL}}} \stackrel{df}{=} f^{\perp}$, where $f^{\perp}$ is the *strict extension* of the function f, i.e., $f^{\perp}(a) \stackrel{df}{=} f(a)$, provided that $a \neq \perp$ and $f(a)$ was defined, and equals $\perp$, otherwise.

- *Assignable values*: for any value sort $F(\lambda, s)$, in $\mathbb{L}^{\mathrm{PFOL}}$, the set of assignable values $|V^{\mathbb{G}^{\mathrm{PFOL}}}|_{F(\lambda,s)}$ is obtained from $|\mathbb{G}^{\mathrm{PFOL}}|_{F(\lambda,s)}$ by removing the "undefined constant".

The first example is in a sense more interesting, because it shows that a distinction between "values' and "assignable values" may be already present in the logic, whereas the second example merely shows how we can cope with the "deficiencies" of our *total* meta-framework.

6 Context Parchment Morphisms

6.1 Definition

If we think of context parchments as a way of *presenting* logics, their *morphisms* intuitively express the idea that one logic is *built over* another one.[4]

Definition 10. Let $\mathfrak{P}$, and $\mathfrak{P}'$ be context parchments. A morphism from $\mathfrak{P}$ to $\mathfrak{P}'$ consists of:

- a functor $\Phi : \mathbf{Sig} \to \mathbf{Sig}'$,
- a natural transformation $\alpha : \Phi; \mathbf{Lan}' \dot{\to} \mathbf{Lan}$,
- a natural transformation $\beta : \mathbf{Mod} \dot{\to} \Phi^{op}; \mathbf{Mod}'$,
- a meta-signature morphism $l : \mathbb{L}' \to \mathbb{L}$,
- a $\mathbb{L}'$-homomorphism $g : \mathbb{G}' \to \mathbb{G}|_l$

such that for any $\Sigma \in \mathbf{Sig}$, and any model $M \in \mathbf{Mod}_\Sigma$:

$$\begin{array}{ccc}
\mathbf{Lan}_\Sigma & \xleftarrow{\alpha_\Sigma} & \mathbf{Lan}'_{\Phi(\Sigma)} \\
\Big\downarrow {\scriptstyle \mathbf{mod}_\Sigma(M)} & & \Big\downarrow {\scriptstyle \mathbf{mod}'_{\Phi(\Sigma)}(\beta_\Sigma(M))} \\
\mathbb{L} & \xleftarrow[l]{} & \mathbb{L}'
\end{array}$$

Context parchments, together with their morphisms (with an obvious definition of composition), constitute a category, which we shall denote by $\mathfrak{ConPar}$.

6.2 Examples

In this section, we shall give some examples of context parchment morphisms. We shall use context parchments introduced in section 5.3 for this purpose.

[4] It is probably a good idea to stress at this point, that this is by no means the only possible way of "understanding" context parchment morphisms. It is the one which comes from "standard" examples.

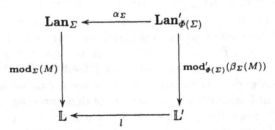

"Syntactic restrictions". Before introducing context parchments FOL, $MSOL$, and SOL, for first-order, monadic second-order, and second-order logics, we mentioned that all of them are based on a *common relational base*—relational signatures—as signatures, and relational structures—as models. The essential difference between these logics, comes from the allowed *variable range.* As we have seen, the variable range of FOL is "contained" in the variable range for MSOL, which in turn is "contained" in the variable range for the (full) second-order logic—SOL. These relationships are expressed by the existence of appropriate context parchment morphisms: $(\Phi, \alpha, \beta, l, g)$: MSOL $\rightarrow$ FOL, and $(\Phi, \alpha', \beta, l', g')$: SOL $\rightarrow$ MSOL. In both cases Φ, is the identity functor on **RelSig**, and β is the identity natural transformation. The *language components*—α's and l's, are appropriate "inclusions". Translations between semantic universes—the g's, are identity homomorphisms, which stresses the fact, that all the differences between these three logics are "forced" by syntax. This may justify the name we have chosen for the morphisms—"syntactic restrictions".

"Semantic restrictions". The situation is somewhat different, with the remaining two examples introduced in section 5.3—the weak second-order logic WSOL, and partial first-order logic PFOL. Let us start with the case of WSOL. Although the obvious morphism WSOL $\rightarrow$ FOL, may be seen as a "syntactic restriction", the relationship between WSOL, and the two other second-order logics—MSOL, and SOL, have more "semantic" flavour. A morphism $(\Phi, \alpha', \beta, l', g'')$: SOL $\rightarrow$ WSOL, is defined *almost* like the one from SOL, to MSOL, but the homomorphism of the semantic universes—g'', is no longer the identity. The reason is very simple—the *assignable values* for WSOL, are properly included in the assignable values of the l'-reduct of the semantic universe for SOL. A very similar situation occurs in the case of a morphism $(\Phi, \mathbf{id}, \beta, \mathbf{id}, g''')$: MSOL $\rightarrow$ WSOL.

Slightly more interesting example of a "semantic restriction", is the relationship between PFOL, and FOL, expressed by existence of a context parchment morphism PFOL $\rightarrow$ FOL. The signature functor, translates *partial* relational signatures to *ordinary* relational signatures, by simply "forgetting" the partial operations. The natural transformation between *language functors,* consists of appropriate inclusions. *Partial* relational structures, are mapped to *total* ones, by forgetting the interpretation for the partial operations. The *universal language* for FOL is simply included in the one for PFOL. The interesting part, is the semantic universe homomorphism. Unlike the g's in the previous examples, it has to do some (very simple) "encoding"—for all the "functional" sorts (i.e., sorts of the form $F(w, s)$), the homomorphism translates every total function $f : w \rightarrow s$, to its *strict extension,* $f^{\perp} : w^{\perp} \xrightarrow{s} s^{\perp}$.

Expressing relationships between logics is only one possible application of the concept of context parchment morphism. Also interesting, and perhaps even more useful from "practical" point of view, is the possibility of *modular construction* of logics, by "putting context parchments together".

6.3 Categories of Context Parchments

Using universal constructions in categories of parchments, and λ-parchments, for modular construction of logics, has been advocated in [10], and [11]. In the case of context parchments, situation is very similar. Due to the lack of space, we shall not go into the details, but instead—try to sketch the main ideas, and, where possible refer to the corresponding parts of [10], and [11].

Using Prop. 4, Prop. 7, and Prop. 8, and almost repeating the proof of Theorem 12, from [11], we can show, that:

Proposition 11. *The category of context parchments* $\mathfrak{ConPar}$*, is complete.* □

This means, that we can "put context parchments together" using categorical limits. As it has been pointed out in [10], we should not expect, that the "categorical nonsense" will always give the "expected" parchment as a result. Sometimes we have to "repair" the result slightly. The same remains true for context parchments as well.

In section 5.2, we have seen, that every *logical* context parchment $\mathfrak{P}$, generates a context institution $\mathcal{I}(\mathfrak{P})$. In [13], a notion a *context institution morphism* has been introduced. Context institutions and their morphisms constitute a category $\mathfrak{ConIns}$. A natural question arises—whether the construction of $\mathcal{I}(\mathfrak{P})$, extends to context parchment morphisms between logical context parchments? As it turns out, in general, the answer is "no". The same phenomenon occurs in the case of ordinary parchments—[10], and λ-parchments—[11]. We have to impose some additional requirements on the *semantic universe homomorphism* g. First, as in [10], and [11], we have to assume that g is a *monomorphism*, which not only *preserves*, but also *reflects*, the *distinguished truth values*. In addition, g must also *reflect* the *assignable values*. A context parchment morphism satisfying the above requirements will be called *logical*. Logical context parchments, and logical context parchment morphisms constitute a category $\mathfrak{LConPar}$. The construction $\mathcal{I}(\mathfrak{P})$, extends then, to a functor $\mathcal{I} : \mathfrak{ConIns} \to \mathfrak{LConPar}$[5]

Observe, that if $\mathfrak{p} : \mathfrak{P} \to \mathfrak{P}'$ is logical, then there is a bijection between the corresponding sets of assignable values. This means, that morphisms from SOL, to WSOL, and from MSOL to WSOL, discussed in section 6.2, are *not* logical. This is actually quite fortunate, as otherwise, by existence of the appropriate context institution morphism, we could have expected for example, that a WSOL-formula of the form $\forall X.\phi$, where X is a "set"-variable, is true under a WSOL valuation v, if and only if it is true under "the same" valuation in SOL (or MSOL), which is not too reasonable.

Unfortunately, the category of logical context parchments $\mathfrak{LConPar}$, is not complete, so every time we want to combine some logical context parchments together, we first have to take a limit of the corresponding diagram in $\mathfrak{ConPar}$,

[5] Actually, the definition of context institution morphism has to be generalized slightly—instead of the requirement, that carrier of a "translated" model *equals* the translation of its carrier, we have to require that they are *isomorphic*.

and then, possibly, "massage" the result to make it logical (examples from cf. [10], may be easily adopted).

One of the advantages of context parchments is that they are usually much simpler, than the corresponding ordinary parchments, or λ-parchments (the Reader may wish to compare the examples from section 5.3, and examples given in [10], and [11]). The only exception are *ground logics*, i.e., logics without variables, where all three notions essentially coincide.

There are however constructions which can be expressed at the level of context parchments, but cannot be performed at the level of ordinary, nor λ-parchments. Let us give a simple example.

Let $\mathbb{B}$, denote a context parchment defined as follows:

- $\mathbf{Sig}^{\mathbb{B}} \stackrel{df}{=} \mathbb{1}$, where $\mathbb{1}$, is a one-object category, with the only object denoted by "$\bullet$",
- $\mathbf{Lan}^{\mathbb{B}}_{\bullet} = \mathbb{L}^{\mathbb{B}} \stackrel{df}{=} \langle\{\star\}, \emptyset, \emptyset, \emptyset\rangle$
- $\mathbf{Mod}^{\mathbb{B}} : \mathbb{1} \to \mathbf{Class}$, $\mathbf{Mod}^{\mathbb{B}}_{\bullet} \stackrel{df}{=} \{\bullet\}$,
- $mod^{\mathbb{B}}_{\bullet}(\bullet) \stackrel{df}{=} \mathbf{id}_{\star}$,
- $|\mathbb{G}^{\mathbb{B}}|_{\star} \stackrel{df}{=} \{\mathrm{tt}, \mathrm{ff}\}$, $D^{\mathbb{B}} \stackrel{df}{=} \{\mathrm{tt}\}$.

The parchment $\mathbb{B}$, is a parchment of *classical truth values*. Of course, we can extend it, by adding some *connectives*—**Con**, and/or *quantifiers*, and obtain a context parchment which we shall denote by $\mathbb{B}^{\forall,\mathbf{Con}}$. For example, if we take any context parchment $\mathfrak{P}$, "based on $\mathbb{B}$", i.e., such that $|\mathbb{G}^{\mathfrak{P}}| = \{\mathrm{tt}, \mathrm{ff}\}$, and such that the set of generalized operations in $\mathbb{L}^{\mathfrak{P}}$ is empty, then by taking a pullback in $\mathfrak{ConPar}$, of the diagram:

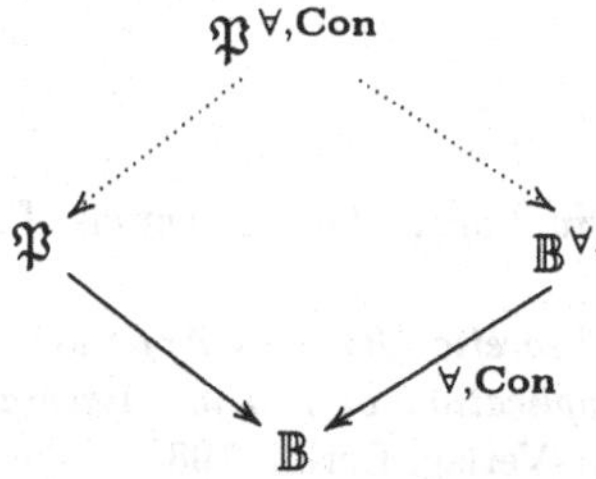

with empty set of *connectives*—**Con**, we obtain a context parchment $\mathfrak{P}^{\forall}$—which is a "universal closure" of $\mathfrak{P}$[6].

Applying the above construction for $\mathfrak{P} = \mathrm{EL}$, where EL is a context parchment for equational logic, and with $\mathbf{Con} \stackrel{df}{=} \{\neg, \wedge\}$, gives $\mathrm{FOL}^{=}$—a context parchment for first-order logic with equality, as a result.

7 Conclusions and Future Work

In the paper, we introduced notions of *context parchment*, and *context parchment morphism*, which are meant to play the same role with respect to *context insti-*

[6] Note, that $\mathfrak{P}^{\forall}$ will always be *logical*.

tutions, as ordinary *parchments*, or *λ-parchments* and their morphisms—with respect to *institutions*. We illustrated the notions with several examples, and have shown, that any *logical* context parchment generates a context institution. The construction can also be extended to *logical context parchment morphisms.*

The category of context parchments and their morphisms—$\mathfrak{ConPar}$, can be used for modular construction of logics. Since $\mathfrak{ConPar}$ is complete, we can use categorical limits to "put context parchments together". Although, parchments and λ-parchments can be used for this purpose as well, context parchments, apart from generating context institutions, are usually simpler and more "e-conomical". By "economy" we mean that certain features of the logic, like for example quantifiers, can be be effectively presented at the level of context parchments, whereas at the level of parchments and λ-parchments they must be fully defined *internally*.

By a more clear "separation" of the *algebra of truth values*, from the rest of the parchment, context parchments make it possible to define operations like *quantifier closure*, or *closure wrt. a set of connectives* for an arbitrary context parchment "based" on a given space of *truth values*.

We hope, that context parchments can be used for defining, and combining *inference systems* for logics.

Recently, a new notion of *model-theoretic parchment*, has been proposed in [12]. Model-theoretic parchments eliminate the need for the *universal language* and *semantic universe*, and provide a natural framework for describing "non-classical" logics, such as modal logic for example. We believe, that it should be fairly straightforward to define a "context version" of model-theoretic parchments.

References

1. J. Barwise. *Axioms for abstract model theory.* Annals of Mathematical Logic, vol. 7, 1974, pp. 221-265.
2. P. Burmeister. *A Model Theoretic Oriented Approach to Partial Algebras. Introduction to Theory and Application of Partial Algebras – Part I.* Mathematical Research Vol. 32, Akademie-Verlag, Berlin, 1986.
3. R. Diaconescu, J. Goguen, and P. Stefaneas. *Logical Support for Modularisation.* In G. Huet and G. Plotkin, editors, *Logical Environments*, pp. 83-130, Cambridge University Press, 1993.
4. H.D. Ebbinghaus, J. Flum, and W. Thomas. *Mathematical Logic*, Springer-Verlag, 1984.
5. A.A. Fraenkel, Y. Bar-Hillel, and A. Levy. *Foundations of Set Theory*, second revised edition. Studies in Logic and the Foundations of Mathematics, Vol. 67, North-Holland, 1973.
6. J.A. Goguen and R.M. Burstall. *Introducing Institutions.* In Proceedings, Logics of Programming Workshop, LNCS 164, pp. 221-256, Springer-Verlag, 1984.
7. J.A. Goguen and R.M. Burstall. *A study in the foundations of programming methodology: Specifications, institutions, charters and parchments.* In D. Pitt, S. Abramsky, A Poigné, and D. Rydeheard, editors, Proceedings, Conference on

Category Theory and Computer Programming, LNCS 240, pp. 313-333, Springer-Verlag, 1986.

8. J.A. Goguen and R.M. Burstall. *Institutions: Abstract model theory for specification and programming.* Journal of the Association for Computing Machinery, vol. 39, pp. 95-146, 1992.
9. H. Herrlich, G.E. Strecker. *Category Theory*, second edition. Heldermann Verlag, Berlin, 1979.
10. T. Mossakowski. *Using limits of parchments to systematically construct institutions of partial algebras.* In Magne Haveraaen and Olaf Owe and Ole-Johan Dahl, editors, *Recent Trends in Data Type Specifications. 11th Workshop on Specification of Abstract Data Types joint with the 8th general COMPASS workshop. Oslo, Norway, September 1995. Selected papers.* LNCS, vol. 1130, pp. 379-393, Springer-Verlag, 1996.
11. T. Mossakowski, A. Tarlecki, and W. Pawłowski. *Combining and representing logical systems* . In E. Moggi, editor, *Category Theory in Computer Science.* LNCS, Springer-Verlag 1997. To appear.
12. T. Mossakowski, A. Tarlecki, and W. Pawłowski. *Combining and representing logical systems using model-theoretic parchments.* Presented at 12th Workshop on *Algebraic Development Techniques WADT'97*, Tarquinia, June 1997.
13. W. Pawłowski. *Context Institutions.* In Magne Haveraaen and Olaf Owe and Ole-Johan Dahl, editors, *Recent Trends in Data Type Specifications. 11th Workshop on Specification of Abstract Data Types joint with the 8th general COMPASS workshop. Oslo, Norway, September 1995. Selected papers.* LNCS, vol. 1130, pp. 436-457, Springer-Verlag, 1996.
14. H. Rasiowa, R. Sikorski. *The Mathematics of Metamathematics.* Państwowe Wydawnictwo Naukowe, Warszawa, 1963.
15. R. Sikorski. *Algebra of Formalized Languages.* Colloquium Mathematicum, Vol. IX, 1962, pp. 1-31.
16. A. Tarlecki, R. Burstall, and J. Goguen. *Some fundamental algebraic tools for the semantics of computation, part 3: Indexed categories.* Theoretical Computer Science, 91, pp. 239-264, 1991.
17. U. Wolter, R. Wessäly, M. Klar, and F. Cornelius. *Four Institutions: A Unified Presentation of Logical Systems for Specification.* Bericht-Nr. 94-24, Technische Universität Berlin, 1994.

Verifying a Compiler Optimization for Multi-Threaded Java

Bernhard Reus, Alexander Knapp, Pietro Cenciarelli, and Martin Wirsing

Ludwig–Maximilians–Universität München
{reus,knapp,cenciare,wirsing}@informatik.uni-muenchen.de

Abstract. The specification for the object-oriented concurrent language Java [3] is rather loose with respect to the interaction of shared memory and the local working memories of different threads. This allows maximal freedom in the language implementation. Such freedom is reflected in the semantics provided in [2], where threads-memory interaction is formalized in terms of structures called *event spaces*. Two kinds of memories are described in the Java specification: a "normal" memory and a more liberal one, where values can sometimes be stored even before they are produced as results of computation. Here we compare two structural operational semantics of a sublanguage of Java modelling the two types of memory. The two semantics share the same set of operational rules but put different requirements (expressed as first order theories) on the notion of event space. We prove a result which is informally stated in [3]: the two semantics coincide for properly synchronized programs. This shows the applicability of a new technique for combining structural operational semantics and first order specification of process behaviour.

1 Introduction

A concurrent program consists of multiple tasks that are or behave as if they were executed all at the same time. Such tasks can be implemented using *threads* (short for "threads of execution"), which are sequences of instructions that run independently within the encompassing program. The object-oriented language Java supports thread programming (see e.g. [1], [4]).

Java threads share a common memory, but keep working copies of shared variables in private working memories. It is only when leaving a synchronized block that *must* a thread copy the content of its working memory in the main memory. However, possible implementations of the run-time system *may* choose to update the value of a variable in the main memory as soon as a thread makes an assignment to its working copy of that variable. The Java language specification [3] leaves freedom to the implementation in that respect.

A particular implementation technique is also discussed in [3], where a value can be stored by a thread in the main memory before such value is produced by the computation. This is called a *prescient* store action [3, §17.8]. The only restriction is that between the prescient store and the matching assignment nothing "bad" happens, e.g. no other thread reads illegitimately the prescient value. Consider the following code example.

```
o.b = 2;
for(o.a = 1; o.a < 10; o.a = o.a + 1)
  o.b = o.a + o.b;
```

where `o` is (a reference to) an object with two attributes `a` and `b` of type `int`. When executed the value of `o.a` can only be stored after `o.a` has been assigned a new value, i.e. after it was incremented. If prescient store operations are permitted, then it would be also legal to store the value 10 for `o.a` in advance, i.e. before the loop is entered, excluding thereby that any other thread can load `o.a` before the end of the loop.

The rearrangement of store operations can be used to speed up programs when updating of variables is split into a thread action (called *Store*) and a memory action (*Write*). The global memory can concurrently provide the value of a pre-stored variable while a second thread waits for it. Re-grouping the store-operations might also optimize memory access itself.

In [2] we present a structural operational semantics (in the style of [5]) of a nontrivial sub-language of Java which includes dynamic creation of objects, blocks, and synchronization of threads. The notion of *event space* is introduced in that paper to formalize the communication protocol between shared memory and threads. Event spaces correspond roughly to *configurations* in Winskel's *event structures* [6], which are used for denotational semantics of concurrent languages.

Here we exploit the flexibility of the approach proposed in [2], where the operational semantics is given parametrically in the notion of event space, and compare two language implementations which share the same set of operational rules. The implementations are obtained by imposing different requirements on event spaces, so that prescient stores are possible in one case and impossible in the other. Such requirements are expressed in simple first order clauses. In this framework we prove that prescient and nonprescient semantics coincide for properly synchronized programs, that is programs where any two threads are not allowed to write a variable into the global memory without synchronization (*race conditions* [4]). This property was only informally stated in [3, §17.8]. We also provide an example where prescient store actions for non-properly synchronized programs lead to inconsistent memory contents.

The contribution of the paper is twofold: on the one hand it provides welcome formal confirmation of the intuitive correctness of certain compiler optimization techniques; on the other hand it shows the applicability of an innovative technique for combining structural operational semantics and first order axiomatisation of process behaviour.

The paper is organized as follows: Section 2 recapitulates the definition of event spaces from [2]. These are used in Section 3 for the SOS-rules. Next, the axiomatization of event spaces is changed (Section 4) in order to allow for prescient stores and prescient operational semantics is defined in Section 5. It is then proven, in Section 6, that for properly synchronized programs the extension is conservative w.r.t. the old semantics.

2 Event Spaces

The execution of a Java program comprises many *threads* of computation running in parallel. Threads exchange information by operating on values and objects residing in a shared *main memory*. As explained in the Java language specification [3], each thread also has a private *working memory* in which it keeps its own working copy of variables that it must use or assign. As the thread executes a program, it operates on these working copies. The main memory contains the master copy of each variable. There are rules about when a thread is permitted or required to transfer the contents of its working copy of a variable into the master copy or vice versa. Moreover, there are rules which regulate the *locking* and *unlocking* of objects, by means of which threads synchronize with each other. These rules are given in [3, Chapter 17] and have been formalized in [2] as "well-formedness" conditions for structures called *event spaces*. We summarize their definition and usage.

Event spaces will be included in the configurations of multi-threaded Java to constrain the applicability of certain operational rules. Additionally, they will be used to model the working memories of all threads. The main memory is modeled as an abstract store that can be thought of as mapping addresses of instance variables (left-values, from a semantic domain *LVal*) of objects (from a semantic domain *Obj*) to values or object references (right-values, from a semantic domain *RVal*).

In accord with [3], the terms *Use*, *Assign*, *Load*, *Store*, *Read*, *Write*, *Lock*, and *Unlock* are used here to name actions which describe the activity of the memories during the execution of a Java program. *Use* and *Assign* denote the above mentioned actions of the private working memory. *Read* and *Load* are used for a loosely coupled copying of data from the main memory to a working memory and dually *Store* and *Write* are used for copying data from a working memory to the main memory (as mentioned in the introduction).

We let the metavariable A (possibly indexed) stand for a generic action name. Moreover, we let B range over the set of thread actions and C over the set of memory actions, that is: $B \in \{Use, Assign, Load, Store, Lock, Unlock\}$, $C \in \{Read, Write, Lock, Unlock\}$.

Let *Thread_id* be a set of thread identifiers. An *action* is either a 4-tuple of the form (A, θ, l, v) where $A \in \{Assign, Store, Read\}$, $\theta \in$ *Thread_id*, $l \in$ *LVal* and $v \in$ *RVal*, or a triple (A, θ, l), where θ and l are as above and $A \in \{Use, Load, Write\}$, or a triple (A, θ, o), where $A \in \{Lock, Unlock\}$ and $o \in$ *Obj*.

Events are instances of actions, which we think of as happening at different times during execution. We use the same tuple notation for actions and their instances (the context clarifies which one is meant) and let a, b, c stand for either. Sometimes we omit components of an action or event: we may write e.g. $(Read, l)$ for $(Read, \theta, l, v)$ when θ and v are not relevant.

An *event space* is a poset of events (thought of as occurring in the given order) in which every chain can be enumerated monotonically with respect to the arithmetical ordering $0 \leq 1 \leq 2 \leq \ldots$ of natural numbers, and which satisfies the conditions (17.2.1–17.6.2') of Table 1. These conditions, which formalize directly the rules of [3, Chapter 17], are expressed by clauses of the form:

$$\forall \boldsymbol{a} \in \eta \,.\, (\Phi \Rightarrow ((\exists \boldsymbol{b}_1 \in \eta \,.\, \Psi_1) \vee (\exists \boldsymbol{b}_2 \in \eta \,.\, \Psi_2) \vee \ldots\ (\exists \boldsymbol{b}_n \in \eta \,.\, \Psi_n)))$$

where $\boldsymbol{a}$ and $\boldsymbol{b}_i$ are lists of events, η is an event space and $\forall \boldsymbol{a} \in \eta \,.\, \Phi$ means that Φ holds for all tuples of events in η matching the elements of $\boldsymbol{a}$ (and similarly for $\exists \boldsymbol{b} \in \eta \,.\, \Psi$). Such statements are abbreviated by adopting the following conventions: quantification over $\boldsymbol{a}$ is left implicit when all events in $\boldsymbol{a}$ appear in Φ; quantification over $\boldsymbol{b}_i$ is left implicit when all events in $\boldsymbol{b}_i$ appear in Ψ_i. Moreover, a rule of the form $\forall \boldsymbol{a} \in \eta \,.\, (\mathit{true} \Rightarrow \ldots)$ is written $\boldsymbol{a} \Rightarrow (\ldots)$. The term $(A, \theta, x)_n$ denotes the n-th occurrence of (A, θ, x) in a given space, if such an event exists, and is undefined otherwise.

We include the origin of each rule from [3, Chapter 17] and refer to [3] and [2] for more detail. For instance, rule (17.2.1) says that actions performed by any thread are totally ordered and (17.2.2) that so are the actions performed by the main memory for any variable or object. Similarly, rules (17.6.2) and (17.6.2') say that a lock action acts as if it flushes all variables from the thread's working memory, i.e. before use they must be assigned or loaded from main memory.

A *complete event space* is an event space that additionally fulfills the axioms (17.2.6) and (17.2.7) such that any *Read* and *Store* events are "completed" by corresponding *Load* and *Write* events.

A new event $a = (A, \theta, x)$ is adjoined to an event space η by extending the execution order as follows: if A is a thread action, then $b \leq a$ for all instances b of (B, θ) in η; if a is a main memory action, then $c \leq a$ for all instances c of (C, x) in η. Moreover, if A is *Load* then $c \leq a$ for all instances c of $(\mathit{Read}, \theta, x)$ in η, and if A is *Write* then $c \leq a$ for all instances c of $(\mathit{Store}, \theta, x)$ in η. The term $\eta \oplus a$ denotes the space thus obtained, provided it obeys the above rules, and it is otherwise undefined. If it is defined then $\eta \oplus a\!\downarrow$ yields true and false otherwise. If η is an event space and $\boldsymbol{a} = (a_1, a_2, \ldots a_n)$ is a sequence of events, we write $\eta \oplus \boldsymbol{a}$ for $\eta \oplus a_1 \oplus a_2 \oplus \cdots \oplus a_n$ and analogously $\eta \oplus \boldsymbol{a}\!\downarrow$.

3 Operational Semantics

We briefly recapitulate the structural operational semantics of multi-threaded Java in [2]. We restrict ourselves to those parts that are relevant for the "prescient" compiler optimization.

Objects are kept in the main memory. We use a semantic domain *Store* that is abstractly given by the following five semantic functions: $\mathit{upd} : \mathit{LVal} \times \mathit{RVal} \times \mathit{Store} \to \mathit{Store}$ updates a given store; we write $\mu[l \mapsto v]$ for $\mathit{upd}(l, v, \mu)$. The function $\mathit{lval} : \mathit{Obj} \times \mathit{Identifier} \times \mathit{Store} \rightharpoonup \mathit{LVal}$ retrieves the left-value of an instance variable (such as of `o.a`). Analogously, $\mathit{rval} : \mathit{LVal} \times \mathit{Store} \rightharpoonup \mathit{RVal}$ retrieves the right-value of a left-value; we write $\mu(l)$ for $\mathit{rval}(l, \mu)$. New objects are allocated by $\mathit{new}_C : \mathit{Store} \to \mathit{Obj} \times \mathit{Store}$. This family of functions is indexed by class types $C \in \mathit{ClassType}$. For a given store μ, $\mathit{new}_C(\mu)$ yields an object o and a store μ' such that μ' extends μ where $\mu'(\mathit{lval}(o, i, \mu'))$ is defined for any identifier i ranging over all instance variables of the class C.

$$(B,\theta),(B',\theta) \Rightarrow (B,\theta) \leq (B',\theta) \vee (B',\theta) \leq (B,\theta) \tag{17.2.1}$$

$$(C,x),(C',x) \Rightarrow (C,x) \leq (C',x) \vee (C',x) \leq (C,x) \tag{17.2.2}$$

$$(Assign,\theta,l) \leq (Load,\theta,l) \Rightarrow (Assign,\theta,l) \leq (Store,\theta,l) \leq (Load,\theta,l) \tag{17.3.2}$$

$$(Store,\theta,l)_m < (Store,\theta,l)_n \Rightarrow \\ (Store,\theta,l)_m \leq (Assign,\theta,l) \leq (Store,\theta,l)_n \tag{17.3.3}$$

$$(Use,\theta,l) \Rightarrow (Assign,\theta,l) \leq (Use,\theta,l) \vee (Load,\theta,l) \leq (Use,\theta,l) \tag{17.3.4}$$

$$(Store,\theta,l) \Rightarrow (Assign,\theta,l) \leq (Store,\theta,l) \tag{17.3.5}$$

$$(Assign,\theta,l,v)_n \leq (Store,\theta,l,v') \Rightarrow \\ v = v' \vee (Assign,\theta,l,v)_n < (Assign,\theta,l)_m \leq (Store,\theta,l,v') \tag{17.1}$$

$$(Load,\theta,l)_n \Rightarrow (Read,\theta,l)_n \leq (Load,\theta,l)_n \tag{17.3.6}$$

$$(Write,\theta,l)_n \Rightarrow (Store,\theta,l)_n \leq (Write,\theta,l)_n \tag{17.3.7}$$

$$(Store,\theta,l)_m \leq (Load,\theta,l)_n \Rightarrow (Write,\theta,l)_m \leq (Read,\theta,l)_n \tag{17.3.8}$$

$$(Lock,\theta,o)_n \leq (Lock,\theta',o) \wedge \theta \neq \theta' \Rightarrow (Unlock,\theta,o)_n \leq (Lock,\theta',o) \tag{17.5.1}$$

$$(Unlock,\theta,o)_n \Rightarrow (Lock,\theta,o)_n \leq (Unlock,\theta,o)_n \tag{17.5.2}$$

$$(Assign,\theta,l) \leq (Unlock,\theta) \Rightarrow \\ (Assign,\theta,l) \leq (Store,\theta,l)_n \leq (Write,\theta,l)_n \leq (Unlock,\theta) \tag{17.6.1}$$

$$(Lock,\theta) \leq (Use,\theta,l) \Rightarrow \\ (Lock,\theta) \leq (Assign,\theta,l) \leq (Use,\theta,l) \vee \\ (Lock,\theta) \leq (Read,\theta,l)_n \leq (Load,\theta,l)_n \leq (Use,\theta,l) \tag{17.6.2}$$

$$(Lock,\theta) \leq (Store,\theta,l) \Rightarrow (Lock,\theta) \leq (Assign,\theta,l) \leq (Store,\theta,l) \tag{17.6.2'}$$

$$(Read,\theta,l)_n \Rightarrow (Load,\theta,l)_n \tag{17.2.6}$$

$$(Store,\theta,l)_n \Rightarrow (Write,\theta,l)_n \tag{17.2.7}$$

Table 1. Event space axioms

The local variables of a block are kept in a stack of environments. *Environments*, denoted *Env*, are pairs (I, ρ) of declared identifiers $I \subseteq \mathit{Identifier} \cup \{\mathit{this}\}$ and a map $\rho : I \rightharpoonup \mathit{RVal}$ representing the values they (possibly) have. Environments are also used to store the information on which object's code is currently being executed ($\rho(\mathit{this})$). An environment ρ is updated as usual by $\rho[i \mapsto v]$. The empty environment is denoted by $\rho_\emptyset$. Let *S-Stack* be the domain of (single-threaded) *stacks of environments*. The empty stack is written $\sigma_\emptyset$. The operation $\mathit{push} : \mathit{Env} \times \textit{S-Stack} \rightarrow \textit{S-Stack}$ is the usual one on stacks. We use the operations $\sigma[i \mapsto v]$ for updating stacks, and $\sigma(i)$ for retrieving values. Each thread of execution of a Java program has its own stack. We call $\textit{M-Stack} = \textit{Thread\_id} \rightharpoonup \textit{S-Stack}$ the domain of *multi-threaded stacks*, ranged over by σ. Given $\sigma \in \textit{M-Stack}$, the multi-threaded stacks $\mathit{push}(\theta, \rho, \sigma)$, $\sigma[\theta, i \mapsto v]$ map θ' to $\sigma(\theta')$ when $\theta \neq \theta'$, and otherwise map θ respectively to $\mathit{push}(\rho, \sigma(\theta))$, $\sigma(\theta)[i \mapsto v]$. Note that an additional operation is necessary for extending (stacks of) environments when dealing with local variable declarations, but those are not addressed in this paper (cf. [2]). We also write $\sigma(\theta, i)$ instead of $\sigma(\theta)(i)$.

The operational semantics works on a set *M-Term* of *multi-threaded abstract terms* that contain *single-threaded abstract terms* from a set *S-Term*. We let the metavariable t range over *S-Term*. To each syntactic category of Java we associate a homonymous category of abstract terms. The well-typed terms of Java are mapped to abstract terms of corresponding category by a translation $(\_)^\circ$, which we leave implicit when no confusion arises. Abstract blocks are terms of the form $\{t\}_{(I,\rho)}$ where the source I of the environment (I,ρ) contains the local variables of the block. A multi-threaded abstract term T is a set of pairs (θ,t), where $\theta \in$ *Thread_id* and $t \in$ *S-Term* and no distinct elements of T bear the same thread identifier. Multi-threaded abstract terms $\{(\theta_1,t_1),(\theta_2,t_2),\ldots\}$ are written as lists $(\theta_1,t_1) \mid (\theta_2,t_2) \mid \ldots$ and pairs (θ,t) are written t when θ is irrelevant.

The *configurations* of multi-threaded Java are 4-tuples (T,η,σ,μ) consisting of an M-term T, an event space η, an M-stack σ, and a store μ. The operational semantics is the binary relation $\longrightarrow$ on configurations inductively defined by the rules that follow. In the rule schemes in Tables 2–4, the metavariables range as follows: $i \in$ *Identifier*, $k \in$ *Identifier* $\cup$ *LVal*, $l \in$ *LVal*, $o \in$ *Obj*, $e \in$ *Expression*, $v \in$ *RVal*, $s \in$ *Statement*, $b \in$ *BlockStatement*, $B \in$ *BlockStatement**, $q \in$ *Block*, $t \in$ *S-Term*, and $T \in$ *M-Term*. Stacks, event spaces, and stores are *omitted* when they are not relevant.

We write $store_\eta(\theta,l)$ for the oldest unwritten value of l stored by θ in η. More formally: let an event $(Store,\theta,l)_n$ in η be called *unwritten* if $(Write,\theta,l)_n$ is undefined in η; then, $store_\eta(\theta,l) = v$ if there exists an unwritten $(Store,\theta,l,v)_n$ such that for any unwritten $(Store,\theta,l)_m$ we have $n \leq m$; if no such a *Store* event exists, $store_\eta(\theta,l)$ is undefined. Similarly, we write $rval_\eta(\theta,l)$ for the latest value of l assigned or loaded and read by θ in η.

[assign1]	$\dfrac{e_1 \longrightarrow e_2}{e_1 = e \longrightarrow e_2 = e}$	[assign2]	$\dfrac{e_1 \longrightarrow e_2}{k = e_1 \longrightarrow k = e_2}$
[assign3']	$(\theta, l = v), \eta \longrightarrow (\theta, v), \eta \oplus (Assign, \theta, l, v)$		
[assign4']	$(\theta, i = v), \sigma \longrightarrow (\theta, v), \sigma[\theta, i \mapsto v]$		
[binop1]	$\dfrac{e_1 \longrightarrow e_2}{e_1 \ \mathbf{op}\ e \longrightarrow e_2 \ \mathbf{op}\ e}$	[binop2]	$\dfrac{e_1 \longrightarrow e_2}{v \ \mathbf{op}\ e_1 \longrightarrow v \ \mathbf{op}\ e_2}$
[binop3]	$v_1 \ \mathbf{op}\ v_2 \longrightarrow v_1 \ op\ v_2$	[pth]	$(e) \longrightarrow e$
[access1]	$\dfrac{e_1 \longrightarrow e_2}{e_1 . i \longrightarrow e_2 . i}$	[access2]	$o.i, \mu \longrightarrow lval(o,i,\mu), \mu$
[this]	$(\theta, \mathtt{this}), \sigma \longrightarrow (\theta, \sigma(\theta, this)), \sigma$	[new]	$\mathtt{new}\ C\,(\,), \mu \longrightarrow new_C(\mu)$
[val']	$(\theta, l), \eta \longrightarrow (\theta, rval_\eta(\theta, l)), \eta \oplus (Use, \theta, l)$		
[var']	$(\theta, i), \sigma \longrightarrow (\theta, \sigma(\theta, i)), \sigma$		

Table 2. Expressions

[statseq1]	$\dfrac{b_1 \longrightarrow b_2}{b_1\,B \longrightarrow b_2\,B}$	[statseq2]	$\dfrac{b, \mu_1 \longrightarrow \mu_2}{b\,B, \mu_1 \longrightarrow B, \mu_2}$
[expstat1]	$\dfrac{e_1 \longrightarrow e_2}{e_1\,; \longrightarrow e_2\,;}$	[expstat2]	$\dfrac{e, \mu_1 \longrightarrow v, \mu_2}{e\,;\,, \mu_1 \longrightarrow \mu_2}$
[skip]	$;\,, \sigma \longrightarrow \sigma$	[block1]	$\{\,\}_\rho\,, \sigma \longrightarrow \sigma$

[block2']
$$\frac{(\theta, B_1), push(\theta, \rho_1, \sigma_1) \longrightarrow (\theta, B_2), push(\theta, \rho_2, \sigma_2)}{(\theta, \{B_1\}_{\rho_1}), \sigma_1 \longrightarrow (\theta, \{B_2\}_{\rho_2}), \sigma_2}$$

[if1]
$$\frac{e_1 \longrightarrow e_2}{\mathtt{if}(e_1)\ s \longrightarrow \mathtt{if}(e_2)\ s}$$

[if2]	$\mathtt{if}(\mathit{true})\ s \longrightarrow s$	[if3]	$\mathtt{if}(\mathit{false})\ s, \mu \longrightarrow \mu$

[while]
$$\mathtt{while}(e)\ s \longrightarrow \mathtt{if}(e)\ \{\ s\ \mathtt{while}(e)\ s\ \}$$

[for]
$$\mathtt{for}(e_1;\ e_2;\ e_3)\ s \longrightarrow \{\ e_1;\ \mathtt{while}(e_2)\ \{\ s\ e_3;\ \}\ \}$$

Table 3. Statements

[synchro1]
$$\frac{e_1 \longrightarrow e_2}{\mathtt{synchronized}(e_1)\ q \longrightarrow \mathtt{synchronized}(e_2)\ q}$$

[synchro2]
$$\frac{q_1 \longrightarrow q_2}{\mathtt{synchronized}(o)\ q_1 \longrightarrow \mathtt{synchronized}(o)\ q_2}$$

[lock]
$$\frac{(\theta, e), \eta_1 \longrightarrow (\theta, o), \eta_2}{(\theta, \mathtt{synchronized}(e)\ q), \eta_1 \longrightarrow (\theta, \mathtt{synchronized}(o)\ q), \eta_2 \oplus (\mathit{Lock}, \theta, o)}$$

[unlock]
$$(\theta, \mathtt{synchronized}(o)\ \{\,\}_\rho), \eta \longrightarrow \eta \oplus (\mathit{Unlock}, \theta, o)$$

[read]
$$T, \eta, \mu \longrightarrow T, \eta \oplus (\mathit{Read}, \theta, l, \mu(l)), \mu$$

[load]
$$T, \eta \longrightarrow T, \eta \oplus (\mathit{Load}, \theta, l)$$

[store]
$$T, \eta \longrightarrow T, \eta \oplus (\mathit{Store}, \theta, l, v)$$

[write]
$$T, \eta, \mu \longrightarrow T, \eta \oplus (\mathit{Write}, \theta, l), \mu[l \mapsto \mathit{store}_\eta(\theta, l)]$$

[par]
$$\frac{t_1 \longrightarrow t_2}{t_1 \,|\, T \longrightarrow t_2 \,|\, T}$$

Table 4. Multi-threaded Java

The rules [assign3', val', lock, unlock, read, load, store, write] make use of the well-formedness conditions of event spaces via the $\oplus$. The rules [read, load, store, write] are spontaneous in the sense that they do not depend on T. The [store] rule additionally "guesses" the value of the last *Assign*; its correctness is ensured by axiom (17.1). *Synchronization*, i.e. mutual exclusion, is handled by [synchro1, synchro2, lock, unlock], by [par] sequential computations are lifted to multi-threaded ones.

4 Prescient Event Spaces

The *prescient* store actions are introduced in [3, 17.8] as follows: " ... the *store* [of variable V by thread T] action [is allowed] to instead occur before the *assign* action, if the following rule restrictions are obeyed:

- If the *store* occurs, the *assign* is bound to occur. ...
- No *lock* action intervenes between the relocated *store* and the *assign*.
- No *load* of V intervenes between the relocated *store* and the *assign*.
- No other *store* of V intervenes between the relocated *store* and the *assign*.
- The *store* action sends to the main memory the value that the *assign* action will put into the working memory of thread T.

The last property inspires us to call such an early *store* action *prescient*: ... "

The specification above seems to assume that it is known which *Store* events are prescient and which prescient *Store* event is matched by which *Assign* event. We do not assume such knowledge but adopt a more general approach introducing so-called complete labellings. These labellings are not necessarily unique but it is always possible to infer a complete labelling at run time. It will turn out, however, that the semantics is independent of the choice of complete labellings, see Corollary 6.

In order to define the new *prescient event spaces* we proceed as follows:

First, we have to add new relations (cf. axioms (17.2.1), (17.2.2)) between certain actions of different threads in order to be able to formalize the preconditions of the second, third, and fourth requirement above. *Assign*, *Load* or *Store* actions for the same variable and *Lock* actions must be comparable. To this end let $D = \{Assign, Load, Store\}$, then we stipulate:

$$(D,\theta,l),(D,\theta',l) \Rightarrow (D,\theta,l) \leq (D,\theta',l) \vee (D,\theta',l) \leq (D,\theta,l)$$

Since *Store* and *Lock* events are already comparable, by transitivity also *Lock* and D actions are comparable.

Second, rules (17.3.3), (17.3.5), (17.1), and (17.6.2') are now used for the definition of a predicate *prescient* on event spaces and *Store* events yielding true iff a *Store* is necessarily prescient. We define $prescient_\eta((Store,\theta,l)_n)$ to be valid if one of the rules in Table 5 holds. Note that η is usually omitted if it is clear from the context.

Rules (P1–P4) simply tell that a *Store* event which does not obey old rules (17.3.3), (17.3.5), (17.1), or (17.6.2') is necessarily prescient. Rule (P5) is sound because if there is only one $(Assign,\theta,l,v)$ between two stores and the first is *prescient*, then by re-arranging the prescient *Store* two *Store* events would follow each other without a triggering *Assign* in between, which contradicts the old semantics.

Third, keep rules (17.2.1), (17.2.2), (17.3.4), (17.3.6), (17.3.7), (17.3.8), (17.5.1), (17.5.2), and (17.6.2).

$$(Store,\theta,l)_m \leq (Store,\theta,l)_n \not\Rightarrow (Store,\theta,l)_m \leq (Assign,\theta,l) \leq (Store,\theta,l)_n \quad \text{(P1)}$$

$$(Store,\theta,l)_n \not\Rightarrow (Assign,\theta,l) \leq (Store,\theta,l)_n \quad \text{(P2)}$$

$$(Assign,\theta,l,v')_m \leq (Store,\theta,l,v)_n \not\Rightarrow v = v' \vee (Assign,\theta,l,v')_m \leq (Assign,\theta,l)_k \leq (Store,\theta,l,v)_n \quad \text{(P3)}$$

$$(Lock,\theta) \leq (Store,\theta,l)_n \not\Rightarrow (Lock,\theta) \leq (Assign,\theta,l) \leq (Store,\theta,l)_n \quad \text{(P4)}$$

$$(Store,\theta,l)_m \leq (Assign,\theta,l)_k \leq (Assign,\theta,l)_{k'} \leq (Store,\theta,l)_n \wedge prescient((Store,\theta,l)_m) \Rightarrow k = k' \quad \text{(P5)}$$

Table 5. Rules for *prescient*

Fourth, adapt rule (17.3.2) as follows, allowing prescient *Stores* on the right hand side of an implication:

$$\begin{aligned}&(Assign,\theta,l,v) \leq (Load,\theta,l) \Rightarrow\\ &\quad((Assign,\theta,l) \leq (Store,\theta,l) \leq (Load,\theta,l)) \vee\\ &\quad((Store,\theta,l,v) \leq (Assign,\theta,l,v) \leq (Load,\theta,l) \wedge prescient(Store,\theta,l,v))\end{aligned}$$

and rule (17.6.1) as follows:

$$\begin{aligned}&(Assign,\theta,l,v) \leq (Unlock,\theta) \Rightarrow\\ &\quad(Assign,\theta,l) \leq (Store,\theta,l)_n \leq (Write,\theta,l)_n \leq (Unlock,\theta)) \vee\\ &\quad((Store,\theta,l,v)_n \leq (Assign,\theta,l,v) \leq (Unlock,\theta) \wedge\\ &\quad(Write,\theta,l)_n \leq (Unlock,\theta) \wedge prescient((Store,\theta,l,v)_n))\end{aligned}$$

Finally, we need an additional rule corresponding to the second, third, and fourth requirements in the citation at top of Section 4. We add a new rule scheme: for any $a \in \{(Lock), (Load,l), (Store,l)\}$:

$$\begin{aligned}&(Store,\theta,l,v)_n < a \wedge prescient((Store,\theta,l,v)_n) \Rightarrow\\ &\quad(Store,\theta,l,v)_n \leq (Assign,\theta,l,v) \leq a\end{aligned} \quad (17.8)$$

Next, we redefine the operation $\oplus$ on *prescient event spaces*: A new event a is adjoined to a prescient event space η as in the case for old event spaces, but one additional condition. Let $A \in \{Store, Assign, Load\}$. If $a = (A,\theta,l)$ and $b = (A,\theta',l) \in \eta$ then $b \leq a$. Also, the term $\eta \oplus a$ denotes the space thus obtained, provided it obeys the above rules for prescient event spaces, and it is otherwise undefined.

Analogously to the predicate *prescient* one can also define a predicate *non_prescient* which contains only *Stores* that are necessarily non-prescient. We define *non_prescient*$((Store,\theta,l)_m)$ on an (implicitly) given event space to be true if one of the rules of Table 6 is fulfilled.

Rule (NP2) is the dual of (P5). Moreover, rule (NP2) is raised by (17.3.3) and (NP3) by new rule (17.8). Observe also that the predicate *prescient* propagates from past to present whereas *non_prescient* is computed in the opposite direction. Note that $\neg$ *non_prescient*(B) is *not* equivalent to *prescient*(B)

$$\neg\exists(Assign,\theta,l,v)\,.\,(Store,\theta,l)_m \leq (Assign,\theta,l,v) \tag{NP1}$$

$$(Store,\theta,l)_m \leq (Assign,\theta,l)_k \leq (Assign,\theta,l)_{k'} \leq (Store,\theta,l)_n \wedge \quad non\_prescient((Store,\theta,l)_n) \Rightarrow k = k' \tag{NP2}$$

$$\forall a \in \{(Lock),(Load,l),(Store,l)\}\,.\,(Store,\theta,l,v)_m < a \Rightarrow \quad \neg\exists(Assign,\theta,l,v)\,.\,(Store,\theta,l,v)_m \leq (Assign,\theta,l,v) \leq a \tag{NP3}$$

Table 6. Rules for *non_prescient*

and hence also *prescient*(B) $\vee$ *non_prescient*(B) does not always hold and *prescient*(B) $\wedge$ *non_prescient*(B) is not always false.

A prescient event space η is called *consistently complete* if it is complete and for no instance of a *Store*, say s, we have that $prescient_\eta(s) \wedge non\_prescient_\eta(s)$. Note that it makes only sense for the final event space of a reduction sequence to be consistently complete (as for complete). During execution, the matching *Assign* for a prescient *Store* might not have happened and therefore the corresponding *Store* would be considered *non_prescient*, which might lead to a contradiction. A consistently complete event space fulfills the first and last requirement in [3, §17.8] (see top of Section 4), because a prescient *Store* would otherwise have no matching *Assign* and hence by rule (NP1) contradict consistently completeness.

There might be a *Store* event s in a given event space for which neither *prescient*(s) nor *non_prescient*(s) is derivable. In this case one needs a "labelling" of *Store* events, i.e. a predicate fixing whether a *Store* shall be considered prescient or not. More formally, a *labelling* for a prescient event space is a predicate ℓ on *Store* events such that it obeys rules (L1–L3) in Table 7.

$$prescient(s) \Rightarrow \ell(s) \tag{L1}$$

$$non\_prescient(s) \Rightarrow \neg\ell(s) \tag{L2}$$

$$((Store,\theta,l)_m \leq (Assign,\theta,l)_k \leq (Assign,\theta,l)_{k'} \leq (Store,\theta,l)_n \Rightarrow k = k') \Rightarrow \quad (\ell((Store,\theta,l)_m) \Rightarrow \ell((Store,\theta,l)_n)) \tag{L3}$$

$$prescient(s) \Rightarrow p^*(s) \tag{PC1}$$

$$p(s) \Rightarrow p^*(s) \tag{PC2}$$

$$((Store,\theta,l)_m \leq (Assign,\theta,l)_k \leq (Assign,\theta,l)_{k'} \leq (Store,\theta,l)_n \Rightarrow k = k') \Rightarrow \quad (p^*((Store,\theta,l)_m) \Rightarrow p^*((Store,\theta,l)_n)) \tag{PC3}$$

Table 7. Rules for labelling and prescient closure

Rule (L3) implies that $\neg\ell$ is closed under (NP2).

Let p be any binary predicate on event spaces and *Store* events (where we usually omit the event space argument). Then we define the *prescient closure* of p, the binary predicate p^*, inductively by rules (PC1–PC3) of Table 7.

Lemma 1. *For any consistently complete event space one can give a labelling.*

Proof. Choose a p such that $\neg(p(s) \wedge \mathit{non\_prescient}(s))$ holds for any *Store* event s. This is possible since the event space is consistently complete; for example, $p = \mathit{prescient}$ (or equivalently $p = \mathit{false}$) will do. It remains to prove that p^* is a labelling: rules (L1) and (L3) hold by (PC1) and (PC3), respectively. In order to show rule (L2) prove by induction on the derivation of $p^*(s)$ that $\mathit{non\_prescient}(s) \wedge p^*(s)$ leads to a contradiction. In the (PC1)-case one needs consistently completeness and in the (PC2)-case the assumption on p.

For a consistently complete prescient event space with a labelling the *Assign* events matching the prescient *Stores* can also be singled out as follows: Let ℓ be a labelling on an prescient event space η. A *matching* (labelling of *Assigns*) on ℓ and η, m_ℓ, is a predicate on the *Assign* events of η fulfilling the three axioms in Table 8.

$$\begin{aligned}
&\forall a \in \{(\mathit{Lock}), (\mathit{Load}, l), (\mathit{Store}, l)\} \,.\, (\mathit{Store}, \theta, l, v) < a \wedge \ell((\mathit{Store}, \theta, l, v)) \Rightarrow \\
&\qquad (\mathit{Store}, \theta, l, v) \leq (\mathit{Assign}, \theta, l, v) \leq a \wedge m_\ell((\mathit{Assign}, \theta, l, v)) \\
&(\mathit{Store}, \theta, l, v) \wedge \ell((\mathit{Store}, \theta, l, v)) \Rightarrow \\
&\qquad (\mathit{Store}, \theta, l, v) \leq (\mathit{Assign}, \theta, l, v) \wedge m_\ell((\mathit{Assign}, \theta, l, v)) \\
&(\mathit{Store}, \theta, l, v)_k \leq (\mathit{Assign}, \theta, l, v)_m < (\mathit{Assign}, \theta, l, v)_n \wedge \\
&\qquad\qquad \ell((\mathit{Store}, \theta, l, v)_k) \wedge m_\ell((\mathit{Assign}, \theta, l, v)_m) \wedge m_\ell((\mathit{Assign}, \theta, l, v)_n) \Rightarrow \\
&\qquad (\mathit{Store}, \theta, l, v)_k \leq (\mathit{Assign}, \theta, l, v)_m \leq (\mathit{Store}, \theta, l, v)_{k'} \leq (\mathit{Assign}, \theta, l, v)_n \wedge \\
&\qquad\qquad\qquad\qquad\qquad\qquad \ell((\mathit{Store}, \theta, l, v)_{k'})
\end{aligned}$$

Table 8. Rules for matching

It is easily checked that the following predicate fulfills the axioms for matchings.

$$\begin{aligned}
&\hat{m}_\ell((\mathit{Assign}, \theta, l, v)_m) \Leftrightarrow \\
&\qquad \exists(\mathit{Store}, \theta, l, v).\, (\mathit{Store}, \theta, l, v) \leq (\mathit{Assign}, \theta, l, v)_m \wedge \ell((\mathit{Store}, \theta, l, v)) \wedge \\
&\qquad \neg\exists(\mathit{Assign}, \theta, l, v)_n \,.\, (\mathit{Store}, \theta, l, v) \leq (\mathit{Assign}, \theta, l, v)_n < (\mathit{Assign}, \theta, l, v)_m
\end{aligned}$$

A *complete labelling* is a pair consisting of a labelling and a matching for this labelling.

For the sake of simplicity we assume in the rest of the paper that a complete labelling is always given and exhibited in form of special action names, i.e. *pStore* and *pAssign*. If $\mathit{prescient}(\mathit{Store}, \theta, l, v)$ holds then $(\mathit{Store}, \theta, l, v)$ is denoted $(\mathit{pStore}, \theta, l, v)$ and analogously for the matching *Assign* we use *pAssign*.

5 Prescient Operational Semantics

We obtain the prescient operational semantics from the old semantics of Section 3 just by switching from the event spaces of Section 2 to the prescient event spaces of Section 4 keeping the operational rules untouched.

For the prescient operational semantics we write $\longrightarrow\!\!\triangleright$. Moreover, let $\mathit{Conf}_\triangleright$ denote the set of configurations with prescient event spaces, and $\mathit{Conf}_\blacktriangleright$ those according to the definition $\longrightarrow$ of Section 2.

Lemma 2. *Any event space η (obeying the old rules) is also a prescient event space, thus any old configuration is a new configuration, i.e. $\mathit{Conf}_{\blacktriangleright} \subseteq \mathit{Conf}_{\triangleright}$, and any reduction $\Gamma \longrightarrow \Gamma'$ is also a prescient one, i.e. $\Gamma \mathrel{-\!\!\triangleright} \Gamma'$ holds as well.*

Proof. Assume η is an event space satisfying the old rules. By a simple induction, $\mathit{prescient}_{\eta}(s)$ never holds for any *Store* event s in η. Thus η is a prescient event space because the new rules form a subset of the old rules. Since the configurations only differ in the event space definition and the rules of the semantics are not changed at all, the other claims of the lemma now hold trivially.

Since we use labellings our operational semantics is very liberal. It accepts reductions using *Store* events even if it is not clear during execution whether this *Store* event is meant to be prescient or not. In such a case, however, the prescient *Store* is not done as early as possible. Therefore, in practical cases, any *Store* which is not immediately recognized by the rules (P1–P5) can be considered nonprescient. This corresponds to the prescient closure false^* (cf. Lemma 1) meaning that the labelling is computed at run time. By definition also $\hat{m}_{\mathit{false}^*}$ is computable at run time, thus a complete labelling is, too.

6 Prescient Semantics is conservative

The relation between the "normal" and the "prescient" semantics is described in [3, §17.8] as follows: "The purpose of this relaxation is to allow optimizing Java compilers to perform certain kinds of code rearrangements that preserve the semantics of properly synchronized programs but might be caught in the act of performing memory actions out of order by programs that are not properly synchronized."

This has to be formalized in the sequel. The following notation, exemplified for $\longrightarrow$ only, will be used analogously for all kinds of arrows: $\stackrel{r}{\longrightarrow}$ denotes a one-step reduction with rule r; if $e = (r_1, \ldots, r_n)$ is a list of rules then $\stackrel{e}{\longrightarrow}$ denotes $\stackrel{r_1}{\longrightarrow} \ldots \stackrel{r_n}{\longrightarrow}$; if the list is irrelevant we write $\longrightarrow^*$. For rules that change the event space we often decorate arrows with actions instead of rule names as the latter are ambiguous.

First, we observe that $\mathrel{-\!\!\triangleright}$ and $\longrightarrow$ can not be bisimilar by definition since $\mathrel{-\!\!\triangleright}$ permits *Store*-actions where $\longrightarrow$ does not. But $\mathrel{-\!\!\triangleright}$ cannot even be bisimilar to the reflexive closure of $\longrightarrow$, since simulating a $(\mathit{pStore}, \theta, l)$ and the following *Writes* by void steps leads to inequivalent configurations (since the main memories will contain different values for l).

As a prerequisite for a simulation relation of type $\mathit{Conf}_{\blacktriangleright} \times \mathit{Conf}_{\triangleright}$, we define an equivalence on prescient configurations $\sim \subseteq \mathit{Conf}_{\triangleright} \times \mathit{Conf}_{\triangleright}$ as follows:

$$(T, \eta, \sigma, \mu) \sim (T', \eta', \sigma', \mu') \iff T = T' \wedge \sigma = \sigma' \wedge (T, \eta, \sigma, \mu) \downarrow (T', \eta', \sigma', \mu')$$

$$(T, \eta, \sigma, \mu) \downarrow (T', \eta', \sigma', \mu') \iff \forall \boldsymbol{a} \,.\, \eta \oplus \boldsymbol{a}\!\downarrow \Leftrightarrow \eta' \oplus \boldsymbol{a}\!\downarrow \wedge$$

$$\forall e.\, (T, \eta, \sigma, \mu) \stackrel{e}{\mathrel{-\!\!\triangleright}}{}^{c} (T_1, \eta_1, \sigma_1, \mu_1) \wedge (T', \eta', \sigma', \mu') \stackrel{e}{\mathrel{-\!\!\triangleright}}{}^{c} (T_2, \eta_2, \sigma_2, \mu_2) \Rightarrow \mu_1 = \mu_2$$

where a is any sequence of actions, e is a sequence of rules and $(T, \sigma, \eta, \mu) \longrightarrow^{c} (T', \sigma', \eta', \mu')$ if $(T, \sigma, \eta, \mu) \longrightarrow^{*} (T', \sigma', \eta', \mu')$ such that η' is complete.

This equivalence relation is obviously preserved by the rules of the semantics:

Lemma 3. *The relation $\sim$ is an equivalence relation such that if $\Gamma_1 \sim \Gamma_2$ then $\Gamma_1 \xrightarrow{r} \Gamma_1'$ iff $\Gamma_2 \xrightarrow{r} \Gamma_2'$ for any rule r, and if such a reduction r exists then $\Gamma_1' \sim \Gamma_2'$ holds.*

In order to establish a bisimulation result, we must delay all the operations which are possible due to a $(pStore, \theta, l, v)$ until the matching *pAssign* event.

But that will not work for all kinds of programs. Consider the following example:

$$(\theta,\ \{\ \texttt{synchronized}(o)\ \{\ l = v;\ \}_{\rho_\emptyset}\ \}_{\rho_\emptyset})\ \mid\ (\theta',\ \{\ l = v';\ \}_{\rho_\emptyset})$$

Its execution may give rise to a sequence of computation steps which contains the following complete subsequence of actions:

$$(Lock, \theta, o), (Assign, \theta, l, v), (Store, \theta, l, v), (pStore, \theta', l, v'),$$
$$(Write, \theta', l), (Write, \theta, l), (Unlock, \theta, o), (pAssign, \theta', l, v')$$

In a simulation the $(Store, \theta', l, v')$ is illegal w.r.t. to the old event space definition and can only be simulated by a void (i.e. delaying) step as well as the following *Write*. Now the $(Write, \theta, l)$ is bound to occur before the *Unlock* and therefore also $(Store, \theta, l, v)$. Finally, after the *Assign* we must recover the pending prescient $(Store, \theta', l, v')$ and its corresponding $(Write, \theta', l)$. According to this simulation l has value v' in the global memory, but the reduction via $\longrightarrow$ yields v for l. Thus, both end-configurations are not equivalent, a contradiction.

Therefore, we have to restrict ourselves to "properly synchronized" programs. A multi-threaded program T is called *properly synchronized* if for any configuration $(T', \eta', \sigma', \mu')$ such that $(T, \eta, \sigma_\emptyset, \emptyset) \longrightarrow^{*} (T', \eta', \sigma', \mu')$ and $(Write, \theta_1, l, v_1) \leq (Write, \theta_2, l, v_2)$ in η' there is a $(Lock, \theta_3, o)$ in η' such that $(Write, \theta_1, l, v_1) \leq (Lock, \theta_3, o) \leq (Write, \theta_2, l, v_2)$. To be "properly synchronized" is a semantical (and rather intricate) property which for a program is hard to tell in advance. A sufficient condition for "properly synchronizedness" is the syntactic criterion that in a program shared variables may only be written in synchronized blocks. It is clear, that in any execution sequence two *Write* actions must then be separated by the corresponding *Lock*.

In the sequel Δ (possibly with annotations) stands for configurations in $Conf_{\blacktriangleright}$ and Γ for new configurations in $Conf_{\triangleright}$. Recall that any old configuration is also a valid one in the new sense by Lemma 2. According to the observations above, we define a new reduction relation $\rightarrowtail : (Conf_{\blacktriangleright} \times E^*) \times (Conf_{\blacktriangleright} \times E^*)$ where $E = \{(pStore), (Write), (Read)\}$ by the rules of Table 6. Note that we do not need to treat $(Load)$ events (cf. rule (17.8)). The corresponding $\rightarrowtail$-configurations (Δ, e) consist of an old configuration $\Delta \in Conf_{\blacktriangleright}$ plus a list of "pending" events e. Appending an event a at the end of a list e is written $e \circ a$. An additional operation $split_{\theta,l}(e)$ is needed. Given a list of events e it yields a pair of lists (e_l, e') where both are sublists of e; e_l is obtained from e by extracting all $(pStore, \theta, l)$, $(Write, \theta, l)$ and $(Read, \theta', l)$ events simultaneously changing a $(pStore, \theta, l)$ into $(Store, \theta, l)$, and e' is e_l's complement w.r.t. e.

$$(\Delta, e) \overset{(pStore,\theta,l,v)}{\rhd\!\!\!\longrightarrow} (\Delta, e \circ (pStore, \theta, l, v)) \qquad (\text{red}_s)$$

$$(\Delta, e) \overset{(Write,\theta,l)}{\rhd\!\!\!\longrightarrow} (\Delta, e \circ (Write, \theta, l)) \ \text{ if } \ (pStore, \theta, l, v) \in e \qquad (\text{red}_w)$$

$$(\Delta, e) \overset{(Read,\theta',l,v)}{\rhd\!\!\!\longrightarrow} (\Delta, e \circ (Read, \theta', l, v)) \ \text{ if } \ (Write, \theta, l) \in e \qquad (\text{red}_r)$$

$$(\Delta, e) \overset{(pAssign,\theta,l,v)}{\rhd\!\!\!\longrightarrow} (\Delta', e') \ \text{ if } \ split_{\theta,l}(e) = (e_l, e') \wedge \Delta \xrightarrow{(Assign,\theta,l,v)} \Delta_1 \xrightarrow{e_l} \Delta' \qquad (\text{red}_a)$$

$$(\Delta, e) \overset{r}{\rhd\!\!\!\longrightarrow} (\Delta', e) \text{ for any other case } r \text{ if } \Delta \xrightarrow{r} \Delta' \qquad (\text{red}_d)$$

Table 9. Rules for the simulating reduction relation

To relate configurations of $\longrightarrow\!\!\!\rhd$ and $\rhd\!\!\!\longrightarrow$ reductions the simulation relation $\approx \subseteq Conf_{\rhd} \times (Conf_{\blacktriangleright} \times E^*)$ is defined as follows:

$$\Gamma \approx (\Delta, e) \quad \text{if, and only if,} \quad \Delta \overset{e}{\longrightarrow\!\!\!\rhd} \Gamma_\Delta \wedge \Gamma_\Delta \sim \Gamma$$

i.e. Γ is equivalent to (Δ, e) if Γ is equivalent to the completion of Δ, usually called Γ_Δ, by executing the pending events in e. Note that $\longrightarrow\!\!\!\rhd$ is used here for the sequence of events e, as e may contain prescient *Store* events.

Below we use the following notation of a commuting diagram

$$\begin{array}{ccc} \Gamma & \longrightarrow & \Gamma_1 \\ \downarrow & \sim & \downarrow \\ \Gamma_3 & \longrightarrow & \Gamma_2 \end{array}$$

stating that $\Gamma \longrightarrow \Gamma_1 \longrightarrow \Gamma_2$ and $\Gamma \longrightarrow \Gamma_3 \longrightarrow \Gamma_2'$ and $\Gamma_2 \sim \Gamma_2'$. This notation is also used for any other kind of arrows.

Lemma 4. *If $\Gamma \approx (\Delta, e)$ and $\Gamma \overset{r}{\longrightarrow\!\!\!\rhd} \Gamma'$, where r is as in case* (red$_\text{d}$) *and Γ stems from a properly synchronized program, then $\Delta \xrightarrow{r} \Delta'$ and the diagram*

$$\begin{array}{ccccc} \Delta & \overset{e}{\longrightarrow\!\!\!\rhd} & \Gamma_\Delta & \sim & \Gamma \\ \downarrow r & & r \downarrow & & \downarrow r \\ \Delta' & \overset{e}{\longrightarrow\!\!\!\rhd} & \Gamma'_\Delta & \sim & \Gamma' \end{array}$$

commutes, hence in particular $\Gamma \approx (\Delta, e) \overset{r}{\rhd\!\!\!\longrightarrow} (\Delta', e) \approx \Gamma'$ holds.

Proof. (sketched) By definition of $\rhd\!\!\!\longrightarrow$ we have $\Delta \xrightarrow{r} \Delta'$ as we consider case (red$_\text{d}$). Next, we have to check that r does not depend on e, such that commutation is possible. Proof is by inspecting the relevant laws for event spaces: rules

(17.3.2), (17.3.4), (17.6.2) refer to *Load* events which are not possible as long as e contains a corresponding *pStore*, (17.3.7) is not relevant as matching *Writes* are treated in (red_w). Thus, we are left with (17.6.1). Cases, however, where *Store* and *Write* in e allow r to be an *Unlock* are excluded by the rules for labellings.

To prove that the diagram commutes it suffices by definition of $\sim$ to show that the same actions are executed, but maybe in different order. We have to ensure that *Write* events of the same variable from different threads are not re-ordered. But this could only happen if $r = (\mathit{Write}, \theta, l)$ and another $(\mathit{Write}, \theta', l) \in e$ which is impossible since only properly synchronized programs are considered.

Theorem 5. *For properly synchronized programs the relation* $\approx$ *is a simulation relation of* $\longrightarrow\!\!\!\triangleright$ *and* $\triangleright\!\!\!\longrightarrow$, *i.e. if* $\Gamma \xrightarrow{r}\!\!\triangleright \Gamma'$ *during the execution of such a program and* $\Gamma \approx (\Delta, e)$ *then there is a* (Δ', e') *such that* $(\Delta, e) \triangleright\!\!\xrightarrow{r} (\Delta', e')$ *and* $\Gamma' \approx (\Delta', e')$.

Proof. Assume $\Gamma \approx (\Delta, e)$, i.e. $\Delta \xrightarrow{e}\!\!\triangleright \Gamma_\Delta \sim \Gamma$. We do a case analysis for $\Gamma \xrightarrow{r}\!\!\triangleright \Gamma'$:
Case $\Gamma \xrightarrow{Write}\!\!\triangleright \Gamma'$: if $(\mathit{pStore}, \theta, l) \in e$ then it holds that $(\Delta, e) \triangleright\!\!\xrightarrow{r} (\Delta, e \circ r)$ by (red_w). Moreover, by Lemma 3, $\Gamma' \approx (\Delta, e \circ r)$.

If $(\mathit{pStore}, \theta, l) \notin e$ then by Lemma 4, $(\Delta, e) \triangleright\!\!\xrightarrow{r} (\Delta', e')$ and $\Gamma' \approx (\Delta', e)$.
Case $\Gamma \xrightarrow{pAssign}\!\!\triangleright \Gamma'$. Let $\mathit{split}_{\theta,l}(e) = (e_l, e')$. Since an *Assign* is always possible, assume that $\Delta \xrightarrow{(Assign,\theta,l,v)} \Delta_1$. Now every action in e_l becomes legal for the old semantics, so we can further assume $\Delta_1 \xrightarrow{e_l} \Delta'$, such that $(\Delta, e) \triangleright\!\!\xrightarrow{r} (\Delta', e')$. One can prove analogously to Lemma 4 that the left rectangle in

$$
\begin{array}{ccccc}
\Delta & \xrightarrow{e}\!\!\triangleright & \Gamma_\Delta & \sim & \Gamma \\
{\scriptstyle (Assign,\theta,l,v)}\big\downarrow & & \big\downarrow{\scriptstyle r} & & \big\downarrow{\scriptstyle r} \\
\Delta_1 & \sim & & & \\
{\scriptstyle e_l}\big\downarrow & & & & \\
\Delta' & \xrightarrow{e'}\!\!\triangleright & \Gamma'_\Delta & \sim & \Gamma'
\end{array}
$$

commutes; the right rectangle commutes by Lemma 3, thus $(\Delta, e) \triangleright\!\!\xrightarrow{r} (\Delta', e')$ and $\Gamma' \approx (\Delta', e')$.

For *pStore* and *Read* one proceeds as for *Write*, all other cases follow from Lemma 4.

Our main result is the following corollary which states that the prescient semantics is conservative, i.e. any prescient execution sequence of a properly synchronized program can be simulated by a "normal" execution of Java.

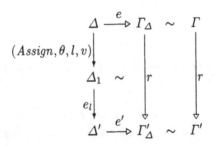

Corollary 6. *Given* $\Gamma \in \mathit{Conf}_\triangleright$ *from a properly synchronized program and* $\Delta \in \mathit{Conf}_\blacktriangleright$, *if* $\Gamma \sim \Delta$ *and* $\Gamma \longrightarrow\!\!\triangleright^* \Gamma'$ *such that the event space* $\eta_{\Gamma'}$ *of* Γ' *is consistently*

complete, then for any complete labelling of $\eta_{\Gamma'}$ *there is a reduction sequence* $\Delta \longrightarrow^* \Delta'$ *such that* $\Gamma' \sim \Delta'$.

Moreover, if two different complete labellings yield two different reduction sequences $\Delta \longrightarrow^* \Delta'_1$ *and* $\Delta \longrightarrow^* \Delta'_2$, *then still* $\Delta'_1 \sim \Delta'_2$ *holds.*

Proof. First, observe that if $\Gamma \sim \Delta$ then $\Gamma \approx (\Delta, \varepsilon)$. By a simple induction on the length of the derivation by Theorem 5, we get $(\Delta, \varepsilon) \rhd\!\!\!\longrightarrow^* (\Delta', e)$ and $\Gamma' \approx (\Delta', e)$. Now $e = \varepsilon$ follows from the fact that Γ' is consistently complete which entails that all prescient stores are matched by an *Assign* such that e must be empty in the end. From $e = \varepsilon$ we immediately get $\Gamma' \sim \Delta'$. Also from $(\Delta, \varepsilon) \rhd\!\!\!\longrightarrow^* (\Delta', \varepsilon)$ we can strip off a derivation $\Delta \longrightarrow^* \Delta'$ by definition of $\rhd\!\!\!\longrightarrow$.

The second claim follows just by transitivity of $\sim$ as $\Delta'_1 \sim \Gamma' \sim \Delta'_2$.

For our running example we can conclude that the corollary is applicable if all threads write o exclusively in `synchronized` blocks.

7 Conclusion

We have presented an event space semantics for multi-threaded Java with prescient stores. The informal statements in [3, §17.8] have been formalized and proven completely. In fact, the main motivation for this work was to understand what they meant. Correspondingly, we presented an operational semantics for prescient stores by just refining the axioms of the event space, leaving untouched the laws of the operational semantics. This demonstrates the flexibility of the event space approach.

Future work will include the extension of the treated language, e.g. `wait` and `notify`, exceptions, method calls, and the application of the semantics to correctness proofs of Java programs.

Acknowledgement: We used Paul Taylor's `diagram.sty`.

References

1. Ken Arnold and James Gosling. *The Java Programming Language.* Addison–Wesley, Reading, Mass., 1996.
2. Pietro Cenciarelli, Alexander Knapp, Bernhard Reus, and Martin Wirsing. From Sequential to Multi-Threaded Java: An Event-Based Operational Semantics. In *Proc. 6th Int. Conf. Algebraic Methodology and Software Technology*, Lect. Notes Comp. Sci., Berlin, 1997. Springer. To appear.
3. James Gosling, Bill Joy, and Guy Steele. *The Java Language Specification.* Addison–Wesley, Reading, Mass., 1996.
4. Doug Lea. *Concurrent Programming in Java.* Addison–Wesley, Reading, Mass., 1997.
5. Gordon D. Plotkin. Structural Operational Semantics (Lecture notes). Technical Report DAIMI FN–19, Aarhus University, 1981 (repr. 1991).
6. Glynn Winskel. An Introduction to Event Structures. In Jacobus W. de Bakker, editor, *Linear Time, Branching Time and Partial Order in Logics and Models for Concurrency*, volume 354 of *Lect. Notes Comp. Sci.*, Berlin, 1988. Springer.

Categories of Relational Structures

Michał Walicki
Institute of Informatics
University of Bergen
michal@ii.uib.no

Marcin Białasik
Institute of Computer Science
Polish Academy of Sciences
marcinb@ipipan.waw.pl

Abstract. We characterise compositional homomorphims of relational structures. A study of three categories of such structures – viewed as multialgebras – reveals the one with the most desirable properties. We study also analogous categories with homomorphisms mapping elements to sets (thus being relations). Finally, we indicate some consequences of our results for partial algebras which are special case of multialgebras.

1 Introduction

In the study of universal algebra, the central place occupies the pair of "dual" notions of congruence and homomorphism: every congruence on an algebra induces a homomorphism into a quotient and every homomorphism induces a congruence on the source algebra. Categorical approach attempts to express *all* (internal) properties of algebras in (external) terms of homomorphisms. When passing to relational structures, however, the close correspondence of these internal and external aspects seems to get lost.

The most common generalisation of the definition of homomorphism to relational structures says that a set function $\phi : \underline{A} \to \underline{B}$, where both sets are equipped with respective relations $R^A \subseteq \underline{A}^n$ and $R^B \subseteq \underline{B}^n$, is a (weak) homomorphism iff

$$\langle x_1 ... x_n \rangle \in R^A \;\Rightarrow \langle \phi(x_1) ... \phi(x_n) \rangle \in R^B \tag{1.1}$$

Now *any* equivalence on $\underline{A}$ gives rise to a weak homomorphism and, conversely, a weak homomorphism induces, in general, only an equivalence relation on $\underline{A}$. Hence this homomorphism does not capture the notion of congruence and this is just one example of an internal property of relational structures that cannot be accounted for by relational homomorphisms (in various variants). Probably for this reason, the early literature on homomorphisms of relations is extremely meagre [20, 24] and most work on relations concerns the study of relation algebras, various relational operators and their axiomatizations. Although in recent years several authors begun studying relational structures and their homomorphisms in various contexts, a general treatement of relational homomorphisms is still missing. This growing interest is reflected in numerous suggestions on how the definition of relational homomorphism could be specialized to obtain a more useful notion. This issue is our main objective.

In a more concise, relational notation, (1.1) is written as $R^A; \phi \subseteq \phi; R^B$. This somehow presupposes that R is a binary relation (of course, a homomorphism is such a relation, too) since composition has a standard definition only for binary relations. There seems to be no generally accepted definition of composition of relations of arbitrary arities. In the following, we will compose arbitrary relations (within the structures), like R above, with binary relations (obtained from homomorphisms between the structures). We choose to define the composition of relations $R^A \subseteq \underline{A}^{n+1}$, resp. $R^B \subseteq \underline{B}^{n+1}$ with a binary relation $\phi \subseteq \underline{A} \times \underline{B}$ as the relations on $\underline{A}^n \times \underline{B}$, as follows:

$$\begin{aligned} \langle a_1 ... a_n, b\rangle \in R^A;\phi &\Leftrightarrow \exists a \in \underline{A} : \langle a_1 ... a_n, a\rangle \in R^A \ \wedge \ \langle a, b\rangle \in \phi \\ \langle a_1 ... a_n, b\rangle \in \phi;R^B &\Leftrightarrow \exists b_1 ... b_n \in \underline{B} : \langle b_1 ... b_n, b\rangle \in R^B \ \wedge \ \langle a_i, b_i\rangle \in \phi \end{aligned} \tag{1.2}$$

This definition is certainly not the only possible one.[1] The reason for this choice (which, hopefully, will become convincing later) is our intension to treat relations in an algebraic way. It allows us to view relations as set-valued functions and turns relational structures into algebraic ones (*algebras of complexes* from [15, 16]). In particular, it admits composition of relations of arbitrary arities (analogous to composition of functions) with binary relations as a special case.

Now, table 1 presents a sample of proposed definitions of relational homomorphisms gathered from [20, 11, 7, 18, 23, 3, 21, 22]. It uses binary relations but with the above definition (1.2) it may be used for relations R of arbitrary arity (notation is explained at the end of this Introduction). The names are taken from the articles introducing the respective definitions and they themselves should suffice to ilustrate the existing confusion.

homomorphism ϕ		relational def.	logical def. $\forall x, y:$
1.	weak	$\phi^-;R^A;\phi \subseteq R^B$	$R^A(x,y) \Rightarrow R^B(\phi(x),\phi(y))$
2.	loose	$R^A;\phi \subseteq \phi;R^B$	1.
3.	full	$\phi^-;R^A;\phi = \phi^-;\phi;R^B;\phi^-;\phi$	$\exists x',y' : R^A(x',y') \Leftrightarrow R^B(\phi(x),\phi(y))$
4.	'strong'	$\phi^-;R^A;\phi \supseteq \phi^-;\phi;R^B;\phi^-;\phi$	$\exists x',y' : R^A(x',y') \Leftarrow R^B(\phi(x),\phi(y))$
5.	outdegree	$R^A;\phi = \phi;R^B;\phi^-;\phi$	$\exists x' : R^A(x',y) \Leftrightarrow R^B(\phi(x),\phi(y))$
6.	indegree	$\phi^-;R^A = \phi^-;\phi;R^B;\phi^-$	$\exists y' : R^A(x,y') \Leftrightarrow R^B(\phi(x),\phi(y))$
7.	'very strong'	$\phi;\phi^-;R^A;\phi \supseteq \phi;R^B$	$\exists x',y' : R^A(x',y') \Leftarrow R^B(\phi(x),y)$
8.	regular	5. & 6.	5. & 6.
9.	closed	$R^A;\phi \supseteq \phi;R^B$	$\exists y' : R^A(x,y') \Leftarrow R^B(\phi(x),y)$
10.	strong	$R^A = \phi;R^B;\phi^-$	$R^A(x,y) \Leftrightarrow R^B(\phi(x),\phi(y))$
11.	tight	$R^A;\phi = \phi;R^B$	2. & 9.

– primed symbol z' denotes some element such that $\phi(z') = \phi(z)$

Table 1. Some definitions of relational homomorphisms

This paper is an attempt to bring some order into this chaos. Given the combinatorial possibilities of defining homomorphisms of relational structures, a complete classification seems hardly possible. The very issue of the "criteria of usefulness" of various definitions, depending on the intended applications, may be debatable. Nevertheless, we hope that approaching the problem from a more algebraic perspective may bring at least some clarification. Instead of listing and defending new definitions we have chosen compositionality and the elementary properties of the resulting categories as the basis for comparison. We believe these to be

[1] It can be seen as standard composition of binary relations if we view a tuple $\langle a_1 ... a_n, x\rangle$ as a pair $\langle \overline{a}, x\rangle$ and let $\langle \overline{a}, \overline{b}\rangle \in \phi \Leftrightarrow \langle a_i, b_i\rangle \in \phi$ for all $1 \leq i \leq n$.

important properties and our results should be useful at least to those who share this belief.

Section 2 addresses the question of composition of homomorphisms of relations. We show that there are exactly 9 such homomorphisms which are closed under composition and give their characterization – in fact, most of the suggested defintions, like most of those in table 1, do *not* enjoy this property which we believe is crucial. The section ends with some remarks on the weaker notions of congruence related to these compositional homomorphisms. In section 3 we introduce *multialgebras* which are relational structures with composition of relations of arbitrary arities defined in the way (1.2) reflecting the traditional algebraic way of composing functions. We study epi-mono factorisation, (finite) completeness and co-completeness of three categories of such structures. Two of them turn out to have extremely poor structure indicating that the respective notions of homomorphisms may be less usefull than the remaining one. Then, in section 4 we study the analogous categories with homomorphisms being themselves relations. Such notions have been occassionally introduced in the literature but our results here do not look promising for their further use. Section 5 contains some remarks on the consequences of the results from section 3 for partial algebras which are special cases of multialgebras.

Most proofs are rather straightforward, involving basic facts from relational calculus and category theory. The paper presents a comprehensive view of several results and space limitations do not allow us to include even some counter-examples, not to mention the full proofs. The details can be found in [2].

Notation. In addition to the standard notions of algebraic signature, structure, etc., we will use their analogues for relational structures. A *relational signature* is a pair $\langle \mathcal{S}, \mathcal{R} \rangle$ where $\mathcal{S}$ is a set (of sort symbols) and $\mathcal{R}$ is a set of relation symbols with given arities (also called *type* of relation). A *relational Σ-structure* is a pair $A = \langle |A|, \mathcal{R}^A \rangle$, where $|A|$ is an $\mathcal{S}$-sorted set called a *carrier* and $\mathcal{R}^A$ is a set of relations, such that for each $[R_i : s_1 \times ... \times s_n] \in \mathcal{R}$ there is $R_i^A \subseteq |A|_{s_1} \times ... \times |A|_{s_n}$. [2]

We will study extensively algebras whose carriers are power sets. For such an algebra A with $|A| = \wp(\underline{A})$, the set $\underline{A}$ is called the *underlying set.* Given a function $f : \underline{A} \to B$, we will use its additive pointwise extension without making it explicit in the notation – for any $X \subseteq \underline{A}$, $f(X)$ means $\bigcup_{x \in X} f(x)$. We do not make explicit the distinction between elements and 1-element sets: if $|A| = \wp(\underline{A})$ and $a \in \underline{A}$, we write $a \in A$ meaning $\{a\} \in A$.

Readers should be wary about the notion of a homomorphism. For relational structures, it is a mapping between carriers with some additional constraints. It is just the same for ordinary algebras. Confusion may arise in connection with power set algebras, where various types of mappings are possible, including the ones between underlying sets or carriers (power sets). We will usually discuss them separately but they do come together on some occasions.

The source of a homomorphism $\psi : A \to B$ between two structures will be the

---

[2] To simplify the notation, we limit ourselves to single sorted structures claiming the results carry over to the multi-sorted case.

underlying set $\underline{A}$ (rather than the carrier $|A|$), and its *kernel* a relation $\sim \subseteq \underline{A} \times \underline{A}$ such that $a \sim b$ iff $\psi(a) = \psi(b)$. The equivalence classes under an equivalence $\sim$ are denoted $[x]$. Composition is written in diagrammatic order as $f;g$ for $g(f(\_))$. For a binary relation/function ϕ, ϕ^- denotes its inverse $= \{\langle y,x\rangle : \langle x,y\rangle \in \phi\}$

2 Homomorphisms of relational structures

Confronted with a tremendous number of possible definitions of a homomorphism, we believe that the property of being compositional may serve as an important feature distinguishing the more "relevant" ones. Theorem 2.2, which is the main result of this section, gives an exhaustive characterization of compositional definitions. We give only one counter-example showing non-compositionality of full homomorphisms from table 1 which were considered in [20, 18] as *the* homomorphisms between relations. In a more special form, they also appear in the study of partial algebras [6].

Example 2.1 *Let A, B, C be structures with one relation R.*

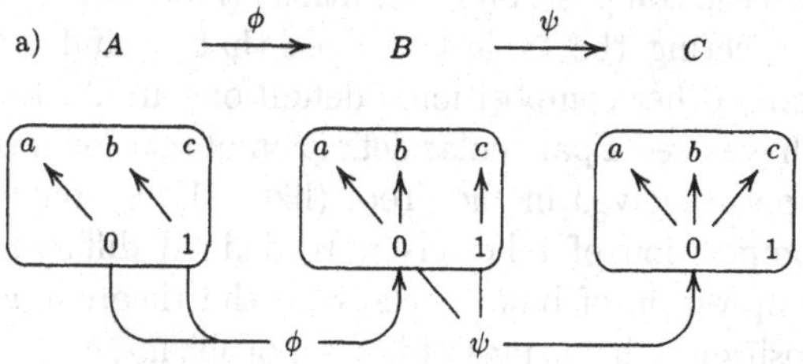

Both ϕ and ψ are full but $\phi;\psi$ is not. Although $\phi;\psi(0) = 0$ and $\langle 0,c\rangle \in R^C$, there is no $x \in A$ in the pre-image of c, such that $\langle 0,x\rangle \in R^A$.

We assume a fixed relational signature, with R ranging over all relation symbols, and consider definitions of homomorphisms $\phi : A \to B$ of the form

$$\Delta[\phi] \quad \Leftrightarrow \quad l_1[\phi]; R^A; r_1[\phi] \quad \bowtie \quad l_2[\phi]; R^B; r_2[\phi] \tag{2.3}$$

where $l[\_]$'s and $r[\_]$'s are relational expressions (using only composition and inverse), and $\bowtie$ is one of the set-relations $\{=, \subseteq, \supseteq\}$. A definition is *compositional* iff for all $\phi : A \to B$, $\psi : B \to C$, we have $\Delta[\phi]$ & $\Delta[\psi] \Rightarrow \Delta[\phi;\psi]$, i.e.:

$$\begin{aligned} & l_1[\phi]; R^A; r_1[\phi] \bowtie l_2[\phi]; R^B; r_2[\phi] \;\; \& \\ & l_1[\psi]; R^B; r_1[\psi] \bowtie l_2[\psi]; R^C; r_2[\psi] \\ & \qquad \Rightarrow l_1[\phi;\psi]; R^A; r_1[\phi;\psi] \;\; \bowtie \;\; l_2[\phi;\psi]; R^C; r_2[\phi;\psi] \end{aligned} \tag{2.4}$$

Theorem 2.2 *A definition is compositional iff it is equivalent to one of the following forms (where $\bowtie \in \{=, \subseteq, \supseteq\}$ and $\triangleright \in \{=, \supseteq\}$):*

$$\begin{array}{llll} 1) & R^A;\phi \bowtie \phi;R^B & 2) & \phi^-;R^A;\phi \triangleright R^B \\ 3) & \phi^-;R^A \triangleright R^B;\phi^- & 4) & R^A \triangleright \phi;R^B;\phi^- \end{array} \tag{2.5}$$

The proof of the "if"part is an easy check that 1)–4) satisfy (2.4). In fact, this holds for *any transitive* set-relation $\bowtie$. E.g., for 3) we verify:

$$\begin{array}{rl} & \phi^-; R^A \bowtie R^B; \phi^- \quad \& \quad \psi^-; R^B \bowtie R^C; \psi^- \\ \Rightarrow & \psi^-; \phi^-; R^A \bowtie \psi^-; R^B; \phi^- \ \& \ \psi^-; R^B; \phi^- \bowtie R^C; \psi^-; \phi^- \\ & \Rightarrow (\phi;\psi)^-; R^A \bowtie R^C; (\phi;\psi)^- \end{array}$$

The "only if" part is more tedious but amounts to a simple induction on the complexity of the expressions $l_i[\_]$ and $r_i[\_]$ using the following facts. Since homomorphisms are functions we have

$$a)\ \phi^-;\phi;\phi^- = \phi^- \qquad b)\ \phi;\phi^-;\phi = \phi \qquad c)\ \phi^-;\phi = id_{\phi[\underline{A}]} \tag{2.6}$$

One of the three possibilities for case 1) in the Theorem 2.2 is $R^A; \phi \subseteq \phi; R^B$. In fact, this case subsumes (or better, is equivalent to) *all* other definitions using $\subseteq$, also those not conforming to the compositional formats from the theorem!

Claim 2.3 *For any definition of the form* $\Gamma[\phi] \Leftrightarrow l_1[\phi]; R^A; r_1[\phi] \subseteq l_2[\phi]; R^B; r_2[\phi]$, *we have* $\Gamma[\phi] \Leftrightarrow R^A; \phi \subseteq \phi; R^B$.

Thus, there are 9 basic compositional definitions (more can be obtained by their conjunctions). Inspecting the table 1, we see that 1. and 2. define the same notion, and the only other compositional definitions are 9., 10. and 11.

Although we have used a particular definition of relational composition (1.2), all counter-examples involved in the proof (like 2.1) use only binary relations. Thus, even if composition of relations were defined differently, as long as it subsumes the composition of binary relations, the theorem gives the maximal number of compositional definitions of homomorphisms.

On the other hand, one might probably come up with other forms of homomorphism definition that are not covered by (2.3), e.g., allowing complementation. However, all commonly used forms do conform to this format. Some authors consider certain modifications of the definitions from table 1, for example, requiring surjectivity. With this restriction, full, outdegree and indegree homomorphisms (3,5,6) do compose. But this is just a round-about way of enforcing the equality $\phi^-;\phi = id_B$ (instead of the more limited case c) of (2.6)), and leads, respectively, to special cases of 2), 1) and 3) from the Theorem 2.2.

2.1 Congruences on relational structures

Congruences of relational and power structures were studied in [3, 4, 5]. The latter works study primarily lifting of various properties of a structure to its power structure without focusing on the notion of homomorphism.

As observed before, any equivalence gives rise to a (weak) homomorphism. However, the more specific definitions from theorem 2.2 may lead to more specific relations. We consider first equational definitions from the theorem, i.e.:

$$\begin{array}{llll} 1) & R^A;\phi = \phi; R^B & 2) & \phi^-; R^A; \phi = R^B \\ 3) & \phi^-; R^A = R^B; \phi^- & 4) & R^A = \phi; R^B; \phi^- \end{array} \tag{2.7}$$

and characterize these kernels which turn out to be not merely equivalences but congruences of a sort. First, we define relational quotient structures.

Definition 2.4 *Given a relational structure* $A = \langle \underline{A}, R_1^A, R_2^A ... \rangle$ *and an equivalence* $\sim \subseteq \underline{A} \times \underline{A}$*, a quotient structure* $A/_\sim = Q$ *is defined by* $\underline{Q} = \{[a] : a \in \underline{A}\}$ *and* $R_i^Q = \phi^-; R_i^A; \phi$*, where* $\phi : \underline{A} \to \underline{Q}$ *maps* $a \mapsto [a]$.

Claim 2.5 *Let* $\sim$ *be an equivalence on* A *and* Q, ϕ *be as in definition 2.4.*

if $\sim$ satisfies		then
2)		$\phi^-; R^A; \phi = R^Q$
1)	$\sim; R^A; \sim = R^A; \sim$	$R^A; \phi = \phi; R^Q$
3)	$\sim; R^A; \sim = \sim; R^A$	$\phi^-; R^A = R^Q; \phi^-$
4)	$\sim; R^A; \sim = R^A$	$R^A = \phi; R^Q; \phi^-$

In 1), 3) and 4) the relation $\sim$ has a flavour of a congruence:

1) called a *tight* congruence, can be stated as: $\forall a_1...a_n, b, a_1'...a_n' \; \exists b' \sim b :$
$R^A(a_1...a_n, b) \wedge a_1 \sim a_1'...a_n \sim a_n' \Rightarrow R^A(a_1'...a_n', b')$
3) yields a dual condition: $\forall a_1...a_n, b, b' \; \exists a_1' \sim a_1...a_n' \sim a_n :$
$R^A(a_1...a_n, b) \wedge b' \sim b \Rightarrow R^A(a_1'...a_n', b')$;
4) is strongest: $\forall a_1...a_n, b, a_1'...a_n', b' :$
$R^A(a_1...a_n, b) \wedge a_1' \sim a_1...a_n' \sim a_n \wedge b' \sim b \Rightarrow R^A(a_1'...a_n', b')$.

On the other hand, for any (at least weak) homomorphism we have the converse:

Claim 2.6 *Given a homomorphism* $\phi : A \to B$*, let* $\sim$ *be the kernel of* ϕ

	if $\phi : A \to B$ satisfies	then $\sim$ is an equivalence and
1)	$R^A; \phi = \phi; R^B$	$\sim; R^A; \sim = R^A; \sim$
3)	$\phi^-; R^A = R^B; \phi^-$	$\sim; R^A; \sim = \sim; R^A$
4)	$R^A = \phi; R^B; \phi^-$	$\sim; R^A; \sim = R^A$

There is no line for condition 2) since $\phi^-; R^A; \phi = R^B$ obviously implies that $\sim$ is an equivalence but, in fact, this follows for any function ϕ.

This isn't the strongest formulation. For instance, 1) implies $R^A; \phi = \phi; R^B; \phi^-; \phi$ which is sufficient to get the respective property of $\sim$. In general, since $\sim$ is induced from the image of A under ϕ, restricting the homomorphisms on the R^B-side to this image (by $\phi^-; \phi$) will yield the same properties of $\sim$.

Similar results do not follow for homomorphisms defined by $\supseteq$ in place of $=$ in (2.7). We can uniformly replace $=$ by $\supseteq$ in proposition 2.6, but then the statements in the right column are trivial for any mapping ϕ. If the target algebra is total then the kernel may retain the flavor of congruence. However, in general such homomorphisms induce only an equivalence relation.

3 Multialgebras

Multialgebras can be described as relational structures with a specific composition of relations of arbitrary arities. This issue offers several choices leading to possibly general and complicated solutions (see e.g. [9, 26]). Definition (1.2) was motivated by the wish to view relations as set-valued functions where the last,

n-th argument of an n-ary relation corresponds to an element of the result set obtained by applying the set-valued function to the first $n-1$ arguments. This view appears in [24], was elaborated in [15, 16], then in [25] and re-emerged recently in the algebraic approaches to nondeterminism [17, 14, 1, 28, 29]. It is based on the obvious isomorphism between the set-valued operations and relations:

$$\begin{array}{c} [A_1 \times ... \times A_n \to \wp(A)] \simeq [A_1 \times ... \times A_n \to [A \to Bool]] \simeq \\ [A_1 \times ... \times A_n \times A \to Bool] \simeq \wp(A_1 \times ... \times A_n \times A) \end{array} \quad (3.8)$$

Thus composition of relations becomes naturally the composition of the respective set-valued functions given by additive extension to sets. We may now talk about similarity types, or signatures, for multialgebras in the way entirely analogous to standard algebras. Another consequence of this change of perspective is that the structure of the derived operators ceases to be simply a Boolean algebra relatively independent from the actual multialgebra.[3] In fact, they become related to the signature in the same way as in classical universal algebra. Some universal algebraic aspects of power structures are addressed in [12, 10, 4, 19, 5].

Definition 3.1 *Let $\Sigma = \langle \mathcal{S}, \mathcal{F} \rangle$ be a signature. A Σ-multialgebra A is given by:*

- *a carrier $|A| = \{|A|_s\}_{s \in \mathcal{S}}$, where for each $s \in \mathcal{S}$, $|A|_s = \wp(\underline{A}_s)$ of some* underlying set *$\underline{A}_s$, with the obvious embedding $\underline{A}_s \hookrightarrow \wp(\underline{A}_s)$;*
- *a function $f^A : \underline{A}_{s_1} \times ... \times \underline{A}_{s_n} \to \wp(\underline{A}_s)$ for each $[f : s_1 \times ... \times s_n \to s] \in \mathcal{F}$, with composition defined through additive extension to sets, i.e. $f^A(X_1, ..., X_n) = \bigcup_{x_i \in X_i} f^A(x_1, ..., x_n)$.*

Although the carrier of a multialgebra is a power set, and hence a Boolean algebra, this isn't reflected in the signature containing only the "declared" operations. This does not change the actual structures but has some implications for the possible homomorphisms (which do not have to preserve the set operations). This also distinguishes multialgebras from *algebras of complexes* from [15].

We'll consider categories of multialgebras with homomorphisms corresponding to the condition 1) of theorem 2.2. Their multialgebraic form is as follows.

Definition 3.2 *A homomorphism $\phi : A \to B$ is a mapping $\phi : \underline{A} \to \underline{B}$ of one of the three modes*

$$\begin{array}{rcl} \textit{weak} & - & \text{when } \phi(f^A(a_1...a_n)) \subseteq f^B(\phi(a_1)...\phi(a_n)) \\ \textit{closed} & - & \text{when } \phi(f^A(a_1...a_n)) \supseteq f^B(\phi(a_1)...\phi(a_n)) \\ \textit{tight} & - & \text{when } \phi(f^A(a_1...a_n)) = f^B(\phi(a_1)...\phi(a_n)) \end{array}$$

Multialgebras are "partial" in the sense that operations may return empty set of values. By the pointwise extension of operations, they are strict in all arguments. Notice also that we allow empty carriers, i.e. $\wp(\emptyset) = \{\emptyset\}$ is a possible carrier of a multialgebra. Thus we will often refer to three special multialgebras, namely: "empty" with carrier $\wp(\emptyset)$, "unit" with carrier $\wp(\{\bullet\})$ and all operations returning $\{\bullet\}$, and "e-unit" with carrier $\wp(\{\bullet\})$ and all operations returning $\emptyset$.

[3] Cohn [7], p.204 sees this as *the* reason for the lacking interest in relational homomorphisms.

Obviously, a homomorphism $\phi : A \to B$ maps not only the underlying sets $\phi : \underline{A} \to \underline{B}$ but also the whole carriers (through its unique extension), i.e. $\phi : \wp(\underline{A}) \to \wp(\underline{B})$. When the source algebra A is "more partial" than the target B (meaning the operations more often return $\emptyset$), there will be neither a tight nor a closed homomorphism, but often a weak one.

An arbitray category of Σ-multialgebras is denoted $\mathsf{MAlg}(\Sigma)$, possibly with a subscript $_{\mathsf{W}}$, $_{\mathsf{T}}$, $_{\mathsf{C}}$ indicating the mode. These three cannot be mixed – e.g. a composition of a weak and a closed homomorphism may yield a mapping which is of none of the three modes. The following fact is common to all three categories.

Claim 3.3 *Let ϕ be a homomorphism in any category $\mathsf{MAlg}(\Sigma)$*
1) ϕ is epi (mono) iff it is surjective (injective) on the underlying set.
2) ϕ is an isomorphism iff it is tight, surjective and injective.

That the definition of tight homomorphisms is closest to the classical definition, together with 2), might suggest that $\mathsf{MAlg_T}(\Sigma)$ is *the* appropriate category to work with. However, as we will see, this category is much poorer then $\mathsf{MAlg_W}(\Sigma)$.

3.1 Epi-mono factorisation

We show that all three categories have epi-mono factorisations and construct factorisation systems for $\mathsf{MAlg_W}(\Sigma)$ and $\mathsf{MAlg_T}(\Sigma)$.

A *quotient* of a multialgebra A wrt. an *equivalence* $\sim$ (denoted $A/_\sim$) was defined in 2.4: it is a multialgebra Q such that $\underline{Q} = \{[a] : a \in \underline{A}\}$ and $f^Q([a_1]...[a_n]) = \bigcup_{a'_i \in [a_i]} [f^A(a'_1...a'_n)]$ for any function f.[4] Recall also that a kernel of a (weak) homomorphism is an equivalence and, more importantly, the mapping $\phi : A \to A/_\sim$ given by $\phi(a) = [a]$ is a weak homomorphism. We also have the following implication for equivalences $\sim_1 \subseteq \sim_2$ which, however, cannot be reversed:

Claim 3.4 *If $\sim_1 \subseteq \sim_2 \subseteq \underline{A} \times \underline{A}$, there is a weak morphism $\phi : A/_{\sim_1} \to A/_{\sim_2}$.*

In multialgebraic setting a tight congruence $\sim$ is an equivalence such that for all $f \in \Sigma$ we have $\forall a_1...a_n, b_1...b_n : a_i \sim b_i \Rightarrow f^A(a_1...a_n) \sim f^A(b_1...b_n)$, with $\sim$ on sets defined by the Egli-Milner extension : $A \sim B \Leftrightarrow \forall a \in A\ \exists b \in B : a \sim b \wedge \forall b \in B\ \exists a \in A : a \sim b$. We then have the familiar facts: if $\sim$ is a tight congruence on A then $\phi : A \to A/_\sim$ defined by $m \mapsto [m]$ is a tight homomorphism; and kernel of any tight homomorphism is a tight congruence.

Theorem 3.5 [**of homomorphisms**] *Given a homomorphism $\phi : A \to B$, its kernel $\sim$, and $Q = A/_\sim$, define $\psi_1 : A \to Q$ and $\psi_2 : Q \to B$ as follows: $\psi_1(a) = [a]$ and $\psi_2([a]) = \phi(a)$. Then ψ_1 is epi, ψ_2 is mono, and*
1) if ϕ is weak then both ψ_1 and ψ_2 are weak.
2) if ϕ is tight then both ψ_1 and ψ_2 are tight;
3) if ϕ is closed then ψ_1 is weak and ψ_2 is closed;

[4] Observe that, if we perform this construction on a usual algebra with $\sim$ being an equivalence but not a congruence, we obtain a well-defined multialgebra.

To obtain epi-mono factorisation for closed homomorphisms we need a different notion of quotient: a *closed quotient* of a multialgebra A wrt. an equivalence $\sim$ (denoted $A/^{c}_{\sim}$) is a multialgebra Q such that $\underline{Q} = \{[a] : a \in \underline{A}\}$ and $f^Q([a_1]...[a_n]) = \bigcap_{a'_i \in [a_i]}[f^A(a'_1...a'_n)]$ for any function f.

Theorem 3.6 *With the notation from theorem 3.5, except for* Q *which now denotes* $A/^{c}_{\sim}$*: if* ϕ *is closed then* ψ_1 *is a closed epi and* ψ_2 *is a closed mono.*

In order to obtain factorisation systems we need precise characterization of the homomorphisms into a quotient. This presents some problems since the actual construction can be subsumed by many different definitions – mere surjectivity, for example, is not sufficient. Similar difficulties emerge with partial algebras which are special cases of multialgebras. Two possible definitions are given below:

Definition 3.7 *We say that* $\eta : A \to B$ *is*

- full *iff* $\forall f \in \Sigma, a_1...a_n \in \underline{A} : f^B(\phi(a_1)...\phi(a_n)) \cap \phi[A] = \bigcup_{a'_i \in [a_i]} \phi(f^A(a'_1...a'_n))$.
- fully-tight *iff* $\forall f \in \Sigma,\ b_1...b_n \in \underline{B} : f^B(b_1...b_n) = \bigcup_{a_i \in \phi^-(b_i)} \phi(f^A(a_1...a_n))$.

Obviously, tight $\Rightarrow$ fully-tight $\Rightarrow$ full, while (full and surjective) $\Rightarrow$ fully-tight, but none of these implications can be reversed. Full are not compositional and to repair this, one might, in addition, require surjectivity. However, fully-tight can be defined by $\phi^-; R^A_f; \phi = R^B_f$ (R_f is the relation for f), and so by theorem 2.2, are compositional. We suggest that this is *the* right definition – being compositional, it also yields the left factor of a factorisation system in $\mathsf{MAlg}_\mathsf{W}(\Sigma)$.

Theorem 3.8 $\langle \mathcal{E}, \mathcal{M} \rangle$ *is a factorisation system for*

- $\mathsf{MAlg}_\mathsf{W}(\Sigma)$ *with* $\mathcal{M}$ *all monomorphisms and* $\mathcal{E}$ *all full epimorphisms.*
- $\mathsf{MAlg}_\mathsf{T}(\Sigma)$ *with* $\mathcal{M}$ *all monomorphisms and* $\mathcal{E}$ *all epimorphisms.*

In $\mathsf{MAlg}_\mathsf{C}(\Sigma)$ homomorphisms into the *closed* quotient can be characterised dually to fully-tight morphisms (with intesection instead of union) but this class is not closed under composition, and hence does not lead to a factorisation system.

3.2 Subobjects

Defining, as usual, a *subobject* to be an equivalence class of monomorphisms, we obtain three different notions of weak, tight and closed subobjects in the respective categories. In plain multialgebraic terms: B is a *tight* (*weak, closed*) subalgebra of A iff $|B| \subseteq |A|$ and $\forall f\ \forall b_1...b_n \in |B| : f^B(b_1...b_n) = f^A(b_1...b_n)$ (with $\subseteq$ and $\supseteq$, respectively).

Claim 3.9 *A homomorphism* $\phi : B \to A$ *induces a subalgebra* $[B]$ *of* A*, which is of the same mode as* ϕ*.*

The construction is straightforward in $\mathsf{MAlg}_\mathsf{T}(\Sigma)$ but in the other two cases involves taking unions, resp., intersections of the images of the result sets in order to endow the image with the apropriate algebraic structure.

The different modes of homomorphisms lead also to different constructions of minimal subalgebras or, more generally, of intersections of subobjects.

Claim 3.10 *For any A, any set* $\{\nu_i : B_i \hookrightarrow A\}$ *of A's subobjects in* $\mathsf{MAlg_W}(\Sigma)$ *(resp.* $\mathsf{MAlg_T}(\Sigma)$*) has an intersection* $(B, \nu) = \bigcap(B_i, \nu_i)$.

We do not have a counterpart of this fact for $\mathsf{MAlg_C}(\Sigma)$ where subalgebras can be constructed, so to speak, in two opposite directions. On the one hand, empty subalgebra will be a closed subobject of a given algebra A. On the other hand, if we keep the same carrier $|B| = |A|$, and define all operations to return the full carrier, we get a closed subalgebra B of A.

3.3 Limits and co-limits in $\mathsf{MAlg}(\Sigma)$.

$\mathsf{MAlg_W}(\Sigma)$ has all (finite) limits and co-limits. The categories $\mathsf{MAlg_T}(\Sigma)$ and $\mathsf{MAlg_C}(\Sigma)$ are neither (finitely) complete nor co-complete. Since these statements involve long lists of specific results, we only summarize them in table 2.

	$\mathsf{MAlg_W}(\Sigma)$	$\mathsf{MAlg_T}(\Sigma)$	$\mathsf{MAlg_C}(\Sigma)$
products	+	–	–
equalizers	+	–	–
finite completeness	+	–	–
initial objects	+	–	–
co-products	+	–	–
co-equalizers	+	+	
finite co-completeness	+	–	–
epi-mono factorisation	+	+	+
factorisation system	+	+	

Table 2. Properties of various $\mathsf{MAlg}(\Sigma)$.

The initial object in $\mathsf{MAlg_W}(\Sigma)$ is not (any form of) term structure but the empty algebra. Nevertheless, the category $\mathsf{MAlg_W}(\Sigma)$ possesses many desirable properties not possessed by the other two categories. It has perhaps been a prevailing opinion among mathematicians interested in the question that weak homomorphisms of relational structures provide the most useful notion. The above results justify and demonstrate this opinion. We have thus studied all the cases of one type of compositional homomorphisms – case 1) from theorem 2.2. We conjecture that the negative results about the tight and closed homomorphisms of this type will also occur for the tight and closed variants of the remaining three types.

4 Homomorphisms as Relations

Since the carrier of a multialgebra is a power set, one might attempt to define a multialgebra as a usual algebra over such a carrier. Such a generality, however, reduces to the considerations of standard algebras. In order to capture the intended (power-)set structure, some additional restrictions have to be imposed, the least of which is monotonicity of the operations within the algebra. (Such a variant, called power-algebras, was used, for instance in [28], for modelling a kind of "call-by-name" passing of nondeterministic parameters.)

We will not consider this alternative here and, instead, will look closer at the alternative notions of homomorphisms between multialgebras. There are three immediate possibilities here; a homomorphism $\phi : A \to B$ may be of three kinds: PP : $\underline{A} \to \underline{B}$ – point-point, PS : $\underline{A} \to \wp(\underline{B})$ – point-set, or SS : $\wp(\underline{A}) \to \wp(\underline{B})$ – set-set. Various modes of the first kind have been considered so far. The last kind, when taken without additional restrictions, treats multialgebras as standard algebras which only "by accident" happen to have power sets as carriers. Of course, a PP-homomorphism induces a PS one by pointwise extension, and a PS-homomorphism induces an SS one by additive extension. I.e., modulo the obvious type-conversion, we have: PP(ϕ) $\Rightarrow$ PS(ϕ) $\Rightarrow$ SS(ϕ). All three kinds can be further combined with the three modes – tight, weak, or closed – leading to a taxonomy which is simple in structure but messy in interrelations.

PS-homomorphisms seem a natural counterpart of the set-valued functions within algebras and have received sporadical attention in the literature. For these reasons, we cast a closer look at the categories PSAlg(Σ), highlighting some points illustrating the lack of equally pleasing categorical properties as those possessed by the category $\mathsf{MAlg}_\mathsf{W}(\Sigma)$.

The only positive result, and the one not present in MAlg(Σ), is that initial objects in $\mathsf{PSAlg}_\mathsf{T}(\Sigma)$ resemble the classical term algebras (theorem 4.7). This seems the only reason for which they attracted some attention, e.g., in [13, 14].

Before proceeding further, we observe the following general fact. Claim 3.3 showed that isomorphism of any mode is actually tight. Although this was stated for the PP-homomorphisms, it does not depend on the kind of homomorphisms:

Claim 4.1 *If $\phi : A \to B$ is an isomorphism then* SS(ϕ) $\Rightarrow$ PS(ϕ) $\Rightarrow$ PP(ϕ).

4.1 The category PSet

As a technical tool, we introduce the category PSet. It is a special case of any of the three PSAlg(Σ) categories where Σ contains only one sort symbol. Thus the negative results for PSet can be immediately applied to PSAlg(Σ). Characterization of epis and monos, as well as non-existence of epi-mono factorisations for PSAlg(Σ) are obtained from the results in this subsection.

Definition 4.2 *The objects of* PSet *are power sets, i.e. for each object S of* Set, *$\wp(S)$ is an object of* PSet. *A morphism $\phi : \wp(S) \to \wp(T)$ is a* Set *morphism $\underline{\phi} : S \to \wp(T)$ extended pointwise, i.e. for $X \in \wp(S) : \underline{\phi}(X) = \bigcup_{x \in X} \underline{\phi}(x)$.*

Claim 4.3 *Let I :* PSet $\to$ Set *be the inclusion and $\wp$:* Set $\to$ PSet *send each set to its power set $S \mapsto \wp(S)$, and each morphism to its pointwise extension to power set. These functors are adjoint $\wp \dashv I$.*

In particular, the initial object (preserved by $\wp$) in PSet will be $\wp(\emptyset) = \{\emptyset\}$. It is initial in the same way as $\emptyset$ is initial in Set. This is also the terminal object in PSet, because morphisms send elements of the source to subsets of the target, and hence the unique morphism from any other set will map all elements to $\emptyset$.

Claim 4.4 PSet *is complete but not co-complete.*

Another negative fact of importance is the lack of epi-mono factorisation shown by a counter-example using the following characterization.

Lemma 4.5 *A morphism $\phi : S \to \wp(T)$ in* PSet *is:*
mono iff pointwise extension is injective: $\forall X, Y \subseteq S : X \neq Y \Rightarrow \phi(X) \neq \phi(Y)$;
epi iff it is surjective on T, *i.e.* $\forall t \in T \exists s \in S : \phi(s) = \{t\}$.

Claim 4.6 PSet *does not have epi-mono factorisation.*

4.2 The categories PSAlg(Σ)

Empty algebra is terminal in each PSAlg(Σ) category. It is also initial in $\mathsf{PSAlg_W}(\Sigma)$. For $\mathsf{PSAlg_T}(\Sigma)$, we have a more general result.

Theorem 4.7 *Let* U : $\mathsf{PSAlg_T}(\Sigma) \to \mathsf{PSet}$ *be the forgetful functor given by* $U(A) \stackrel{\text{def}}{=} \wp(\underline{A})$. *There is an adjunction* $\mathcal{T}_\Sigma \dashv U$.

Unlike for $\mathsf{MAlg_W}(\Sigma)$ where an initial object was the empty algebra, here we obtain essentially standard term algebra $\mathcal{T}_\Sigma(X)$ for each $\wp(X)$, and initial term algebra $\mathcal{T}_\Sigma$. Its initiality depends heavily on the fact that we have PS homomorphisms – they allow us to send a single element $t \in \mathcal{T}_\Sigma$ to a *set* of elements t^A in the target algebra A. In particular, we may have that $t^A = \emptyset$. The construction does not work for $\mathsf{PSAlg_W}(\Sigma)$ or $\mathsf{PSAlg_C}(\Sigma)$.[5]

Both U and $\mathcal{T}_\Sigma$ are faithful (but not full) and so we obtain, as a corollary and using lemma 4.5, the following characterization of epis and monos.

Claim 4.8 *Let* $\phi : M \to N$ *be in* $\mathsf{PSAlg_T}(\Sigma)$ *and* $\underline{\phi} = U(\phi)$ *be its image in* PSet. *Then 1)* ϕ *is mono iff* $\underline{\phi}$ *is mono, and 2)* ϕ *is epi if* $\underline{\phi}$ *is epi.*

The "only if" in 2) does not hold but, to our surprise, we were not able to find a necessary condition for epis in $\mathsf{PSAlg_T}(\Sigma)$.

Table 3 summarizes the properties of the three PS-categories.

	$\mathsf{PSAlg_W}(\Sigma)$	$\mathsf{PSAlg_T}(\Sigma)$	$\mathsf{PSAlg_C}(\Sigma)$
products	+	+	+
equalizers	+	+	–
completeness	+	+	–
initial objects	+	+	–
co-products	+	–	–
co-equalizers	–	–	–
co-completeness	–	–	–
epi-mono factorisation	–	–	–

Table 3. Properties of PSAlg(Σ).

[5] Almost the same construction would give us an adjunction between **Set** and $\mathsf{PSAlg_T}(\Sigma)$. A specific subcategory of $\mathsf{PSAlg_T}(\Sigma)$ (with all ground terms determinisitc in each algebra) was used in [8] for studying nondeterministic automata and recognizable sets. The initiality result for this subcategory was also mentioned there.

5 Partial Algebras

We model a partial function $f : X \to Y$ by a multifunction $\hat{f}$ returning an appropriate singleton set whenever f is defined, and the empty set otherwise. Given a partial algebra A, let $\hat{A}$ be its multialgebraic representation with functions given below ($\mathsf{dom}(f^A)$ is a domain on which f^A is defined):

$$\begin{array}{l} a_1...a_n \in \mathsf{dom}(f^A) \Leftrightarrow f^{\hat{A}}(a_1...a_n) \neq \emptyset \\ a_1...a_n \in \mathsf{dom}(f^A) \Leftrightarrow f^{\hat{A}}(a_1...a_n) = \{f^A(a_1...a_n)\} \end{array} \tag{5.9}$$

We list a few relevant definitions concerning partial algebras.

Definition 5.1 *$\phi : A \to B$ is a* weak *homomorphism iff $\phi(\mathsf{dom}(f^A)) \subseteq \mathsf{dom}(f^B)$ and $\forall a_i \in \mathsf{dom}(f^A) : \phi(f^A(a_i)) = f^B(\phi(a_i))$. It is also* [6]*:*

tight *if* $f^A; \phi = \phi; f^B$ *i.e.* $\mathsf{dom}(f^A) = \phi^-(\mathsf{dom}(f^B))$

full *if* $\forall a_i \in A : f^B(\phi(a_i)) \in \phi[A] \Rightarrow \exists a_i' \in \mathsf{dom}(f^A) : \phi(a_i') = \phi(a_i)$

full injective *if it is injective, and* $f^A = \phi; f^B; \phi^-$

Let $\mathsf{PAlg_W}$ and $\mathsf{PAlg_T}$ be the categories of partial algebras with weak and tight homomorphisms, respectively. Closed homomorphisms rarely appear in literature so we do not consider them here. The functor $\hat{\ } : \mathsf{PAlg}(\Sigma) \to \mathsf{MAlg}(\Sigma)$ sends any partial algebra to a multialgebra $\hat{A}$ as in (5.9); and a morphism $\phi : A \to B$ to the morphism $\hat{\phi} : \hat{A} \to \hat{B}$ given by $\hat{\phi}(a) = \phi(a)$. Thus $\hat{\ }$ is essentially an inclusion (disregarding the conversion of elements into 1-element sets). $\widehat{\mathsf{PAlg_W}(\Sigma)}$ and $\widehat{\mathsf{PAlg_T}(\Sigma)}$ are full subcategories of $\mathsf{MAlg_W}(\Sigma)$, resp. $\mathsf{MAlg_T}(\Sigma)$ with objects being deterministic multialgebras. It is hardly surprising that:

Claim 5.2 *Functors $\hat{\ } : \mathsf{PAlg_W}(\Sigma) \to \mathsf{MAlg_W}(\Sigma)$ and $\hat{\ } : \mathsf{PAlg_T}(\Sigma) \to \mathsf{MAlg_T}(\Sigma)$ are full and faithful.*

Thus $\hat{\ }$ reflect limits and co-limits and many properties of the two categories of partial algebras follow now by revisiting their multialgebraic proofs and checking whether the constructions, when applied to the deterministic multialgebras, yield deterministic objects. One such consequence is that any set of subobjects of an algebra in $\mathsf{PAlg_W}(\Sigma)$, resp. $\mathsf{PAlg_T}(\Sigma)$, has an intersection, in particular, the intersection of a set of tight subalgebras is a tight subalgebra.

Yet another consequence is that factorisation systems for $\mathsf{MAlg_W}(\Sigma)$, and $\mathsf{MAlg_T}(\Sigma)$, specialise to factorisation systems for $\mathsf{PAlg_W}(\Sigma)$ and $\mathsf{PAlg_T}(\Sigma)$. The definition 5.1 of full homomorphisms is a special case of the relational definition and the multialgebraic one 3.7. Thus, there is a one-to-one correspondence: ϕ is full-and-surjective iff $\hat{\phi}$ is full(y-tight) epi. Both kinds compose.

As argued after definition 3.7, the simplest characterisation of compositional homomorphisms into quotients is $\phi^-; R^A; \phi = R^B$. It specialises to partial algebras yielding a notion stronger than full and weaker than full-and-surjective. The latter, used as the left factor of factorisation system, satisfy this condition

[6] The literature on partial algebras calls tight homomorphisms *closed* and the full injective ones *relative-injective.* 'Full injective' is equivalent to 'full-*and*-injective'.

and can be safely replaced by such homomorphisms. Similarly, the constructions of quotients and associated congruences from $\mathsf{MAlg}_{\mathsf{W}}(\Sigma)$ and $\mathsf{MAlg}_{\mathsf{T}}(\Sigma)$ specialise to respective partial algebraic notions. Nevertheless, $\mathsf{MAlg}(\Sigma)$ being more general do not contain all the information relevant in the study of partial algebras. Various characterizations, numerous factorisation systems, etc. for the latter cannot be obtained from the former.

6 Informal Summary

According to theorem 2.2 there are only 9 compositional homomorphisms of relations: 1 weak (subset), 4 tight (equality) and 4 closed (superset). A closer study of three of them (together with the 3 respective PS-homomorphisms) shows that closed homomorphisms yield categories with very poor structural properties and one may expect that this to be the case for the remaining three closed homomorphisms. Thus, we conjecture that, accepting compositionality and the categorical properties we have studied among the adequate criteria for choice of the homomorphisms, there are at most 5 "useful" possibilities. Among them, the 4 tight ones can be expected to have significant weaknesses, exemplified by the one case, $\mathsf{MAlg}_{\mathsf{T}}(\Sigma)$, which has been given a detailed treatement here. But even if one still does not want to work with weak homomorphisms, the results of this paper can provide a useful tool preventing one from looking for new, idiosyncratic notions serving only very peculiar purposes.

The suggested definition of composition of relations of arbitrary arities turns relational structures into algebras – namely, multialgebras. These provide a convenient way for algebraic study of power set structures, relational structures, as well as for modeling phenomena like partiality and nondeterminism.

1) Multialgebras do not give one a grasp on the individual elements – ground terms may denote sets. This causes trouble in many situations, in particular, when constructing unique homomorphisms. Two homomorphisms may reflect the Σ-algebraic structure in the same way but still be different. E.g., algebras with carriers $\underline{A} = c^A = \{1,2\} = c^B = \underline{B}$, even with tight homomorphisms, will have "only one" homomorphism in the sense that $\phi(c^A) = c^B$ but, actually, there are two $\phi_1 = id$ and $\phi_2(1) = 2,\ \phi_2(2) = 1$.

2) Many (counter-)examples are based on empty signatures or signatures with only one constant. It seems that some nice results follow only if Σ is non-void (like relevant initial objects in $\mathsf{PSAlg}_{\mathsf{T}}(\Sigma)$), while others only if it is void.

2.1) Weak homomorphisms seem to work best in combining these two poles – $\mathsf{MAlg}_{\mathsf{W}}(\Sigma)$ and $\mathsf{PSAlg}_{\mathsf{W}}(\Sigma)$ had most desirable properties. However, this happens at the cost that many categorical constructions simply yield degenerated cases – empty algebras or algebras where (all) operations return empty set.

2.2) The mere look at the tables 2-3 could suggest that the weak categories $\mathsf{MAlg}_{\mathsf{W}}(\Sigma)$, resp. $\mathsf{PSAlg}_{\mathsf{W}}(\Sigma)$, are most interesting. In fact, the corresponding tight categories $\mathsf{MAlg}_{\mathsf{T}}(\Sigma)$, resp. $\mathsf{PSAlg}_{\mathsf{T}}(\Sigma)$, lack many of the desirable properties possessed by these weak ones.

However, consulting the actual proofs and constructions we observe that, in the situations when some construction works in the tight categories, it often yields a more natural – less degenerate and closer to the classical – result than the corresponding construction in the weak categories. Nonexistence of some constructions in the tight categories is often a side-effect of excluding the degenerate cases admissibile in the weak categories.

3) There is a certain sense of correspondence between the categories $\mathsf{MAlg_W}(\Sigma)$ and $\mathsf{PSAlg_T}(\Sigma)$. The weak PP-homomorphisms allow the target to have "larger result sets" than the source. Similar effect is achieved in PS-homomorphisms in that a single element from the source can be mapped on a set of elements in the target. It all depends on what homomorphisms are supposed to be used for, but assuming that this kind of "reducing the result sets" in the source is intended, there seems to be little point in combining PS-homomorphisms (which do that) with the weakness requirement (which does the same again).

In fact, $\mathsf{PSAlg_T}(\Sigma)$ seems to be the most interesting among the PS-categories, in particular, due to the natural homomorphism condition (involving set equality) and the existence of interesting initial objects (not present in any other category we have studied). However, it is not only easier to work with PP-homomorphisms – the category $\mathsf{MAlg_W}(\Sigma)$ had, in addition to initial objects, factorisation system and all (finite) co-limits. These may be strengths making it the most attractive of all the structures we have reviewed.

References

[1] Białasik, M. and Konikowska, B. Reasoning with nondeterministic specifications. Tech. Rep. 793, Polish Academy of Sciences, Institute of CS. (1995).

[2] Białasik, M. and Walicki, M. Relations, multialgebras and homomorphisms. Tech. Rep. 838, Polish Academy of Sciences, Institute of CS. (1997).

[3] Boyd, J. P. Relational homomorphisms. *Social Networks 14*, 163–186. (1992).

[4] Brink, C. Power structures. *Algebra Universalis 30*, 177–216. (1993).

[5] Brink, C., Jacobs, D., Netle, K., and Sekran, R. Generalized quotient algebras and power algebras. (1997), [unpublished].

[6] Burmeister, P. *A Model Theoretic Oriented Approach to Partial Algebras.* Akademie-Verlag, Berlin. (1986).

[7] Cohn, P. M. *Universal Algebra.* D. Reidel Publishing Company. (1981), [series "Mathematics and Its Applications", vol. 6].

[8] Elienberg, S. and Wright, J. Automata in general algebras. *Information and Control 11*, 425–470. (1967).

[9] Glenn, P. Identification of certain structures as split opfibrations over Δ^{op}. (1997), [to appear in Journal of Pure and Applied Algebra].

[10] Goldblatt, R. Varieties of complex algebras. *Annals of Pure and Applied Logic 44*, 173–242. (1989).

[11] Grätzer, G. *Universal Algebra.* Springer. (1968).

[12] Grätzer, G. and Whitney, S. Infinitary varieties of structures closed under the formation of complex structures. *Colloq. Math. 48.* (1984).

[13] Hussmann, H. Nondeterministic algebraic specifications and nonconfluent term rewriting. In *Algebraic and Logic Programming.* LNCS vol. 343, Springer. (1988).

[14] Hussmann, H. *Nondeterminism in Algebraic Specifications and Algebraic Programs.* Birkhäuser. (1993).

[15] JÓNSSON, B. AND TARSKI, A. Boolean algebras with operators I. *American J. Mathematics 73*, 891–939. (1951).

[16] JÓNSSON, B. AND TARSKI, A. Boolean algebras with operators II. *American J. Mathematics 74*, 127–162. (1952).

[17] KAPUR, D. *Towards a Theory of Abstract Data Types.* Ph. D. thesis, Laboratory for CS, MIT. (1980).

[18] ŁOŚ, J. Homomorphisms of relations. (1985), [manuscript, Warszawa].

[19] MADARÁSZ, R. Remarks on power structures. *Algebra Universalis 34*, 2, 179–184. (1995).

[20] MOSTOWSKI, A. *Mathematical Logic.* Warszawa-Wrocław. (1948), [in Polish].

[21] NIPKOW, T. Non-deterministic data types: models and implementations. *Acta Informatica 22*, 629–661. (1986).

[22] NIPKOW, T. *Observing non-deterministic data types.* LNCS vol. 332, (1987).

[23] PATTISON, P. The analysis of semigroups of multirelational systems. *J. Mathematical Psychology 25*, 87–117. (1982).

[24] PICKERT, G. Bemerkungen zum homomorphie-begriff. *Mathematische Zeitschrift 53.* (1950).

[25] PICKETT, H. Homomorphisms and subalgebras of multialgebras. *Pacific J. of Mathematics 21*, 327–342. (1967).

[26] TOPENTCHAROV, V. V. Composition générale des relations. *Algebra Universalis 30*, 119–139. (1993).

[27] WALICKI, M. AND BROY, M. Structured specifications and implementation of nondeterministic data types. *Nordic Journal of Computing 2*, 358–395. (1995).

[28] WALICKI, M. AND MELDAL, S. *Multialgebras, power algebras and complete calculi of identities and inclusions.* LNCS vol. 906, Springer. (1995).

[29] WALICKI, M. AND MELDAL, S. Algebraic approaches to nondeterminism – an overview. *ACM Computing Surveys*, vol. 29, no. 1 (1997).

Author Index

Lecture Notes in Computer Science

For information about Vols. 1–1300

please contact your bookseller or Springer-Verlag

Vol. 1301: M. Jazayeri, H. Schauer (Eds.), Software Engineering - ESEC/FSE'97. Proceedings, 1997. XIII, 532 pages. 1997.

Vol. 1302: P. Van Hentenryck (Ed.), Static Analysis. Proceedings, 1997. X, 413 pages. 1997.

Vol. 1303: G. Brewka, C. Habel, B. Nebel (Eds.), KI-97: Advances in Artificial Intelligence. Proceedings, 1997. XI, 413 pages. 1997. (Subseries LNAI).

Vol. 1304: W. Luk, P.Y.K. Cheung, M. Glesner (Eds.), Field-Programmable Logic and Applications. Proceedings, 1997. XI, 503 pages. 1997.

Vol. 1305: D. Corne, J.L. Shapiro (Eds.), Evolutionary Computing. Proceedings, 1997. X, 307 pages. 1997.

Vol. 1306: C. Leung (Ed.), Visual Information Systems. X, 274 pages. 1997.

Vol. 1307: R. Kompe, Prosody in Speech Understanding Systems. XIX, 357 pages. 1997. (Subseries LNAI).

Vol. 1308: A. Hameurlain, A M. Tjoa (Eds.), Database and Expert Systems Applications. Proceedings, 1997. XVII, 688 pages. 1997.

Vol. 1309: R. Steinmetz, L.C. Wolf (Eds.), Interactive Distributed Multimedia Systems and Telecommunication Services. Proceedings, 1997. XIII, 466 pages. 1997.

Vol. 1310: A. Del Bimbo (Ed.), Image Analysis and Processing. Proceedings, 1997. Volume I. XXII, 722 pages. 1997.

Vol. 1311: A. Del Bimbo (Ed.), Image Analysis and Processing. Proceedings, 1997. Volume II. XXII, 794 pages. 1997.

Vol. 1312: A. Geppert, M. Berndtsson (Eds.), Rules in Database Systems. Proceedings, 1997. VII, 214 pages. 1997.

Vol. 1313: J. Fitzgerald, C.B. Jones, P. Lucas (Eds.), FME '97: Industrial Applications and Strengthened Foundations of Formal Methods. Proceedings, 1997. XIII, 685 pages. 1997.

Vol. 1314: S. Muggleton (Ed.), Inductive Logic Programming. Proceedings, 1996. VIII, 397 pages. 1997. (Subseries LNAI).

Vol. 1315: G. Sommer, J.J. Koenderink (Eds.), Algebraic Frames for the Perception-Action Cycle. Proceedings, 1997. VIII, 395 pages. 1997.

Vol. 1316: M. Li, A. Maruoka (Eds.), Algorithmic Learning Theory. Proceedings, 1997. XI, 461 pages. 1997. (Subseries LNAI).

Vol. 1317: M. Leman (Ed.), Music, Gestalt, and Computing. IX, 524 pages. 1997. (Subseries LNAI).

Vol. 1318: R. Hirschfeld (Ed.), Financial Cryptography. Proceedings, 1997. XI, 409 pages. 1997.

Vol. 1319: E. Plaza, R. Benjamins (Eds.), Knowledge Acquisition, Modeling and Management. Proceedings, 1997. XI, 389 pages. 1997. (Subseries LNAI).

Vol. 1320: M. Mavronicolas, P. Tsigas (Eds.), Distributed Algorithms. Proceedings, 1997. X, 333 pages. 1997.

Vol. 1321: M. Lenzerini (Ed.), AI*IA 97: Advances in Artificial Intelligence. Proceedings, 1997. XII, 459 pages. 1997. (Subseries LNAI).

Vol. 1322: H. Hußmann, Formal Foundations for Software Engineering Methods. X, 286 pages. 1997.

Vol. 1323: E. Costa, A. Cardoso (Eds.), Progress in Artificial Intelligence. Proceedings, 1997. XIV, 393 pages. 1997. (Subseries LNAI).

Vol. 1324: C. Peters, C. Thanos (Eds.), Research and Advanced Technology for Digital Libraries. Proceedings, 1997. X, 423 pages. 1997.

Vol. 1325: Z.W. Raś, A. Skowron (Eds.), Foundations of Intelligent Systems. Proceedings, 1997. XI, 630 pages. 1997. (Subseries LNAI).

Vol. 1326: C. Nicholas, J. Mayfield (Eds.), Intelligent Hypertext. XIV, 182 pages. 1997.

Vol. 1327: W. Gerstner, A. Germond, M. Hasler, J.-D. Nicoud (Eds.), Artificial Neural Networks – ICANN '97. Proceedings, 1997. XIX, 1274 pages. 1997.

Vol. 1328: C. Retoré (Ed.), Logical Aspects of Computational Linguistics. Proceedings, 1996. VIII, 435 pages. 1997. (Subseries LNAI).

Vol. 1329: S.C. Hirtle, A.U. Frank (Eds.), Spatial Information Theory. Proceedings, 1997. XIV, 511 pages. 1997.

Vol. 1330: G. Smolka (Ed.), Principles and Practice of Constraint Programming – CP 97. Proceedings, 1997. XII, 563 pages. 1997.

Vol. 1331: D. W. Embley, R. C. Goldstein (Eds.), Conceptual Modeling – ER '97. Proceedings, 1997. XV, 479 pages. 1997.

Vol. 1332: M. Bubak, J. Dongarra, J. Waśniewski (Eds.), Recent Advances in Parallel Virtual Machine and Message Passing Interface. Proceedings, 1997. XV, 518 pages. 1997.

Vol. 1333: F. Pichler. R.Moreno-Díaz (Eds.), Computer Aided Systems Theory – EUROCAST'97. Proceedings, 1997. XII, 626 pages. 1997.

Vol. 1334: Y. Han, T. Okamoto, S. Qing (Eds.), Information and Communications Security. Proceedings, 1997. X, 484 pages. 1997.

Vol. 1335: R.H. Möhring (Ed.), Graph-Theoretic Concepts in Computer Science. Proceedings, 1997. X, 376 pages. 1997.

Vol. 1336: C. Polychronopoulos, K. Joe, K. Araki, M. Amamiya (Eds.), High Performance Computing. Proceedings, 1997. XII, 416 pages. 1997.

Vol. 1337: C. Freksa, M. Jantzen, R. Valk (Eds.), Foundations of Computer Science. XII, 515 pages. 1997.

Vol. 1338: F. Plášil, K.G. Jeffery (Eds.), SOFSEM'97: Theory and Practice of Informatics. Proceedings, 1997. XIV, 571 pages. 1997.

Vol. 1339: N.A. Murshed, F. Bortolozzi (Eds.), Advances in Document Image Analysis. Proceedings, 1997. IX, 345 pages. 1997.

Vol. 1340: M. van Kreveld, J. Nievergelt, T. Roos, P. Widmayer (Eds.), Algorithmic Foundations of Geographic Information Systems. XIV, 287 pages. 1997.

Vol. 1341: F. Bry, R. Ramakrishnan, K. Ramamohanarao (Eds.), Deductive and Object-Oriented Databases. Proceedings, 1997. XIV, 430 pages. 1997.

Vol. 1342: A. Sattar (Ed.), Advanced Topics in Artificial Intelligence. Proceedings, 1997. XVII, 516 pages. 1997. (Subseries LNAI).

Vol. 1343: Y. Ishikawa, R.R. Oldehoeft, J.V.W. Reynders, M. Tholburn (Eds.), Scientific Computing in Object-Oriented Parallel Environments. Proceedings, 1997. XI, 295 pages. 1997.

Vol. 1344: C. Ausnit-Hood, K.A. Johnson, R.G. Pettit, IV, S.B. Opdahl (Eds.), Ada 95 – Quality and Style. XV, 292 pages. 1997.

Vol. 1345: R.K. Shyamasundar, K. Ueda (Eds.), Advances in Computing Science - ASIAN'97. Proceedings, 1997. XIII, 387 pages. 1997.

Vol. 1346: S. Ramesh, G. Sivakumar (Eds.), Foundations of Software Technology and Theoretical Computer Science. Proceedings, 1997. XI, 343 pages. 1997.

Vol. 1347: E. Ahronovitz, C. Fiorio (Eds.), Discrete Geometry for Computer Imagery. Proceedings, 1997. X, 255 pages. 1997.

Vol. 1348: S. Steel, R. Alami (Eds.), Recent Advances in AI Planning. Proceedings, 1997. IX, 454 pages. 1997. (Subseries LNAI).

Vol. 1349: M. Johnson (Ed.), Algebraic Methodology and Software Technology. Proceedings, 1997. X, 594 pages. 1997.

Vol. 1350: H.W. Leong, H. Imai, S. Jain (Eds.), Algorithms and Computation. Proceedings, 1997. XV, 426 pages. 1997.

Vol. 1351: R. Chin, T.-C. Pong (Eds.), Computer Vision – ACCV'98. Proceedings Vol. I, 1998. XXIV, 761 pages. 1997.

Vol. 1352: R. Chin, T.-C. Pong (Eds.), Computer Vision – ACCV'98. Proceedings Vol. II, 1998. XXIV, 757 pages. 1997.

Vol. 1353: G. BiBattista (Ed.), Graph Drawing. Proceedings, 1997. XII, 448 pages. 1997.

Vol. 1354: O. Burkart, Automatic Verification of Sequential Infinite-State Processes. X, 163 pages. 1997.

Vol. 1355: M. Darnell (Ed.), Cryptography and Coding. Proceedings, 1997. IX, 335 pages. 1997.

Vol. 1356: A. Danthine, Ch. Diot (Eds.), From Multimedia Services to Network Services. Proceedings, 1997. XII, 180 pages. 1997.

Vol. 1357: J. Bosch, S. Mitchell (Eds.), Object-Oriented Technology. Proceedings, 1997. XIV, 555 pages. 1998.

Vol. 1358: B. Thalheim, L. Libkin (Eds.), Semantics in Databases. XI, 265 pages. 1998.

Vol. 1360: D. Wang (Ed.), Automated Deduction in Geometry. Proceedings, 1996. VII, 235 pages. 1998. (Subseries LNAI).

Vol. 1361: B. Christianson, B. Crispo, M. Lomas, M. Roe (Eds.), Security Protocols. Proceedings, 1997. VIII, 217 pages. 1998.

Vol. 1362: D.K. Panda, C.B. Stunkel (Eds.), Network-Based Parallel Computing. Proceedings, 1998. X, 247 pages. 1998.

Vol. 1363: J.-K. Hao, E. Lutton, E. Ronald, M. Schoenauer, D. Snyers (Eds.), Artificial Evolution. XI, 349 pages. 1998.

Vol. 1364: W. Conen, G. Neumann (Eds.), Coordination Technology for Collaborative Applications. VIII, 282 pages. 1998.

Vol. 1365: M.P. Singh, A. Rao, M.J. Wooldridge (Eds.), Intelligent Agents IV. Proceedings, 1997. XII, 351 pages. 1998. (Subseries LNAI).

Vol. 1367: E.W. Mayr, H.J. Prömel, A. Steger (Eds.), Lectures on Proof Verification and Approximation Algorithms. XII, 344 pages. 1998.

Vol. 1368: Y. Masunaga, T. Katayama, M. Tsukamoto (Eds.), Worldwide Computing and Its Applications — WWCA'98. Proceedings, 1998. XIV, 473 pages. 1998.

Vol. 1370: N.A. Streitz, S. Konomi, H.-J. Burkhardt (Eds.), Cooperative Buildings. Proceedings, 1998. XI, 267 pages. 1998.

Vol. 1372: S. Vaudenay (Ed.), Fast Software Encryption. Proceedings, 1998. VIII, 297 pages. 1998.

Vol. 1373: M. Morvan, C. Meinel, D. Krob (Eds.), STACS 98. Proceedings, 1998. XV, 630 pages. 1998.

Vol. 1375: R. D. Hersch, J. André, H. Brown (Eds.), Electronic Publishing, Artistic Imaging, and Digital Typography. Proceedings, 1998. XIII, 575 pages. 1998.

Vol. 1376: F. Parisi Presicce (Ed.), Recent Trends in Algebraic Development Techniques. Proceedings, 1997. VIII, 435 pages. 1998.

Vol. 1377: H.-J. Schek, F. Saltor, I. Ramos, G. Alonso (Eds.), Advances in Database Technology – EDBT'98. Proceedings, 1998. XII, 515 pages. 1998.

Vol. 1378: M. Nivat (Ed.), Foundations of Software Science and Computation Structures. Proceedings, 1998. X, 289 pages. 1998.

Vol. 1379: T. Nipkow (Ed.), Rewriting Techniques and Applications. Proceedings, 1998. X, 343 pages. 1998.

Vol. 1380: C.L. Lucchesi, A.V. Moura (Eds.), LATIN'98: Theoretical Informatics. Proceedings, 1998. XI, 391 pages. 1998.

Vol. 1381: C. Hankin (Ed.), Programming Languages and Systems. Proceedings, 1998. X, 283 pages. 1998.

Vol. 1382: E. Astesiano (Ed.), Fundamental Approaches to Software Engineering. Proceedings, 1998. XII, 331 pages. 1998.

Vol. 1383: K. Koskimies (Ed.), Compiler Construction. Proceedings, 1998. X, 309 pages. 1998.